COLLOIDAL DISPERSIONS

COLLOIDAL DISPERSIONS

SUSPENSIONS, EMULSIONS, AND FOAMS

Ian D. Morrison
Sydney Ross

A John Wiley & Sons, Inc., Publication

Cover image courtesy of Eric R. Weeks, Department of Physics, Emory University.

This book is printed on acid-free paper. ∞

Copyright © 2002 by John Wiley and Sons, Inc., New York. All rights reserved.

Published simultaneously in Canada.

No part of this publication may be reproduced, stored in a retrieval system or transmitted in any form or by any means, electronic, mechanical, photocopying, recording, scanning or otherwise, except as permitted under Sections 107 or 108 of the 1976 United States Copyright Act, without either the prior written permission of the Publisher, or authorization through payment of the appropriate per-copy fee to the Copyright Clearance Center, 222 Rosewood Drive, Danvers, MA 01923, (978) 750-8400, fax (978) 750-4744. Requests to the Publisher for permission should be addressed to the Permissions Department, John Wiley & Sons, Inc., 605 Third Avenue, New York, NY 10158-0012, (212) 850-6011, fax (212) 850-6008, E-mail: PERMREQ@WILEY.COM.

For ordering and customer service, call 1-800-CALL-WILEY.

Library of Congress Cataloging-in-Publication Data is available.

ISBN 0-471-17625-7

Printed in the United States of America.

10 9 8 7 6 5 4 3 2 1

A low man goes on adding one to one.
His hundred's soon hit:
A high man, aiming at a million,
Misses an unit.

BROWNING, "A Grammarian's Funeral"

This book is dedicated to the memory of a high man

FREDERICK MAYHEW FOWKES

by two grateful learners.

CONTENTS

Preface		xxi
Acknowledgments		xxiii
Introduction		xxv

1 Optical Properties: Light Scattering — 1
- 1.1 Rayleigh Theory — 4
- 1.2 Molecular Weights by Laser Light Scattering — 5
- 1.3 Mie Theory — 7
- 1.4 Fraunhofer Diffraction — 9
- 1.5 Quasi-elastic Light Scattering — 10
- 1.6 Laser Velocimetry — 13
- References — 14

2 Rheology — 16
- 2.1 The Motion of Disperse-Phase Systems — 16
  - 2.1.a Newtonian Flow — 17
  - 2.1.b Poiseuille Flow — 19
  - 2.1.c Elasticity — 20
  - 2.1.d Plastic Flow and Shear-Thinning Flow — 21
  - 2.1.e Dilatancy — 25
  - 2.1.f Summary of Common Rheograms — 27
  - 2.1.g Thixotropy and Rheopexy — 29
  - 2.1.h Viscoelasticity — 30
  - 2.1.i Units and Dimensions — 33
- 2.2 Instruments to Measure Viscosity — 34
  - 2.2.a Capillary and Other Tube Viscosimeters — 34
  - 2.2.b Falling-Ball and Rising-Bubble Viscosimeters — 37
  - 2.2.c Falling-Needle Viscosimeters — 38
  - 2.2.d Vibrating Probe Viscosimeters — 38
  - 2.2.e Piston Viscosimeters — 39
  - 2.2.f Microbalance Viscosimeters — 39
  - 2.2.g Rheometers — 39
    - 2.2.g(i) The Brookfield Viscosimeter — 41
    - 2.2.g(ii) Coaxial Cylindrical Rheometers — 41

viii   CONTENTS

		2.2.g(iii)	Cone-and-Plate Rheometers	42
		2.2.g(iv)	Squeezing Flow Rheometers	44
	2.3	Manufacturers of Rheometers		45
	References			46

3 Kinetic and Statistical Properties 48

  3.1 Kinetic Properties 48
      3.1.a Motion of Particles by an Applied Force 48
      3.1.b Motion of Particles by Gravity 49
      3.1.c Motion of Particles by Centrifugation 50
      3.1.d Motion of Particles by an Applied Electric Field 51
      3.1.e Motion of Particles by Viscous Forces 51
      3.1.f Motion of Particles by Ultrasound 52
  3.2 Statistical Properties 54
      3.2.a Brownian Motion 54
      3.2.b Sedimentation Equilibrium 57
      3.2.c Colligative Properties 58
      3.2.d Fractals 58
  References 60

4 Particle Sizing 62

  4.1 Types of Distributions 63
  4.2 Sampling and Sample Preparation 66
  4.3 Sizing by Flow 67
      4.3.a Sieving 67
      4.3.b Filtration 69
      4.3.c Sedimentation 70
          4.3.c(i) Theory 70
          4.3.c(ii) Detectors 71
      4.3.d Elutriators, Impactors, and Cyclones 73
      4.3.e Centrifugation 74
          4.3.e(i) Disk Centrifuge 75
          4.3.e(ii) Uniform-Start Centrifuge 76
      4.3.f Ultracentrifugation 76
      4.3.g Capillary Hydrodynamic Chromatography 77
      4.3.h Field-Flow Fractionation 78
          4.3.h(i) Sedimentation Field-Flow Fractionation 78
          4.3.h(ii) Thermal Field-Flow Fractionation 79

4.4	Sizing by Single-Particle Detection		81
	4.4.a	Optical Microscope	81
	4.4.b	Image Analysis	81
	4.4.c	Ultramicroscopy	82
	4.4.d	Electron Microscopy	83
	4.4.e	Time-of-Transition Method	83
	4.4.f	Particle Counting: Photozone Detection	84
	4.4.g	Particle Counting: Electrozone Detection	86
4.5	Sizing by Light Scattering		87
	4.5.a	Rayleigh Scattering	88
	4.5.b	Turbidity	89
	4.5.c	Mie Scattering	90
	4.5.d	Fraunhofer Diffraction	90
	4.5.e	Polarization Effects	92
	4.5.f	Quasi-Elastic Light Scattering (QELS)	92
	4.5.g	Diffusing-Wave Spectroscopy	93
	4.5.h	Light Scattering as a Detector	94
4.6	Sizing by Acoustics		94
	4.6.a	Scattering and Absorption	94
	4.6.b	Speed of Sound	96
	4.6.c	Electroacoustics	98
4.7	Surface Area by Adsorption		98
	4.7.a	Multilayer Gas Adsorption	99
	4.7.b	Monolayer Gas Adsorption	101
	4.7.c	Pore-Volume Analysis	102
	4.7.d	Adsorption from Solution	103
4.8	Indirect Methods		105
4.9	Comparisons of Techniques		106
4.10	Manufacturers		109
	References		111

5 Processing Methods for Making Emulsions and Suspensions — 117

5.1	Dry Grinding (Micronizing)		118
5.2	High-Speed Stirrers		118
	5.2.a	Blade Stirrers	118
	5.2.b	Rotor-Stator Dispersers	120
5.3	High-shear Mills		121
	5.3.a	Colloid Mills	121
	5.3.b	Homogenizers	122
	5.3.c	Ultrasonic Dispersers	123

	5.4	Electrostatic Dispersing	123
	5.5	Impact Mills	124
		5.5.a Ball and Jar Mills	124
		5.5.b Vibratory Mills	125
		5.5.c Attritors	126
		5.5.d Sand or Other Small-Media Mills	127
	5.6	Mills for High-viscosity Dispersions	129
		5.6.a Roll Mills	129
		5.6.b Heavy-Duty Mixers	130
	5.7	Chemical Processing Aids	130
	5.8	Manufacturers of Milling Equipment	132
		References	134
6	**Liquid Surfaces and Interfaces**	**136**
	6.1	Molecular Theory of Surface and Interfacial Tension	136
	6.2	Thermodynamics of Surfaces and Interfaces	140
		6.2.a Coalescence of Droplets	141
		6.2.b Spreading of One Liquid on Another	142
		6.2.c Encapsulation of One Liquid by Another	142
		6.2.d Works of Adhesion and Cohesion	144
		6.2.e Free Energy of Emersion	146
	6.3	Thermodynamics of Curved Interfaces	146
	6.4	Sessile and Pendent Drops	149
	6.5	The Kelvin Equation	150
	6.6	Ostwald Ripening	151
	6.7	Capillary or Laplace Flow	152
	6.8	Marangoni Flow	153
		6.8.a Processes Dependent on Marangoni Flow	154
		6.8.b Thermally Induced Marangoni Flow	155
		References	155
7	**Liquid/Solid Interfaces**	**157**
	7.1	Thermodynamics of Liquids in Contact with Solids	157
	7.2	Degrees of Liquid/Solid Interaction	160
	7.3	Floating Particles	161
	7.4	Initial and Final Spreading Coefficients	162
	7.5	Determination of Spreading Pressures by Adsorption Techniques	167
	7.6	Dynamics of Spreading	168

	7.7	Final Spreading	169
	7.8	Detergency	170
	7.9	Thin-Film Coatings	171
	7.10	Electrical Charges	173
		References	174

8 Theories of Surface and Interfacial Energies — 177

- 8.1 Forces of Attraction Between Molecules — 177
- 8.2 The Componental Theory of Interfacial Tension — 179
  - 8.2.a Dispersion Force Contributions to Interfacial Tension — 180
  - 8.2.b Acid–Base Contributions to Interfacial Tension — 185
- 8.3 Dispersion Energies of Solid Substrates — 190
- 8.4 Acid–Base Character of Solid Substrates — 196
  - References — 197

9 Experimental Methods of Capillarity — 200

- 9.1 Surface and Interfacial Tensions — 200
  - 9.1.a Capillary Height — 200
  - 9.1.b Drop Weight — 201
  - 9.1.c Wilhelmy Plate — 201
  - 9.1.d The Du Noüy Ring — 202
  - 9.1.e Sessile and Pendent Drop — 202
  - 9.1.f Maximum Pull on a Rod — 203
  - 9.1.g Maximum Bubble Pressure — 205
  - 9.1.h Spinning Drop — 206
  - 9.1.i Vibrating Jet — 207
- 9.2 Methods to Measure Contact Angle — 208
  - 9.2.a On Flat Surfaces — 208
    - 9.2.a(i) The Contact Angle Goniometer — 208
    - 9.2.a(ii) Interfacial Meniscus — 209
  - 9.2.b On Fibers: The Fiber Balance — 209
  - 9.2.c On Powders — 210
    - 9.2.c(i) The Bartell Cell — 210
    - 9.2.c(ii) The Washburn Equation — 210
    - 9.2.c(iii) By Centrifugation — 212
    - 9.2.c(iv) By Compression on a Langmuir Balance — 213
    - 9.2.c(v) By Heats of Immersion — 213
- 9.3 Manufacturers of Equipment — 214
  - References — 215

10 Wetting of Irregular Surfaces — 218

- 10.1 Introduction — 218
- 10.2 Irregularities Due to Roughness: Wenzel Equation — 218
- 10.3 Irregularities Due to Chemical Heterogeneity — 219
- 10.4 Irregularities Due to Pores — 220
- 10.5 Irregularities Due to Scratches — 221
- 10.6 Jamin Effect in Capillaries — 222
- 10.7 True Contact Angles on Rough Surfaces — 224
- 10.8 Techniques — 224
  - 10.8.a The Wilhelmy Method (Plate or Fiber) for Measuring Contact Angles on Irregular Surfaces — 224
  - 10.8.b Goniometer Methods for Measuring Contact Angle Hysteresis — 226
  - References — 226

11 Surface-Active Solutes — 228

- 11.1 Introduction — 228
- 11.2 Range of Solutes from Lipophilic to Hydrophilic — 230
- 11.3 Types: Anionic, Cationic, and Nonionic — 233
- References — 236

12 Physical Properties of Insoluble Monolayers — 237

- 12.1 Observations — 237
- 12.2 Viscoelasticity of Insoluble Monolayers — 242
- 12.3 Built-up Films: Langmuir–Blodgett Films — 243
- References — 244

13 Aqueous Solutions of Surface-Active Solutes — 246

- 13.1 Adsorption — 246
- 13.2 Lundelius's Rule and the Ferguson Effect — 249
- 13.3 Micellization and Solubilization — 250
  - 13.3.a Self-Assembly by Spontaneous Association — 250
  - 13.3.b Methods to Determine Critical Micelle Concentration — 253
  - 13.3.c Equilibrium or Phase Separation? — 254
  - 13.3.d Effects of Varying Molecular Structure — 257
  - 13.3.e Mixtures of Homologs — 261
  - 13.3.f Solubilization by Micelles (Microemulsions) — 261
- 13.4 The Krafft Point — 264
- 13.5 Elasticity of Surface Films — 265
- 13.6 Dynamic Surface Tensions — 268

	13.7	Ultralow Dynamic Surface Tension	270
	13.8	Phase Behavior of Surface-Active Solutes	271
	13.9	The Effects of Geometric Packing	273
		References	276

14 Surface Activity in Nonpolar Media — 279

- 14.1 The Inverse Micelle — 279
- 14.2 Adsorption From Nonpolar Solution — 280
- 14.3 Electrical Charges in Nonpolar Media — 282
- 14.4 Theories of Bubble Rise in Viscous Media — 287
  - 14.4.a Rise in Pure Liquids: Hadamard Regime — 288
  - 14.4.b Effect of Surface-Active Solutes: Stokes Regime — 289
- References — 290

15 Thermodynamics of Adsorption from Solution — 292

- 15.1 The Gibbs Adsorption from Solution — 292
  - 15.1.a Two-Component Systems — 294
  - 15.1.b Three-Component Systems — 296
- 15.2 The Surface Tension Isotherm — 300
  - 15.2.a Adsorption Isotherm: Langmuir — 301
  - 15.2.b Equation of State: Frumkin — 301
- 15.3 Standard-State Free Energies — 302
- 15.4 Traube's Rule — 305
- References — 307

16 The Relation of Capillarity to Phase Diagrams — 308

- 16.1 Introduction — 308
- 16.2 Regular Solution Theory — 311
- 16.3 Critical Point Wetting — 313
- References — 314

17 Electrical Charges in Dispersions — 316

- 17.1 Ions and Charged Particles in Dispersions — 316
  - 17.1.a Origin of Charges at Surfaces in Aqueous Media — 317
  - 17.1.b Point of Zero Charge in Aqueous Media — 317
  - 17.1.c Origin of Charges at Surfaces in Nonpolar Media — 318
  - 17.1.d Preferential Adsorption — 321
  - 17.1.e Ions Near Charged Particles; the Electrical Double Layer — 324

| | | 17.1.e(i) | Gouy–Chapman theory | 324 |
| | | 17.1.e(ii) | Stern Theory | 329 |

17.2 Electrokinetic Phenomena ... 331
17.3 Zeta Potential ... 333
17.4 Measurements of Zeta Potential ... 334
    17.4.a Electrophoresis ... 334
        17.4.a(i) Microelectrophoresis ... 335
        17.4.a(ii) Laser Doppler Microelectrophoresis ... 337
        17.4.a(iii) Phase-Angle Light Scattering ... 338
        17.4.a(iv) Electrophoresis at Low Conductivity ... 339
    17.4.b Electro-osmosis ... 339
    17.4.c Streaming Current and Streaming Potential Techniques ... 340
    17.4.d Electroacoustic Techniques ... 341
    17.4.e Charge-to-Mass Ratio in Low-Conductivity Media ... 343
17.5 Electrical Effects in Dry Powders ... 343
17.6 Manufacturers of Equipment to Measure Particle Charge ... 347
References ... 348

18 Forces of Attraction Between Particles ... 352

18.1 The Forces of Attraction Between Molecules ... 352
18.2 Dispersion Forces of Attraction Between Particles ... 354
18.3 Theoretical Approaches ... 355
    18.3.a Molecular Approach: Hamaker Theory ... 355
        18.3.a(i) The Effects of Intervening Substances ... 356
        18.3.a(ii) Values of Hamaker Constants for Some Common Materials ... 357
        18.3.a(iii) The Hamaker Equation for Different Geometries ... 357
        18.3.a(iv) Dispersion-Force Contributions to Interfacial Tensions ... 362
        18.3.a(v) The Effect of Retardation ... 363
        18.3.a(vi) A Criticism of the Molecular Approach ... 364
    18.3.b Molar Approach: Lifshitz Theory ... 364
        18.3.b(i) Two Half-Spaces, Nonretarded Interaction ... 365
        18.3.b(ii) Method of Ninham and Parsegian ... 365
        18.3.b(iii) Two Half-Spaces, Including Retardation ... 369

		18.3.b(iv) Lifshitz Theory for Dispersions in Ionic Solutions	369
18.4	Other Methods to Obtain Hamaker Constants		370
	18.4.a	Direct Measurements of Forces of Attraction	370
	18.4.b	Indirect Measurements of the Forces of Attraction	370
	References		371

19 Forces of Repulsion — 374

19.1	Electrostatic Repulsion Between Flat Plates		374
19.2	Electrostatic Repulsion Between Spheres		377
	19.2.a	The Derjaguin Approximation	377
	19.2.b	The Debye–Hückel Approximation	378
	19.2.c	Constant Surface-Charge Density	379
19.3	Repulsion by Polymer Layers		379
	References		381

20 Dispersion Stability — 383

20.1	Lyophilic and Lyophobic Dispersions		383
20.2	Kinetics of Coagulation		383
	20.2.a	Rates of Flocculation	384
	20.2.b	Rates of Rapid Flocculation	386
	20.2.c	Interparticle Forces from Rates of Flocculation	386
20.3	Electrocratic Repulsion Versus Dispersion-Force Attraction		387
	20.3.a	DLVO Theory: Two Flat Plates	388
	20.3.b	DLVO Theory: Two Equal Spheres	389
	20.3.c	Example Application of DLVO Theory	390
	20.3.d	Critical Coagulation Concentration	391
	20.3.e	Schulze–Hardy Rule	393
	20.3.f	Secondary Minimum	394
20.4	Colloid Stability and Complex Ion Chemistry		394
20.5	Electrostatic Stability in Nonpolar Media		396
20.6	Theory of Steric Stabilization		398
	References		400

21 Polymeric Stabilization — 402

21.1	Adsorption from Solution by Solids		402
	21.1.a	Adsorption of Polymers	403
	21.1.b	Acidic or Basic Character of Solid Substrates	406

	21.1.c	Traube's Rule Revisited	409
	21.1.d	Rates of Adsorption on Solid Surfaces	409
21.2		Stabilization and the Phase Diagram	411
21.3		Effect of Free Polymer on Dispersion Stability	413
21.4		Steric versus Electrostatic Stabilization	414
21.5		Electrosteric Stabilization	415
21.6		Block Copolymers	416
		References	418

22 Emulsions   420

22.1		Definitions and Glossary of Terms	420
22.2		Determination of Emulsion Type	420
22.3		Interfacial Tension	423
22.4		Coalescence of Emulsion Droplets	423
22.5		Bancroft's Rule and Its Exceptions	424
22.6		Amphipathic Particles as Emulsion Stabilizers	426
22.7		The HLB Scale	429
22.8		The Phase Inversion Temperature (PIT)	432
22.9		Mechanical Properties of the Interface	434
22.10		Variation with Concentration of Internal Phase	436
	22.10.a	Rheology of Emulsions	437
	22.10.b	Electrical Conductivity of Emulsions	439
22.11		Measurement of Emulsion Stability	441
22.12		Making Emulsions	443
	22.12.a	The Method of Phase Inversion	443
	22.12.b	Phase-Inversion Temperature (PIT) Method	444
	22.12.c	Condensation Methods	444
	22.12.d	Intermittent Milling	444
	22.12.e	Electric Emulsification	445
	22.12.f	Special Methods	445
22.13		Breaking Emulsions	446
	22.13.a	Mechanical Demulsification	446
	22.13.b	Thermal Demulsification	446
	22.13.c	Chemical Demulsification	447
22.14		Microemulsions and Miniemulsions	448
22.15		Emulsion Polymerization	451
		References	452

23 Foams   456

23.1	Properties of Foams	456
23.2	Geometry of Bubbles and Foams	456

23.3		Equation of State of Foam	459
23.4		Theories of Foam Stability	460
	23.4.a	Thermodynamic Stability	460
	23.4.b	Gibbs-Marangoni Effect	462
	23.4.c	Mutual Repulsion of Overlapping Double Layers	463
	23.4.d	Drainage of Foams	464
	23.4.e	Elasticity of Lamellae	466
	23.4.f	Enhanced Viscosity or Rigidity	467
	23.4.g	Foam Boosters in the Presence of Oil Drops	470
	23.4.h	Black Films	470
	23.4.i	Influence of Liquid-Crystal Phases	471
	23.4.j	Effect of Dispersed Particles on Foam Stability	473
	23.4.k	Foaminess and Phase Diagrams	474
	23.4.l	Evanescent Foams and Stable Foams	480
23.5		Mechanisms of Antifoaming Action	480
	23.5.a	Impairment of Foam Stability	480
	23.5.b	Surface and Interfacial Tension Relations	481
	23.5.c	Foam Inhibition: Aqueous Systems	483
	23.5.d	Compounded Foam Inhibitors	485
	23.5.e	Silicone Foam Inhibitors	486
	23.5.f	Hydrophobic Particles	486
23.6		Methods of Measuring Foam Properties	489
	23.6.a	Stability	489
	23.6.b	Surface Rheology	491
		References	493

24 Technology of Suspensions — 499

24.1		Dispersing Agents	499
24.2		Experimental Techniques for Suspensions	500
	24.2.a	Adsorption from Solution	500
	24.2.b	Preferential Adsorption	502
	24.2.c	Adsorption of Dyes	504
	24.2.d	Surface-Tension Titrations	506
	24.2.e	Flocculated and Deflocculated Suspensions	506
	24.2.f	Sediment Volumes	508
	24.2.g	Ultrasonic Techniques	510
		24.2.g(i) Ultrasonic Spectroscopy of Suspensions	510
		24.2.g(ii) Electroacoustic Techniques	512
24.3		Rheology of Suspensions	512

	24.3.a	Dilute Suspensions	513
	24.3.b	Flocculated Suspensions	513
	24.3.c	Deflocculated Suspensions	514
	24.3.d	Concentrated, Deflocculated Suspensions	515
	24.3.e	Pigments in Polymer Solutions	516
	24.3.f	Rheology for Thin-Film Coatings	520
	24.3.g	Particles as Thickeners	520
	24.3.h	Powder Flow	522
24.4	Flocculation of Suspensions		523
	24.4.a	Double Layer Compression	523
	24.4.b	Flocculation by Neutral Polymers	524
	24.4.c	Flocculation by Polyelectrolytes	525
	24.4.d	Flocculation versus Stabilization	525
	24.4.e	Heteroflocculation	526
24.5	Nucleation and Crystal Growth		526
	References		528

25 Special Systems — 532

25.1	Model Colloidal Systems		532
	25.1.a	Particles of Uniform Size	532
	25.1.b	Surfaces of Uniform Composition	536
		25.1.b(i) Soap Films	536
		25.1.b(ii) Homotattic Solid Surfaces	537
	25.1.c	Colloidal Crystals	538
25.2	Clouds		539
25.3	Smokes		539
25.4	Lubricating Greases		539
25.5	Glues		540
25.6	Inks for Xerography		540
25.7	Electrophoretic Displays		543
25.8	Nanoparticles and Nanotechnology		545
	25.8.a	Synthesis	546
	25.8.b	Assembly	547
	25.8.c	Self-Assembly	547
	25.8.d	Applications	548
	References		548

26 Appendices — 554

26.1	Physical Constants	554
26.2	Units	554
26.3	Mathematical Formulae Used in Text	555

26.4	Electrostatic and Induction Contributions to Intermolecular Potential Energies	556
26.5	Electric properties of representative molecules	557
26.6	Lifetimes of Contributors to Colloid and Interface Science	557
26.7	Awards	561
	ACS Award in Colloid or Surface Chemistry	561
	Arthur W. Adamson Award for Distinguished Service in the Advancement of Surface Science	562

Bibliography 563

Index 589

PREFACE

This book was originally intended to be the second edition of our book *Colloidal Systems and Interfaces* published in 1988. That book was closely related to a four-day short course on emulsions and dispersions that had been taught by Professor S. Ross and Professor F. M. Fowkes since 1967, joined by Dr. I. D. Morrison in 1985, and by 1988 had gathered some 2000 alumni. Since the death of Professor Fowkes, our two-day short course has been sponsored by the American Chemical Society's Department of Continuing Education under the title *Dispersion of Fine Powders in Liquids* and our three-day short course was sponsored by Rensselaer Polytechnic Institute under the title *Colloid Chemistry Applied to Industrial R&D*. The ACS course has been delivered over fifty times to groups of participants varying in number from fifteen to thirty, as well as in-house to larger groups, bringing the total to some 3000 alumni.

This book, like the first, is intended for the industrial chemist or chemical engineer who may not have had a formal university course in colloid and interface chemistry but finds that the nature of the problems that must be solved necessitates the rapid acquisition of some knowledge of that subject. The major step in solving a problem is to define it. If this step is not well considered, the enterprise is sick. We hope to display the armory of concepts and techniques that are available in this discipline, so that investigators may orient their thinking along lines already laid down by the experience of previous workers. Every topic we broach is treated at greater length in monographs and reviews. We do no more than outline its nature, define its terms, explain its elementary concepts, and direct the reader to sources of fuller information. Our book therefore is an index of related topics, by means of which the enquirer, with a specific problem in mind, may hope to find the appropriate context to help formulate it. A great body of organized knowledge is at hand, but many who could use it are only vaguely aware of its existence or are intimidated by its bulk and impenetrability. This book is a guide to those so perplexed.

An outgrowth of colloid science is surface science, which has grown remarkably since the 1960s when ultrahigh vacuum systems made clean solid substrates available for adsorption studies. Further advances in instrumentation, such as scanning–tunneling microscopy and atomic-force spectroscopy have provided additional probes for the study of solid surfaces. In the same way that polymer science developed as a separate outgrowth from colloid science, so now surface science has reached its independent status with its own techniques, journals, monographs, and textbooks. The subject is so large that its inclusion here could not be other than an inadequate treatment.

The behavioral phenomena of foams, emulsions, and suspensions are almost always very complicated. To understand all the details involved in these phenomena requires a more advanced knowledge than what is provided in just the chapters headed "Foams," "Emulsions," and "Suspensions." The explanations of the phenomena are, therefore, materially facilitated by a preliminary account of the general principles of the science.

Considerable additions of material pertaining to recent advances in the various subjects treated, and some deletions of mathematical derivations, have been made from our first book. Our experience has been that the students attending short courses are more familiar with surface properties than with particulate and collective properties. These topics have been given greater attention in this second book. The title of the book has been changed to better describe its contents.

We are on the threshold of another scientific revolution brought about by nanotechnology. Properties of matter measured on a scale of nanometers are now within our reach. Even as we write, new techniques are being developed, soon to be reported in the specialized journals of this field, to remove guesswork from our theories, and to advance our knowledge of phenomena. As the pace of development quickens, so does the rate at which current techniques and even current modes of thought become obsolescent. Today's knowledge may be only of historical interest tomorrow. Even experience may become irrelevant in a mere 20 years, long before a young scientist has reached the end of active life. There is nothing else for it but the prolongation of studenthood throughout one's whole career, by attending short courses and habitual reading. Now more than ever before may it be said that the art is long and life is brief.

Ian D. Morrison
Sydney Ross

ACKNOWLEDGMENTS

The digital image of the Weissenberg effect for Figure 2-14 was taken by Mr. R. Welch and Dr. J. Bico from the Non-Newtonian Fluid Dynamics Research Group of Prof. G. McKinley at MIT. The fluid is 500 ppm of high molecular weight polystyrene in an oligomeric styrene resin with a relaxation time of approximately 3 s.

The data used in Figure 3-2 was supplied by Mr. Douglas Martin and Mr. Martin Forstner from the Biophysics Group of Prof. Josef Käs at the University of Texas in Austin.

Digital images for Figure 4-3 were supplied by M. André Charron and are used with permission of Huntsman Tioxide and The Federation of Societies for Coating Technology.

The pictures for Figure 7-11 were supplied by Prof. Vera Žutić of the Rudjer Bosšković Institute, Zagreb and are used with permission of the American Chemical Society.

Figures 17-13 and 17-14 were calculated with *Mobility* ©, a computer program purchased from the University of Melbourne, and used with permission of the Royal Society of Chemistry.

Figures 13-6, 20-7 and 20-8 are used with permission of Harcourt, Inc.

Figures 22-6 is used with permission of the Italian Chemical Society.

INTRODUCTION

This book is addressed to industrial scientists and engineers. The students who attend two- and three-day short courses taught by the authors are typical. They are trained in chemistry or chemical engineering, all with undergraduate degrees, a few with graduate degrees. Their primary assignments are not directly related to colloid chemistry and few have had any training in colloid chemistry. However, as is so often true in industrial research, development, and engineering, these assignments require a practical understanding of the general principles of colloid chemistry.

This book is a handbook, a resource for the explanation of important colloidal phenomena, for summaries of practical theories, for references to more detailed accounts, and for leads to suppliers and manufacturers. This book does not contain mathematical derivations so often required of students. These can be found by reference. The book does not explain all the newest research topics. It does contain important ideas to help solve problems and to design new products.

Almost all interesting materials and processes include multiple phases. When those phases are finely divided, as in paints, cosmetics, pharmaceuticals, reinforced plastics and rubbers, foods, ceramics, and all life forms, colloidal properties become significant. Inclusion of some of the topics discussed in this book provides a useful addition to the training of students.

After writing the tenth volume of *The Story of Civilization* to 1987, Will and Ariel Durant reread their texts to propose a list of the important lessons of history.[1] Similarly, after rereading this text we propose the following as the important lessons of colloid chemistry.

Particulate Phenomena

1. The scattering of light depends on the ratio of the particle size to the wavelength of incident light. For particles less than the wavelength of light Lord Rayleigh derived a relation between the intensity of light scattered, I, the particle radius, a, and the wavelength of the incident light, λ. This relation shows that large particles scatter more light than small ones and that blue light is more strongly scattered than red.

$$I \sim \frac{a^6}{\lambda^4}$$

---

[1] Durant, W.; Durant, A. *The Lessons of History*; Simon and Schuster: New York; 1968. The lessons were on the earth, biology, race, character, morals, religion, economics, socialism, government, war, growth and decay, and progress.

2. A significant attractive force between dispersed particles arises from the coordination of electron motion between particles. This attractive force is the common cause of the flocculation of dispersions. These forces are variously called van der Waals forces, London forces, Lifshitz forces, and dispersion — forces. The magnitudes of these forces vary with the polarizability and densities of the materials, e.g., metals attract each other more strongly than insulators.
3. Dispersions of solid or liquid particles in liquids can be stabilized against flocculation when the particles have enough electric charge. This mode of stabilization is called electrocratic. The magnitude of electrostatic repulsion is reduced with increasing ionic strength. The balance between electrostatic repulsion and dispersion force attraction is described by the DLVO theory. This theory rationalizes the Schulze-Hardy rule that the concentration of added electrolyte needed to flocculate an electrostatically stabilized dispersion varies inversely with the sixth power of the valence of counterions.
4. Dispersions of solid or liquid particles in liquids can be stabilized against flocculation when surface coatings of polymers are thick enough. This mode of stabilization is called steric. The dispersion is stable under the same conditions of temperature, pressure, and solvent composition as the polymer is when in solution.
5. Bancroft's empirical rule is that the emulsifier is more soluble in the continuous phase. Exceptions to Bancroft's rule occur only with weak emulsifiers.
6. Mathematical descriptions based on theories, confirmed by observations, show that the effectiveness of electrocratic stabilization increases the larger the dispersed particle, whereas the effectiveness of steric stability is greater the smaller the dispersed particle. This beneficent arrangement of natural cause and effect allows the stabilization of dispersions containing a wide range of particle sizes by means of polymers containing one or more ionic charge, thus combining electrocratic and steric stabilization. The two can also be combined by means of two different stabilizers.

Surface Phenomena

1. Increasing the surface area raises the potential energy of the system. The surface tension at the liquid/air interface is numerically equal to the increase in Helmholtz free energy per unit of area. The interfacial tension at the liquid/liquid interface is numerically equal to the increase in Helmholtz free energy per unit area at that interface. The interfacial Helmholtz free energies per unit area at the solid/vapor and solid/liquid interfaces are not so readily measured, but are assumed to exist there by analogy. The Young–Dupré equation is based on this assumption and the success of Fowkes' treatment of contact angles confirms that equation when applied to smooth, uniform, solid substrates.
2. The magnitudes of surface and interfacial tensions depend on molecular attractions within the liquid or interactions across the interface, respectively.

At the liquid/vapor surface the larger the intermolecular attractions within the liquid, the greater is the surface tension; at the liquid/liquid interface the larger the mutual interaction between the immiscible liquids, the smaller is the interfacial tension.

3. When the surface energy of the solid/vapor interface is less than the sum of the liquid/vapor interface plus the solid/liquid interface, then a sessile drop of the liquid displays a finite contact angle against the solid that is independent of the size of the drop. This is described by the Young-Dupré equation, which was proved experimentally for smooth, uniform, solid substrates by the success of Fowkes' equation when applied to Zisman's low-energy, uniform, solid surfaces. Modern instrumentation allows surface roughness to be taken into account and so allows the application of the Young–Dupré equation to extend beyond its former limitations.

4. The pressure changes across a curved interface and is equal to the product of interfacial energy and the curvature of the interface. Laplace derived the relation between the pressure change, Δp, the interfacial tension, σ, and the two principal radii of curvature, R_1 and R_2:

$$\Delta p = \sigma \left(\frac{1}{R_1} + \frac{1}{R_2} \right)$$

5. Adsorption lowers the potential energy of a surface, liquid or solid. The magnitude of this decrease on a liquid substrate is measured by the decrease of surface tension. On a solid substrate, it can be calculated from the adsorption isotherm by means of the Gibbs adsorption equation. Any factor, such as a change of composition, concentration, or temperature, tending to decrease solubility promotes surface activity (that is, any approach to a phase boundary in the phase diagram of the system.)

Complex phenomena are best understood when the most important aspects are identified first. These general lessons provide a solid starting point. The rest of this book leads the way to a deeper understanding.

Specialized dictionaries for colloid chemistry: Becher, P. *Dictionary of colloid and surface science*. Marcel Dekker: New York; 1990. Schramm, L.L. *The language of colloid and interface science — A dictionary of terms*. American Chemical Society: Washington, DC 1993. Schramm, L.L. *Dictionary of colloid and interface science*. John Wiley & Sons: New York; 2001.

COLLOIDAL DISPERSIONS

1 Optical Properties: Light Scattering

The optical properties of a disperse system depend strongly on its degree of subdivision. This is obvious when we consider the optical properties of humid air, then the clouds that form from the humid air, then the rain that comes from the clouds, and finally the puddles that form from the rain. All four are dispersions of water in air. The humid air is a molecular dispersion of water in air, angstroms in diameter, the clouds are drops of water, microns in diameter, the raindrops millimeters in diameter, and the puddles are meters in diameter. The humid air is essentially transparent, the clouds white and opaque, the rain looks like little lenses, and the puddles are again largely transparent.

A similar dependence of optical properties on the degree of dispersion is found when transparent glass windows are shattered. The transparent pane is transformed into an opaque pile of shattered pieces. Common table salt is opaque white in the shaker and transparent when added to water. All sorts of solutions are transparent until heated or cooled so that the solute precipitates. Titania particles about a quarter of micron in diameter are used to make paper and paint opaque. Titania particles, an order of magnitude smaller, are used in clear lotions such as sun screens.

In 1869 Tyndall produced artificial mists (aerosols) and demonstrated that a light beam is made visible when particles intercept the beam.[1] He used the absence of the effect as a sensitive test for the cleanliness of air and of solutions, and to this day it remains the simplest test to differentiate between suspensions and solutions. He also found that the scattering power of a given mass of suspended particles increases the larger the particle size. This phenomenon is the most sensitive indicator of the onset of flocculation of sols and of precipitation. He showed that the scattered light at right angles to the incident beam is completely linearly polarized, and this led him to suggest that the light of the sky, which is polarized at right angles to the sun, could be accounted for by scattering in the higher regions of the atmosphere.

No quantitative theory of light scattering was possible before Maxwell's discovery of the electromagnetic nature of light and the formulation of Maxwell's equations. The electromagnetic nature of light is that light can be described as a simultaneous traveling electric field and a traveling magnetic field. How light travels through matter is determined by the dielectric and magnetic properties of that matter. When light travels from one medium to another, its direction and

OPTICAL PROPERTIES: LIGHT SCATTERING

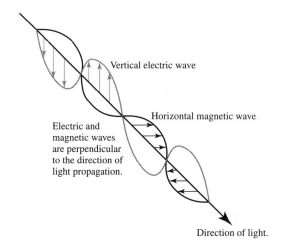

Figure 1.1 A light wave is comprised of traveling electric and magnetic waves.[2]

velocity change. As light travels through a prism it is bent and separated into its components. This redirection of light is called light refraction.

Figure 1.1 gives a schematic of Maxwell's description of light. The beam of light is a traveling electric and a traveling magnetic wave. The electric and magnetic fields are perpendicular to each other. The wavelength of the two traveling waves is the wavelength of the light. The electric and magnetic fields can rotate around the direction of propagation. If the rotation of the electric and magnetic fields is random, the light is called unpolarized. If the direction of the electric and magnetic fields remain constant, then the light is called polarized. Light from an incandescent bulb is unpolarized; light from a laser is polarized.

An electromagnetic wave impinging on matter induces oscillating electric and magnetic motion within atoms, thereby generating secondary electromagnetic waves, which are emitted in all directions and are observed as scattered light. The scattering center could be an atom or molecule, or a solid or liquid particle.

Scattered light is not light that refracts off a particle, but rather light that is generated in the particle by the motion of electrons. Since the intensity of the scattered light increases with the number and size of the scattering centers, particles have much greater scattering power than atoms or molecules.

A consequence of the electromagnetic theory of light is that its velocity, c, in a homogeneous medium is related to the dielectric constant, ε_0, and magnetic permeability of the medium, μ_0, by the expression

$$c = (\varepsilon_0 \mu_0)^{-1/2} \tag{1.1}$$

Substituting numerical values for free space gives the speed of light in a vacuum.

$$c = [(4\pi \times 10^{-7} \text{ W/A} - \text{m})(8.9 \times 10^{-12} C^2/\text{N} - \text{m}^2)]^{-1/2}$$
$$= 3.0 \times 10^8 \text{ m/s} \tag{1.2}$$

The differences in dielectric and magnetic properties at the interface between phases, such as the interface between a particle and the liquid in which it is dispersed, determines how the light is affected as it moves across the interface. A direct consequence of Maxwell's theory is that all scattering phenomena of light are shown to arise from inhomogeneities.

The scattering of light by particles and drops depends on the size and shape of the particles, the dielectric and magnetic properties of the particles and the dielectric and magnetic properties of the medium. Most common materials have small magnetic polarizabilities and the difference between the magnetic properties of the two phases is small. Therefore the magnetic properties of particles are usually ignored in calculating optical properties.

The scattering of light by particles and drops depends on the size and shape of the particles and drops and the differences of the dielectric properties of the two phases. The dielectric properties of a material are related to its refractive indexes by the Kramers–Kronig relations. In general the refractive index of a material is comprised of two components, one real and one imaginary. The real part of the refractive index is commonly measured with a refractometer. The imaginary part of the refractive index is related to the absorption coefficient. Both these components vary with the wavelength of light.

A simple consequence of Maxwell's theory is that no light is scattered by a particle or drop when it has the same refractive index as the medium. Small differences in refractive indexes result in only weak scattering. Many cosmetics and personal hygiene products are formulated so as to produce small differences between dispersed phases and the continuous medium in order to appear clear. On the other hand, titania, with its high refractive index, is used in paints, inks, and paper coatings to produce maximum scattering and hence opacity.

The dependence of scattering on particle size and shape is more complex. By far the most calculations of the scattering properties of disperse systems are done assuming that the scattering particle or drop is a sphere. The high symmetry of a sphere reduces the complexity of the calculations.

Unfortunately, few practical systems contain only spherical dispersed phases. The usual approach is to ignore the influence of shape on the scattering. This is a poor approximation.

Visible light varies in wavelength from the near infrared with a wavelength of about 700 nm to the edge of the ultraviolet with a wavelength of about 400 nm. This range is a narrow portion of the electromagnetic spectrum. Since the wavelengths of visible light are near the size of colloidal particles, most scattering experiments in colloid science use visible light, especially the light from a helium–neon laser with a wavelength of 632.8 nm, although in principle the sources from x-rays with wavelengths less than 0.1 nm to microwaves with wavelengths about 1 cm could be used.

If we consider only the scattering by spheres, the mathematics of light scattering can be simplified to three size ranges. These size ranges are set by the ratio of the wavelength of the incident light to the size of the particle. Particles

4 OPTICAL PROPERTIES: LIGHT SCATTERING

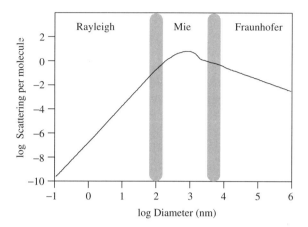

Figure 1.2 Scattering of visible light per water molecule.[3]

small compared to the wavelength of light are treated by Rayleigh theory; particles about the same size as the wavelength of light are treated by Mie theory; particles larger than the wavelength of the light are treated by Fraunhofer theory, often called Fraunhofer diffraction (See Figure 1.2).

1.1 RAYLEIGH THEORY

For particles of size much less than the wavelength of light, the electric field traversing the particle is uniform throughout the particle at any instant. The upper limit of particle radius for this to hold is generally taken to be of the order of 20–30 nm. Polymer solutions, micelles, and nanoparticles obtained by nucleation are within this range. This condition of uniform electric field and the further assumption of transparent particles allowed Rayleigh to apply Maxwell's equations to the propagation of light through a medium containing small spherical particles, treated as discontinuities in refractive index, to obtain the relation (as subsequently corrected)

$$I_u = \frac{8\pi^4 a^6}{r^2 \lambda^4} \left(\frac{n^2 - 1}{n^2 + 2}\right)^2 (1 + \cos^2 \theta) \tag{1.3}$$

where I_u is the scattered intensity from a single spherical particle of radius a, in the direction θ, at a distance r from the particle, per steradian, when illuminated by unpolarized light of unity intensity of wavelength λ (in that medium), and n is the ratio of the refractive index of the particle to that of the medium. Rayleigh's light-scattering equation has the following important corollaries:

(a) When the refractive index of the particle is the same as that of the medium $n = 1$, no light is scattered.

(b) The intensity of the scattered light increases as the sixth power of the radius of the particle, a; therefore, in a polydisperse system, the intensity of the scattered light is dominated by the larger particles.

(c) The intensity of the scattered light is inversely proportional to the fourth power of the wavelength of incident light λ; therefore, blue light with its shorter wavelength is scattered more than red light.

The blue of the sky comes from the scattering of sunlight by air molecules. The simplest observational evidence that dust is irrelevant is provided by blue skies in an extremely clean region, such as Antarctica or Greenland or by observations in airplanes: the sky becomes a deeper blue as the plane climbs to ever higher altitudes. The blue constituents of the scattered light are more intense than the red. The sky appears blue because scattered blue light comes to us from all quarters of the heavens; in the absence of an atmosphere no scattering takes place and the heavens would appear black except for direct observation of the sun, moon, and stars. Another possibly equally significant factor for the blue of the sky is the preferential absorption of red light by oxygen molecules which can be traced to the O—O stretching vibration, with a fundamental wavelength of 777 nm.[4] The blue of distant mountains, as mentioned in Scott's *Marmion*, "Far in the distant Cheviots *blue*," is caused by the intervention of a glowing blue atmosphere.

While the blue of the sky is fully accounted for by air molecules, the colors at sunrise and sunset arise from larger particles. The yellowish-red of dawning day and of sunsets depend on the optical thickness along the line of sight. For a fixed loading of particles, this optical thickness increases the greater the number of particles. In early morning, the atmosphere is stable (temperature inversion), there is little mixing and the concentration of particles is high. As the day progresses, heating of the ground destroys the inversion, there is mixing, and the particle concentration decreases. Therefore sunrises are brighter than sunsets, or so one would suppose. For a thorough explanation of sky colors at an elementary level, see Bohren and Fraser.[5]

Water is not colorless but slightly blue-green when observed by looking at white light transmitted through a long tube filled with water. The color is noticeable at a pipe length of one meter and is more intense with longer lengths of tube. This intrinsic color of water can be traced to the O—H stretching vibration with a fundamental wavelength around 3000 nm (near IR absorption.) The intrinsic color of ice is more difficult to detect by eye, but its absorption spectrum is very like that of water with a slight shift towards the red caused by more extensive hydrogen bonding.[4]

1.2 MOLECULAR WEIGHTS BY LASER LIGHT SCATTERING

Rayleigh also obtained an expression for the scattering of light by particles of any size or shape as long as their refractive index is close to that of the medium. These conditions are well suited to the study of polymers in solution. The observed quantities are the intensity of scattered light above that scattered by the pure solvent, I_{ex}, as a function of polymer concentration, c, and of scattering angle, θ, and the change in the refractive index of the solution as a function of concentration, $(\partial n/\partial c)_{T,p}$. These data are extrapolated to zero concentration and to zero scattering angle, with the following limiting relations to obtain the weight-average molecular weight M_w and the mean squared radius of gyration $\langle r_g^2 \rangle$.[6,7,8]

$$\frac{Kc}{R(\theta,c)} = \frac{1}{M_w}\left[1 + \frac{16\pi^2 \langle r_g^2 \rangle \sin^2(\theta/2)}{3\lambda^2}\right] \quad \text{as } c \to 0 \quad (1.4)$$

$$\frac{Kc}{R(\theta,c)} = \frac{1}{M_w} + 2Bc \quad \text{as } \theta \to 0 \quad (1.5)$$

Combining Equations (1.4) and (1.5) gives

$$\lim_{\substack{\theta \to 0 \\ c \to 0}} \frac{Kc}{R(\theta,c)} = \frac{1}{M_w} \quad (1.6)$$

where

$$K = \frac{4\pi^2 n_0^2}{\lambda_0^4 N_0}\left(\frac{\partial n}{\partial c}\right)_{T,p}^2 \quad (1.7)$$

and

$$R(\theta,c) = r^2 I_{ex}/I_0 = \text{the Rayleigh ratio}$$

B = virial coefficient
I_0 = incident light intensity
λ_0 = light wavelength *in vacuo*
N_0 = Avogadro's number
n = refractive index of the solution
n_0 = refractive index of the solvent
r = distance from sample to detector

Thus the molecular weight, M_w, can be evaluated from Eq. (1.6), the virial coefficient, B, can be evaluated from Eq. (1.5), and the mean squared radius of gyration, $\langle r_g^2 \rangle$, evaluated from Eq. (1.4). This evaluations have been made simple with a Zimm plot[6] in which $Kc/R(\theta,c)$ is plotted versus $\sin^2(\theta/2) + gc$ where g is an arbitrary constant chosen to provide a convenient spread of the data. Equipment to perform the necessary measurements of scattering intensity at low concentrations and multiple scattering angles as well as perform the Zimm plot calculations is available from Wyatt Technology (www.wyatt.com).

1.3 MIE THEORY

Most dispersions of practical interest contain larger particles, which may absorb light; hence a more general treatment than Rayleigh theory is needed. The red light of a helium-neon laser with a wavelength of 632.8 nm is the most common light-scattering source. Particles with diameters greater than about 50 nm cannot be considered small compared to the wavelength of the scattering light. The electrons in these particles no longer sense the same magnitude and direction of the electric field all at the same time.

In 1908 Mie developed a more general solution for the scattering of light by spherical particles of any size and any refractive index.[9] His theory was later extended to include cylinders and stratified spheres.[10] The essential result of Mie's treatment is the appearance of many maxima in the intensity of the scattered light as a function of scattering angle; the number and position of these maxima depend on the refractive-index difference and the size of the particle.

Figure 1.3 shows the angular dependence of the scattered intensity for increasing ratios of particle radius to the wavelength of incident light. As the size of the particle increases, the scattered light develops lobes of higher intensity at characteristic angles. For the largest particles, bright forward-scattering lobes develop corresponding to the onset of diffraction effects (Fraunhofer diffraction). The experimental technique is to measure the intensity of light as a function of angle from an angle close to the source of the laser light (backscattering) to an angle nearly directly into the laser beam (forward scattering.) If all the particles are the same diameter then scattering patterns similar to those of Figure 1.3 are obtained. The particle size could easily be deduced by comparing calculated scattering patterns to the experimental data. Monodisperse samples of sulfur sols[11] and monodisperse selenium sols[12,13] were used to confirm Mie's theory.

The scattering diagrams of Figure 1.3 were normalized to unity in the forward direction. When all the scattering functions of Figure 1.3 are included in the same diagram (Figure 1.4) then the strong dependence of the intensity on particle diameter is seen. An increase in diameter from 50 nm to 2000 nm results in a six order of magnitude increase in scattered-light intensity. This means that in polydisperse dispersions, the light scattered from the largest particles easily overwhelms the light scattered from the smaller ones.

If particle properties of a polydisperse system, such as refractive index, distributions of size and shape, and the absorption spectrum are known, the light scattering can be calculated. Efficient computer programs are now available to calculate Mie functions (Bohren and Huffman, 1983). However, the inverse problem, that is, to obtain the distribution of size and shape from the data, is impossible, unless the distribution is very narrow, because of the overlapping of maxima and minima from different sizes of particles.

When uniform dispersions of particles of a size approaching the wavelength of light, such as Tyndall's mists and La Mer's sulfur sols, are illuminated with white light, the scattered light appears colored, the color varying with the angle of observation. The variations of the color with the angle of observation are

8   OPTICAL PROPERTIES: LIGHT SCATTERING

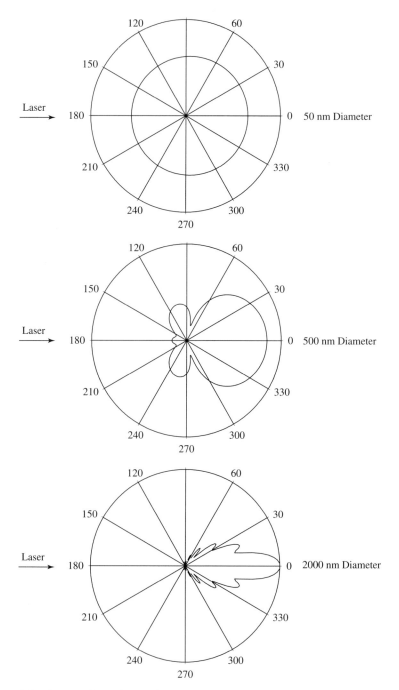

Figure 1.3 Angular dependence of scattered intensity for polystyrene spheres suspended in water. The arrows indicate the direction of the incident laser light. Each diagram has been normalized to have unit intensity scattering in the forward direction.

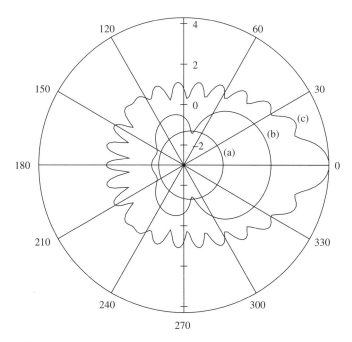

Figure 1.4 The scattering diagrams of polystyrene in water from Figure 1.3 plotted on the same graph. Note the strong dependence of the scattered-light intensity on the size of the particle.

called higher order Tyndall spectra or HOTS.[14] The colors are more brilliant the closer the dispersion is to being monodisperse. Each size of particle generates a characteristic number of bands of color. Particle size is determined by locating the angles of maximum red (or green) intensity and comparing with computed tables.

When the particles are not larger than a few tenths of a wavelength, the scattering is described by Rayleigh theory and the intensity increases rapidly towards the violet. The scattering by such small particles always shows a blue-violet light independent of composition as long as they are insulators. (Nanoparticle gold is red because of surface plasmon resonance.) Mie theory applies to larger particles and predicts that the color also depends on particle size and index of refraction. This explains why the smoke of a cigar or cigarette is blue when puffed immediately into the air but, if it is kept in the mouth first, becomes white. The particles of smoke covered with a layer of water are so much larger.[15] In the same way steam from a cooling tower becomes more and more opaque the farther from the vent.

1.4 FRAUNHOFER DIFFRACTION

When the particle diameter is much larger than the wavelength, the electric and magnetic fields of the incident light attain all possible values throughout

10   OPTICAL PROPERTIES: LIGHT SCATTERING

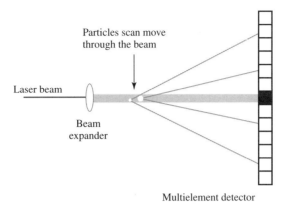

Figure 1.5 The concentric rings of forward-scattered light from particles large compared to the wavelength of light, Fraunhofer diffraction.

the particle. In this limit, the particles act as tiny lenses and the scattering phenomenon is called Fraunhofer diffraction. The light scattered from a monodisperse dispersion of spheres consists of a central bright spot surrounded by concentric dark and bright rings whose intensity diminishes farther from the center of the pattern. The scattering angle at which the first dark ring, or diffraction minimum, occurs depends on the size of the particles; the smaller the particle, the higher the angle of the first dark ring. See Figure 1.5.

When particles are large compared to the wavelength of light, almost all the scattered light is in the forward direction. The simplest optics to detect the intensity and position of the forward-scattered rings of light are concentric rings of light-sensitive detectors such as photodiodes. Commercial detectors used in Fraunhofer diffraction have between thirty-two and a couple of hundred concentric, solid-state detectors (see Table 4.10). The intensity can be measured almost simultaneously from all the detectors, allowing an almost instantaneous determination of the particle size.

The pattern of light scattered from a polydisperse dispersion is the sum of the light scattered from each individual size. Therefore the determination of the polydispersity requires deconvoluting the measured intensity as a function of forward-scattered angle into sums of scattering patterns for monodisperse spheres. This technique is particularly suitable for aerosols as the droplets are always large compared to the wavelength of the light and the droplets are spheres. Furthermore, since the detection and particle size determination is so fast, Fraunhofer diffraction can be used to characterize changing or moving dispersions.

1.5   QUASI-ELASTIC LIGHT SCATTERING

The light-scattering methods described so far use the time-average scattered intensity, usually as a function of scattering angle, sometimes as a function of wavelength. For each, the experimental procedure is to position a detector at some known angle and measure the intensity of light scattered from the dispersion. However, if the light intensity is measured quickly (on the order of milliseconds) the light intensity is found to vary. The source of this variation in intensity has been shown to be due to the Brownian motion of the suspended particles. If all the particles were fixed in place, then the scattered light intensity would be a function of the scattering properties of each particle and their relative position. The relative positions are important because the light scattered from each particle may constructively or destructively interfere with the light scattered from any other particle.

However, the particles are not fixed in position. Brownian motion causes them to diffuse randomly through the dispersion. With the changing relative positions, the light intensity changes; the scattered light intensities from individual particles interfere more or less with each other. Figure 1.6 shows the variation in light intensity for a dispersion of 261 nm diameter polystyrene spheres in water. Figure 1.7 shows the variation in light intensity for a dispersion of 85 nm diameter polystyrene spheres in water.

The light scattered from the larger spheres (Figure 1.6) is seen to vary more slowly than the light scattered from the smaller spheres (Figure 1.7). On the average the smaller spheres move more quickly than the larger spheres, therefore the variation in constructive and destructive interference is more rapid. The time dependence of the fluctuating intensity therefore is a function of particle size. The autocorrelation function* is a mathematical transformation of the intensity data to measure the variability of the intensity. Modern digital autocorrelators are

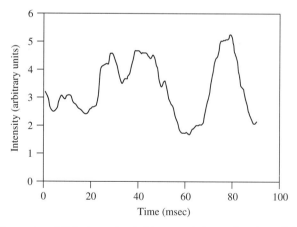

Figure 1.6 The scattered light intensity (arbitrary units) as a function of time for 261 nm polystyrene spheres in water at 25°C. The scattering angle is 40°.

---

*The autocorrelation function is a mathematical transform of data in the same sense as the Fourier transform. The Fourier transform produces a frequency-dependent function. The autocorrelation function produces a time-dependent function.

12 OPTICAL PROPERTIES: LIGHT SCATTERING

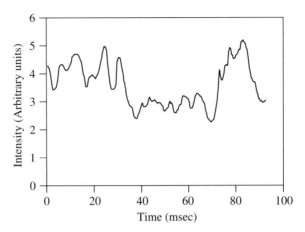

Figure 1.7 The scattered light intensity (arbitrary units) as a function of time for 85 nm diameter polystyrene spheres in water at 25°C. The scattering angle is 40°.

able to calculate the autocorrelation function in real time. For a monodisperse system of spherical particles in Brownian motion, the normalized autocorrelation function, $g^{(1)}$, is a simple exponential decay,

$$g^{(1)}(\tau) = \exp(-\Gamma\tau) \qquad (1.8)$$

where τ is the delay time of the autocorrelation function and Γ is the decay constant given by

$$\Gamma = D\kappa^2 \qquad (1.9)$$

where

$$\kappa = (4n\pi/\lambda)\sin\frac{\theta}{2} \qquad (1.10)$$

D is the Brownian diffusion coefficient of the particles [see Eq. (1.11)], and n, λ, and θ are the refractive index of the medium, wavelength of the light, and scattering angle, respectively.

The autocorrelation of the time dependence of light scattered from 96 nm polystyrene spheres is shown in Figure 1.8. The diffusion coefficient is found by fitting the data to Eq. (1.8) and solving for the diffusion coefficient, Eq. (1.9). The diffusion coefficient D is inversely proportional to particle diameter, as shown by the Einstein equation,

$$D = \frac{kT}{3\pi\eta d} \qquad (1.11)$$

where k is the Boltzmann constant, T is the absolute temperature, η is the viscosity of the liquid, and d is the diameter of the particle. Therefore, by measuring the scattered intensity as a function of time, for example, Figures 1.6 and 1.7,

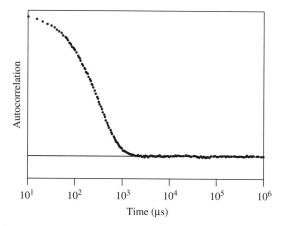

Figure 1.8 The autocorrelation of the time-dependent intensity of scattered light from a monodisperse dispersion of 96 nm diameter polystyrene spheres in water. Scattering angle of 90° and at 25°C.

calculating the autocorrelation of the data, Figure 1.8, fitting the data to an exponential, Eq. (1.8), and solving for the diffusion coefficient, Eq. (1.9), the particle diameter can be calculated by Eq. (1.11). This procedure is so accurate that quasi-elastic light scattering is used to determine the sizes of polystyrene particle standards.

For a polydisperse system of spherical particles in Brownian motion, the normalized autocorrelation function, $g^{(1)}(\tau)$, is an integral of exponentials:

$$g^{(1)}(\tau) = \int_0^\infty F(\Gamma) \exp(-\Gamma \tau) \, d\Gamma \tag{1.12}$$

The advantage of using quasi-elastic light scattering (the time dependence of scattered light) over Mie scattering (the angular dependence of scattered light) is that the subsequent deconvolution of the data for polydisperse samples is less susceptible to experimental fluctuations (see Chapter 4).

1.6 LASER VELOCIMETRY

Laser velocimetry is a light-scattering technique to measure the velocities of suspended particles. The technique is applicable to velocities from a few microns per second, appropriate for low-field electrophoretic measurements to hundreds of meters per second, appropriate for turbulent flow in air measurements. The experimental setup is to split a laser beam into two parts and direct the two beams to cross each other. As a particle passes through this scattering volume it scatters light from both beams. Since the beams hit the particle from different directions, the scattered light from the beams are slightly out of phase. The frequency of this out-of-phase signal is directly proportional to the velocity of the particle in the direction perpendicular to the scattering volume. Three different color lasers with three photodetectors each sensitive only to a particular color may be used to measure all three components of the velocity of particle. This is particularly useful in mapping turbulent flows.

The crossed laser beams create a standing optical fringe whose spacing, δ, is given by

$$\delta = \frac{\lambda}{2\sin(\theta/2)} \qquad (1.13)$$

where λ is the wavelength of the laser and θ is the angle between the two beams. A HeNe laser beam with a wavelength of 633 nm, split and crossed at an angle of 10°, produces a standing optical fringe pattern with a spacing of 3630 nm. The beat frequency, f, of a particle passing through the scattering volume with velocity, u, is

$$f = \frac{u}{\delta} \qquad (1.14)$$

If the velocity of the particle were 10^4 nm/sec, the frequency of the beat signal for the example just given would be 2.75 Hz. Low frequencies are difficult to measure so the experimental frequency is shifted by reflecting one of the beams from a mirror vibrating at a few hundred Hertz, thus moving the signal to a higher frequency. Laser velocimetry is most often used in colloid chemistry for the measurement of electrophoretic velocities. Only a fraction of a second is required for this measurement. Therefore the electric field can be reversed often, thus preventing electrode polarization or reactions. This enables measurements at high electric fields and at lower conductivities.

An interesting application of laser velocimetry is to measure blood flow in the small blood vessels of the microvasculature, such as the low-speed flows associated with nutritional blood flow in capillaries close to the skin surface and flow in the underlying arterioles and venules, part of the regulation of skin temperature. The tissue thickness is typically 1 mm, the capillary diameters 10 μm and the velocities 0.01 to 10 mm/sec. A scanning measurement can be made to map blood flow over an area.[16] Both large areas (a full torso) and small area (part of a finger) can be scanned, enabling the blood flow to be mapped.

REFERENCES

[1] Tyndall, J. On the blue colour of the sky, the polarization of skylight, and on the polarization of light by cloudy matter generally, *Phil. Mag.* **1869**, [5], *37*, 384–394.

[2] Maxwell, J.C. *A treatise on electricity and magnetism*, 3rd ed.; Clarendon Press: Oxford, 1891, Vol. 2, p 439; Dover: New York; 1954.

[3] Bohren, C.F. Optics, Atmospheric, *Encycl. Appl. Phys.* **1995**, *12*, 405–434; Figure 7.

[4] Quickenden, T.; Handon, A. The colour of water and ice, *Chem. in Britain* **2000**, *36*(12), 37–39.

[5] Bohren, C.F.; Fraser, A.B. Colors of the sky, *The Phys. Teacher* **1985**, *23*, 267–272; Bohren, C.F. *Clouds in a glass of beer*; Wiley: New York; 1987.

[6] Kerker, M. *The scattering of light and other electromagnetic radiation*; Academic Press: New York; 1969; pp 433–437.

[7] Kratochvil, P. *Classical light scattering from polymer solutions*; Elsevier: New York; 1987; pp 122–144.

[8] Chu, B. *Laser light scattering, Basic principles and practice*, 2nd ed.; Academic Press: New York; 1991; pp 14–20.

[9] Mie, G. Beiträge zur Optik trüber Medien, speziell kolloidaler Metalösungen (Contributions to the optics of diffusing media, particularly of colloidal metals in solution), *Ann. Phys. (Leipzig)* **1908**, [4], *25*, 377–445.

[10] Kerker loc. cit.

[11] La Mer, V. K. Nucleation in phase transitions, *Ind. Eng. Chem.* **1952**, *44*, 1270–1277.

[12] Dauchot, J.; Watillon, A. Optical properties of selenium sols I. Computation of extinction curves from Mie equations, *J. Colloid Interface Sci.* **1967**, *23*, 62–72.

[13] Watillon, A.; Dauchot, J. Optical properties of selenium sols II. Preparation and particle size distribution, *J. Colloid Interface Sci.* **1968**, *27*, 507–515.

[14] Kerker loc. cit.; pp 324, 396–401.

[15] Minnaert, M. *The nature of light and colour in the open air*; Dover: New York; 1954; pp 236–237.

[16] www.moor.co.uk/laser_doppler_theory.htm

2 Rheology

2.1 THE MOTION OF DISPERSE-PHASE SYSTEMS

The viscosity of a fluid is essentially its internal friction. A body incapable of flow has infinite internal friction; a gas, on the other hand, flows readily because its internal friction is small; liquids are in between. The study of flow, or rheology, provides information about the internal structure of a liquid system; conversely, by controlling the internal structure, flow of a desired quality can be obtained.

For example, consider the flow of liquid between the two concentric cylinders shown in Figure 2.1. If a fluid and an outer cylinder are at rest and an inner cylinder revolves uniformly, a circular motion, communicated to the fluid, will be propagated by decreasing degrees through the fluid to the outer cylinder. The internal friction of the liquid diminishes the velocity of this motion even as it is being communicated to layers of liquid lying farther away.

The shearing stress applied by the inner cylinder is the force per unit of surface area of contact; the rate of shear of the liquid is its velocity gradient in a direction normal to that of the inner cylinder. Newton described these phenomena in the following:

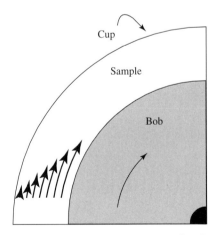

Figure 2.1 Velocity profile of liquid flow around a moving inner cylinder surrounded by a stationary outer cylinder. The arrows represent the direction and magnitude of the fluid flow.

If a fluid and an outer cylinder are at rest and an inner cylinder revolves uniformly, a circular motion, communicated to the fluid, will be propagated by decreasing degrees through the fluid to the outer cylinder.

— *Principia*, Prop. LI, Cor. V.

The flow of a dispersed-phase system reflects the complex interplay between the motion of the liquid and the motion of the particles. The overall motion could be analyzed in stages, first the analysis of how the liquid would flow when free of any dispersed-phase, then the effect of the dispersed phase on the continuous phase, and finally the effect of the motion of the dispersed phases on each other. For example, if the fluid contained between the two concentric cylinders of Figure 2.1 is a pure liquid then a stated motion of the inner cylinder would require a certain shear stress (force per unit area) to be applied to the outer cylinder. If a low concentration of particles were introduced into the liquid between the moving cylinders then the particles would impart to the liquid a small structure. To overcome this structure, a larger shear stress would have to be applied to the inner cylinder to keep it moving at the same velocity. As more and more particles are added to the dispersion, the structure would likely become stronger and stronger; therefore a larger and larger stress would be needed to keep the inner cylinder moving at the same shear rate. The overall flow properties of the dispersed-phase system is a complex interplay between all the components. The rheological experiment is to measure the flow under various controlled conditions and find a consistent explanation of the data.

A key to obtaining useful rheological data is make measurements in geometries where the shear rates are constant. This requires that thin layers of samples be used. Two types of shear are shown in Figure 2.2. For simple and rotational shear, the shear rate is nearly constant across thin layers.

2.1.a Newtonian Flow

The viscosity coefficient η is defined as the ratio of the shearing stress τ (Pa) to the rate of shear $\dot{\gamma}$ (s^{-1}) for one-dimensional flow. If the viscosity coefficient is independent of the rate of shear, the fluid is called Newtonian.

$$\tau = \eta \dot{\gamma} \qquad (2.1)$$

The viscosity η is the ratio of the shear stress (having units of pressure, 1 N/m^2 = 1 Pa) to the shear rate, having units of s^{-1}. Therefore the viscosity has units of Pa-s. The viscosity of water at room temperature is about 1 mPa-s. The viscosities of liquids, emulsions, and dispersions almost always decrease with higher temperatures, because the internal friction decreases as the temperature rises.

The rheology of many simple materials is Newtonian. Examples of Newtonian liquids are pure liquids, solutions of solutes smaller than macromolecules, and

18 RHEOLOGY

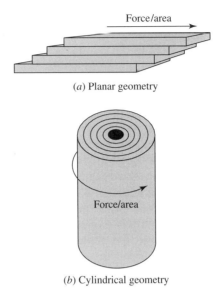

Figure 2.2 Various types of shear. (*a*) Simple shear (planar). (*b*) Rotational shear.

TABLE 2.1 The Viscosity of Some Familiar Materials at Room Temperature[1]

Liquid	Approximate Viscosity (Pa-s)
Glass	10^{40}
Molten glass (500°C)	10^{12}
Bitumen	10^{8}
Molten polymers	10^{3}
Golden syrup	10^{2}
Liquid honey	10^{1}
Glycerol	10^{0}
Olive oil	10^{-1}
Bicycle oil	10^{-2}
Water	10^{-3}
Air	10^{-5}

even some polymer solutions. The viscosity of various materials ranges over many orders of magnitude. Table 2.1 gives the viscosity of some familiar liquids.

Table 2.2 gives shear rates typical of some familiar processes. However, many liquids, including some that are common in everyday life, are non-Newtonian. Examples of these are paints, tomato ketchup, cold creams, ointments, and margarine. The flow of these and systems like them can still be described by a plot of shearing stress against rate of shear (the rheogram), but the result is not a straight

TABLE 2.2   Shear Rates Typical of Some Familiar Materials and Processes[2]

Situation	Typical Range of Shear Rates (s^{-1})	Application
Sedimentation of fine powers in a suspending liquid	$10^{-6}-10^{-4}$	Medicines, paints
Leveling due to surface tension	$10^{-2}-10^{-1}$	Paints, printing inks
Draining under gravity	$10^{-1}-10^{1}$	Painting and coating
Extruding	$10^{0}-10^{2}$	Polymers
Chewing and swallowing	$10^{1}-10^{2}$	Foods
Dip coating	$10^{1}-10^{2}$	Paints, confectionery
Mixing and stirring	$10^{1}-10^{3}$	Manufacturing liquids
Pipe flow	$10^{0}-10^{3}$	Pumping, blood flow
Spraying and brushing	$10^{3}-10^{4}$	Spray-drying, painting, fuel atomization
Rubbing	$10^{4}-10^{5}$	Application of creams to skin
Milling pigments in fluid bases	$10^{3}-10^{5}$	Paints, printing inks
High speed coating	$10^{5}-10^{6}$	Paper
Lubrication	$10^{3}-10^{7}$	Gasoline engines

line through the origin. Gravity is a ubiquitous source of stress on structures. As the sediment formed from a flocculated dispersion is pulled by the forces of gravity, it slowly collapses. The sediment height will decrease with time. The sediment formed by deflocculated particles is tightly packed and the sediment height does not change with time.

2.1.b   Poiseuille Flow

A useful application of Eq. (2.1) is to describe mathematically the volume flow of a liquid through a cylindrical pipe as a function of the pressure drop down the pipe. (Figure 2.3). The rigorous derivation of this problem requires transforming the equations of continuity and motion into cylindrical coordinates.[3] The result is

$$\omega = \frac{\pi \Delta p r^4}{8L\eta} \tag{2.2}$$

Equation (2.2) is due to Poiseuille (1844) and relates the rate of flow, ω, to the applied pressure on the liquid flowing through a pipe, Δp, r, the radius of the pipe, L, the length of the pipe, and η, the viscosity of the fluid. A common use of Eq. (2.2) is to calculate either the volume-flow rate at a given pressure drop or the pressure drop required to produce a given flow rate through a pipe

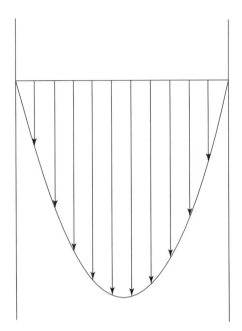

Figure 2.3 Parabolic profile of velocity in a cylindrical pipe.

of known dimensions for a liquid of known viscosity. This is important in the sizing of pipes and pumps for pilot and manufacturing equipment.

The rheogram for the flow of liquid in a pipe may be plotted directly from measured observations as ω versus Δp, and if the liquid is Newtonian, that is, η constant, a straight line through the origin is the result. The slope of the line is $\pi r^4/8L\eta$.

2.1.c Elasticity

Materials that stretch when under stress and return to their original shape when the stress is removed are called elastic. Rubber bands are obvious examples. The behavior of elastic materials cannot be described by Newton's equation, (2.1). When a stress is applied to an elastic material, it stretches a certain distance and then stops. When a stress is applied to a viscous material, it continues to flow. The elasticity of a material is described by Hooke's law (*ut tensio sic vis* or the extension is proportional to the force):

$$\tau = G\varepsilon \tag{2.3}$$

where τ is the stress, ε is the strain (fractional change in length) and G is called the elastic modulus and has units of Pa. Thus if a rod of length L and cross section A is extended by a force F to length $L + X$, the strain is $\varepsilon = X/L$ or extension per unit length and the stress is $\tau = F/A$ or force per unit area.

The modulus, of coefficient of elasticity, is given by $G =$ stress $\div$ strain. The elasticity of a material can be constant over a wide range of stresses so that the elastic modulus is a constant. However, the elasticity of many materials is nonlinear and the elastic modulus varies with the strain.

2.1.d Plastic Flow and Shear-Thinning Flow

Many materials are only elastic as long as the stress does not exceed some critical value. This critical stress is called the yield point. These materials will stretch and return to their original shape when the stress is below the yield point, but will stretch, then flow and be permanently deformed when the stress is higher than the yield point. Such materials are called plastic. An interesting example is provided by the flow of glaciers for which the rheogram is shown on Figure 2.4. The diagram shows the extremely high yield point, approximately 20 kPa, and its extremely high apparent viscosity. The flow behavior of ice was first demonstrated by J. D Forbes in 1842 by planting a line of stakes across the Mer de Glace glacier. On returning the following year he found that the linear array had moved into a parabolic curve.[4,5] The rheological behavior of ice, metals, or monomineral rocks depends on the statistical distributions of the orientations of the crystallographic axes in their different grains, termed its *fabric*. Continuous strain of a polycrystalline material progressively produces a fabric. In general, the crystallographic axes rotate until an orientation allowing easy glide is reached. The situation in a system of dispersed solids is similar in that the flocculation of particles produces a fabric.[6]

Bingham pointed out that the flow behavior of many materials that have a yield point can be approximated on the rheogram as a straight line intercepting the axis at the yield point. Two numerical indices are then available to describe this behavior: the yield point and the plastic viscosity, U, derived from the slope of the straight line. Such materials are named "Bingham bodies." Bingham bodies

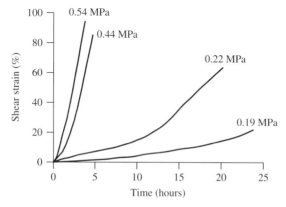

Figure 2.4 Creep versus time of ice at different shear stresses demonstrating plastic flow.[7]

TABLE 2.3 Terms Commonly Used to Describe Flow Properties of Consistency of Pigment Dispersion with Approximate Ranges in Absolute Units[10]

Descriptive Terms Frequently Used	Plastic Viscosity		Yield Value		Example
	Range	Magnitude (Pa-s)	Range	Magnitude (Pa)	
Watery; thin soupy; highly fluid; nontacky	Low	10.01–0.1	Low	0–1	Spraying lacquer; gravure ink
Creamy	Low	0.1–0.5	Medium	5–500	Paints; ketchup
Pasty; stiff; buttery; high-consistency; nonfluid; nontacky; short	Low	0.001–0.05	High	50–500	Textile color pastes; (stipple paints); shaving cream; mayonnaise
Fluid; low tack	Medium	0.5–5	Low	0–100	Black news ink
Buttery; stiff; pasty; salve like; short	Medium	0.5–5	High	100–1,000	Ointments
Long; molasseslike; tacky; highly viscous	High	10–100	Medium	10–1,000	Rotary press ink
Long; heavy bodied; high consistency; tacky	High	10–100	Medium high	100–3,000	Job press; offset inks
Leathery; tough; rubbery sticky	Very high	10^2–10^6	Very high	—	Resin melts; rubber; asphalt

such as spraying lacquer have low yield-point values (0–1 Pa) and low plastic viscosities (0.01–0.10 Pa-s); bodies such as paints or ketchup have medium yield-point values (5–500 Pa) and low plastic viscosities (0.1–0.5 Pa-s); and bodies such as ointments have high yield-point values (100–1000 Pa) and medium to high plastic viscosities (0.5–5 Pa-s). Several examples of combinations of yield-point values and plastic viscosities are listed in Table 2.3.

A gradual transition from Newtonian to plastic flow is often seen as the solids content of a flocculated dispersion is increased. The reason for this is that the higher concentrations of particles create a stronger structure in the dispersion with a correspondingly higher yield value. (Figure 2.5)

The sharp increase in plastic behavior can be seen in the rheograms for a series of methyl cellulose sols (Figure 2.6) The data are reported as the revolution per minute versus the load in grams. These are proportional to shear rate and to shear stress, respectively, therefore the apparent viscosity at any point is the ratio of the load to the rpm. Note that the apparent yield point increases dramatically from 14 to 18 percent solids — a narrow range of concentrations.

A second example of the transition from plastic to Newtonian flow is found in a series of kaolin dispersions to which has been added various amounts of the dispersant, tetrasodium pyrophosphate, TSPP (Figure 2.7). With no dispersant added, the clay slurry has a yield point of almost 20 Pa. The yield point is usually called the apparent yield point because it is the extrapolation of a mathematical fit to the rheogram at high shear rates to zero shear rate. Each addition of the dispersant reduces the apparent yield point, clearly showing that the dispersant is reducing the interparticle forces between the clay platelets. At the highest level of dispersant added, 0.3 weight percent, the kaolin dispersion flows like a Newtonian liquid.

With the advent of ever more sensitive rheometers, fewer and fewer materials are found to be strictly elastic below the yield point but rather deform only slowly. This is generally true for flocculated dispersions. At low shear stresses

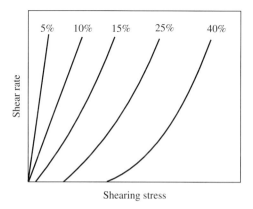

Figure 2.5 Rheograms showing the transition from Newtonian to plastic flow with concentration of flocculated solids in a solid–liquid dispersion.

24 RHEOLOGY

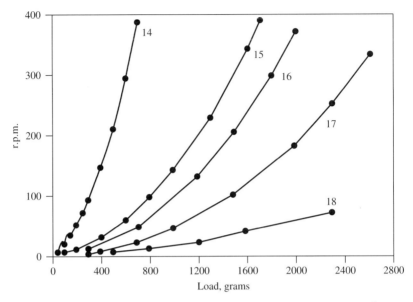

Figure 2.6 Pseudoplastic flow of a series of methyl cellulose sols.[8]

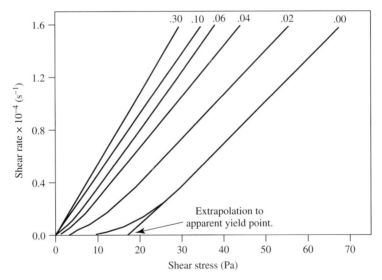

Figure 2.7 Rheograms of 20 weight-percent deionized kaolin slurries at several levels of tetrasodium pyrophosphate addition. The figures on the curves indicate percent TSPP per weight of clay. An extrapolation of the linear region determines an apparent yield point.[9]

the floc structure remains intact. The flocculated dispersion will deform under stress and return to its original position if the stress is removed quickly. However, when the stress is maintained for longer times, the floc structure slowly flows and the dispersion will only return partially to its initial position when the stress is removed. At higher shear stresses, the floc structure is quickly destroyed and the flow approximates that of Newtonian fluid.

The yield point of a Bingham body marks the destruction of an internal static structure such as is found in a flocculated suspension. On shearing beyond the yield point, the structure may degenerate further, which causes a reduction of the apparent viscosity coefficient, an effect described as "shear thinning."

Some plastic materials have vanishingly small yield points but still show "shear thinning." The internal structure of such materials is too fragile to withstand even the slightest shearing stress and gradually disintegrates as further shear is applied. These materials, having no appreciable yield point but an apparent viscosity coefficient that continually decreases with applied stress, are sometimes called "pseudoplastic." Figure 2.8 illustrates the structure of a flocculated dispersion at rest and under flow. The flocculated structure provides an initial resistance to flow. As the shear stresses increase with flow, the structure is gradually destroyed and the flow increases. With the high sensitivity of modern controlled-stress rheometers, yield points down to 1 mPa are detectable. This corresponds to the presence of a flocculated structure too weak to be apparent even by merely pouring the dispersion.

The rheogram of many shear thinning systems is linear on a log–log plot of the viscosity versus shear rate, $\dot{\gamma}$, (or shear stress, τ).

$$\log(\eta) = (1 - n)\log(\dot{\gamma}) + \log(\eta_0) \qquad (2.4)$$

where η is the viscosity, n is called the power law coefficient and η_0 is the viscosity at 1 s^{-1}. The power law coefficient, n, is 1 when the dispersion is Newtonian, and between 0 and 1 when the dispersion is shear thinning. As dispersions are made more and more stable by the addition of a dispersant, the value of the power law coefficient rises from a fraction towards unity. Therefore the power law coefficient is a convenient measure of the effectiveness of dispersants. The flow of liquids through pipes is a strong function of the power law coefficient. When the coefficient is zero, the fluid is Newtonian and the velocity profile is parabolic with the highest fluid velocity in the center of the pipe. For lower values of the power law coefficient, the fluid profile becomes flatter, with the velocity nearly constant in the middle of the tube. This type of flow is called plug flow. For lower and lower values of the power law coefficient the diameter of the "plug" in the middle of the pipe gets larger and larger.

2.1.e Dilatancy

A completely different type of flow behavior is displayed by materials whose resistance to flow increases with increased applied stress. Every child who has

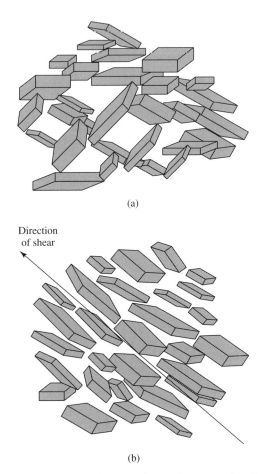

Figure 2.8 A flocculated suspension under rest and under shear.

played with a bucket and spade on the seashore has discovered how difficult it is to stir wet sand. The sand particles are round, approximately all of the same size, and are completely wetted by seawater. The difficulty experienced in stirring this mixture is caused by the obstruction to flow of a closely packed and deflocculated system of particles. Reynolds in 1885 coined the term "dilatancy" to describe this effect. His explanation is that the sand particles, when undisturbed, have settled to a close-packed arrangement and that any disturbance causes rearrangement to a smaller number of nearest neighbors, in which the particles are farther apart. The familiar example that he had in mind was the drying of moist sand when one stands on it. The pressure of the foot expands the underlying sand structure, and the surrounding surface moisture flows into it leaving a visible dry area; on raising the foot, a puddle of water is disclosed as the expanded structure settles back, releasing water as the voids collapse. After a short time the puddle soaks back into the dry area, and all is as before.

THE MOTION OF DISPERSE-PHASE SYSTEMS 27

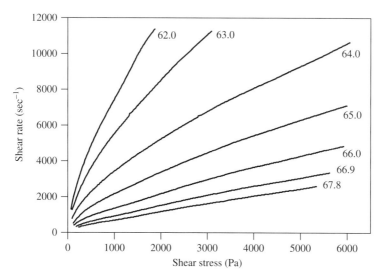

Figure 2.9 Rheograms for a series of curves of deflocculated paper-coating-grade clay, at weight-percent solids indicated, showing development of shear thickening behavior.[11]

The degree of dilatancy of a deflocculated dispersion depends strongly on the volume loading of particles. At high volume loadings, the flow of the dispersion is greatly hindered by the motion of particles interfering with each other. This interference decreases with particle loading. Figure 2.9 shows rheograms for a series of deflocculated paper-coating-grade clay, at weight percent solids indicated, showing development of shear-thickening behavior.[11] The apparent viscosity at any point is the ratio of the shearing stress to the shear rate. The 62% solids dispersion has a low viscosity. As the solids concentration is increased, the initial slopes of the rheograms decrease indicating a higher and higher viscosity. Furthermore, the slope of each rheogram becomes less the faster the dispersion is stirred, indicating a higher apparent viscosity.

A simple demonstration of the effect can be made with a plastic bottle with an extended neck containing sand and water (Figure 2.10). When the container is squeezed, the net volume increases (this is hard to believe!) and the level of liquid in the neck declines. Try it for yourself!

Dilatant mixtures, similar to sand and water, are created by cornstarch and water. A common dilatant fluid is Dow Corning 3179, a silicone compound better known as Silly Putty. Other examples are the polishing compounds used in the semiconductor industry, which require high concentrations of finely divided solids and therefore exhibit dilatant properties.

2.1.f Summary of Common Rheograms

Newtonian flow, plastic flow, shear thinning, and dilatancy are the common kinds of rheograms. These rheograms are shown schematically in Figure 2.11.

28 RHEOLOGY

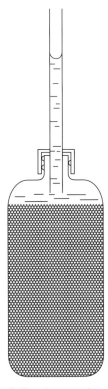

Figure 2.10 A mysterious bottle.

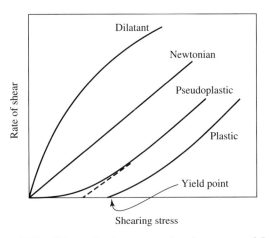

Figure 2.11 Schematic rheograms of various types of flow.

The experimental data are usually taken either by applying a controlled stress and measuring the rate of strain or by applying a controlled rate of strain and measuring the consequent stress. For many materials the resulting rheogram is independent of the rate at which the independent variable is changed. This is true as long as the time the system takes to relax from a perturbed state is small compared to the time of the measurement. If the system does not relax to a new steady state after the application of a change within the time of the measurement, then the rheogram will depend on the time scale of the measurement. This can be important information because many processes involve rapid changes of conditions. Knowing how the dispersion responds is important. A simple example of a rheogram that might depend on the rate of the measurement is that of a flocculated dispersion where the floc might take some time to re-form. The next sections discuss the various time-dependent rheological effects.

2.1.g Thixotropy and Rheopexy

A feature of some shear-thinning liquid systems is that significant periods of time are required for an internal structure to re-form (heal) after it has been broken by an applied stress. The time required may be anything from a few seconds to a few months depending on the nature of the system. Lubricating greases are prone to show this phenomenon. A sample of the grease may be worked energetically, and the rate at which its consistency returns as its structure re-forms is tested periodically. The effect is shown on the rheogram by tracing the variation of the rate of shear as the stress is gradually increased and then reversing the direction of the variation of stress. If the internal structure is reformed slowly, the reverse rheogram curve lies above the ascending curve, as shown in Figure 2.12. Behavior of this description is called thixotropy. The slow healing of a shear-thinned system is measured conveniently by taking the area of the hysteresis loop. This measure of thixotropy depends on the rate of change of the shear stress. A more systematic study of thixotropy is to measure the viscoelasticity of the dispersion. The viscoelastic characterization of a dispersion explicitly accounts for time-dependent phenomena by measuring the elasticity as well as the viscosity.

Thixotropic behavior can also be detected by a decrease in apparent viscosity when a dispersion is stirred at a *constant* shear rate over a period of time. This is the result of a gradual deterioration of structure that is unable to re-form instantly.

Another kind of non-Newtonian behavior can be made manifest by the stirring at constant shear. The viscosity builds up with time, creating a structure that is unable to relax instantly. The effect is the opposite of thixotropy, sometimes called antithixotropy, but Freundlich coined the term "rheopexy" for this phenomenon, commenting that "it is too cumbersome to say 'thixotropic sols that can be solidified by orienting the particles.'"

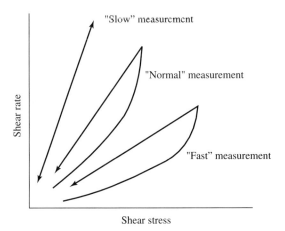

Figure 2.12 Rheograms of a thixotropic system at different sweep rates. The "up" curves are measured at steadily increasing shearing stresses and the down curves are measured at steadily decreasing shearing stresses. The area of the loop is a measure of thixotropic breakdown. The less rapid the cycling the less the thixotropy as more time is given for healing.

2.1.h Viscoelasticity

Viscous flow is described by

$$\eta = \frac{\tau}{\dot{\gamma}} \tag{2.5}$$

where η is called the apparent viscosity at any given rate of shear. Only materials that flow are described by this equation. Viscous fluids eventually take the shape of their container.

The property of elasticity is described by Hooke's equation

$$G = \frac{\tau}{\varepsilon} \tag{2.6}$$

where G is called the apparent elasticity. Elastic materials deform but do not generally take the shape of their container.

The flow behavior of many dispersed-phase systems cannot be described by these equations. These materials do not take the shape of their containers at first, but they eventually flow. They are "solids" at short times but "liquids" at long times.[12] Or they may be "solids" at low stresses but "liquids" at high stresses. This behavior is typical of a piece of plastic. The time required for a material to change from elastic behavior to viscous behavior can vary immensely.

Thixotropy is an example of time-dependent behavior. The most common experimental and mathematical treatment of viscoelasticity is applied to polymers and polymer solutions.[13] The study of the viscoelasticity of complex fluids containing structure-forming materials has adopted concepts from polymer viscoelasticity to elucidate structural information.[14]

Mechanical models of viscoelastic materials are constructed from combinations of viscous components and elastic components. These models resemble electric circuits comprising capacitors and resistors. The elements for viscoelastic behavior are dashpots, the viscous elements obeying Newton's equation, and springs, obeying Hooke's equation. Complex viscoelastic models are built from these elements. Figure 2.13 shows the viscoelastic elements and three simple mechanical models for viscoelastic behavior.

Maxwell postulated that the simplest realistic viscoelastic model[15] is the combination of a viscous component in parallel with an elastic component. The mechanical model is that of a spring stretched through a viscous liquid. The differential equation that describes the motion is

$$\tau = \eta \frac{d\varepsilon}{dt} - \frac{\eta}{GA} \frac{dF}{dt} \qquad (2.7)$$

where G is the elastic modulus and has units of Pa. The viscous term is another form of Newton's equation, (2.1), where the shear rate is replaced by the strain rate

$$\dot{\gamma} = \frac{d\varepsilon}{dt} \qquad (2.8)$$

The strain, ε, is the fractional change in position of a fluid element under stress. The alternate form of Newton's equation is

$$\tau = \eta \frac{d\varepsilon}{dt} \qquad (2.9)$$

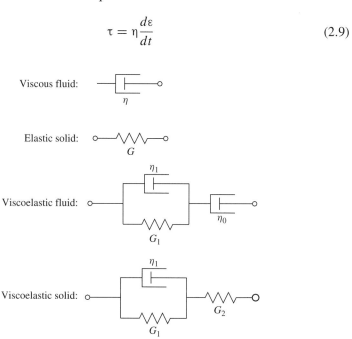

Figure 2.13 The mechanical models for viscoelasticity. (*a*) The viscous fluid. (*b*) The elastic solid. (*c*) The viscoelastic fluid. (*e*) The viscoelastic solid.

Equation (2.7) can be solved for various experiments. For example, if the viscoelastic material is subject to a sudden strain and held in that position ($d\varepsilon/dt = 0$) then the stress decays exponentially as the system relaxes

$$\tau(t) = \tau_0 \exp\left(-\frac{G}{\eta}t\right), \quad (2.10)$$

where τ_0 is the stress at time 0. The ratio of the viscosity to the elastic modulus can be calculated from the slope and intercept of the stress versus time on a log-log plot. The ratio of the viscosity to the elastic modulus is called the characteristic time. The time-dependent stress that accompanies a constant strain is called stress relaxation. This experiment measures the stress relaxation.

If the experiment is to apply a constant stress ($d\tau/dt = 0$) and measure the strain, then Eq. (2.7) gives the strain as a function of time.

$$\varepsilon = \tau_0 \left(\frac{1}{G} + \frac{t}{\eta}\right) \quad (2.11)$$

The strain increases instantaneously by the ratio of the stress to the elastic modulus and then increases linearly with time inversely proportional to the viscosity. When the strain versus time data are plotted on a linear scale, the intercept of the line gives the inverse of the elastic modulus, G, and the slope of the line gives the inverse of the viscosity. The time-dependent strain that accompanies a constant stress is called creep relaxation. This experiment measures the creep relaxation.

When the viscoelastic data do not fit these simple models, more complex models can be proposed and fitted to data. The experiments remain essentially the same; that is, they measure the stress as a function of strain or strain rate or measure the strain or strain rate as a function of a constant applied stress.

A more general technique is to study viscoelastic behavior while the applied stress or strain is changed sinusoidally with time. The frequency of the sinusoidally applied stress or strain is varied. The amplitude of the stress or strain is kept small enough to remain in the linear range of response.

Viscous components are dominant at low frequencies (long times) and elastic components dominate at high frequencies (short times). Therefore the analysis of the flow of a system as a function of the frequency of the applied strain or stress will give the elastic and viscous properties as a function of frequency. This analysis is described in more specialized texts.[16] These techniques work quite well for polymers, polymer solutions, and other solutions of complex materials. The experiments are more difficult for flocculated particles as the floc structures are usually quite brittle. The linear region of viscoelastic behavior is at extremely low strains where the data are difficult to obtain.

Viscoelastic measurements are made in the linear region. Many practical problems are at large strains, far from the linear region. Care needs to be taken in extrapolating conclusions from data taken at small deformations to processes of interest at large deformation.

One further complication arises in the viscoelastic characterization of disperse systems. The usual viscoelastic analysis is based on the approximation that stresses only produce strains in the same direction as the applied stress. This is true when the structures are small, such as for low molecular weight polymers or their solutions. However, when the structures in the dispersion are large, as they are in dispersions of particles or liquid drops, then applied stresses can cause movement, such as rotation, in directions other than the applied stress. The stresses produced in a complex system in directions other than the direction of the applied stress are called normal stresses.

A common manifestation of normal stresses is the "die swell" of molten polymers when the polymer is extruded from a die. The diameter of the extruded polymer is larger than the diameter of the hole in the die. Normal stresses can also be detected as an upward pressure against the rotating cone or plate in a rheometer. The rise of a polymer solution up the shaft of a rotating stirrer is an example of motion caused by normal stresses and is called the Weissenberg effect (Figure 2.14). The rise of the polymer solution is surprising, since centripetal forces would normally cause the spinning fluid to be thrown away from the center of rotation of the liquid.

2.1.i Units and Dimensions

The viscosimetric units and dimensions of the terms described are collected in Table 2.4.

Figure 2.14 The Weissenberg effect. A stirred polymer solution consising of 500 ppm of high-molecular-weight polystyrene in an oligomeric styrene resin rises up the shaft of the stirrer.[17]

TABLE 2.4 Viscosimetric Units[18]

Property	Symbol	Unit	CGS units	SI units	Dimensions
Viscosity (Newtonian)	η	Poise	dyn-s/cm^2	10^{-1} N-s/m^2	$[M^{-1}L^{-1}T^{-1}]$
Kinematic viscosity	St	Stoke	cm^2/s	10^{-4} m^2/s	$[L^2T^{-1}]$
Fluidity	Φ	Rhe	cm^2/dyn-s	10 m^2/N-s	$[M^{-1}LT]$
Plastic viscosity	U	Poise	dyn-s/cm^2	10^{-1} N-s/m^2	$[ML^{-1}T^{-1}]$
Mobility	μ	Rhe	cm^2/dyn-s	10 m^2/N-s	$[M^{-1}LT]$
Yield value	f		dyn/cm^2	10^{-1} N/m^2	$[ML^{-1}T^{-2}]$
Thixotropic breakdown	M		dyn/cm^2-s	10^{-1} N/m^2-s	$[ML^{-1}T^{-3}]$
Rate of shear	$D, \dot{\gamma}$		s^{-1}	s^{-1}	$[T^{-1}]$
Shearing stress	τ		dyn/cm^2	10^{-1} N/m^2	$[ML^{-1}T^{-2}]$
Torque	τ		dyn-cm	10^{-7} N-m	$[ML^{-1}T^{-3}]$
Rate of rotation	ω		rad/s	rad/s	$[T^{-1}]$

2.2 INSTRUMENTS TO MEASURE VISCOSITY

The methods to measure the rheology of a dispersed-phase system can be grouped into two categories: (1) those in which a stress is applied to a sample and the resulting strain or rate of strain is measured and (2) those in which a strain or strain rate is applied to a sample and the stress required to do so is measured. The first group is exemplified by letting a dispersion drain from a container under the influence of gravity and measuring the rate of flow. The second group is exemplified by measuring the torque on a spindle in a dispersion while turning at a fixed rpm. All of these measurements can be made as a function of time and/or temperature. Sometimes these experiments are conducted with an oscillatory motion. The simplest experimental technique is to measure the draining of a liquid through a capillary with the capillary viscosimeter.

2.2.a Capillary and Other Tube Viscosimeters

A complete capillary viscosimeter consists of five essential parts: (1) a fluid reservoir, (2) a capillary of known dimensions, (3) a unit to control and measure the applied pressure, (4) a unit to determine flow rate, and (5) a unit to control temperature. Capillary viscosimeters are designed to make use of Poiseuille's equation, hence they are useful only for Newtonian liquids. The applied stress may be the action of gravity on the liquid or it may be created by pressure of a gas to push the liquid through the capillary. The most common instrument of this type is the Ostwald–Ubbelohde viscosimeter (Figure 2.15). A standard liquid of known viscosity and density is used to calibrate the instrument. Since the viscosity of liquids varies with temperature, the instrument is immersed in

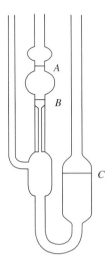

Figure 2.15 Capillary viscosimeter.

a constant-temperature bath. The instrument is filled to the level C with liquid, which is then raised by suction above point A and allowed to flow through the capillary. The time t for the liquid to move from, A to B is measured. This time is proportional to the ratio of viscosity to density, η/ρ. Comparing the unknown 1 with the standard 2 gives

$$\eta_1 = \eta_2 \left(\frac{\rho_1 t_1}{\rho_2 t_2} \right) \qquad (2.12)$$

This instrument is satisfactory for simple liquids and solutions. For different ranges of viscosity, capillaries of different diameters are used; the higher the viscosity, the wider the capillary needed. The capillary viscosimeter is a one-point instrument, so called because it measures a single average applied stress. Other one-point instruments use the time required for a ball, a long needle, or a bubble to pass through the liquid. Instruments that can measure enough points on the rheogram to define its shape are called rheometers rather than viscosimeters. A capillary instrument can be made into a rheometer by varying the applied stresses.

The advantage of the capillary viscosimeter is its simplicity. All that is needed is the viscosimeter, a constant-temperature bath and a stopwatch. A set of capillary viscosimeters is purchased with various capillary diameters, the larger diameters being used for the higher viscosities. The capillary viscosimeter is also the instrument of choice for low-viscosity fluids, particularly pure solvents or dilute solutions. It is not suitable for dispersions or emulsions as sedimentation or creaming and flocculation against the walls are common. Even for polymer solutions care has to taken to be sure the capillary is clean after each measurement.

A particularly useful adaptation of the capillary viscosimeter is produced by Viscotek Corporation. The principle of operation is based on measuring pressure drops caused by the continuous forced flow of solvent and sample through two stainless steel capillary tubes placed in series. The pressure drop across each capillary tube obeys Poiseuille's law,

$$\Delta p = \eta Q R \tag{2.13}$$

where Δp is the pressure drop as measured by a differential pressure transducer, Q is the flow rate, R is the tube resistance, and η is the viscosity. The sample solution is loaded into the sample loop via the syringe pump and then pushed into capillary 2 when the injection valve is turned. A steady-state condition is reached when the sample solution completely fills capillary 2, solvent remaining in capillary 1 at all times. The relative viscosity of the sample solution is determined simply and directly by the ratio of the pressure drops. From the measured relative viscosity, all other solution viscosity measurements can be calculated. The determination of intrinsic viscosity (to obtain polymer molecular weights) is greatly simplified by automatic sample injection as the system is constantly flushed clean with a flow of solvent.

Capillary viscosimeters are also used to measure the flow of liquids at very high shear rates as high pressures can be used to force liquid flow through small capillaries. Data like these are particularly useful in characterizing materials for coating operations where shear rates are often quite high and out of the range of normal rotational rheometers. Commercial capillary viscosimeters are available from Parr Physica which can measure up to a shear rate of 10^6 s^{-1}.

The viscosity of liquids can be measured in a capillary viscosimeter by a second method where the capillary rise is measured rather than the rate of drainage. The height of liquid in the capillary is measured as a function of time. The height, h, rises from rest to the final height, h_∞ exponentially as a function of time, t, according to the relation

$$h(t) = h_\infty \left(1 - \exp(-t/\tau)\right) \tag{2.14}$$

where τ is the characteristic time. The viscosity, η, is given by[19]

$$\eta = \frac{g \rho r^2 \tau}{4 h_\infty} \tag{2.15}$$

where g is the acceleration of gravity, r is the radius of the tube, and ρ is the density of the liquid. This technique has an advantage for some opaque liquids because the advancing liquid front is sometimes easier to detect than a receding one.

In a device to measure viscosities at high pressures and temperatures the sample is placed within a stainless steel torus approximately 0.06 m in diameter, which is mounted on the rim of a wheel of about 2 m in diameter. When the wheel is rotating, the liquid volume is drawn with the wheel in the direction of rotation. At a constant rotational velocity the liquid plug sits at a steady-state displacement, which is viewed through high-pressure windows. The extra torque on the drive shaft caused by the displacement of the liquid is related to the shear stress. Silicone oils of known viscosity in the range of 1–200 mPa-s are used for calibration.[20]

2.2.b Falling-Ball and Rising-Bubble Viscosimeters

Figure 2.16 shows a diagram of a falling-ball viscosimeter. The tube is filled with a liquid of unknown viscosity but known density. A ball of known size and density is allowed to sediment under the influence of gravity and the time required to fall a known distance is measured. The viscosity is determined from Stokes' law,

$$\eta = \frac{2a^2 g (\rho_2 - \rho_1) t}{9d} \qquad (2.16)$$

where η is the viscosity, a is the radius of the ball, g is the acceleration due to gravity, ρ_2 is the density of the ball, ρ_1 is the density of the liquid, and t is the time the ball takes to fall between the marks a distance d apart. The falling-ball viscosimeter is also available with a magnetic sensor at the two marks so that the time can be measured in an opaque liquid. This viscosimeter is a single-point

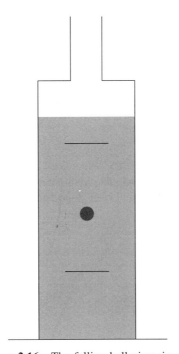

Figure 2.16 The falling-ball viscosimeter.

measurement so that only Newtonian fluids can be thus measured. A falling-ball viscosimeter is available from Parr Physica and Thermo Haake.

One variation of the falling-ball viscosimeter is to fill a tube with the unknown liquid, leave a small, but known volume of air at the top of the tube and seal the tube. When the tube is inverted, the bubble rises through the liquid. The viscosity of the liquid can be determined also by Eq. (2.16) where ρ_1, the density of the air is effectively zero. Sometimes results are reported as the viscosity divided by the density (kinematic viscosity) since this is the experimental measurement,

$$\frac{\eta}{\rho_2} = \frac{2a^2 gt}{9d} \qquad (2.17)$$

where a is the known radius of the bubble. The viscosity of a liquid divided by its density is the kinematic viscosity and has units of Stokes (1 St = 10^{-4} m^2 s^{-1}).

A further enhancement of the technique is to provide tubes filled with liquids of known viscosity and density, hence known kinematic viscosities. The procedure is to fill one tube to the proper level with the liquid of unknown kinematic viscosity, cap it, and invert all of the tubes at the same time. The kinematic viscosity of the unknown is then estimated by interpolating between the tubes that have nearly the same times of bubble rise.

2.2.c Falling-Needle Viscosimeters

The falling-needle viscosimeter consists of a slender hollow cylinder (the needle) with hemispherical ends dropping through a liquid in a cylindrical container. The terminal velocity of the falling needle is measured by timing the travel across a known distance. The time can be measured manually or with a magnetic sensor. The advantages of the falling-needle viscosimeter over the falling-ball viscosimeter are (1) The effects of the wall are explicitly accounted for with the concentric geometries of a needle falling in a cylinder whereas the falling-ball viscosimeter equations assume an infinite fluid. (2) The moving needle is more stable than the falling ball during descent. (3) The density of the needle can easily be changed with the addition of weights within the hollow needle. (4) Because of the concentric geometry, shear rates can be calculated. Non-Newtonian fluids can be characterized by varying the shear rate produced by the falling needle.[21] A falling needle viscosimeter which can easily measure viscosities up to 10^8 mPa-s is available from Stonybrook Scientific.

2.2.d Vibrating Probe Viscosimeters

One on-line technique is to extend a wire loop about six inches long into the process stream. One end of the wire loop is vibrated at 120 Hz with a known applied force. The movement of the wire loop is attenuated by the viscosity of the liquid. The other end of the wire loop is attached to an electromagnetic sensor which detects the amplitude of the motion. The signal from the electromagnetic sensor is proportional to the viscosity. A vibrating probe viscosimeter is available from Automation Products.

Another on-line technique is to insert a vibrating probe about five inches long into a process stream. The rod is vibrated at a constant frequency and a constant amplitude of 1 μm. The power to maintain this constant strain rate is a function of the viscosity of the liquid. The constant strain rate provides the precision of measurement. For greater sensitivity a sphere or thicker rod can be attached to the end of the rod. The viscosimeter can be attached to small chambers for laboratory use, requiring from 10 to 35 mL of liquid depending on the head design and is available from Nametre.

2.2.e Piston Viscosimeters

These in-line viscosimeters operate by withdrawing a small portion of the fluid from a flowing stream into a chamber. The chamber contains a piston which can be moved back and forth magnetically and the time the piston takes to move a known distance is a measure of the viscosity.

The chamber containing the neutrally buoyant piston has two electromagnetic coils. First the lower one is activated to pull the piston downward. The upper coil senses the motion. The downward motion of the piston expels fluid in the chamber around the moving piston. The velocity of the piston is proportional to the viscosity. When the piston has reached the lower end of the chamber, the upper coil is activated to pull the piston upward with a known force. The lower coil senses the speed of the upward moving piston. That motion is impeded by the inward flow of the fluid around the piston. This process of moving the piston with a known electromagnetic force and measuring the velocity of the piston with a second coil takes only a few seconds and automatically refills the chamber with fresh fluid on every cycle. The instrument simultaneously measures the temperature. A piston viscosimeter is available from Cambridge Applied Systems.

2.2.f Microbalance Viscosimeters

This handheld instrument uses only about 10 mL of sample. A drop of liquid is placed on top of a quartz microbalance, which is vibrated horizontally at a fixed high frequency. The oscillating crystal produces a shear wave that travels into the fluid only a few microns. Through concepts developed over many years with surface-wave-acoustics technology, a correlation has been developed that relates the change in performance of the crystal to the viscosity of the fluid. The instrument is most sensitive for low viscosities. A microbalance viscosimeter is available from DICKEY-john Corp.

2.2.g Rheometers

For almost all polymer solutions, surfactant solutions, dispersion, foams, and emulsions, the ratio of shear stress to shear rate is not constant for all shear rates. The reason is simple; the flow of complex fluids depends on the structure of the fluid and that structure changes with shear. A useful understanding of the flow of complex materials requires measuring the viscosity over a range of shear rates. The viscosimeters described above are not adequate to characterize complex fluids completely; either the shear rate is not uniform across the sample when the measurement is made or the viscosity is only measured at one shear rate. (They are, however, quite useful in many practical applications such as quality control.)

Rheometers, equipment to measure viscosity as a function of shear, are designed to provide a constant shear rate throughout the sample and to vary the shear rate over a reasonable range. A constant shear rate is provided by measuring the flow of a thin layer of fluid, generally fluid held between two concentric cylinders or fluid between a cone and a flat plate. Almost all rheometers are designed to use both concentric cylinders and cones and plates. Concentric cylinder geometries are preferred for low-viscosity fluids or volatile solvents. Cone and plate geometries are preferred for high shear rates, high viscosities, and small sample volumes. The cone and plate geometries are much easier to clean.

Rheometers vary in whether they provide a constant shear rate and measure the resulting stress or whether they provide a constant shear stress and measure the shear rate. In constant shear-rate rheometers either one of the concentric cylinders or one of the cones or plates is turned and the stress transmitted to the other is measured. Constant stress rheometers can be constructed so that the stress is applied with either cylinder or with either the cone or the plate and the motion of that part measured. Therefore, in constant stress rheometers the stress and motion can be measured on the same part of the equipment, for example, the bob or the cone. This simplifies the design.

All modern rheometers are computer controlled so that they can be programmed to vary either the shear rate or stress over a wide and continuous range. The rheometers can also vary the time over which the data are taken to look for time effects. Any elasticity in a sample is reflected in time-dependent viscosities. This means, of course, that if the viscosity varies with time, the rheogram cannot be interpreted with Newton's equation but with the more elaborate analysis of viscoelasticity. A common indication of viscoelastic behavior is thixotropy, where the rheogram shows hysteresis over the same range.

Common measurement programs are:

Stress viscometry

Measures the shear rate at a constant applied stress

Constant stress viscometry

Measures the creep and creep recovery at constant applied stress

Constant rate viscometry

Measures the necessary stress to attain a constant shear rate

Oscillation viscometry

Measures the complex viscosity as a function of frequency of applied oscillatory motion

Stress sweep viscometry

Measures the complex viscosity as a function of stress at a fixed frequency.

2.2.g(i) The Brookfield Viscosimeter Possibly the most common viscosimeter in industrial use is the Brookfield viscosimeter. The unit consists of a rotating rod connected to the motor by a torsional spring. Various cylindrical bobs of different diameters can be attached to the rod. The suspended bobs can be turned over a range of rotational velocities. The stress on the bob generated by the motion through the liquid at a known rotational velocity is measured by the deflection of the torsional spring. Figure 2.17 shows the Brookfield viscosimeter in its standard configuration. The shear rate for this geometry is not well defined and the resulting torque measurements give only a indication of the viscosity; however, the technique is simple and hence popular.

More recent models of the Brookfield viscosimeter come with concentric cylinder sample chambers so that the shear rates are better defined (Figure 2.18). In this model, the Brookfield viscosimeter can be used to obtain a viscosity over a range of shear rates. A wide variety of viscosimeters, for both controlled stress and controlled strain, are available from Brookfield Engineering.

2.2.g(ii) Coaxial Cylindrical Rheometers Couette or rotational flow occurs in the annular gap between concentric cylinders, usually called the bob and the cup. The advantage of coaxial rheometers is that these rheometers provide a more constant shear rate across the sample and that they can be operated over a much

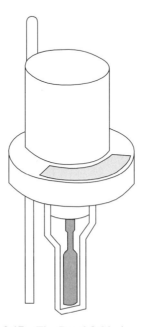

Figure 2.17 The Brookfield viscosimeter.

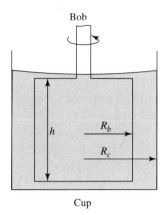

Figure 2.18 The concentric cylinder rheometer. For constant shear-rate measurements, either cylinder is rotated at a constant rate and the stress transmitted to the other cylinder is measured.

wider range of shear rates. The shear rate is varied by controlling the rate of rotation of the bob or the cup, so that the instrument is capable of providing a rheogram. In bob-and-cup rheometers, the driving force is balanced by viscous forces. At steady state the rate of rotation is constant and the shear rate is nearly uniform; the following relation holds:

$$\tau = 4\pi r^2 \eta h \Omega \tag{2.18}$$

where τ is the torque, η is the viscosity coefficient, h is the height of the liquid in the cup, and Ω is the rate of rotation (radians per second.) This relation ignores the effect of the fluid flow at the bottom of the bob. The factor r^2 is calculated from the inner radius of the cup and the outer radius of the bob as follows:

$$r^2 = \frac{R_c^2 R_b^2}{R_c^2 - R_b^2} \tag{2.19}$$

where R_b and R_c are the radii of the bob and cup, respectively. Greater sensitivity can be obtained by using what is called a double Couette, where a hollow bob surrounds a stationary cylinder inside the cup. This gives two annular liquid films.

In the original Couette rheometer (1890) and the MacMichael rheometer (1915), the rate of rotation of the cup is the independent variable and the torque is measured. In the latter the torque is measured as "MacMichael degrees" (300° per revolution.)

2.2.g(iii) **Cone-and-Plate Rheometers** The cone-and-plate rheometer (Figure 2.19) consists essentially of a flat plate and a cone, the apex of which barely misses making contact with the plate. The sample of liquid is inserted

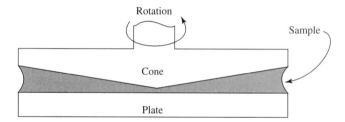

Figure 2.19 Schematic diagram of the cone-and-plate rheometer.[22]

into the narrow gap between the cone and the plate, where it is retained by capillary forces. The cone is almost a flat plate, e.g., 179.7 degrees. The torque resulting from rotating the cone at constant angular velocity is recorded. The design ensures that all portions of the sample are subjected to the same shear rate. Bob-and-cup rheometers do not provide this feature. For Newtonian liquids this difference would not matter because the flow of the liquid is directly proportional to the shearing stress, so for an average shear a proportionately average flow results. Not so for non-Newtonian fluids. If there is a yield point, no flow occurs at shearing stresses less than the yield point. Even at greater shearing stresses the flow is not proportional to the shear. Plug flow and other anomalies begin to appear when non-Newtonian fluids are subjected to shear rates that vary locally. The cone-and-plate rheometer is designed to avoid these anomalies.

From an article by R. McKennell of Ferranti Ltd., Manchester, England,[23]

Theoretically a cone of half-angle of $\theta_0 = 6°$ gives rise to a departure of only 0.35% from shear-rate uniformity. However, if θ_0 is greater than about 4°, errors may arise due to edge effect and, at higher shear rates, due to temperature rise within the fluid. In practice an angle of only 0.3° is used, with an average gap width of about 0.05 mm, requiring a sample of about 0. 1 cc. Cone angles of this order of magnitude simplify the mathematical analysis of non-Newtonian flow data because the whole of the measured sample attains a uniform shear rate and hence a constant apparent viscosity.

Let the rate of rotation be Ω rad/s, the cone half-angle be θ_0, and the gap width be w at radial distance x within the sample of radius R. The linear velocity of a point on the cone is proportional to the radial distance. The width of the gap is also proportional to the radial distance. The rate of shear D is given by

$$D = \frac{\text{linear velocity}}{\text{gap width}} = \frac{\Omega x}{w} (s^{-1}) \qquad (2.20)$$

Therefore,

$$D = \Omega/\theta_0 \quad \text{(i.e., independent of the distance } x\text{)} \qquad (2.21)$$

44   RHEOLOGY

The torque relations and velocity distribution in the cone-and-plate design are analyzed by Bird, Stewart and Lightfoot[24] Table 2.5 shows the fundamental simplicity of the cone-and-plate rheometer compared to the cylindrical rheometer by listing expressions for various rheological properties as a function of the dimensions of the equipment and the experimental values.

TABLE 2.5 Rheological Equations[23]

Property	Couette Rheometer	Cone-and-Plate Rheometer
Viscosity coefficient	$\eta = \tau/(4\pi r^2 h \Omega)$	$\eta = 3\tau/(2\pi R^3 D)$
Rate of shear	$D_{max} = 2\Omega r^2/R_b^2$	
	$D_{min} = 2\Omega r^2/R_c^2$	$D = \Omega/\theta_0$
Plastic viscosity	$U = (\tau - \tau_0)/(4\pi r^2 h \Omega)$	$U = 3(\tau - \tau_0)/(2\pi R^3 D)$
Yield value	$f = U\tau_0/[\ln(R_c/R_b)]$	$f = U\tau_0$
	(($\tau_0 =$ extrapolated) value of torque for $\Omega = 0$)	

Viscoelastic fluids return only partially to their original form when the applied stress is released. When stress is suddenly applied and maintained constant, the resulting strain (or change in form) with time is called a *creep curve*. In *stress relaxation* the sample is brought suddenly to a given deformation and the stress required to maintain this deformation is measured as a function of time. Viscoelastic behavior is important in connection with plastics manufacture, performance of lubricant greases, application of paints, processing of foods, and movement of biological fluids. Most of these systems consist of suspended particles or emulsion droplets of colloidal size; the interactions between such particles give rise to their viscoelastic or other non-Newtonian properties. The observed viscoelasticity of a material, therefore, can give information about how its constituents are spatially distributed and the nature of their interactions.

Materials of different types require instruments appropriate to their properties. In materials that flow readily, elastic strains relax quickly. Instruments that use forced sinusoidal deformations (mechanical spectrometers or oscillatory viscoelastometers) are the most useful to determine the relations between stress, deformation, and time. Small deformations are used so that the microstructure is not destroyed during and by the experiment. A suitable rheological test program is based on the identification of important flow conditions, both for production and use.

2.2.g(iv) Squeezing Flow Rheometers Squeezing flow rheometers measure the pressure generated between a stationary wall and a rotating cam. The shear rate can be varied by changing the cam rotation speed. These flows and pressures are similar to lubricating processes, especially in automobile engine crankshafts, and to flows and pressures in polymer extruders. The rheometers have a cam rotating inside a cylindrical, fluid-filled cavity with a pressure transducer flush-mounted on the inside wall. These rheometers can be mounted inside moving equipment, like gear pumps or extruders with the cam mounted on the rotating shafts for real time measurements. The rotating cam produces a wedge-shaped cavity that fluctuates in gap as the eccentric cam rotates. The maximum in the pressure occurs when the cam is closest to the wall. Cams with dual lobes of different clearances can be used for simultaneous dual shear-rate measurements.[25]

2.3 MANUFACTURERS OF RHEOMETERS

A wide choice of instruments is available for materials exhibiting a wide range of properties. About 40 rheometers are described by Walters (1980).[11]

Company	Telephone/FAX	Internet	Rheometers
ATS Rheosystems 52 Georgetown Rd Bordentown, NJ 08505	(609) 298-2522/ (609) 298-2795	www.atsrheosystems.com	Controlled stress and controlled strain
Automation Products, Inc. 3030 Max Roy St. Houston, TX 77008	(800) 231-2062/ (713) 869-7332	www.dynatrolusa.com	Vibrating rod in-line rheometers
Bohlin Instruments Inc. 1004 Eastpark Blvd. Cranbury, NJ 08512	(609) 655-4447/ (609) 655-1475	www.bohlin.co.uk	Controlled stress and controlled strain
Brinkmann Instruments, Inc. One Cantiague Rd. P.O. Box 1019 Westbury, NY 11590-0207	(800) 645-3050/ (516) 334-7500	www.brinkmann.com	Capillary viscosimeters
Brookfield Engineering Laboratories 11 Commerce Blvd. Middleboro, MA 02346	(800) 628-8139/ (508) 946-6262	www.belgmbh.com	Rotational and controlled shear rate
Camtel, Ltd. 5 Carrington House 37 Upper King St. Royston, Herts SG8 9AZ, UK	+44(0)-176-324-4280 +44(0)-176-324-4290	www.camtel.co.uk	Interfacial rheometer Wilhelmy plate and ring Bartell technique
Cambridge Applied Systems, Inc. 196 Boston Ave. Medford, MA 02155	(781) 393-6500/ (781) 393-6515	www.cambridgeapplied.com	Piston in-line rheometers

46 RHEOLOGY

Company	Telephone/FAX	Internet	Rheometers
Chemcal Electro-Physics Corp. 705 Yorklyn Rd. Hockessin, DE 19707	(800) CEP-CHEM/ (302) 239-4677	www.cep-corp.com	Squeezing flow
DICKEY-john Corp. 5200 DICKEY-john Rd Auburn, IL 62615	(800) 637-2952/ (217) 438-6012	www.djscientific.com	Handheld, quartz crystal microbalance sensor
Nametre 25 Wiggins Ave. Bedford, MA 01730	(781) 275-9660/ (781) 275-9665	www.nametre.com	Vibrating probe in-line rheometers
Paar Physica USA, Inc. 1056 Technology Park Dr. Glen Allen, VA 23060	(800) 688-FLOW/ (804) 266-6176	www.paarphysica.com	Falling ball High-shear capillary Magnetoviscometer Controlled stress and controlled strain
Rheometric Scientific, Inc. One Possumtown Rd. Piscataway, NJ 08854	(732) 560-8550/ (732) 560-7451	www.rheometrics.com	High-shear capillary Elongational Controlled stress and controlled strain
Stony Brook Scientific, Ltd. 51 Hartz Way Secaucus, NJ 07094	(201) 863-2212/ (201) 863-1890	www.stonybrooksci.com	Disposable/Portable high pressure high temperature falling needle
TA Instruments 109 Lukens Dr. New Castle, DE 19720	(302) 427-4000/ (302) 427-4001	www.tainst.com	Controlled stress and controlled strain
Thermo Haake (USA) 53 W. Century Road Paramus, NJ 07652	(201) 265-7865/ (201) 265-1977	www.haake.de	Falling-ball viscosimeter Controlled stress and controlled stain

REFERENCES

[1] Barnes, H.A.; Hutton, J.F.; Walters, K. *An introduction to rheology*; Elsevier: New York; 1989; p 11.

[2] Barnes loc. cit.; p13.

[3] Bird, R.B.; Stewart, W.E.; Lightfoot, E.N. *Transport phenomena*; Wiley: New York; 1960; pp 42–47.

[4] Forbes, J.D. *Occasional papers on the theory of glaciers: now first collected and chronologically arranged. With a prefatory note on the recent progress and present aspect of the theory*; Adam and Charles Black: Edinburgh; 1859; pp 1–12.

[5] Cunningham, F.F. *James David Forbes: Pioneer Scottish glaciologist;* Scottish Academic Press: Edinburgh; 1990; p 329.

[6] Lliboutry, L.A. *Very slow flows of solids*; Kluwer: Dordrecht; 1987; p 80.

[7] Griggs, D.T.; Coles, N.E. Creep of a single crystal of ice, *SIPRE Technical Report* **1954**, *11*, 24.

[8] Fischer, E.K. *Colloidal dispersions*; Wiley: New York; 1950; p 20.

[9] Olivier, J.P.; Sennett, P. Electrokinetic effects in kaolin-water systems. I. The measurement of electrophoretic mobility, *Clays Clay Miner* **1967**, *15*, 345–346.

[10] Fischer loc. cit.; p 156.

[11] Walters, K., Ed. *Rheometry: Industrial applications*; Wiley-RSP: New York; 1980; p 358.

[12] Larson, R.G. *The structure and rheology of complex fluids*; Oxford University Press: New York; 1999; pp 3–4.

[13] Rohn, C.L. *Analytic polymer rheology*; Hanser Publishers: New York; 1995.

[14] Larson, R.G loc. cit.

[15] Rohn loc. cit.; pp 209–210.

[16] Russel, W.B.; Saville, D.A.; Showalter, W.R. *Colloidal dispersions;* Cambridge University Press: New York; 1989.

[17] Barnes loc. cit.; p 154. The photograph courtesy of Mr. R. Welsh and Dr. J. Bico of M.I.J.

[18] Fischer loc. cit.; p 154.

[19] Acton, J.R.; Squire, P.T. *Solving equations with physical understanding*; Adam Hilger: Boston; 1985; p 48.

[20] Johnsen, E.E.; Førdedal, H.; Urdahl, O. A simplified experimental approach for measuring viscosity for water-in-crude-oil emulsions under flowing conditions, *J. Dispersion Sci. Tech.* **2001**, *22*, 33–39.

[21] Park, N.A.; Irvine, T.F., Jr. Measurements of rheological fluid properties with the falling needle viscometer, *Rev. Sci. Instrum.* **1988**, *59*, 2051–2058.

[22] Bird loc. cit.; p 99.

[23] McKennell, R. Cone-plate viscometer, *Anal. Chem.* **1956**, *28*, 1710–1714.

[24] Bird loc. cit.; pp 98–101.

[25] www.cep-corp.com

3 Kinetic and Statistical Properties

Particles and drops in suspension are constantly moving and colliding with each other and with walls and surfaces. Their motion may be caused by outside forces such as gravity and electric fields, bulk fluid motion such as pumping and stirring, or it may be caused by thermal forces that give rise to Brownian motion.

3.1 KINETIC PROPERTIES

The motion of particles caused by external forces such as gravity and centrifugal forces is often used to determine the size of the particles. The motion of particles caused by an externally applied electric force is often used to determine the charge on the particle.

3.1.a Motion of Particles by an Applied Force

Gravitational forces, centrifugal forces, and electric forces can move particles suspended in a fluid. Each of these forces causes particles to accelerate. The velocity of the particles would increase without bounds except that the increasing velocity of the particles produces an increasing viscous drag. When the viscous force balances the force of gravity, or of the centrifuge, or of an applied electric force, the particles move with a steady velocity. This velocity is called the terminal velocity. The key questions are what is the terminal velocity? How do particle properties such as size and charge relate to that velocity? And how long does a particle take to attain that velocity?

Stokes showed that the viscous drag on a sphere is proportional to the velocity of the particle. The drag force is $6\pi\eta a v$, where η is the fluid viscosity, a is the particle radius, and v is the particle velocity. The total force on the particle subject to an applied force is

$$F_{\text{total}} = F_{\text{applied}} - 6\pi\eta a v \tag{3.1}$$

where F_{applied} is the applied force (gravitational, centrifugal, or electrical). Newton's equation is that the total force is equal to the mass of the particle times the acceleration

$$F_{\text{total}} = m\frac{dv}{dt} \tag{3.2}$$

where t is time. Combining these equations and solving gives the velocity as a function of time

$$v = \frac{F_{\text{applied}}}{6\pi\eta a}\left(1 - \exp\left(-\frac{6\pi\eta a}{m}t\right)\right) \qquad (3.3)$$

Therefore the terminal velocity is

$$v_t = \frac{F_{\text{applied}}}{6\pi\eta a} \qquad (3.4)$$

and the characteristic time (the time for the particle to reach $1/e$, 37%, of its final velocity) is

$$t_0 = \frac{m}{6\pi\eta a} \qquad (3.5)$$

The time for velocity to reach 95% of its maximum is $3t_0$ or

$$t_{95\%} = 3t_0 = \frac{m}{2\pi\eta a} \qquad (3.6)$$

For a 1000 nm (1 μm) diameter particle with a density of 1 gm/cm³ in water, this time is about a third of a microsecond. For all practical purposes, submicron particles attain their terminal velocities instantaneously. Larger particles take longer to approach the terminal velocity since the ratio, m/a, varies as the diameter squared. A particle 1 mm in diameter would take about a third of a second to reach terminal velocity.

Note that the terminal velocity scales with the magnitude of the force applied, but the time does not. That is, larger applied fields produce larger terminal velocities but the time it takes the particles to approach that velocity depends only on the particle size and liquid viscosity.

3.1.b Motion of Particles by Gravity

The sedimentation or creaming of particles is caused by gravity. The gravitational force on a particle is the product of its effective mass times the acceleration of gravity. The effective mass of a particle is the product of its volume times the density difference between the particle and the suspending liquid. When the particle is a sphere, the force on the particle is given by

$$F_{\text{gravitational}} = \tfrac{4}{3}\pi a^3 \Delta\rho g \qquad (3.7)$$

where a is the radius of the sphere, $\Delta\rho$ is the density difference between the particle and the suspending liquid, and g is the gravitational acceleration. Obviously if the particle is less dense than the liquid, the particle rises. When the gravitational force is used in the equation for the terminal velocity, Eq. (3.4), the expression for the terminal velocity of a particle as it settles is found.

$$v_{t\text{(gravitational)}} = \frac{2a^2 \Delta \rho g}{9\eta} \quad (3.8)$$

The quadratic dependence of the sedimentation rate on the radius of the particle points out the importance of keeping the particle size small to reduce sedimentation. For example, the terminal velocity for a 5 μm radius, $\rho = 2$ gm/cm^3 particle in water at 20°C is about 5.5×10^{-3} cm/sec. Such a particle takes about 3 minutes to settle 1 cm. A 0.5 μm radius particle takes about 5 hours to fall the same distance. If sedimentation were used to determine particle size and the particle detector were 1 cm from the top of the liquid, then 5 μm particles would reach the detector in 3 minutes, but the 0.5 μm particles would take 5 hours. If the equipment were not meticulously isolated from mechanical and thermal fluctuations over a five-hour period, the detection of the smaller particles would be inaccurate.

3.1.c Motion of Particles by Centrifugation

The centrifugal force of acceleration is given by an expression similar to the expression for gravitational acceleration except that the constant acceleration of gravity is replaced by the acceleration caused by the centrifuge.

$$F_{\text{centrifugal}} = \frac{4}{3}\pi a^3 \Delta\rho R \omega^2 \quad (3.9)$$

where R is the distance from the point of observation to the center of rotation of the centrifuge and ω is the angular velocity in radians per second (1 rpm = $2\pi/60$ radians per second). A particle 10 cm from the center of rotation of a centrifuge running at 10,000 rpm feels an acceleration of 1.1×10^7 cm/sec^2 or 1.1×10^4 times the acceleration due to gravity. The terminal velocity of a particle in a centrifugal field is

$$v_{t\text{(centrifugal)}} = \frac{2a^2 \Delta \rho R \omega^2}{9\eta} \quad (3.10)$$

The same particle will be pushed out from the center of the centrifuge ten thousand times faster than it would settle by the force of gravity. Note that the terminal velocity depends on the distance of the particle from the center of rotation. As the particle moves, it travels farther from the center of the centrifuge and hence picks up speed. In most centrifuges equipped for particle sizing, the particle detector is at a fixed position. Calculation of the time required for a particle of a known size to settle from the liquid surface to the detector must take into account the steady increase in particle velocity as it moves farther from the center of rotation of the

3.1.d Motion of Particles by an Applied Electric Field

The force on a particle induced by an applied electric field is approximately the charge on the particle times the magnitude of the electric field. Corrections to this estimate have to do with the motion of ions in the electrical double layer around the particle and can be neglected for the present calculations. A spherical, electrically insulated particle has an electric charge of

$$Q = 4\pi a D \varepsilon_0 \Phi_0 \qquad (3.11)$$

where D is the dielectric constant of the medium, ε_0, the permittivity of free space, and Φ_0, the surface potential of the particle. The force on the particle created by an applied electric field of V volts across a cell gap of length L is the charge times the electric field.

$$F_{electrical} = \frac{4\pi a D \varepsilon_0 \Phi_0 V}{L} \qquad (3.12)$$

The ratio of the electrical force to the gravitational force is

$$\frac{F_{electrical}}{F_{gravitational}} = \frac{3 D \varepsilon_0 \Phi_0 V}{a^2 \Delta \rho g L} \qquad (3.13)$$

where $\Delta \rho$ is the density difference and g is the acceleration due to gravity. Equation (3.13) shows that the smaller the particle, the greater the influence of electric forces over gravitational forces. Electrophoretic motion of small particles is easier to measure than rate of sedimentation of small particles. A particle, 0.5 µm in radius, 2 gm/cm³ in density, dispersed in water, with 50 mV zeta potential, subjected to an electric field of 25 V/cm experiences an electric force about 100 times the gravitational force.

3.1.e Motion of Particles by Viscous Forces

The derivations above show that when particles are subjected to an applied force, they rapidly attain terminal velocity. Similarly, a particle suspended in a fluid rapidly attains the velocity of the liquid when the viscous drag of liquid motion is the applied force. These results are part of fluid mechanics. In the general analysis of fluid flow, Osborne Reynolds identified a dimensionless grouping of constants that could be used to identify transitions from laminar to turbulent flow. The Reynolds number can be interpreted as the ratio of inertial to viscous forces. For flows with large Reynolds numbers, the inertia of the fluid is so large that it cannot keep up smoothly with sudden changes in fluid velocity and turbulence begins. The Reynolds number, R_e, is given by

$$R_e = \frac{vd\Delta\rho}{\eta} \qquad (3.14)$$

The characteristic length d and the characteristic velocity v are determined by the geometry of the flow. For flow in a tube, d is taken to be the tube diameter and v the average velocity of flow. For flow around a sphere, d is taken to be the diameter of the sphere and v the velocity of the liquid far from the sphere.[1] Thus even in turbulent flow of the liquid the particles follow the flow. Engineering practice has determined that when the Reynolds number is greater than 2000 the fluid flow becomes turbulent.

To provide a scale appropriate for the Reynolds number for the flow of dispersed particles in liquids, the following relation between units is useful.

$$1 = \frac{1\frac{m}{\sec} \cdot 1\ \mu m \cdot 1\frac{gm}{cm^3}}{1\ mPa\text{-}sec} \qquad (3.15)$$

That is, a particle diameter of 1 μm, a density difference of 1 gm/cm^3, moving at 1 m/sec in a viscosity of 1 mPa-s, the viscosity of water at room temperature, has a Reynolds number of 1. Faster velocities, larger particles, larger density differences, and lower viscosities all lead to conditions where the particle does not follow the streamline.

This effect is used in equipment designed to separate particles based on their size, such as impactors and cyclones. The smaller the particles, the faster the flow required to separate them by size. The Reynolds number is also useful in predicting where in a process particles might remove themselves from the fluid flow owing to a sudden change in fluid direction. An example is in the pumping of dispersions through pipes with sudden changes in direction as in around values.

Even when the Reynolds number for the entire flow with a characteristic length of the pipe diameter is large and the macroscopic flow turbulent, the Reynolds number for a particle with a characteristic length of its diameter is much smaller. Under these conditions, the particle follows the flow of the liquid within each turbulent cell while the turbulent cells tumble chaotically down the pipe. This is an important consideration when trying to understand the rate of mixing or the rate of adsorption or the shear stresses on flocs.

3.1.f Motion of Particles by Ultrasound

Newton first proposed that sound is a traveling pressure wave in the medium. The direction of the expansion and compression of the medium is parallel to the direction of propagation of the sound. Since the traveling pressure wave alternately expands and compresses the medium, the propagation of the sound depends on the mechanical and thermal properties of the medium. When the medium contains particles, its mechanical and thermal properties are changed and the propagation of the sound altered. When sound travels through a liquid containing particles the resulting motion of the particles produces a pressure wave normal to the direction of the sound, scattering sound similarly to the scattering of light by particles.

It was Rayleigh's investigation of sound that led him to think about the scattering of light and ultimately to explain the blue of the sky. The range of frequencies of sound is enormous, say from 0.1 to 10^{13} Hz. Simple pressure transducers are capable of sensing sound waves over large portions of these frequencies. This range of frequencies is much larger than the visible frequencies of light. Therefore exploring the absorption and scattering of sound as a function of frequency becomes a powerful tool. Povey[2] describes ultrasonic methods for determining the concentration of alcohols, sugar, oils, and fats, crystal nucleation rates, emulsion stability, particle sizing, and the sizing of bubbles in foams.

Sound travels easily through metal containers and pipe walls; therefore ultrasound techniques are useful to obtain information about dispersions and emulsions even when they cannot be seen. Similarly, ultrasound techniques can be used for opaque and concentrated dispersions.

The concentration of a dispersed phase can be determined by measuring the speed of sound through the dispersion. The technique is relatively simple provided the temperature is either carefully controlled or known. The speed of sound in a pure liquid is given by a Laplace equation[3]

$$v = \sqrt{\frac{E}{\rho}} \qquad (3.16)$$

where E is the elastic modulus and ρ is the density. This equation indicates that the velocity of sound depends solely on the elastic modulus and the density and not on particle or droplet size. The velocity of sound in emulsions and dispersions therefore depends on the average density and the average speed of sound in the constituents of emulsion or dispersion.[4]

$$\frac{1}{E} = \sum_i \Phi_i \frac{1}{E_i} \quad \text{and} \quad \rho = \sum_i \Phi_i \rho_i \qquad (3.17)$$

where Φ_i are the volume fractions of the individual phases. The speed of sound in a dispersed-phase system is measurable. If the speed of sound in the individual components of a two-phase system is known, the volume fraction of the dispersed phase can be calculated by means of Eqs (3.16) and (3.17). When the values are not known a calibration curve of the speed of sound versus the volume fraction of the dispersed phase is the usual approach.

The mathematical description of acoustic scattering is similar in principle to the mathematical description of light scattering. Both types of scattering depend on particle size and shape. Light scattering depends on the dielectric constants

54 KINETIC AND STATISTICAL PROPERTIES

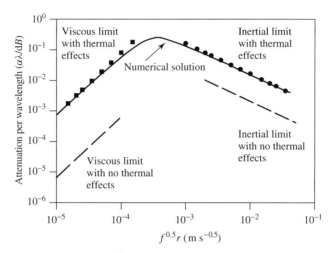

Figure 3.1 Ultrasound attenuation per wavelength as a function of the scaling parameter $f^{0.5}r$ for a 20 vol percent sunflower oil in water emulsion at 30°C where f is the frequency of the sound and r is the particle radius.[6]

of the liquid and particles. Acoustic scattering depends on many more parameters such as densities, acoustic attenuation, viscosity, specific heats, thermal conductivities, and compressibilities. Unfortunately, many of these mechanical and thermal properties are not known for materials of interest.

McClements and Povey[5] first applied scattering theory to a practical emulsion, sunflower oil in water, which forms the basis of many margarine formulations. Pinfield[6] developed simplified approaches to the practical analyses of experimental scattering data (see Figure 3.1). When the frequency is small, the wavelength is long compared to the size of the particle and the attenuation of sound is dominated by viscous effects. When the frequency is large, the attenuation is dominated by inertial effects (i.e., the particles cannot keep up with the high frequency of the sound). In both limits, thermal effects are important. These data and analyses refer to the scattering of acoustic waves whose frequencies vary over four orders of magnitude.

3.2 STATISTICAL PROPERTIES

3.2.a Brownian Motion

The most readily observed statistical property of particulates is Brownian motion. Robert Brown, a Scottish botanist, using a simple microscope, discovered in 1827 that a dispersion of cytoplasmic granules in water showed lively random motions. He subsequently established that the same type of motion could be observed with particles of dead plants, glass, and a wide variety of minerals when ground sufficiently small to be temporarily suspended in water. Among his samples was a chip of stone taken by a relic hunter from an Egyptian sphinx. The motion, therefore, has nothing to do with life. That the effect might be due to convection currents or electric charges was also disproved by later investigators. William Prout, an early proponent of Dalton's atomic theory and the author of Prout's hypothesis that atomic weights are multiples of a subatomic protyle, was also first in his query, "Are the molecular motions of liquids the cause of those motions which solid particles of matter diffused through them sometimes exhibit?" His remarkably prescient suggestion is now known to be correct.[7]

In 1908 Einstein noticed the resemblance between Brownian motion and the hypothetical motion of gas molecules according to the kinetic theory of gases.[8] Working on the supposition that size is the essential difference between suspended particles and gas molecules, Einstein proposed that the mean displacement x of a particle in Brownian motion as a function of time be

$$x = (2Dt)^{1/2} \tag{3.18}$$

where the diffusion coefficient D is given by

$$D = \frac{kT}{3\pi\eta d} \tag{3.19}$$

where η is the viscosity of the medium and d is the diameter of the particle. Equations (3.18) and (3.19) are used extensively in colloid science to estimate the distances materials diffuse with time and to estimate particle diameters from measured diffusion rates.

Equation (3.18) shows that the apparent velocity, $v_{app} = x/t$, of a particle is a function of time; the apparent velocity is larger the shorter the time of observation.

$$v_{app} = \frac{x}{t} = \left(\frac{2D}{t}\right)^{1/2} \tag{3.20}$$

Observed at a distance, particles appear to diffuse slowly, but on microscopic examination the same particles can be seen moving at high velocities. The higher the magnification, the faster the particles appear to be moving. Equation (3.18) gives the total distance moved, albeit by a tortuous path.

From measurements of the diffusion coefficients of molecules in solution, Perrin was able to calculate, by means of Eq. (3.18), the molecular sizes, and from the sizes to calculate Avogadro's number. The agreement of his results with independent determinations was a strong confirmation of the statistical thermodynamics of Boltzmann and showed the connection between Brownian motion and molecular reality, which was Einstein's original hypothesis.[9] Experimental verifications of Einstein's hypothesis were made independently by Svedberg and Perrin, both of whom obtained Nobel prizes in 1926, one in chemistry and the other in physics.

Figure 3.2 shows diffusion data taken from the study of lipids.[10] A few gold-tagged lipids were added to a film of dimyristoyl phophatidylcholine spread on

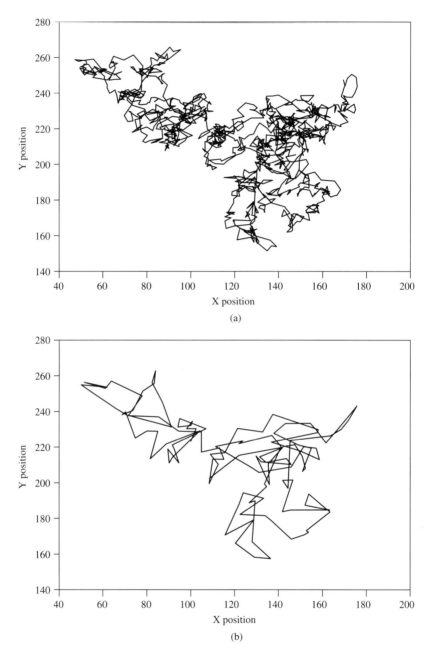

Figure 3.2 The relative distance between two gold-tagged dipalmitoyl phosphoethanolamine molecules diffusing in a monolayer of dimyristoyl phosphatidylcholine on a Langmuir trough. (*a*) The relative positions every 1/3 second connected by straight lines. (*b*) The same data as (*a*) except the relative positions are shown every three seconds.[10] (Data courtesy of the authors.)

a Langmuir trough. The positions of tagged lipids were recorded every 30th of a second. Figure 3.2(a) shows the distance between two particles every third of a second. Figure 3.2(b) uses the same data but shows the distances every three seconds. Figure 3.2 illustrates Einstein's equation: the more frequent the observation of position, the faster the particles appear to be moving.

Nowadays, diffusion coefficients are used to determine particle sizes or polymer molecular weights. They are obtained by measuring the degradation of a sharp boundary between a suspension and its serum. The boundary can be obtained by centrifugation and its rate of spread is followed optically. Diffusion constants can also be obtained from quasi-elastic light-scattering data as described in Chapter 4.

3.2.b Sedimentation Equilibrium

Brownian motion tends to distribute particles uniformly through a dispersion just as molecular diffusion of solutes distributes solute molecules uniformly through a solution.[11] Gravity pulls particles downward just as gravity pulls denser bodies to the earth's surface. Therefore Brownian motion and gravitational sedimentation are opposing forces, one distributing particles throughout the dispersion and the other concentrating particles at the bottom. Dispersions eventually reach a dynamic equilibrium where sedimentation and diffusion are balanced. The concentration of particles as a function of height from the bottom of the dispersion for monodisperse, noninteracting, spherical particles is[12]

$$n = n_0 \exp\left(-\frac{V \Delta \rho g h}{kT}\right) = n_0 \exp\left(-\frac{4\pi a^3 \Delta \rho g}{3kT} h\right) \qquad (3.21)$$

where n is the concentration of particles at any height, h, from the bottom of the dispersion, n_0 is the concentration of particles at the bottom of the dispersions, V is the volume of a particle, $\Delta \rho$ is the density difference between the particle and the liquid, g is the acceleration due to gravity, k is Boltzmann's constant, T is the absolute temperature and a is the radius of the particle. The inverse of the coefficient of h in the exponential in Eq. (3.21) is called the characteristic height, h_0.

$$h_0 = \frac{3kT}{4\pi a^3 \Delta \rho g} \qquad (3.22)$$

or

$$h_0(\mu m) = \frac{0.100}{a^3 \Delta \rho} \qquad (3.23)$$

at 25°C when a is in μm and ρ is in gm/cm^3.

A dispersion of particles 0.5 μm radius, 2 gm/cm^3 particles in water would have a sedimentation height of only about twice their radii. Similar particles only 0.05 μm in radius would form a sediment about 1 mm high. For all practical purposes, such dispersions are completely sedimented.

This result is surprising because common experience is that submicron dispersions remain well dispersed for considerable lengths of time. The source of this apparent stability is not Brownian motion. The Brownian motion only sustains a small concentration of particles above the bottom of the dispersion.

The explanation for the existence of apparently stable dispersions lies in the perturbing influences of thermal fluctuations and mechanical disturbances. Thermal fluctuations, noise, and mechanical perturbations tend to stir dispersions. The smaller the particles, the more susceptible they are to these outside perturbations. Small particles dispersed in a liquid behave like a cloud of dust in air. If a dispersion is kept carefully thermostated and isolated from other external perturbations, sediment forms with a distribution of concentrations as predicted by Eq. (3.21).[13] For practical conditions, extraneous disturbances keep the dispersions mixed. Dispersion stability in the sense of no sedimentation is attained by keeping the particles or droplets small and depending on outside perturbations to keep the dispersion stirred.

3.2.c Colligative Properties

"Colligative properties" usually refer to solutions. These properties are osmotic pressure, freezing point depression, boiling point elevation, and lowering of vapor pressure. All these properties depend, to a first approximation, only upon the concentration of dispersed units, whether the units are atoms, molecules, ions, macromolecules, or dispersed particles. The effects with dispersions are significant when particles are smaller than about 100 nm in diameter and in dispersions free of dissolved species. But these conditions are rare. Kruyt calculated that a typical sol having 1 gm of gold per liter contains 50 mm^3 of gold and the volume of each particle is about 50 μm^3. One liter of the sol therefore contain 10^{18} particles of gold. If a sol with 6.02×10^{23} particles is called a normal solution, this sol is 2×10^{-6} normal and its osmotic pressure amounts to 1/2 mm of water corresponding to a lowering of the freezing point of 4×10^{-6} degree C.[14] Clearly, colloidal sols will show no discernable colligative properties unless they are highly concentrated and severely dialyzed. When particle dispersions are severely dialyzed they can be modeled as a two-component system. These two-component systems have thermodynamic and statistical mechanical properties analogous to two-component solutions including temperature and pressure-dependent phase transitions.[15]

Osmotic pressure and the other colligative properties, however, are useful indices of the concentration of units in both aqueous and nonpolar surfactant solutions. They reveal the existence of associated complexes in solution. For example, self-association can be detected by a decline in the osmotic coefficient of the solution.

3.2.d Fractals

Measurements of Brownian pathways reveal an ever-increasing irregularity with reduction of the time period of observation (see Figure 3.2). Objects that show such irregularity as a function of the scale of measurement are called fractal.[16] Examples of fractal objects are the coastlines of islands and continents, the outlines of snowflakes and flocs, and the surfaces of solids, from carbon blacks to the moon. For example, if the coastline of Britain were measured as a series of different steps of different size, say, from a kilometer step to a centimeter step, the perimeter so found would be larger the smaller the step. In other words there is no single value for the perimeter of a coastline, just as there is no single value for the length of the pathway of a particle in Brownian motion. The effects of the size of the step on the measurements of various coastlines are given in Figure 3.3.

Figure 3.3 reproduces experimental measurements of perimeter length performed on various maps, using shorter and shorter steps. Increasingly precise measurements of a circle stabilize very rapidly near a well-determined value, but an irregular perimeter such as the coastline of Britain shows no such limit. Generally, such data are linear on a log-log plot for a wide range of step sizes. These plots are characterized by their slopes. The fractal dimension is defined

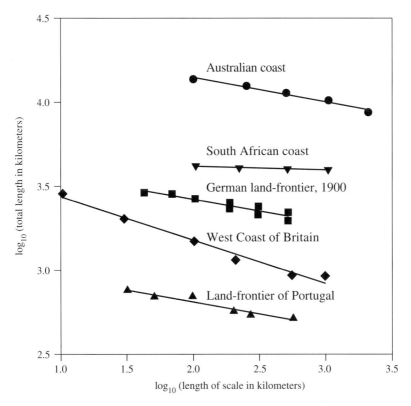

Figure 3.3 Richardson's data on the increase of the perimeters of coastlines with the decrease of the size of the step.[17]

as the slope of the log-log plot plus one. The perfectly smooth circle then has a fractal dimension of unity; irregular perimeters have fractal dimensions larger than unity (but less than two). Irregular or rough surfaces have fractal dimensions between two and three. The more irregular the perimeter or the rougher the surface, the larger the fractal dimension.

Fractal dimensions are used to describe the irregularity of floc shapes and the roughness of surfaces. Digitizing microscopes can be programmed to compute the fractal dimension, which is determined by image analysis as a function of magnification. Another technique is to measure the scattering intensity as a function of scattering angle at low angles. Burns et al.[18] studied the variation of floc structure of uniform polystyrene latex particles using small-angle static light scattering over a range of electrolyte and particle concentrations. This technique allowed them to determine that at high salt levels the particle concentration had little or no effect while at low salt levels (less than 1 m KNO_3) an increase in particle concentration led to a decrease in the fractal dimension. This was attributed to a reduction in the time available for reconfiguring of particles in the flocs caused by the increased frequency of collision. As well as flocs, the fractal morphology of pore structure in microporous membrane filters can be studied by small-angle light scattering.[19] The fractal properties of a substrate have also been studied.[20] The contact angle of a liquid drop on a fractal surface is determined by the fractal dimension, the range of fractal behavior, and the ratio of the composite surface. Fractal surfaces can be super water-repellent (or superwettable) when the surfaces are composed of hydrophobic (or hydrophilic) materials. These authors demonstrate a super water-repellent fractal surface made of alkylketene dimer; a water drop on this surface has a contact angle as large as 174°.

REFERENCES

[1]Bird, R.B.; Stewart, W.E.; Lightfoot, E.N. *Transport phenomena*; Wiley: New York; 1960; pp 107–108.

[2]Povey, M.J. *Ultrasound techniques for fluids characterization*; Academic Press: New York; 1997.

[3]Povey loc. cit.; p 26.

[4]Povey loc. cit.; p 27.

[5]McClements, D.J.; Povey, M.J. Scattering of ultrasound by emulsions, *J. Phys. D: Applied Physics* **1989**, *22*, 38–47.

[6]Pinfield, V.J. Studies *of creaming, flocculation, and crystallization in emulsions: computer modelling and analysis of ultrasound propagation*, PhD Thesis, 1996, Leeds University, U.K.; also, Povey loc. cit.; p 119.

[7]Prout's *Bridgewater treatise*, 4th ed.; Bell and Daldy: London; 1870; p 389. The query does not occur before the 3rd edition of this book. Prout, as an ardent atomist, may even have acquired the idea from Lucretius (ca. 95 to 55 B.C.), who inferred the existence of Brownian motion long before it was discovered (Lucr. *De rer. nat.* **II** ll. 112–141).

[8]Einstein, A. *Investigations on the theory of Brownian motion*; Cowper, A.D., Trans.; Dover: New York; 1956; p 75.

[9] Nye, M.J. *Molecular reality: a perspective on the scientific work of Jean Perrin*; Elsevier: New York; 1972.

[10] Forstner, M.B.; Käs, J.; Martin, D. Single lipid diffusion in Langmuir monolayers, *Langmuir* **2001**, *17*, 567–570.

[11] Einstein loc. cit.; pp 24–28.

[12] van de Ven, T.G.M. *Colloidal hydrodynamics*; Academic Press: New York; 1989; p 271.

[13] Perrin, J.B. Nobel Prize in Physics, 1926 "for his work on the discontinuous structure of matter, and especially for his discovery of sedimentation equilibrium." *Nobel Lectures, Physics, 1922–1941*; Elsevier: New York; 1965.

[14] Kruyt, H.R. *Colloids, a textbook*; van Klooster, H.S., Trans; Wiley: New York; 1930; p 154.

[15] Russel, W.B.; Saville, D.A.; Showalter, W.R. *Colloidal dispersions*; Cambridge University Press: New York; 1989; Chapter 10.

[16] Mandelbrot, B.B. *Fractals: form, chance, and dimension*; W.H. Freeman: San Francisco; 1983.

[17] Mandelbrot loc. cit.; p 32.

[18] Burns, J.L.; Yan, Y.; Jameson, J.; Biggs, S. A light scattering study of the fractal aggregation behavior of a model colloidal system, *Langmuir* **1997**, *13*, 6413–6420.

[19] Cipelletti, L.; Carpineti, M.; Giglio, M. Fractal morphology, spatial order, and pore structure in microporous membrane filters, *Langmuir* **1996**, *12*, 6446–6451.

[20] Onda, T.; Shibuichi, S.; Satoh, N.; Tsujii, K. Super-water-repellent fractal surfaces, *Langmuir* **1996**, *12*, 2125–2127.

4 Particle Sizing

The influence of particle size on macroscopic properties such as the optical and rheological properties is important. A key to choosing the most suitable particle-sizing technique is to find one that is sensitive to those portions of the size distribution that are most significant to pertinent macroscopic properties. For instance, if the optical properties of a suspension are significant, then using an optical technique to measure size distributions is most likely to yield useful information. "Polydisperse" means having more than one size.[1]

In this chapter we discuss various commercially available techniques to measure particle size, dividing the techniques into groups corresponding to the physical property used to determine the size, whether it be flow, single-particle detection, light scattering, acoustics, or adsorption. As the particle size decreases, the specific surface area increases. For particles less than 100 nm, measurements of surface area rather than average particle size may be easier and possibly more useful if the important effects of the particle on the properties of the dispersion depend on interactions at the surface. Detailed monographs devoted to particle-sizing techniques are by Allen,[2] Barth,[3] Jelínek,[4] Kaye,[5] Provder,[6,7,8,9] Orr[10] and Stockham and Fochtman.[11] Some names and addresses of manufacturers of particle-sizing equipment are included at the end of this chapter.

In general, a complete description of all the sizes and the shapes of the particles in a suspension or emulsion is unobtainable. At best the usual particle-sizing techniques give a distribution of "equivalent" sized particles, for instance, a distribution of "equivalent" spheres that settle with the same terminal velocity as the particles in the dispersion. Each of the usual particle-sizing techniques is limited to some size range, which might vary with properties, such as refractive index or density. Particles outside that range are either undetected or produce anomalously high signals. For example, a technique based on measuring changes of electrical resistance as particles pass through a sensing zone may not detect the smallest particles. And a few large particles in a suspension can completely overwhelm most light-scattering measurements. Every technique is sensitive to a certain range of particle sizes and hence each technique provides a different distribution of equivalent sizes for the same sample.

The distribution of particle shapes is even more difficult to obtain. The only simple direct determination of shape is by microscopy. A general rule is to look at the sample through a microscope before any other particle-sizing technique is tried. The information thus easily obtainable is invaluable. Knowing the approximate shape of the particles being sized can be a useful aid in choosing an appropriate technique.

Two sets of values are necessary to specify a particle size distribution: a measure of size of a particle and a measure of the quantity of particles near that size. Particle size can be specified by the radius, diameter, or, if a unique one exists, length of side. The measure usually obtained is one of minimum or maximum chord, fiber length, equivalent diameter from the terminal settling velocity, projected perimeter or diameter of the equivalent circle, or equivalent optical or electrical diameter. The total quantity of particles in each size range is additive and may be expressed as number, volume, mass, or surface area.

4.1 TYPES OF DISTRIBUTIONS

The distribution of particle sizes is sometimes expressed as a histogram, that is, the number (or mass) of particles with equivalent sizes within some range. Some particle-sizing techniques, for example, sieving, produce size distributions as histograms directly. The distribution of particle sizes can also be expressed as a probability function such that the quantity of particles within a small range of sizes is the area for that range under the probability curve. An alternative form to express a distribution is cumulative, in which the ordinate of each point represents the total of all particles with sizes less than the abscissa of that point (see Figure 4.1). When particle-size distributions are plotted as probability functions, the fraction of particles occurring in a certain size range is measured as the area under the curve for that range. When the ordinate is measured in numbers per unit size, the size distribution is called a number distribution. When the ordinate is measured in mass per unit size, the size distribution is called a mass distribution. For instance, particle sizing by sedimentation produces mass distributions directly; particle sizing by electrozone detection produces number distributions directly.

The significant difference between mass distributions and number distributions is illustrated by determining a "biting insect" index as the average mass of horseflies and chiggers per unit land area.[12] A decrease of such an index by 25% could be accompanied by a terrific increase in discomfort. For example, if the mass-heavy horsefly population were to be reduced or exterminated, while at the same time the population of the tiny chiggers tripled, the mass of biting insects per unit land area would indeed be less, but the actual number of biting insects per unit land area would be significantly larger.

Sometimes the probability size distribution of a sample can be adequately described by a Gaussian (normal) distribution function

$$\Phi_g(d) = \frac{1}{\sigma\sqrt{2\pi}} \exp\left[-\frac{(d-d_0)^2}{2\sigma^2}\right] \tag{4.1}$$

where $\Phi(d)$ is the probability density, d is the diameter, d_0 is the mean diameter, and σ is the standard deviation. If the distribution is given as a number density, that is, the number of particles per unit range of particle diameters, it is called

64 PARTICLE SIZING

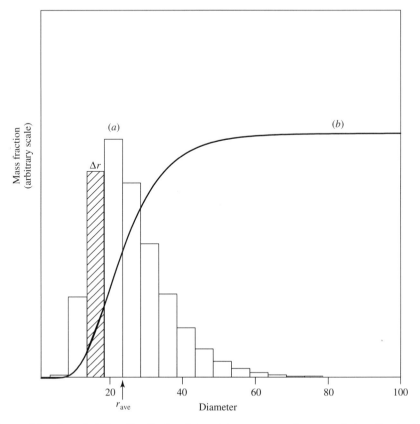

Figure 4.1 A probability distribution (*a*) and its corresponding cumulative distribution (*b*). The hatched area in the probability distribution is the fraction of particles within the range Δr. The average size is shown at r_{ave}. Cumulative distributions start at zero and end at 100%. The midpoint of the cumulative distribution is the average particle size.

a number distribution, and d_0 and σ are the number mean size and the number standard deviation. If the distribution is given as a mass density, that is, the mass of particles per unit range of particle diameters, it is called a mass distribution; and d_0 and σ are the mass mean size and mass standard deviation. If the number distribution is a normal distribution, the mass distribution is not and vice versa.

When a dispersion is obtained by comminution (as is generally the case), then the particle size distribution is often adequately described by the log-normal distribution function

$$\Phi_{\ln}(d) = \frac{1}{\ln \sigma_{\ln} \sqrt{2\pi}} \exp\left[-\frac{\ln^2(d/d_0)}{2 \ln^2 \sigma_{\ln}}\right] \tag{4.2}$$

where d is the diameter, d_0 is the geometric mean diameter, and $\sigma_{\ln}$ is the geometric standard deviation. If the distribution is given as a number density, then the distribution is a number distribution, and d_0 and $\sigma_{\ln}$ are the number geometric mean size and the number geometric standard deviation. If the distribution is given as a mass density, it is a mass distribution, and d_0 and $\sigma_{\ln}$ are the mass geometric mean size and the mass geometric standard deviation. If the number distribution is log-normal, then the mass distribution is so too, with the same geometric standard deviation and vice versa. The relation between the geometric mean for a number distribution, d_n, and the geometric mean for a mass distribution, d_w, is

$$\ln d_n = \ln d_w - 3 \ln^2 \sigma_{\ln} \qquad (4.3)$$

When reporting an average size and standard deviation for a particle size distribution, it is clearly important to state whether the distribution is normal or log-normal, number or mass. Figure 4.2 is a schematic to show the difference between a number distribution and a mass distribution. The mode of the mass distribution is always larger, sometimes much larger, than the mode of the number distribution. A common error is to compare the average particle size of a given powder as obtained by different techniques without taking into account whether they produce mass or number distributions. For a thorough discussion of the statistics of small particles, including graphical techniques, see Herdan.[13]

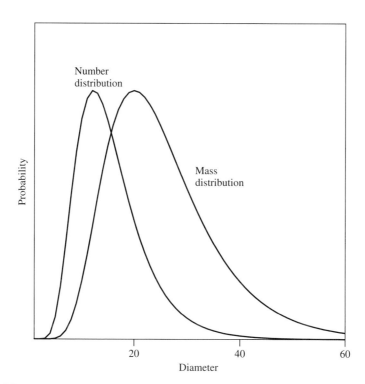

Figure 4.2 A schematic comparing the number and the mass distribution for a given sample.

4.2 SAMPLING AND SAMPLE PREPARATION

Errors in sampling procedure are most common when sizing dry powders. In typical dispersions, the particle sizes are so small and the samples so easily mixed that sampling is usually not a problem. For a detailed discussion of sampling techniques and considerations see Kaye.[5]

Sample preparation is a wider concern. Almost all the particle-sizing techniques require that the dispersion be dilute. Dispersions cannot be diluted without changing some properties of which the most important is stability. When a dispersion is diluted adsorbed species will desorb to establish a new adsorption equilibrium, which may leave the concentration of adsorbate too low to maintain stability with resultant flocculation. The particle-size information obtained on a flocculated dispersion is of no use in understanding the particle-size information about the original dispersion. (See Figure 4.3.)

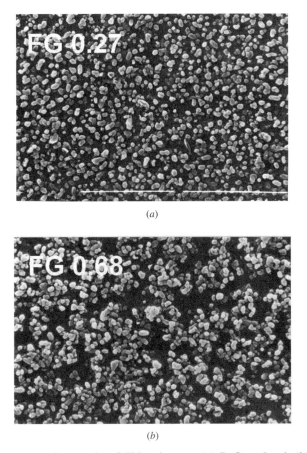

Figure 4.3 Electron micrographs of TiO_2 pigment. (*a*) Deflocculated. (*b*) Flocculated. (Courtesy of Tioxide, Ltd.)

The preferred technique is to isolate some of the dispersion serum, generally by centrifugation, and to use that serum to dilute the dispersion. Thus the adsorption equilibrium remains constant.

An alternative technique is to add one of the many known dispersants when diluting a sample for analysis. Useful for this purpose are Bernhardt's table of over 250 different solids, liquids, and dispersants.[14] The recommended approach is to measure particle size over a period of time to determine that the dispersion is not flocculating.

Micromeritics Corporation has distributed a suggestion for the analysis of magnetic particles that is also generally useful.[15] A sample is mixed well enough with about 1 cc of pure commercial honey to produce a homogeneous dispersion. A drop is placed on a microscope slide. A second slide is placed on top and pressed to form a thin film of the dispersion between the two slides. The top slide is pulled across the lower slide thus shearing the dispersion. This compression and shearing is repeated until the dispersion looks deflocculated under the microscope. The two slides are finally separated and the honey dispersion from one slide is washed off to form a dilute suspension sufficiently stable for quick analysis.

Some light-scattering techniques are applicable at particle concentrations so low that the rate of flocculation is slow compared to the rate of data collection.

4.3 SIZING BY FLOW

Coarse particles may be sized by taking advantage of how different particles respond to fluid motion. The simplest instrument is the sieve, where large particles cannot follow the liquid through a small hole. Sedimentation and centrifugation are also examples of separation by flow as large particles fall though a liquid faster than small ones. These techniques give mass averages and mass distributions.

4.3.a Sieving

A powder can be split into fractions of known size by using a stack of sieves of different mesh. Standard woven-wire sieves are made with openings, called the mesh, in the range of 37–5660 μm. Below, 100 μm variations in aperture size for woven-wire sieves are less discriminating. Micromesh sieves, made by photoetching or electroforming, extend the range downwards to 5 μm.[16] Sieves can be calibrated with standard glass microspheres.[17] Sieving can be "wet" or "dry" according to the medium. "Sifting" is the term used to denote horizontal agitation of the powder over the sieve surface. Sieve surfaces "blind" when particles are jammed into the apertures of the sieve. To minimize blinding, the sieves are vibrated vigorously. Too vigorous a vibration can cause particle attrition and hence a misleading size distribution. Rotation or tapping are the usual forms of agitation to reduce blinding of sieve apertures. The Gilsonic system holds the

68 PARTICLE SIZING

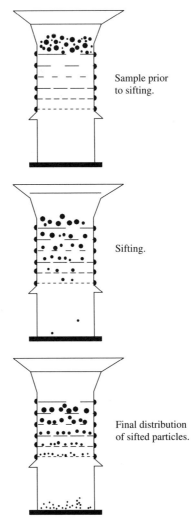

Figure 4.4 Schematic representation of three stages of sieving.[18]

sieves stationary while moving the powder vertically through the sieves with acoustic energy.[16]

Figure 4.4 represents schematically the stages of sieving. A known mass of powder is placed on the upper (coarsest mesh) sieve. The mass of powder is lifted either by air, water pressure or agitation. Gradually the smaller particles work their way down the stack until some final distribution of masses on the various sieves is obtained. The mass of powder left on each sieve as a function of the mesh is the size distribution histogram.

For irregularly shaped particles, the apparent size of the particle varies with its aspect ratio and the length of time spent sieving. Ideally, irregularly shaped

TABLE 4.1 Aperture Dimensions for Wire-Woven Sieves Conforming to the U.S. Sieve Series[19]

Screen Number	Width of Opening (μm)	Screen Number	Width of Opening (μm)
3.5	5660	40	420
4	4760	45	350
5	4000	50	297
6	3360	60	250
7	2830	70	210
8	2380	80	177
10	2000	100	149
12	1680	120	125
14	1410	140	105
16	1190	170	88
18	1000	200	74
20	840	230	62
25	710	270	53
30	590	325	44
35	500	400	37

particles are caught on sieves with a mesh size just smaller than the second largest dimension. (Consider a needle: It will ultimately pass through a sieve whose mesh is just greater than its diameter.)

The American Society for the Testing of Materials (ASTM) specifies standard sieve sizes and procedures (Table 4.1).

To overcome particle–particle agglomeration, pancaking, or electrostatics, powders are often sieved with water. The water can be used as a spray to prevent blinding of the sieves. The disadvantages of using water are that the smaller mesh sizes offer considerable resistance to water and suspended fines are difficult to recover from large volumes of wash water. Various sieves are available with combinations of shaking, tapping, sonic vibration, rocking, and air and water sprays.

Nushart has published a summary of ASTM standards on sieve analysis for specific materials or industries, arguing that sieving is to be included in the portfolio of sizing techniques even though it "isn't equipped with an array of LEDs or a digital timer."[20]

4.3.b Filtration

If smaller and smaller particles can be separated from each other by sieves with smaller and smaller orifices, then using filters with smaller and smaller pores would seem to be a useful method to separate even smaller and smaller particles. Unfortunately, this is not the case. The small holes in filters are actually long tubes (or tortuous paths through a fiber bed) where the transversing particles collide frequently with the walls. The attraction of small particles to filter walls is often larger than inertial forces so that the particles stick and are not separated by size.

In fact, filters with nominal pore diameters of several microns will easily collect submicron particles. This must be born in mind when using submicron particle-sizing equipment. Filtering through a micron-sized filter does not just eliminate the particles greater than one micron, it also eliminates an unknown proportion of submicron particles. The size distribution determined from the filtrate is not indicative of the original size distribution, even the submicron part.

4.3.c Sedimentation

Sedimentation is the most obvious sign that a dispersion contains large particles. No dispersant can negate the pull of gravity. The rate of sedimentation decreases for smaller particles or higher viscosities but is never eliminated. At low concentrations, particles settle independently; at higher concentrations they interfere with one another while settling. Theory requires particles to settle independently. For small particles, say less than a micron, keeping the liquid quiescent is difficult as even small variations of temperature can cause sufficient liquid motion to mask sedimentation. Sedimentation is a good method to determine particle diameters for large particles in dilute suspensions at low viscosities.

4.3.c(i) Theory Sedimentation methods depend on an application of Stokes' law for the viscous drag on a moving spherical particle. At constant velocity, the frictional force F_f is proportional to the radius a, the velocity v, and the viscosity η of the medium. The proportionality constant for conditions of laminar flow is equal to 6π

$$F_f = 6\pi a v \eta \tag{4.4}$$

The gravitational force F_g acting on a sphere of radius a is

$$F_g = \tfrac{4}{3}\pi a^3 (\rho_2 - \rho_1) g \tag{4.5}$$

Under conditions of steady state,

$$F_f = F_g \tag{4.6}$$

Therefore $6\pi a v \eta = \tfrac{4}{3}\pi a^3 (\rho_2 - \rho_1) g$ or

$$v = \frac{2a^2 (\rho_2 - \rho_1) g}{9\eta} \tag{4.7}$$

Equation (4.7) is called Stokes' law. If the particle is not a sphere, then a is the radius of the equivalent sphere, and the size distribution is given in terms of equivalent spherical radii.[21] If h is the distance from the top of the liquid to the sensing zone, then the time for all particles of radius a to have settled past the sensing zone is t_a where

$$t_a = \frac{9\eta h}{2a^2(\rho_2 - \rho_1)g} \tag{4.8}$$

The commonly used techniques start with a homogeneous sample of particles dispersed in a liquid. The simplest method is to sample the concentration of the dispersion at a fixed height at different times. Initially the composition at that height remains constant, with as many particles falling into the sensing zone as falling out of it. At the time when the concentration first changes, then the size of particles that have cleared the volume above the sensing zone is

$$a = \sqrt{\frac{9\eta h}{2t_a(\rho_2 - \rho_1)g}} \tag{4.9}$$

Therefore, concentration-versus-time data can be simply transformed into a concentration-versus-size curve by means of Eq. (4.9). This produces the cumulative size distribution. The probability distribution is the derivative of the cumulative distribution. Obtaining the derivative of data introduces error, hence the probability distribution is less accurate than the cumulative distribution based on sedimentation. The detector can be moved up the sedimentation column to decrease the time required for analysis.

4.3.c(ii) Detectors Particles can be detected in various ways. In the Andreason pipette, samples of the suspension at a determined height are drawn off at various times and analyzed for percent solids (Figure 4.5). Any decrease in the percent solids at a time t is due to a loss of particles with Stokes' velocity greater than h/t. In the Cahn sedimentation balance the increase in mass of particles deposited on a balance pan is monitored continuously. The mass on the pan increases linearly with time until the largest particles have all settled to the depth of the pan. The mass on the pan then increases more and more slowly corresponding to the deposition of smaller and smaller particles. In the Micromeritics SediGraph or the Quantachrome Microscan, the solids concentration is measured continuously by the absorption of x-rays. The advantage of using x-rays as the light source is that the wavelength of x-rays is so short that scattering and diffraction effects are insignificant for particles in this size range.[22] The mass of particles in the beam is determined by Beer's law of absorption. The disadvantage is that the particles must absorb x-rays. Many biological and polymeric materials cannot be sized by this method.

A new detection technique is to measure the speed of sound through the sedimenting (or creaming) dispersion.[23] The speed of sound varies linearly with the mass fraction of dispersed phase. The speed of sound can easily be calibrated for dispersions of known concentration. The sample cell must be held in a water bath, as air is a strong attenuator of sound. A great advantage of this detector is that the sample can be completely opaque (as is often the case in industrial

72 PARTICLE SIZING

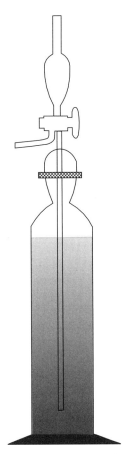

Figure 4.5 In the Andreason sedimentation pipette, the cylinder is filled with homogeneous suspension, the sampling pipette inserted, and the level h of the suspension recorded. Samples of the suspension are removed from the 0-cm level by means of an aspirator at a time t. Any decrease in concentration of suspension is owing to the loss of particles with Stokes sedimentation velocities greater than h/t.[24]

dispersions). The detectors can even be mounted on the walls of pipes and used to monitor the concentration in the passing stream without interfering with the flow and without degradation of the detector.

Sedimentation techniques are limited at the lower sizes by thermal gradients, which disturb steady sedimentation. For narrow tubes, wall effects are significant, especially if the walls are not vertical.[25] The analysis depends on each particle falling independently, which is only true for dilute suspensions. The upper limit of concentration is of the order of 1 to 2% by volume but must be established in each case by making measurements at lower concentrations. When concentrations are too high, particles cluster, tend to settle as a unit, and hence appear to be larger than they really are.

4.3.d Elutriators, Impactors, and Cyclones

Elutriators, impactors, and cyclones separate particles of different size by means of a flow of liquid or gas. The higher the velocity of the fluid, the smaller the size of particle that follows the flow. Particle sizing by the flow of a gas or liquid requires handling large volumes. These techniques are often used for preparative separations; they are therefore better suited for pilot and manufacturing operations than for laboratory work.[26] Klumpar has published a review of commercial equipment and selection criteria for air clarification.[27]

Elutriation is the separation of different size particles by the upwards flow of a liquid. The simplest application is to provide a steady flow of liquid up a vertical column containing a dispersion. Particles larger than a critical size, called the "cut" size, settle by gravitational sedimentation more quickly than they are carried up by the viscous drag of the fluid. Small particles are carried out the top of the column by the liquid flow. This separates the particles into two populations. The technique is often used to purify a dispersion, either eliminating the fines when they are carried off by the flowing fluid or eliminating the oversize particles by leaving them behind in the elutriation chamber. A series of elutriation columns with steadily decreasing diameters to provide faster flow of liquid is used to separate a dispersion into several ranges of particle size.

The capturing of particles on plates (called collectors) placed directly in the flow of a gaseous dispersion is called impaction (see Figure 4.6). Small particles follow the flow around the collector; larger particles impact it and stick. Impaction is not an accurate sizing technique because the flow of gas is turbulent and the velocity varies dramatically at different positions. A series of impactors is called a cascade impactor, in which the air velocity is increased at each succeeding stage to trap particles of progressively smaller size. The instrument is calibrated with standards to provide a measurement of particle size. It is useful for sizing dry powders but is less accurate than sieving or elutriation. Obtaining a representative sample from a gaseous dispersion is a problem because large flowing streams are usually not homogeneous. The collectors are sometimes greased to hold collected particles and rotated to prevent uneven buildup on the collector.

Cyclones are conical chambers where a flow of gas containing particles enters tangentially from one side, whirls around until it exits through an opening in the top of the chamber. The whirling flow of gas accelerates as it approaches the top of the chamber. The larger the particle, the less able it is to follow the gas flow as it circulates toward the center exit. The large particles hit the walls of the cyclone, slide to the bottom, and are caught in a bag. Cyclonic separation is a particularly efficient method to separate fines. Careful engineering of a series of cyclones enables the production of powders within narrow size limits. Often fine particles in a powdered product cause handling or safety concerns so that their separation produces a more useful product. Cyclones are used to separate the xerographic toner used in copiers and printers where large toner particles produce print defects and small toner particles are a health hazard.

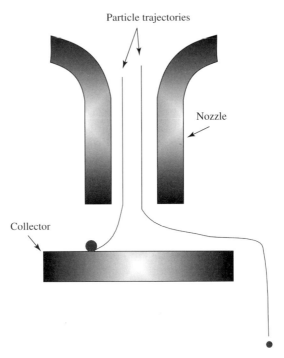

Figure 4.6 A single-stage impactor showing the difference in flow behavior between a large and a small particle.

4.3.e Centrifugation

Gravitational sedimentation as a technique to measure particle size is limited to coarse particles, but centrifugation allows the range to be extended to smaller sizes. It is also useful for larger particles whose density does not differ much from that of the dispersion medium. In terms of terrestrial gravitation the efficiency E of a centrifuge with rotor arm of length R centimeters, moving at ω radians per second, in terms of multiples of terrestrial gravitation is

$$E = \frac{R\omega^2}{g} \tag{4.10}$$

Efficiency thus depends on the angular acceleration, $R\omega^2$. Centrifuge efficiency should be in the range of 100–20,000. An ultracentrifuge, driven by compressed air, of which part drives an air turbine and another part forms an air bearing, is in the range of 200,000–400,000.

In a centrifuge particles do not sediment at constant velocity as in gravitational sedimentation, but they accelerate as they move farther and farther from the axis of rotation. The time t required for a particle of diameter d to move from an initial axial position r_0 to a final axial position r is

$$t = \frac{18\eta \ln(r/r_0)}{(\rho_2 - \rho_1)\omega^2 d^2} \tag{4.11}$$

where η is the viscosity of the fluid, $(\rho_2 - \rho_1)$ is the density difference between the particle and the fluid, and ω is the angular velocity of the centrifuge. The time required for the analysis can be adjusted by varying the angular velocity. The method is absolute and needs no calibration.

4.3.e(i) Disk Centrifuge The commercial instrument first in common use to measure particle sedimentation by centrifugation was the Joyce Loebl disk centrifuge (Figure 4.7(b)).[29] Today disk centrifuges are available from Brookhaven Instruments, CPS, Horiba, and Shimadzu. The rotor is a transparent, hollow plastic disk that is filled, while spinning, with the dispersion medium, or serum. The dispersion is then injected as a second concentric layer situated nearer the center. This technique is referred to as a line start. The dispersion has a higher density than the serum; therefore simply to place the dispersion on top of the serum would lead to the dispersion streaming through the serum as a continuous phase rather than as individual particles. To overcome this hydrodynamic instability the serum is overcoated with a series of layers of gradually increasing density before the dispersion is added. These density-gradient techniques are complex.[30]

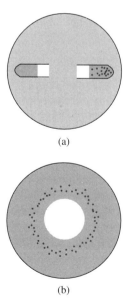

Figure 4.7 Two types of disk centrifuges. (*a*) A disk centrifuge containing the sample in one cuvette and a blank in the second. (*b*) A disk centrifuge in which the sample is injected on top of the liquid inside a spinning disk. Both centrifuge designs have a stationary optical sensor near the outside edge of the spinning disk which detects changes in light absorption.[28]

Particles settle toward the circumference and are detected by the absorption of light at a known distance from the starting line. Equation (4.11) is the relation between the time at which particles are detected and the equivalent spherical radii of those particles. The line start method of centrifugation produces a volume size distribution directly. Mie-scattering theory can be used to convert light absorption to particle volume concentration. Disk centrifuges are used extensively for aqueous dispersions. Additional care has to be taken with nonpolar dispersions: first, that the dispersion solvent does not affect the disk material and second, that static electricity does not build up on the spinning disk surface, causing the particles to stick to the walls.

4.3.e(ii) Uniform-Start Centrifuge A second type of disk centrifuge to determine size distribution uses the typical laboratory centrifuge design, a solid metal disk with machined slots to hold glass, quartz, or polycarbonate cells. The disk for particle sizing holds only two cells, one for the dispersion and one for a reference. One cell is filled with a known volume of the dispersion and one cell is filled with the clear serum and serves to act as the optical reference and as a counterbalance weight. An optical detector, mounted at a known distance from the center of rotation, is used to measure the optical density of the dispersion as a function of time. Generally the light source is a small diode laser and corrections for scattering are applied to the optical data to convert them to particle volume concentration.[31] This centrifugal technique is called "uniform start" as the particle concentration is uniform initially in contrast to the technique described above, the "line start." Some literature refers to this equipment as the cuvette photocentrifuge.

As centrifugal forces cause particles to sediment down the dispersion cuvette, the particle concentration detected gradually drops. At any instant the concentration detected is the concentration of all particles smaller in size than the particle that has had just enough time to sediment from the top of the dispersion to the detection zone. Uniform-start centrifugation reports the cumulative mass distribution directly. The probability distribution is obtained by differentiating the data, a procedure that always introduces error. Therefore uniform-start centrifugation is less accurate than line-start centrifugation. The advantage of uniform-start centrifugation is that the experimental procedure is much easier. Rotational speeds up to 10,000 rpm produce centrifugal forces of 9000 times gravity.

Hoffman has demonstrated the line-start technique with the cuvette photocentrifuge but this requires additional equipment not in wide use.[32]

4.3.f Ultracentrifugation

In 1926 two Nobel prizes were awarded for advances in colloid chemistry: the chemistry prize to Svedberg for "his work on disperse systems" and especially for his invention of the ultracentrifuge and the physics prize to Perrin for "for his work on the discontinuous structure of matter, and especially for his discovery of sedimentation equilibrium." The experimental work of these two men confirmed

Boltzmann's statistical mechanical and Einstein's theory of Brownian motion. These were enough to establish experimentally the size of molecules whose very existence had not been convincingly established until their work.

The ultracentrifuge has become a widely used tool in the study of micellar solutions and macromolecular solutions in general. The equipment can be used to measure sedimentation velocities as described above and to measure sedimentation equilibration. The latter is the distribution of micelles or macromolecules under the influence of strong centrifugal forces at long times. At equilibrium the concentration of micelles or macromolecules has reached an exponential distribution similar to that described by the sedimentation-equilibrium equation, (3.21)

$$c = c_0 \exp\left[-\left(1 - \frac{\rho_1}{\rho_2}\right) \frac{M}{2RT} \omega^2 h^2\right] \quad (4.12)$$

where ρ_1 and ρ_2 are densities of the solvent and polymer respectively, M is the molecular weight of the polymer, ω is the angular velocity of the centrifuge, h is the height above the bottom of the sample where the concentration is c_0. Measurements of sedimentation velocities use rotor speeds up to about 60,000 rpm while the sedimentation equilibria are measured at about 10,000 rpm.[33]

4.3.g Capillary Hydrodynamic Chromatography

When a suspension is injected into a stream of liquid flowing through an open tube or capillary, the particles separate according to size, the largest particles eluting first. Colloidal particles traveling through the capillary will be carried along with the flow of liquid while simultaneously diffusing by Brownian motion radially across the tube. The liquid flow is slowest at the walls of the channel and as particles diffuse nearer the wall, they slow down. The smaller a particle, the closer it can approach the walls of the channel, and the slower it moves. Hence the smaller reside longer in the capillary. The mathematical theory is given by DosRamos and Silebi.[34] All the particles actually move through the column faster than the *average* fluid velocity (Figure 4.8). The separation of particles depends on their size but not their density nor chemical composition.

The equipment for capillary hydrodynamic chromatography (CHDC), also named capillary hydrodynamic fractionation (CHDF), is essentially the same as that used in liquid chromatography, the major difference being the use of inert (nonsticking) walls in the former. If the particles are attracted to the walls, the possibility of separation based on size is lost. The complexity of phenomena in CHDC requires that calibration standards be used. Commercially available equipment uses ultraviolet absorption to detect eluting particles. The measured absorption must be corrected for scattering effects. The commercial unit from Matec Applied Sciences is designed for aqueous dispersions and has a size range from 15 nm to just over a micron. Because the particle fractions are monodisperse as they exit the unit, size-measuring techniques such as quasielastic light scattering or low-angle light scattering can be added as detectors and make the technique absolute.

78 PARTICLE SIZING

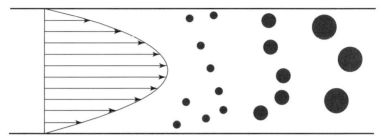

Figure 4.8 The rheological model of separation by hydrodynamic chromatography. The average velocity of the larger particles confined to the center of the tube is greater than the average velocity of the smaller particles, which can diffuse closer to the wall where the rate of fluid flow is less.[35]

4.3.h Field-Flow Fractionation

Field-flow fractionation (FFF) combines capillary hydrodynamic chromatography and field-driven techniques such as centrifugation and electrophoresis. Figure 4.9 shows the general structure of cells used for FFF. The capillary chamber is narrow but wide and the liquid flow is laminar. The applied field in FFF is normal to the fluid flow. The particles are sized according to how they respond to the applied field. The two most popular commercial versions of FFF are sedimentation FFF and thermal FFF.[36] A bibliography for these and other techniques such as electrical FFF[37] and steric FFF is available.[38,39] Sedimentation FFF separates particles based on how they respond to a gravitational or centrifugal field. Thermal FFF separates particles based on how they respond to a thermal gradient. The various FFF techniques have been used to analyze and separate molecules from 10^3 to 10^{16} Da and particles from 0.001 μm to 100 μm. Equipment for field flow fractionation is available from FFFractionation.[40]

4.3.h(i) Sedimentation Field-Flow Fractionation Sedimentation FFF is a particle-sizing technique that combines the flow separation used in capillary hydrodynamic chromatography and the size separation in centrifugation. The sample (generally 5 μL of a 1% dispersion) is injected into a spinning annular channel through which serum is flowing. The larger particles are thrown to the outer edge of the channel by centrifugal force. The liquid at the outer edges of the channel flows more slowly than that in the center, so that the larger particles at the outside move more slowly than the smaller particles in the center. By adjusting the flow rate and the centrifugal force, the separation of the particles by size can be optimized. Figure 4.10 and Figure 4.11 show the size distributions for submicron particles and supramicron particles. Figure 4.11 shows the presence of supramicron sized clusters (up to six) of monodisperse latex beads. No ensemble technique, for example, light scattering, can provide as good a resolution as FFF. The mass of the particles as a function of size is easily measured with an ultraviolet detector on the exit port of the channel. An excellent detection method is to use quasi-elastic light scattering to measure the size of the monodisperse

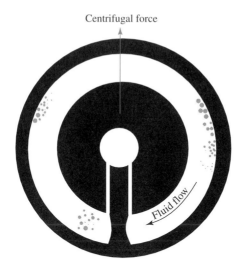

Figure 4.9 The cell for field-flow fractionation showing progressive stages of separation.

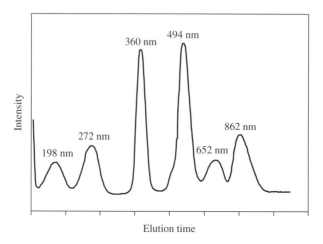

Figure 4.10 The fractionation of submicron polystyrene beads by centrifugation FFF. (Courtesy of FFFractionation, Inc.)

particles eluting at any instant.[41] The method needs no calibration because the laminar flow of liquid in the channel and the centrifugal forces can all be calculated. The engineering challenge is to form tight seals between the spinning channel and the stationary inlet and outlet ports. The commercial equipment is designed for aqueous dispersions.

4.3.h(ii) ***Thermal Field-Flow Fractionation*** Thermal FFF is a technique that separates particles or polymer molecules based on the difference between their

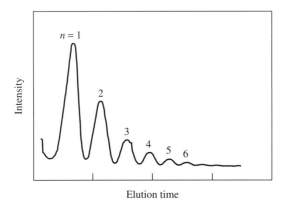

Figure 4.11 The fractionation of supramicron clusters of monodisperse PMMA (polymethylmethacrylate) beads by centrifugation FFF. n is the number of particles in the cluster. (Courtesy of FFFractionation, Inc.)

thermal diffusion, or thermophoresis, which depends on chemical composition, and on Brownian diffusion which depends only on size. A thermal gradient is created between the walls of the cell. Thermophoresis results in the accumulation of particles (or polymers) at the cold wall. Brownian diffusion moves particles away from the wall. The competition between the two types of diffusion results in a separation based on chemical composition and size. The device is constructed with two specially coated and polished thermally conductive bars clamped over a thin Mylar spacer from which a center channel volume only 127 µm (5 mils) thick has been removed. The upper plate is kept hot and the lower plate is cooled (see Figure 4.12). Thermal FFF does not have the problems associated with gel-permeation chromatography such as surface adsorption, shear degradation, size exclusion range, and so on. This technique works effectively for almost any molecular weight, polymer chemistry or solvent.

Historically, thermal FFF was developed to characterize polymers in solution, for example, to determine the high molecular-weight fraction or gel content of polymers. This is otherwise a difficult value to obtain and is important to explain

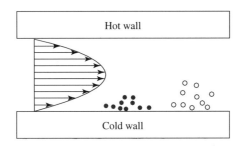

Figure 4.12 A thermal field-flow fractionator from FFFractionation, Inc.

variations in the rheology and stability of polymer-stabilized dispersions, especially of commercial polymers with their batch-to-batch variability. Thermal FFF has been used to separate particles of similar size, stabilized with different surface chemistries.[42,36] The technique is capable of separating particles by both size and surface composition.

4.4 SIZING BY SINGLE-PARTICLE DETECTION

The following techniques isolate and measure single particles. All these methods respond to individual particles, hence the size distributions are number distributions. Allen has written a thorough review of these techniques, including a table with schematics of eight techniques.[43] We include here a description of the important ones for which commercial equipment is readily available.

4.4.a Optical Microscope

The microscope is the basic instrument to study particles; it is the only method by which individual particles are observed directly. It is the first instrument of choice if the material is within its range, because it is accepted as the reference technique. The evidence of the microscope has a high degree of credibility. As well as giving the particle size, it provides information about shape, crystal structure, whether the particles are primary or aggregated, and whether different compounds are in the sample. Care has to be taken to obtain a representative sample, both in what is put under the microscope and in what is taken from the image, and to count a statistically significant number of particles. The short depth of field adds to the sampling problem. Modern image analyzers are capable of doing the counting and subsequent calculations automatically.

The optical microscope will often detect but cannot measure particles below the ~2 μm limit because of diffraction. Specimen contrast often limits the utility of the optical microscope, although special methods such as phase contrast, dark-field illumination, and Nomarski interference are often used to advantage. Specimen contrast for nonabsorbing particles decreases as the refractive index of the particles approaches that of the medium. Optical methods of all types fail for nonabsorbing particles when the refractive index of the particle and the dispersing medium match. To get optimum results, proper techniques must be used: See McCrone and Delly[44] for practical information. This reference also gives detailed techniques for particle identification as well as size and shape analysis.

4.4.b Image Analysis

Image analysis software is available from essentially every manufacturer of optical microscopes or electron microscopes. The characterization of the size distribution of dispersed particles as viewed through a microscope or contained on a photomicrograph requires an accurate estimate of the population in each size group. An automated system should be used to eliminate operator bias and to enable counting large numbers of particles. The idea is quite simple: create a computer program that finds the edges of particles, marks the perimeter of those particles and selects systematically a size parameter such as minimum chord length, average chord length, diameter of inscribed circle or circumscribed circle, and so on. The difficulty comes in making a systematic decision about when a viewed particle comprises more than one smaller particle partially coincident by accident of sampling. If the particle concentration is too high in the field of view, two errors are introduced into the results: first, undercounting of small particles and second, the overcounting of large particles. This has serious consequences in health-related examinations where the fraction of small particles is most critical. To minimize these errors it is common to keep the coverage of particles on the viewing surface, a microscope image or a picture, to less than 5%. Monte Carlo simulations indicate that the particle density needs to be much less than this.[45] Rather than compensate for coincidences by some empirical rule, the dispersion should be sampled at low enough concentration that coincidence of particles is negligible. If there is doubt in the image, retake it at lower concentrations rather than try to resolve the doubt.

Fully automated particle size and shape analysis can be made by allowing the suspension to flow into a wide, shallow stream in such a way that the largest area of the particle is oriented toward a video camera. Images of individual particles are captured, digitized, and analyzed. Image analysis calculates the area and perimeter of each image and the particles can then be classified into categories such as uniparticle, biparticle, aggregate, and so on.[46]

4.4.c Ultramicroscopy

The ultramicroscope was invented by Zsigmondy.[47] With this instrument Zsigmondy was able to show clearly that colloidal particles, particularly gold, in dispersion exist as a separate solid phase and not as a solution. For this invention he was awarded the 1925 Nobel Prize in Chemistry. In essence the ultramicroscope is an improvement on the visual observations by Faraday and Tyndall of the scattering of light by suspended particles. Ultramicroscopy extends the useful range of light microscopy to particles not visible in bright-light illumination. The ultramicroscope employs dark-field illumination in which a beam of light illuminates the sample at right angles to the viewing direction. The low intensity of light scattered by particles at right angles to the direction of illumination can be detected against the dark background. The observer sees a bright flash of light indicating the presence of a particle, but the observer cannot see a true image of the particle and hence cannot determine its size. Counting the number of particles in a known volume of known concentration gives the number average size. Ultramicroscopy has been replaced by more modern methods of size analysis.

4.4.d Electron Microscopy

Electron microscopic methods share with the optical microscope the advantages of direct observation and pictorial representation (imaging.) The major disadvantages are the effects of high vacuum and the high temperature created by impingement of the electron beam on the sample, which inevitably alter its state either by flocculation or coalescence. The preparation of the sample and its imaging are tedious and many pictures have to be taken. The limited depth of field of electron microscopes and artifacts caused by taking microtome cross sections introduce sampling errors. In order to be observed, materials have to be opaque to the electron beam, especially compared to other constituents in the dispersion, such as stabilizers or polymer matrices. The technique is both equipment intensive and labor intensive and so is expensive to run. The degree of magnification in the electron microscope is unknown, so that reference materials are required for calibration. In spite of these defects, the electron microscope can give reliable information about submicron particles with sufficient care.

4.4.e Time-of-Transition Method

The time-of-transition method of particle sizing is a single-particle counting technique used for dilute dispersions and aerosols. The technique is illustrated in Figure 4.13. A laser beam is passed through a wedge prism which refracts the beam by a small angle. The prism is rotated at a constant frequency. A lens focuses the beam. The rotating, focused laser beam writes a circle in space. Particles are presented to the beam in a variety of ways: in flowing or stirred liquid or air, on a moving microscope slide, or sedimenting. A photodiode is placed directly behind the particles, perpendicular to the optical axis. The signal on the photodiode is lower during the time the beam is crossing the particle. When the speed of the scanning laser beam and the time that the light signal is reduced is known, the length of the light beam crossing the particle can be calculated. Numerous corrections can be applied to calculate the size of the particle and many particles can be counted to get a size distribution. In principle, measurements could be made down to about 300 nm, but the limit for current configurations is about 500 nm.[48]

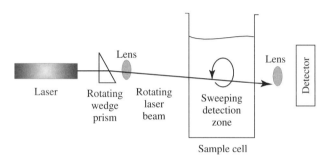

Figure 4.13 The basic optics for the time-of-transition particle-sizing technique.[48]

84 PARTICLE SIZING

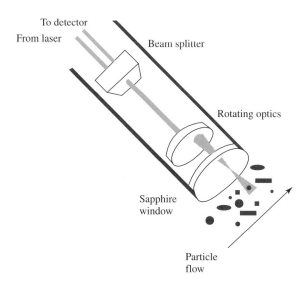

Figure 4.14 Schematic of the detection optics of the Lasentec particle-sizing system.[49]

Lasentec manufactures a line of particle-sizing equipment that is essentially a variation on the time-of-transition method particularly useful for pilot and manufacturing applications.[49] A laser beam from a probe is highly focused and rapidly scanned across particles or droplets suspended in a liquid. As the beam intersects a particle, light reflects back into the probe and the duration of the reflected pulse is recorded and sorted into size channels. (See Figure 4.14.) Typically, 1000 to 20,000 pulses are counted per second. The duration of the reflected pulse is not simply related to the particle diameter since the scanning laser pulse may cross the particle on any path, some longer, some shorter. An algorithm is applied to the data to convert it to a number distribution. The probe is robust and can be inserted into all kinds of process streams such as pipelines, reactor vessels, and tanks, which makes the technique generally useful. The size distributions are output in real time so that the equipment can be used to study changing systems. The lower size limit of detection of particles does not extend as far as that of the optical microscope.

4.4.f Particle Counting: Photozone Detection

Single-particle counting by photozone detection works by measuring the decrease in light intensity as a single particle passes through a narrow area of uniform illumination in the flow channel of a sensor. The drop in light intensity is proportional to the cross-sectional area of the particle. The general name for such instruments is photozone counter (Figure 4.15). This sensing technique is not restricted to aqueous systems; it may be used for nonpolar dispersions and for aerosols. These instruments are inherently inaccurate for particles less than one

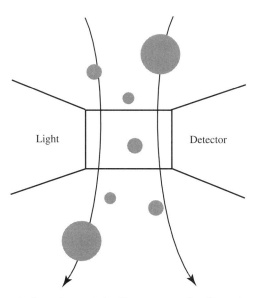

Figure 4.15 The optical sensing and the flow pattern of a dispersion in a typical photozone counter.[50]

micron in diameter, that is, near the wavelength of visible light, as the light absorption is then not a simple function of cross-sectional area because of scattering. Note that photozone detection depends on the change of magnitude of the intensity of light as a particle passes through the illuminated zone, while the time-of-transition methods depend on the length of the light signal as the rotating light beam passes over a particle.

To detect smaller particles, smaller sensing volumes are used. The restriction is that small sensing volumes can be plugged by large particles. The lower limit of detection is 0.2–0.4 μm when conditions of detection and analysis are optimum, but the practical lower limit is about 1.5 μm. Typical particle-counting rates approach 10,000 particles per second. The commercially available equipment has built-in algorithms and mechanical devices to minimize the effects of plugging the sensing zone. When larger sensing volumes are used, sensitivity to smaller particles is lost.

A significant improvement in the quality of data obtained from photozone detection is made by carefully controlling the concentration of particles passing through the sensing zone. This is accomplished by using an autodiluter filled with filtered serum which is used to adjust automatically the particle concentration as the dispersion flows into the sensing volume.[51] The operator starts with a concentrated dispersion and relies on the equipment to dilute the dispersion to the optimum concentration to obtain as high a particle-counting rate as possible but which would be still dilute enough to avoid false signals from particle coincidence in the photozone. This equipment is available from Particle Sizing Systems.

Pelssers et al.[52,53] have designed an instrument in which a flowing stream of the dispersion is hydrodynamically focused within a concentric flow of serum to form a narrow particle stream about 5 μm in diameter. The small stream of dispersion is illuminated with a laser beam about 30-μm wide. This flow and illumination defines a small sampling volume. The light scattered at a low angle, 5°, is detected and analyzed to give the particle size. This procedure allows an accurate determination of particle size to much lower sizes than the usual photozone detection.

4.4.g Particle Counting: Electrozone Detection

Electrozone detection of suspended particles uses the change in electrical resistance when a single particle passes through a sensing zone (see Figure 4.16.) The equipment is often referred to as a Coulter counter because the Coulter Corporation (now Beckman Coulter) produced the first commercial equipment, although other manufactures now produce similar equipment. The change in resistance as a particle passes through the sensing zone is proportional to the volume of the particle so that the technique reports a number distribution of volumes. Up to a million particles can be counted in a short time, providing high precision in particle-size distribution. It is a popular and well established technique for the size range from 0.6 μm to 1200 μm, a lower limit slightly lower than optical-counting techniques. A number of different size sensors are used to optimize sensitivity for different size ranges. The shape of the signal is analyzed to reject the effect of multiple particles in the sensing zone. Electrozone detection is only applicable to aqueous systems, because nonpolar media are not sufficiently conducting. The electrozone counter is also called the resistazone counter in some literature.

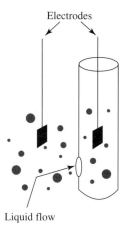

Figure 4.16 Schematic of the sensing zone of an electrozone detector.[54]

The Coulter counter is commonly used to count the number of red and white blood cells for medical examinations. Red blood cells are about 8 μm in diameter and; quite monodisperse; white cells are about 12–15 μm in diameter. Blood is highly conductive and so is a perfect medium for electrozone detection. The adhesion of blood cells to cell surfaces is directly related to the formation of doublets in the serum.[55] Adhesion to walls is decreased by the addition of glutaraldehyde and increased by the addition of dextran and fibronectin. The ratio of doublets to singlets as measured by electrozone detection follows the same trend and makes a simple screening tool.

4.5 SIZING BY LIGHT SCATTERING

The scattering of light is a strong function of particle size. This is illustrated in Figure 4.17 for the scattering of light per molecule as the water cluster size increases from molecules to large drops of rain.[56] Scattering by a molecule that belongs to a cloud droplet is about 10^9 times greater than scattering by an isolated molecule. The initial linear increase in scattering when plotted on a log-log scale is in the regime of Rayleigh scattering. The slope of the line is three as predicted by Rayleigh theory. Above a few hundreds of nanometers, the wavelength of visible light, the scattering per molecule does not increase as much. This is the beginning of the Mie-scattering regime. The linear regime of negative slope beyond a few thousand nanometers is the Fraunhofer-diffraction regime. The beginnings and endings of these regimes vary with the refractive index and extinction coefficient of the scatterer, but the trends are the same. If the data in Figure 4.17 were plotted as the intensity per particle rather than the intensity per molecule, the slope everywhere would be increased by three. This would emphasize the strong dependence of the light scattered by the largest particles

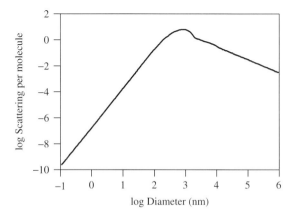

Figure 4.17 The intensity of scattered light per molecule of water as the cluster size increases.[56]

in the dispersion. A special issue of *Applied Optics* features 36 papers on topics related to particle sizing by optical methods.[57]

Laser light is preferred for most scattering measurements, as it is monochromatic, stable and intense. The great advantage of all light-scattering techniques is that particle-size measurements can be made in any solvent, because the sample need only be held in a transparent container. The disadvantages of all light-scattering techniques are that dispersions must be dilute and that the effects of particle shape (other than spherical) are not easily estimated.

4.5.a Rayleigh Scattering

An immediate indication of the presence of particles in a medium is the appearance of turbidity. The analytic technique is called nephelometry or the measure of cloudiness. Turbidity is an important tool in water treatment and purification, in monitoring bacterial growth in microbiological research, in quality control for foods, milk, cheese, and beverage manufacture, in determination of sulfates and silica concentrations, and in sensing nucleation or the onset of precipitation. The turbidity of a dispersion is measured by the attenuation of light passing through a thickness d,

$$I = I_0 \exp(-\tau d) \qquad (4.13)$$

where I_0 is the intensity of the incident beam, I is the intensity of the attenuated beam, and τ is the turbidity coefficient. Turbidity measurements are useful to detect particles that do not absorb light and are less than about 100 nm in diameter, that is, so-called Rayleigh scatterers. All the light lost from the incident beam is by scattering. If the dispersion is monodisperse, the following relation, which gives particle diameter, d, as a function of the ratio of the turbidity τ to the mass concentration of the particles, c, is obtained,[58]

$$d^3 = \frac{\lambda^4}{4\pi^4 \rho} \left[\frac{n^2 + 2}{n^2 - 1} \right]^2 \left(\frac{\tau}{c} \right) \qquad (4.14)$$

where n is the ratio of the refractive indices of the dispersed phase to that of the medium, λ is the wavelength of the light in the medium, ρ is the density of the particles. Flocculation of a fixed mass of dispersed material, as measured by d, can be followed by turbidity measurements, at least within the limits of Rayleigh scattering. Commonly, white light rather than monochromatic light is used and the sample is polydisperse; nevertheless, turbidity gives a light-scattering mean volume. Rayleigh scattering increases as the sixth power of the particle size, so that large particles greatly influence the average size (See Section 1.1).

The term "laser-light scattering" is a designation for a number of different techniques. Outside the field of particle-size analysis, it usually refers to the use of Rayleigh-scattering theory to obtain the molecular weight and radius of gyration of polymers in solution. (See Section 1.2.) A general review of light scattering as an analytic tool is given by Phillies and Billmeyer.[59]

4.5.b Turbidity

Commercially available turbidimeters are calibrated in turbidity units (TU). Originally, one TU was equal to the turbidity caused by 1 ppm of a standard suspended silica. Turbidity is often measured today by applied nephelometry, which measures light scattered by particles at right angles to the incident beam (Figure 4.18). The standard is one nephelometric turbidity unit (NTU). Standards are available from equipment manufacturers. These standards are aqueous dispersions of formazin, a polymer formed by the condensation reaction between hydrazine sulfate and hexamethylenetetramine.[60]

Solutions that contain no second phase do not scatter light so their turbidity is zero. An easy test to decide whether an unknown sample has any dispersed material in it or if all the components are dissolved is to shine a laser pointer through the sample and look for the red Tyndall beam. If the beam is visible, the sample contains a second phase. This simple test can be carried a little further. If the beam appears about the same brightness from all angles of observation, the dispersed material has a size less than about 0.1 µm or so. If the beam looks much brighter in the forward-scattering direction (i.e., looking somewhat back toward the laser pointer), the particles are large compared to the wavelength of the laser light and hence are greater than a micron or two.

Turbidimetric measurements are useful to study the onset of precipitation. When all the components are in one phase no light is scattered and the turbidity is zero. When any of the components begin to precipitate and form a second phase, they will scatter light and the turbidity increases. This same test can be applied to study rates of flocculation if the initial particles are small. Small particles scatter little light and the turbidity is slight. As flocs form the cluster sizes increase, they scatter light more strongly, and turbidity increases. This is useful to determine which additives flocculate a dispersion or over what temperature range a dispersion is stable. Numerous examples of the use of turbidity can be found in the literature.[61] An extensive bibliography of particle sizing using turbidity has been given by Kourti and MacGregor.[62]

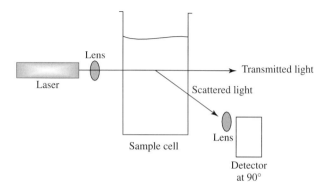

Figure 4.18 The optics of an apparatus used to measure particle size by nephelometry. The light scattered by the particles in the sample cell is detected at 90°.

4.5.c Mie Scattering

Particle-size distributions of suspended particles are measured by the scattering of light in either of two modes: The first is to measure the average intensity of the scattered light as a function of scattering angle (Mie scattering); the second is to measure the time dependence of the scattered light at a fixed scattering angle (quasi-elastic light scattering.) For monodisperse particles either mode of measurement is sufficient to determine particle size. Quasi-elastic light scattering is more useful for polydisperse dispersions.

Figure 1.4 shows the logarithm of the intensity of light scattered from polystyrene particles in water of three sizes, 50 nm, 500 nm, and 2000 nm in diameter, as a function of scattering angle. (The illumination is from the left at $180°$. The forward scattering is to the right at $0°$.) The particle size of monodisperse dispersions in the Mie scattering size range can be easily identified by the distinctive shapes of the scattering functions like these. The intensity of scattered light is measured at as many scattering angles as possible by mounting a single detector on a goniometer or using multiple detectors. Some instruments use more than one wavelength of incident light.

For polydisperse suspensions, however, the analysis is more difficult. The scattered light from a polydisperse system is the superposition of the scattering patterns from each different particle size in the dispersion. The scattering pattern for each size is unique, but the convolution of overlapping angle-dependent scattering functions tends to mask the differences. This means that experimental data, which tend to appear smooth, are actually composed of intricate patterns for each size. The deconvolution of smooth data into a sum of intricate patterns is mathematically ill-conditioned and many equivalent solutions can be found. The analysis is therefore uncertain.

Mie calculations are based on spherical symmetry or the nearly spherical such as ellipsoids. Greater asymmetry or surface roughness is not taken into account and invalidates the application of the theory to most solid particles, especially those created by attrition. Hofer et al.[63] have obtained good results on fat emulsions in water, where the refractive indexes are known and the droplets are spherical, by measuring the scattering at 54 angles.

4.5.d Fraunhofer Diffraction

Particles significantly larger than the wavelength of incident light (usually a helium–neon laser beam of wavelength 632.8 m) act as tiny lenses and diffract light in the same way as does a ruled grating. The phenomenon is called Fraunhofer diffraction, whether it is caused by particles or by a ruled grating. The intensity of light scattered is proportional to the particle volume, and the scattering angle is inversely proportional to the diameter. Neither the intensity of the scattered light nor the angle of scattering is sensitive to particle shape. The intensity of light is measured as a function of angle near the forward direction (low angles.) The central beam is blocked. The original Microtrac Analyzer from

Leeds & Northrup sampled the scattered light through a series of rotating disks with pinholes at different radial positions.

Modern instruments use a series of concentric photosensitive rings on a plane perpendicular to the laser beam. For a monodisperse system the data are described by Airy functions. Figure 4.19 shows the Airy function for one size of particles in a plane but must be imagined in three dimensions as concentric rings of light. The smaller the particle size the more expanded are the rings, approaching Mie scattering in appearance and ultimately requiring a completely different style of detector.

For polydisperse systems the data are combinations of Airy functions which can be deconvoluted by least-squares analysis to give volume distributions. Commercial instruments use flow-through cells and thus are able to monitor flocculation or comminution continuously.[64] Fraunhofer diffraction is particularly suited for application to aerosols and emulsions. An aerosol sprayed across the detection zone of the laser beam results in a nearly instantaneous display of the droplet-size distribution.

The earlier laser diffraction instruments measured light scattered from 0.03° to 3°, a dynamic range of 100:1, and were expected to cover particle sizes from approximately 8 μm to about 800 μm. (A dynamic range of 100:1 is not a specific range; the specific range depends on the degree of magnification.) Later instruments have a dynamic range of several hundred to one. Bott and Hart showed that including two Fourier lenses extends the range farther.[65] This improvement is included in laser diffraction instruments from the Beckman-Coulter Company.

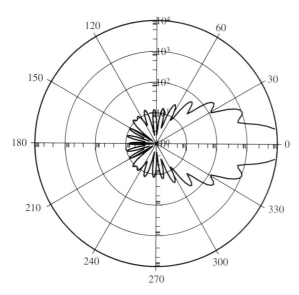

Figure 4.19 A typical Fraunhofer diffraction pattern. The intensity is seen to be a strong function of scattering angle.

4.5.e Polarization Effects

The oscillating electric field of polarized light waves interacting with a particle induces oscillating dipoles therein whose direction is the same as that of the polarization. The oscillating dipoles radiate light in all directions but in the case of molecules or tiny particles not in the direction of polarization. Larger particles scatter more light in that excluded direction. The usual light-scattering instruments measure only light perpendicular to the excluded direction.* By including measurements in the excluded direction and all the way around, more information about particle size can be obtained. A source of white light (all wavelengths, all polarizations) with a series of wavelength and polarization filters on a carousel provides data at various wavelengths and various angles of polarization.[65] These measurements are called polarization intensity differential scattering (PIDS) and instrumentation is commercially available.

4.5.f Quasi-Elastic Light Scattering (QELS)

Light scattering as analyzed by Rayleigh and Mie measures the time-average intensity. In the 1960s it was noticed that the time fluctuations of the intensity could be used to obtain information about the motion of dispersed particles or macromolecules in solution. The motion is Brownian movement. The light scattered by a moving particle is Doppler shifted by a tiny amount with respect to the incident frequency. Each moving particle scatters light at a slightly different frequency depending on its velocity with respect to the incident beam. The superposition of all the frequencies produces a time varying signal with a scale of the order of milliseconds or even microseconds. Brownian motion is inversely proportional to particle size. The time dependence of the fluctuating intensity therefore is a function of particle size. The autocorrelation function of these fluctuations in intensity, a mathematical transformation, has a simple dependence on D and hence on particle size (See Section 1.5).

Commercial instruments take the input from a photomultiplier tube and calculate the autocorrelation function versus delay time automatically. For a monodisperse system of spherical particles in Brownian motion, the autocorrelation function, $g^{(1)}$, is an exponential decay

$$g^{(1)}(\tau) = \exp\left(-D\left(\frac{4n\pi \sin(\theta/2)}{\lambda}\right)^2 \tau\right) \qquad (4.15)$$

where D is the particle diffusion coefficient, τ is the delay time, n, λ, and θ are the refractive index of the medium, wavelength of the light, and scattering

---

* Lasers are generally manufactured so that the polarization is vertical with respect to the feet on the laser. Mounting the photodetectors on the plane of the laser means that only light scattered perpendicular to the polarization of the laser beam is analyzed. Some lasers are sold with the polarization horizontal. Personal experience has shown that this causes great confusion in the building and aligning of light-scattering equipment.

angle, respectively. The diffusion coefficient D is inversely proportional to particle size, as shown by the Einstein equation (see Equation 1.11). A plot of the autocorrelation function versus the decay time is an exponential decay with a slope proportional to the diffusion coefficient and hence inversely proportional to the size of the particle. This technique is so simple and so accurate that it is used to measure reference standards., even for the electron microscope.

For a polydisperse system of spherical particles in Brownian motion, the normalized autocorrelation function, is an integral of exponentials. The mathematical problem is to invert the integral to obtain the distribution of decay constants. Each decay constant corresponds to a particle size through its dependence on the diffusion coefficient. The problem has been solved by the method of cumulants,[66] by the method of regularization,[67] and by a nonnegatively constrained optimization.[68,69] The technique of quasi-elastic light scattering is preferable to the classical Rayleigh–Mie procedure, as the simplicity of Eq. (4.15) results in a more tractable mathematical problem.

Most of the common commercial instruments analyze data taken at only one scattering angle, 90°, but, of course, more dependable results are obtained by combining data taken at many angles.[70] Commercial instruments are available to measure the autocorrelation function and to do the complete mathematical analysis of the data. A review of the application of quasi-elastic light scattering for the determination of particle size distributions has been given by Finsy including round-robin test results.[71]

4.5.g Diffusing-Wave Spectroscopy

All the light scattering techniques discussed above depend on the particle concentration being low enough, or the path length short enough, that incident photons are scattered by only one particle before reaching the detector. However, progress toward obtaining useful information from dispersions at high-volume loadings has been made by the development of a technique called diffusing wave or photon-migration spectroscopy. Light that travels through a concentrated dispersion is multiply scattered. When the concentration of particles is high enough, the scattering is completely random and the properties of the scattered light can be described by diffusive spreading similar to a random walk. The intensity of the scattered light varies with time as the interparticle spacing changes, a phenomenon similar to quasi-elastic light scattering but at shorter time intervals. The fundamental difference is that the diffusion-wave signal is most sensitive to the smallest motions either in terms of the smallest distances or the shortest times, as these are the most correlated.[72] That makes the technique suitable for studying interparticle forces in colloidal suspensions, or measuring displacements down to the angstrom scale, or for studying particle motion other than diffusion, such as in rheological experiments. The phase shift and attenuation can be analyzed to give the particle-size distribution[73] or the structure of gels.[74] The required apparatus is the same as for QELS except simpler because the alignment of the optics is not critical. The technique is new enough that studies are mostly academic.

4.5.h Light Scattering as a Detector

Many particle-sizing techniques are based on physically separating polydisperse samples into nearly monodisperse ranges. Examples are capillary hydrodynamics and field-flow fractionation. Light-scattering techniques are most accurate when the samples are monodisperse. Therefore, using either classical or quasi-elastic light-scattering techniques as the detector provides highly reliable information.

Field-flow fractionation, which separates dispersions into nearly monodisperse fractions, has been combined with multiangle light scattering to create a technique of high resolution for particle sizing or for determining the molecular weight of polymers.[75]

QELS has been used as a detector on chromatographic instruments such as capillary hydrodynamic fractionation, CHDF, from Matec Applied Sciences and field-flow fractionation from FFFractionation.

A fast technique to measure molecular weights and radii of gyration of polymers is light scattering with multiple laser diode detectors (MALS). When one of these detectors is replaced by a quasi-elastic light-scattering detector, which measures time-dependent intensity, the combined equipment (MALS-QELS) allows a simultaneous determination of absolute molecular weight, radius of gyration, and the radius of gyration. These results can be obtained for molecules ranging from thousands to hundreds of millions of Daltons. (www.wyatt.com)

4.6 SIZING BY ACOUSTICS

4.6.a Scattering and Absorption

Light is used routinely to measure the size of particles in suspension. One limitation of optical techniques is that the scattered light needs to reach the photodetector with a minimal amount of multiple scattering or attenuation. This requires that measurements be made in dilute suspension or with the detector placed within the sample. The suspension has to be substantially colorless. Many industrially important dispersions are neither dilute nor clear. Much recent research has gone into measuring particle size by means of sound. The first obvious advantage of an acoustic-sizing technique is that it can be used with opaque dispersions, even through metal walls. One can imagine applications where the acoustic source is mounted on the outside of a pipe or reactor and the detector is mounted on the opposite side.

A second advantage of acoustic measurements is that it is easier to account for multiple scattering and consequently the technique can be used over a very wide range of concentrations, up to the random monodisperse packing limit of about 60%.

A third advantage of acoustic sizing is that the wavelength can be varied over four orders of magnitude. The commercially available instruments allow access to a total of about eight orders of magnitude in frequency. The range is selected depending on the materials of the sample.[76] Light scattering, on the other hand, is rarely done with any wavelengths outside the visible range of 400–700 m, which is much less than one order of magnitude. The analysis of light-scattering data depends strongly on the ratio of the size of the particle to the wavelength of light. This is also true for acoustic scattering, and the wavelength of sound waves can be varied over such a wide range that the frequency can generally be chosen to correspond to the conditions of some useful theory. Again, light-scattering data usually have the form of light intensity at a constant wavelength as a function of scattering angle or time. Acoustic data are usually acoustic absorption at a constant scattering angle as a function of wavelength. This means that the dynamic range for acoustic measurement, that is, the particle-size range over which size measurements can be made by a single instrument, is 0.01 to 1000 μm.[77] Often both the volume fraction and the particle size of the sample are unknown, especially in flowing systems. Use of the frequency-dependent change in sound velocity and attenuation give enough data to establish both concentration and size.[78]

Povey's book (1997) on ultrasonic characterization techniques is the best available general text. He presents a table of comparison of ultrasound and light scattering for aqueous dispersions (see Table 4.2).

Table 4.2 includes one of the major disadvantages of acoustic measurements for particle sizing: The acoustic signal depends on so many properties of the materials of the dispersion. Thermophysical properties include heat capacities, thermal conductivities, compressibilities, the velocity of sound, the acoustic absorption and their temperature dependencies. These properties are often not known for many materials in dispersions. Another disadvantage of acoustic measurements is that air bubbles absorb acoustic energy so any dispersion containing bubbles will be "opaque" to sound.

Nevertheless, ultrasound absorption has been used quite successfully in a number of commercially important applications.[80] Dispersions that have to be diluted

TABLE 4.2 Features of Ultrasound that Distinguish It from Light Scattering in Aqueous Colloids at 20°C[79]

Ultrasound	Light
Transducers are phase sensitive	Transducers are phase insensitive
Wavelength between μm and cm	Wavelength between 0.5 and 1 μm
Frequency between 0.1 and 10^{13} Hz	Frequency between 3×10^{16} and 6×10^{16} Hz
Coherence between pulses	No coherence between pulses
Responds to elastic, thermophysical, and density properties	Responds to dielectric and permeability properties
Particle motion parallel to the direction of propagation; no polarization	Field displacement perpendicular to direction of propagation; polarization is therefore possible
Propagates through optically opaque materials	Sample dilution is normally required

for measurement by light scattering or other techniques need not be diluted for acoustic measurements. An example is the growth of emulsion drops by Ostwald ripening. Figure 4.20 shows the ultrasonic attenuation spectra of an O/W emulsion as a function of time after homogenization. Figure 4.21 shows the corresponding droplet size distributions as a function of time. The average size increases by Ostwald ripening.

Manufacturers need to make the equipment as easy to use as possible. One simplification for acoustic applications is to make as many approximations about the various thermophysical parameters and processes as possible so that the user need not provide so many hard-to-get numbers. This simplification, of course, is better for some systems than others. Since the analysis software is usually proprietary, the user is never quite sure of the results. Relatively few academic publications are available.[81]

4.6.b Speed of Sound

Accurate measurements of sound velocity are more readily made than accurate measurements of the absorption of sound, so that much thought has gone into discovering what properties of multiphase dispersions can be deduced from the speed of sound alone. The most immediately useful quantity that can be determined from the speed of sound is the concentration of particles or droplets in a

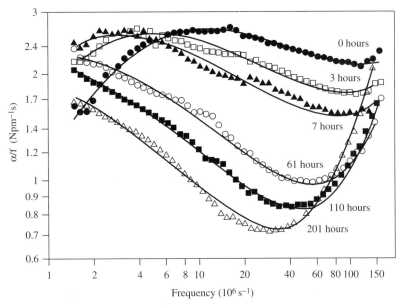

Figure 4.20 Ultrasonic attenuation spectra of 5 wt percent *n*-decane O/W emulsions stabilized by 20 mM sodium dodecyl sulfate, measured at various times after homogenization. Symbols represent measured values and the lines represent best-fit size distributions.[80]

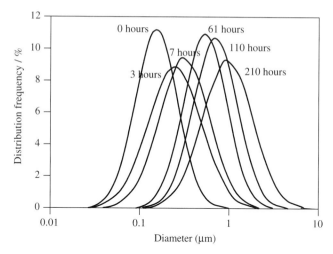

Figure 4.21 Droplet-size distributions of the 5 wt percent n-decane O/W emulsions described in Figure 4.20.[80]

dispersion. The concentration of dispersed phase can be determined from about 1% to randomly close-packed dispersions and this measurement can be made through the opaque walls of pipes and tanks (See Section 3.1.f).

The speed of sound is measured by time of flight, that is, by measuring the length of time taken for a signal to go from source to detector. This is analogous to counting the time between seeing the flash of lightning and the moment the sound is heard. The mathematical analysis is due to Urick[82] who tested the theory with kaolin dispersions in water up to 40 volume percent, emulsions of xylene in water and water in xylene over the entire range of compositions, and of horse blood diluted by its plasma. The velocity measurements made over a range of concentrations for emulsions are the same whether the morphology is W/O or O/W indicating that the velocity of sound is determined by the composition only. The constitutive relations are

$$v = \frac{1}{\sqrt{\kappa \rho}}, \quad \kappa = \sum_j \Phi_j \kappa_j, \quad \rho = \sum_j \Phi_j \rho_j \qquad (4.16)$$

where v is the measured velocity of sound, Φ_j is the volume fraction of component j, κ_j is the adiabatic compressibility of component j and ρ_j is the density of component j. For two-phase systems the volume fractions can be calculated from the measured velocity of sound and the compressibilities and densities of the components. More likely, however, the compressibilities and densities are not known, so an adequate procedure is to make a calibration curve of the speed of sound of the dispersion as a function of volume fraction of dispersed phase. This is easy to do. The major precautions are two: first, compressibilities of liquids are strong functions of temperature so that temperature control is necessary and second, even small quantities of dissolved air can also change compressibilities, especially of water, so that care has to be taken with the conditions of the measurement and the establishment of the calibration curve. Once these precautions are taken, however, measuring the speed of sound through a dispersion is a simple way to get the local volume fraction. Measurement of sound velocity is useful to determine changes in concentration with position, such as occur with settling or creaming. Povey[23] gives examples of results on aqueous solutions of alcohol, sugar, edible oils and fats, and cells, as well as O/W margarine, chocolate, crystallization and other phase changes. Equipment to measure the speed of sound through dispersions is available from Applied Sonics.

4.6.c Electroacoustics

In the first issue of the *Journal of Chemical Physics*, Debye proposed a technique to determine how many molecules of water are intimately connected with different ions in solution by measuring the mass of the hydrated ions.[83] An acoustic signal is passed through the ionic solution. If the frequency of the acoustic signal is high enough that the inertia of the ions is significant, then the motion of the anions and cations differ according to their masses. This difference in motion appears as an electric potential. The converse is also true. If a high-frequency electric field (order of megahertz) is passed through an ionic solution, acoustic signals are generated that depend on the masses of the ions: at low frequencies the ions follow the electric field and produce an acoustic signal. At higher frequencies, the inertia of the ions prevents them from staying in phase with the electric field and the resulting acoustic signal is out of phase and lower in amplitude.

These concepts have been extended to measuring particle size.[84,85,86,87,88] The magnitude of the generated acoustic signal is proportional to the mass concentration of suspended, charged particles. Therefore the signal is greater if the concentration is greater. The acoustic signal can pass through an opaque suspension so that dilution is not needed. The acoustic signal decreases as the frequency of the electric field is increased. From these data, the size of the particles in the suspension can be deduced. An additional advantage of this method is that the charge on the particles may be determined at the same time since the magnitude of the acoustic signal depends on the mass fraction of the particles and the charge per particle.[89] At low frequencies the acoustic signal depends on particle charge; at higher frequencies the acoustic signal depends on particle charge and particle size.

4.7 SURFACE AREA BY ADSORPTION

The specific surface area of a powdered solid is an indirect expression of the particle size. Adsorption techniques deliver the specific surface area rather than particle size. Although specific surface area can be calculated from a particle size distribution, the converse cannot be done. But if the particles are taken as uniform spheres, cubes, or other isometric solids, an average particle size can be obtained from the relation $d = 6/(\rho A)$, where ρ is the density and A is the specific surface area (area/mass). The average radius obtained in this way is the ratio of volume to area, or d^3/d^2, known as a 3:2 average to distinguish it from the ratio of area to diameter, known as a 2:1 average. The determination of surface area of a solid is more accurate the smaller the particle size and is more appropriate when interfacial properties rather than particulate properties are in question.

4.7.a Multilayer Gas Adsorption

The adsorption isotherm relates the amount of adsorbate per gram of adsorbent to each equilibrium pressure of the gas at constant temperature (Figure 4.22). Adsorbate molecules at small pressures are present as a monolayer; at greater pressures, second and higher molecular layers are formed. The transition from the first to the second layer is usually accompanied by a sharp decrease in the binding energy of the adsorbate, as the first layer is directly held by the substrate whereas succeeding layers are farther away and in direct contact only with adsorbed molecules. This transition, therefore, is often marked by a change of slope of the adsorption isotherm. The inflection point of the isotherm occurs approximately at the pressure where the substrate is covered with a fully compressed monolayer, designated n_m. A fully compressed monolayer is an artificial concept, which is never actually attained, because succeeding layers form before the first is saturated. The adsorption isotherm may be determined either by gradual addition of gas[90] or by continuous flow of an adsorbable gas in an inert carrier gas.[91,92]

A theory designed to evaluate n_m from adsorption data was proposed by Brunauer, Emmett and Teller.[93] It is usually referred to as the BET theory. It supposes that the first layer of adsorbate is laid down according to the Langmuir

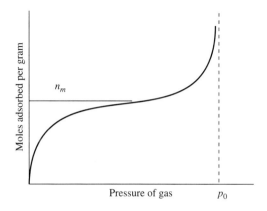

Figure 4.22 A simple multilayer adsorption isotherm typical of nitrogen at 77 K.

equation

$$p = \frac{K\theta}{1-\theta} \tag{4.17}$$

where p is the equilibrium pressure, θ is the degree of coverage, given by n/n_m, where n is the number of moles of adsorbate per gram of adsorbent at pressure p, and K is a constant. The following assumptions underlie the derivation of this equation:

(a) The adsorption is only monomolecular.
(b) The adsorption is localized on specific sites.
(c) The energy of adsorption on the solid substrate is uniform and characteristic of the solid and the adsorbate.
(d) No adsorbate–adsorbate attraction exists.

The BET theory goes on to postulate that simultaneous Langmuir-type adsorptions take place in successive layers. Further assumptions are required to simplify the mathematical treatment, namely, that the energies of adsorption in the second and succeeding layers of adsorbate are also uniform, and that all of these are equal to the heat of liquefaction of the adsorbate. The final expression of the BET theory is

$$n = \frac{n_m C p}{(p - p_0)[1 + (C - 1)(p/p_0)]} \tag{4.18}$$

where p_0 is the saturation vapor pressure of the adsorbate, n_m, and C are constants, which are determined from the data either graphically or by a linear least-squares fit: n_m has the same significance in the BET theory as it has in the Langmuir theory.

The BET theory is the basis of a method to measure the specific surface area of a solid, using the value of n_m, and Eq. (4.25) with $a = 1$. This method has been experimentally shown to be sufficiently accurate for most practical purposes and has successfully met a requirement for a rapid estimate of particle size. This has made possible great advances in both theoretical and practical studies of surface chemistry and physics. We must not, however, confuse the triumphs of the BET theory in this practical application with its claims as a model of the adsorption process. In this latter respect, it is obviously far from satisfactory. The Langmuir model with its unjustifiable assumptions is the basis of the BET model, and the additional assumptions introduced in the BET derivation merely add further improbabilities to the total picture. Nevertheless the following admonition is pertinent:[94]

> From the practical point of view, it is a matter of secondary importance whether our theories and assumptions are correct, if only they guide us to results in accord with facts .... By their aid we can foresee the results of combinations of causes which would otherwise elude us.
>
> — A. W. Rucker, 1901

4.7.b Monolayer Gas Adsorption

At low temperatures a more realistic model than that of Langmuir is to assume that the adsorbate behaves as a nonideal two-dimensional gas. The virial equation of state of such a gas is

$$\frac{FA}{kT} = 1 + \frac{nB}{A} + \frac{n^2C}{A^2} + \cdots \tag{4.19}$$

where F is the two-dimensional (spreading) pressure, A/n is the area per molecule, k is the Boltzmann constant, and B and C are the second and third virial coefficients. The adsorption isotherm equation can be obtained from any equation of state by means of the Gibbs theorem in the form

$$\int d\ln p = \frac{1}{kT} \int \frac{A}{n} dF \tag{4.20}$$

Solving Eq. (4.19) for dF and then substituting in Eq. (4.20) gives

$$\ln p = \ln \frac{n}{A} + \frac{2nB}{A} + \frac{3n^2C}{A^2} + \cdots + \ln K \tag{4.21}$$

where $\ln K$ is the integration constant. Equation (4.21) has been shown to describe the adsorption isotherms of argon on graphitized carbon black in the monolayer region.[95]

Few substrates, of which graphitized carbon black, however, is one, are sufficiently uniform in terms of their adsorptive energies for adsorption of a gas to proceed according to the description given by Eq. (4.21). Heterogeneity of adsorptive energies is the rule rather than the exception. Unless substrate heterogeneity is taken into account in the analytical treatment of the adsorption isotherm, the model lacks an essential feature and the evaluation of n_m and consequently of the specific surface area, is open to doubt.

The variation in adsorptive potential causes the density of the adsorbed molecules (molecules per unit area) to vary across the surface; hence the adsorbate cannot be considered as a single surface phase. An analysis can proceed, however, if the adsorbate is treated as a set of independent surface phases, each with its own homogeneous density. This entails that the substrate be regarded as composed of different unisorptic patches. (A liquid surface is structureless and isoenergetic; a perfect crystalline surface is homotattic; both are unisorptic.) A realistic model of an heterogeneous substrate can be based on the following premises:

1. The whole adsorptive substrate is made up of homotattic patches, each with a different adsorptive energy for the adsorbate, and on each of which monolayer adsorption can be described by an adsorption isotherm equation, such as Eq. (4.21), which has been established experimentally for a homotattic substrate.

2. Elemental patches are filled simultaneously, though not to the same density, subject to the condition that the adsorbed phase on each patch has the same chemical potential. These elemental patches are assumed to be sufficiently large for boundary effects to be unimportant.
3. The number of moles adsorbed on the whole heterogeneous substrate is obtained by summing the individual values of the number of moles adsorbed on each patch, n_i, over all the patches.

$$n(p) = \sum_i n_i(p) = \sum_i A_i f\left(\frac{p}{K_i}\right) \quad (4.22)$$

where A_i is the area of the ith patch and $f(p/K_i)$ is found by solving Eq. (4.21) for n_i/A_i. If the heterogeneity of the substrate can be approximated by a continuous distribution function, then

$$n(p) = A \int dK\, P(K) f\left(\frac{p}{K}\right) \quad (4.23)$$

where A is the specific surface area (m^2/g); $P(K)$ is the probability of a patch having the Henry's law constant K.

The distribution of the K's is a description of the heterogeneity of adsorptive energies of the substrate. Simple techniques for fitting experimental data to Eq. (4.22) or (4.23) are available (CAEDMON).[96]

4.7.c Pore-Volume Analysis

Mercury porosimetry is routinely used to evaluate pore-volume and pore-size distributions of porous solids and particles in a packed bed. The essential idea is to use a nonwetting liquid, such as mercury, and to measure the volume of mercury that is forced into pores as a function of pressure. The radius of the pore is calculated by Laplace's equation [Eq. (6.46)] from the applied pressure, the surface tension of mercury, and an estimate of the mercury/solid contact angle. The pressure required to push mercury into void spaces depends on particle size and packing; therefore, void-volume distribution curves for granular or powdered materials, which are related to their size distributions, are readily obtained while the porosity of the powder is being measured. The pressure at which mercury enters interstitial or void space is much less than that required to penetrate porous particles, and so the effect of particle size is readily distinguished from that of pores. Lowell and Shields[91,92] give a detailed description of mercury porosimetry.

The porosity and permeability of a porous material depend on grain size and shape, and packing. Applications range from predicting the leaching of soil contaminants to the transport efficiency in heterogeneous catalysis. Mercury porosimetry is also used to characterize granular and tableted pharmaceuticals and coated and uncoated papers. Degradation processes in concrete, cement and other construction materials can be studied by porosimetry.[97]

When testing particles or powdered materials, the mercury at low pressures first intrudes into the interstitial or void space of the particles before the particles themselves are penetrated. The pressure at which these void spaces are penetrated is a function of particle size and packing. Therefore in running samples in the porosimeter, a void-volume distribution is always obtained that is characteristic of the particle size-distribution.[98]

4.7.d Adsorption from Solution

Adsorption from solution as a procedure to measure specific surface area is much simpler than any technique that requires vacuum apparatus and is sufficiently rapid to make it attractive for routine determinations, quick estimates, or quality control. Adsorption from solution is a common phenomenon, taking place in dyeing, tanning, decolorization, lubrication, catalysis, flotation, and detergency. Frequently used as adsorbates for determination of specific surface area are fatty acids, dyes, and molecules containing a radioactive isotope. In using this method, two quantities are to be ascertained: the monolayer capacity of the substrate and the area of the molecule in a close-packed monolayer. Both these determinations are subject to error, even more than in gas adsorption.[99] Giles and Nakhwa list the attributes that would be required for a solute to give reliable results for specific surface area:[100]

(a) It should be highly polar to ensure adsorption by polar solids.
(b) It should have hydrophobic properties to permit its adsorption by nonpolar solids.
(c) It should be a small molecule, preferably planar.
(d) It should not be highly surface active, as micelle formation at the surface is undesirable.
(e) It should be colored for ease of analysis.
(f) It should be readily soluble in water for convenience in use but also soluble in nonpolar solvents so that it can be used with water-soluble solids.

Giles, D'Silva and Trivedi[101] give details of a number of dyes recommended for surface area determinations (Table 4.3). When used to examine a wide variety of oxide, carbon, and halide surfaces, ranging in area from 2 to 100 m²/g, these dyes give results lying within 10% of those obtained by nitrogen-gas adsorption.

The adsorption isotherms of these dyes on many solids appear to be Langmuir-type,

$$\frac{n}{n_m} = \theta = \frac{c_2/K}{1 + c_2/K} \qquad (4.24)$$

where n is the amount of solute taken up per gram of the adsorbent when present at concentration c_2 in the solution; n_m is the number of moles of solute in

TABLE 4.3 Details of Recommended Dyes[102]

Name	C.I. Number	Molecular Area Flat σ_0,(nm^2)	Aggregation Number a	Molar Extinction Coefficient ($\times 10^{-4}$)	λ_{max} (nm)
Anionic dyes					
Orange II	15510	1.20	3.0	2.4	480
Naphthalene Red J.	15620	1.50	3.6	2.1	500
Cationic dyes					
Methylene Blue BP	52015	1.20	2.0	4.0	610
Crystal Violet BP	42558	2.25	3.6	3.6	590
Victoria Pure Lake Blue BO	42598	2.70	9.0	6.8	620

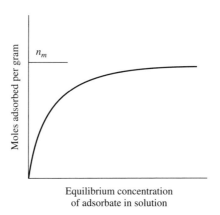

Figure 4.23 The Langmuir adsorption isotherm.

the saturated monolayer per gram of solid and K is related to the solute–solid adsorption potential. Equation (4.24) is illustrated in Figure 4.23.

The evaluation of n_m is obtained from a plot of $1/n$ versus $1/c_2$, which has an intercept of $1/n_m$ and a slope of K/n_m. Besides the value of n_m the calculation of the surface area of the adsorbent requires values for the cross-sectional area per molecule and the aggregation number of molecules flat-wise on top of each other on the substrates. Molecular areas are measured from molecular models and are for the smallest enclosing rectangle. Aggregation numbers are determined by various methods, including comparison with the BET nitrogen area and measurements of dye monolayers on water. The surface area of the adsorbent is then derived as

$$A = \frac{n_m \sigma_0 N_0}{a} \quad (4.25)$$

where N_0 is Avogadro's number, σ_0 is the cross-sectional area per molecule, and a is the aggregation number of the dye.

In spite of the uncertainties in the technique, its extreme simplicity makes it the method of choice for many practical problems. The major experimental consideration is to be able to adjust the concentration of dye and the concentration of particles in suspension so that detection is sensitive enough at the initial and final concentrations. Quite often the absorption spectra of the dye in solution and the dye adsorbed are different enough that the concentration of one can be determined in the presence of the other. When this is the case, the particles do not need to be separated from the serum before analysis.[103,104] Once the proper ranges have been established (usually by trial and error) the method is quite satisfactory and may be used for process control.

Another application of dye adsorption is in the determination of the endpoint of the digestion of a precipitate (Ostwald ripening) at a desired particle size. Samples are withdrawn periodically and filtered with reduced pressure. A fixed volume of dye solution is passed through the cake. If the particles are below the desired size, the dye is completely adsorbed and the filtrate is colorless. At the desired particle size, the filtrate shows traces of dye and further digestion is stopped by removing the precipitate from its supernatant liquor. The method can be entirely empirical, the correct amount of dye to be used from tests on the acceptable product having been established by experience.

4.8 INDIRECT METHODS

The purpose of the Hegman gauge is to determine the size of the coarsest particle. The Hegman gauge is a precision-machined steel plate with a center ditchlike channel tapered from a slight to a deep depression. A sample of the wet dispersion is placed in the tapered channel and leveled with a doctor blade. Particle size is determined by holding the gauge up to the light and identifying the point at which coarse particles begin to appear. The standard channel is tapered from 0 to 5 thousandths of an inch and the depth marked by an engraved scale.[105]

A simple empirical technique to estimate whether a dispersion being milled has reached an appropriate small size is to make a thin coating on a reflecting substrate such as a strip of aluminum painted white or a transparent substrate such as glass. As the particle size is reduced, the smaller particles are more uniformly distributed through the dispersion, and the dispersion will make a more opaque coating.[106] Similarly, many organic pigments change in hue as their particle size is reduced so that a simple colorimetric measurement can sometimes be used to specify the end of a milling process.[107]

4.9 COMPARISONS OF TECHNIQUES

Technique	Greatest Advantage(s)	Greatest Disadvantage(s)	Approximate Size Range	Number or Mass Average	Cumulative or Size Distribution	Aqueous/ Nonpolar
Acoustic absorption	High volume loadings; opaque containers	Dependent on many particle properties often unknown	0.01–1000 μm	Mass	Cumulative	Both
Capillary hydrodynamics	High resolution	Injection of dilute sample	0.01–1 μm	Mass	Size distribution	Mostly aqueous
Centrifugation — line start	High resolution	Tedious sample preparation	0.05–5 μm	Mass	Size distribution	Mostly aqueous
Centrifugation — uniform start	Generally useful	Lower resolution than line start	0.05–5 μm	Mass	Cumulative	Both
Electroacoustics	High volume loadings	Relatively untested	0.01–10 μm	Mass	Cumulative	Both
Elutriators	Inexpensive	Large volumes of water	>5 μm	Mass	Separates into two populations	Aqueous
FFF — centrifugation	High resolution	Aqueous only	0.05–100 μm	Mass	Size distribution	Aqueous

FFF — thermal	Gel content determination	Relatively untested	$5 \times 10^3 - 10^8$ Da	Mass	Size distribution
Gas adsorption	Automation	Sample preparation effects	>0.01 m²/gm	Number	Average only	Dry
Impactors and cyclones	Large dry samples	Separates, don't size	>5 μm	Mass	Separates into two populations	Dry
Light scattering, diffraction	Real time data acquisition, perfect for sprays	Theory does not apply to most samples	2–4000 μm	Mass	Size distribution	Both
Light scattering, low angle	Quantitative for molecular weight	Interference by larger particles	<0.1 μm	Mass	Cumulative	Both
Light scattering, Mie	Simple equipment	Only useful for nearly monodisperse	0.1–2 μm	Mass	Cumulative	Both
Light scattering, QELS	Wide size range and sample types	Limited distribution information	2 nm–2 μm	Mass	Cumulative	Both

Continued

107

4.9 (*Continued*)

Technique	Greatest Advantage(s)	Greatest Disadvantage(s)	Approximate Size Range	Number or Mass Average	Cumulative or Size Distribution	Aqueous/ Nonpolar
Microscopy	Shape and composition information	Tedious data acquisition	>2 µm	Number	Size distribution	Both and dry
Particle counting, optical detection	Dust contamination	Low concentrations	>2 µm	Number	Size distribution	Both
Particle counting, photozone detection	Wide sample compatibility	Less sensitive than electrozone detection	>2 µm	Number	Size distribution	Both
Particle counting, electrozone detection	Biological samples (blood); ubiquitous	High salt concentrations required	>0.5 µm	Number	Size distribution	Aqueous
Sedimentation	Simple apparatus	Slow	>1 µm	Mass	Cumulative	Both
Sieving	Simple apparatus	Large samples Slow	>20 µm	Mass	Size distribution	Aqueous
Turbidity	Simple apparatus	Single point	<0.1 µm	Mass	Single point	Both

4.10 MANUFACTURERS

Company	Telephone/FAX	Internet	Technique
Applied Sonics, Inc. 10092 South Stratford Rd. Littleton, CO 80126	(303) 471-0269 (303) 471-7300	—	Sedimentation, detection of concentration by speed of sound.
ATM Corporation 645 S. 94th Place Milwaukee, WI 53214-1206	(800) 511-2096 (414) 453-1038	www.atmcorporation.com	Sieving
Beckman-Coulter, Inc. 4300 N. Harbor Blvd. P.O. Box 3100 Fullerton, CA 92834-3100	(714) 871-4848 (714) 773-8283	www.beckmancoulter.com	Electrozone counting Laser diffraction QELS Porosimetry and surface area Sedimentation
Brookhaven Instruments Corporation 750 Blue Point Rd. Holtsville, NY 11742-1896	(631) 758-3200 (631) 758-3255	www.bic.com	Disk centrifuge, optical and x-ray detection QELS Fiber optic QELS Porosimetry
Climent Instruments Co. 1320 W. Colton Ave. Redlands, CA 92374	(909) 793-2788 (909) 793-1738	www.climet.com	Particle counting with optical detection
Colloidal Dynamics, Inc. 11 Knight St. Building E18 Warwick, RI 02886	(401) 738-5515 (401) 738-5542	www.colloidal-dynamics.com	Particle size, concentration, and charge by acoustics
CPS 7349 SE Seagate Lane Stuart, FL 34997	(800) 796-5641 (561) 221-7893	www.gwi.net/~sfitzpat	Disk centrifuge
Dantec Systems Corp. 495 Dotzert Court Waterloo, Ont. N2L 6A7	(800) 265-2757 (519) 885-4300	www.dantecmt.com	Particle velocimetry
Dispersion Technology, Inc. 3 Hillside Ave. Mount Kisco, NY 10549	(914) 241-4791 (914) 241-4842	www.dispersion.com dispersi@dispersion.com	Particle size and charge by acoustics
FFFractionation, Inc. 4797 South West Ridge Blvd. Salt Lake City, UT 84118-8429	(801) 955-7550 (801) 955-7553	www.fffract.com	Field flow fractionation techniques

(Continued)

110   PARTICLE SIZING

Company	Telephone/FAX	Internet	Technique
Gilson Company, Inc. P.O. Box 200 Lewis Center, OH 43035-7298	(740) 548-7298 (740) 548-5314	www.globalgilson.com	Sedimentation Sieving
Hach Company P.O. Box 389 Loveland, CO 80539-0389	(800) 227-4242 (970) 669-2932	www.hach.com	Turbidity
Hiden Analytical, Ltd 420 Europa Blvd Warrington, WA5 5UN, UK	+44 (0) 192-544-5225 +44 (0) 192-541-6518	www.hidden.co.uk	Gas adsorption
Horiba Instruments, Inc. 1021 Duryea Ave. Irvine, CA 92714	(949) 250-4811 (714) 250-0924	www.horiba.com	Laser diffraction Mie scattering Disk centrifuge Porosimetry Gas adsorption
Hosokawa Micron Powder Systems 10 Chatham Rd Summit, NJ 07901	(908) 273-6360 (908) 273-7432	www.hosokamicron.com	Classification Sieving
Lasentec 15224 NE 95th St. Redmond, WA 98052	(800) LASENTEC (425) 881-8964	www.lasentec.com	Online optical detection

Malvern Instruments
10 Southville Rd
Southborough,
MA 01772
	(508) 480-0200
	(508) 460-9692
		www.malvern.de	Laser diffraction
			QELS
			Acoustic
			  absorption
			Zeta potential
			Electrozone
			  detection

Matec Applied Sciences
56 Hudson St.
Northboro,
MA 01532
	(508) 393-0155
	(508) 393-5476
		www.matec.com	Capillary
			  hydrodynamics
			Zeta potential
			Acoustic
			  absorption

Micromeritics
  Instrument Corp.
One Micromeritics
Dr Norcross,
GA 30093-1877
	(770) 662-3633
	(770) 662-3696
		www.mircomeritics.com	Electrozone
			  counting
			Laser diffraction
			Porosimetry
			Gas adsorption
			Sedimentation

Microtrac, Inc. 148 Keystone Dr. Montgomeryville, PA 18939	(888) 643-5880 (no FAX)	www.microtrac.com	Laser diffraction QELS

Company	Telephone/FAX	Internet	Technique
Pacific Scientific Instruments 481 California Ave. Grants Pass, OR 97526	(800) 866-7889 (541) 479-3057	www.pacsiinst.com	Particle counting with optical detection
Particle Sizing Systems 75 Aero Camino Suite B Santa Barbara, CA 93117	(805) 968-1497 (805) 969-0361	www.pssnicomp.com	QELS Particle counting with optical detection Electrophoretic light scattering
Precision Detectors, Inc. 10 Forge Park Franklin, MA 02038	(800) 472-6934 (508) 520-8772	www.lightscatter.com	Low angle light scattering QELS
Quantachrome Corp. 1900 Corporate Dr. Boynton Beach, FL 33426	(800) 989-2476 (561) 732-9888	www.quantachrome.com	Sedimentation Porosimetry Mie scattering Laser diffraction
Shimadzu Scientific Instruments, Inc. 7102 Riverwood Dr Columbia, MD 21046	(800) 477-1227 (401) 381-1222	www.shimadzu.com	Centrifugation Laser diffraction
Sympatec, Inc. Princeton Service Center 3490 U.S. Route 1 Princeton, NJ 08540-5706	(609) 734-0404 (609) 734-0777	www.sympatec.com	Laser diffraction Acoustic absorption
TSI, Inc. P.O. Box 64394 St. Paul, MN 55164	(651) 483-0900 (651) 490-2748	www.tsi.com	Size by time of flight Low pressure impactors
Wyatt Technology Corp. 30 S. La Patera Ln, B-7 Santa Barbara, CA 93117	(805) 681-9009 (805) 681-0123	www.wyatt.com	Low angle light scattering QELS

REFERENCES

[1] Gibbons, R.A. Polydispersity, *Nature*, **1963**, *200*, 665–666.

[2] Allen, T. *Particle size measurement*; Chapman and Hall: New York; 1st ed., 1968; 2nd ed., 1977; 3rd ed., 1981; 4th ed. 1990; 5th ed. in 2 vols. 1997.

[3] Barth, H.G., Ed. *Modern methods of particle size analysis*; Wiley: New York; 1984.

[4] Jelínek, Z.K. *Particle size analysis*; Wiley: New York; 1970.

[5] Kaye, B.H. *Direct characterization of fine particles*; Wiley: New York; 1981.

[6] Provder, T., Ed. *Size exclusion chromatography (GPC)*; ACS Symp. Ser. *138*; ACS: Washington, DC; 1980.

[7] Provder, T., Ed. *Particle size distribution I, assessment and characterization*; ACS Symp. Ser. 332; ACS: Washington, DC; 1987.

[8] Provder, T., Ed. *Particle size distribution II, assessment and characterization*; ACS Symp. Ser. 472; ACS: Washington, DC; 1991.

[9] Provder, T., Ed. *Particle size distribution III, assessment and characterization*; ACS Symp. Ser. 693; ACS: Washington, DC; 1998.

[10] Orr, C. *Determination of particle size in Encyclopedia of Emulsion Technology*; Becher, P., Ed.; Marcel Dekker: New York; 1988; Vol. 3; pp 137–169.

[11] Stockman, J.D.; Fochtman, E.G., Eds. *Particle size analysis*; Ann Arbor Science Publishers: Ann Arbor, MI; 1977.

[12] Phillips, D.H. Letter to the Editor, *Chem. Eng. News* May *19*, **1997**, 7.

[13] Herdan, G. *Small particle statistics, an account of statistical methods for the investigation of finely divided materials*; Elsevier: New York; 1953.

[14] Bernhardt, C. Preparation of suspensions for particle size analysis, methodical recommendations, liquids, and dispersing agents, *Adv. Colloid Interface Sci.* **1988**, *29*, 79–139.

[15] Karuhn, R. Dispersion and analysis of magnetic particles with the Elzone, *The mircroReport*; Mircromeritics Corporation: Norcross, GA; 2000, *11*(1), 8–9.

[16] Rideal, G. Absolute precision in particle size analysis, *American Laboratory* **1996** *28*, 46–53.

[17] Rideal, G.R.; Storey, J. Morris, T.R. Sieve calibration — a new simple but high precision approach, *Part. Part. Sys.* **2000**, *17*, 1–7.

[18] Allen loc. cit.; p 179.

[19] Kaye loc. cit.; p 59.

[20] Nushart, S. Sieve analysis: Making your numbers come out right, *Powder Bulk Eng.* **1987**, 1.

[21] Siebert, P. C. Simple sedimentation methods, including the Andreason pipette and the Cahn sedimentation balance, in *Particle size analysis*; Stockman, *loc. cit.* pp 45–55.

[22] Olivier, J.P.; Hickin, G.K.; Orr, C., Jr. Rapid, automatic particle size analysis in the subsieve range, *Part. Tech.* **1970/71**, *4*, 257–263.

[23] Povey, M.J.W. *Ultrasonic techniques for fluids characterization*; Academic Press: New York; 1997.

[24] Jelínek loc. cit.; p 77.

[25] Kaye loc. cit.; p 102.

[26] Kaye loc. cit.; pp 227–270.

[27] Klumpar, I.V. Air classification, Part I — Equipment and selection, *Powder Bulk Eng.* **1987**, *1*, 42–58.

[28] Stockham loc. cit.; p 81.

[29] Puretz, J. Centrifugal particle size analysis and the Joyce–Loebl disc centrifuge, in Stockman loc. cit.; pp 77–87.

[30] Allen loc. cit.; p 320.

[31] Bowen, P.; Hérard, C.; Humphry-Baker, R.; Satoé, E. Accurate submicron particle size measurement of alumina and quartz powders using a cuvette photocentrifuge, *Powder Tech.* **1994**, *81*, 235–240.

[32] Hoffman, R.L. An improved technique for particle size measurement, *J. Colloid Interface Sci.* **1991**, *143*, 232–251.

[33] Vold, R.D.; Vold, M.J. *Colloid and interface chemistry*; Addison Wesley: Reading, MA; 1983; pp 458–465; Hunter, R.J. *Foundations of colloid science*, 2nd ed; Oxford University Press: New York; 2001; pp 122–123.

[34] DosRamos, J.G.; Silebi, C.A. An analysis of the separation of submicron particles by capillary hydrodynamic fractionation (CHDF), *J. Colloid Interface Sci.* **1989**, *133*, 302–320.

[35] Barth loc. cit.; p 279.

[36] Jeon, S.J.; Schimpf, M.E. Thermal field flow fractionation of colloidal particles, *ACS Symp. Ser.* **1998**, *693*, 182–195.

[37] Palkar, S.A.; Murphy, R.E.; Schure, M.R. Charge and hydrophobicity fractionation of colloidal-size polymers using electrical field-flow fractionation and liquid chromatography, *ACS Symp. Ser.* **1998**, *693*, 196–206.

[38] Barman, B.N. FFF Applications, A comprehensive compilation of important field-flow fractionation (FFF) results; *FFFRC: Department of Chemistry*, University of Utah, Salt Lake City, UT 84112; 1989.

[39] See also the collection of papers on field-flow fractionation in *ACS Sypm. Ser.* **1991**, *473*, 198–262.

[40] FFFractionation, Inc. Salt Lake City, UT.

[41] Caldwell, K.D.; Li, J. Emulsion characterization by the combined sedimentation field-flow fractionation-photon correlation spectroscopy methods, *J. Colloid Interface Sci.* **1989**, *132*, 256–268.

[42] Ratanathanawongs, S.K.; Shiundu, P.M.; Giddings, J.C. Size and compositional studies of core-shell latexes using flow and thermal field-flow fractionation, *Colloid Surf., A: Physicochemical Eng. Aspects* **1995**, *105*, 243–250.

[43] Allen, T. Stream scanning methods of particle size measurement; in Allen loc. cit.; 5th ed., Vol. 1; Chapter 7.

[44] McCrone, W.C.; Delly, J.G. *The particle size atlas*; Ann Arbor Science: Ann Arbor, MI; 1993; www.mcri.org.

[45] Kaye, B.H. Operational protocols for efficient characterization of arrays of deposited fineparticles by robotic image analysis systems, *ACS Symp. Ser.* **1991**, *472*, 354–371.

[46] Available from Malvern Instruments.

[47] Zsigmondy, R. *Colloids and the ultramicroscope*; Alexander, J., Trans.; Wiley: New York, 1909.

[48] Weiner, B.B.; Tscharnuter, W.W.; Karasikov, N. Improvements in accuracy and speed using the time-of-transition method and dynamic image analysis for particle sizing, *ACS Symp. Ser.* **1998**, *693*, 88–102.

[49] www.lasentec.com

[50] Allen loc. cit.; p 253.

⁵¹Nicoli, D.F.; Hasapidis, K.; O'Hagan, P.; McKenzie, D.C.; Wu, J.S.; Chang, Y.J.; Schade, B.E.H. High-resolution particle size analysis of mostly submicrometer dispersions and emulsions by simultaneous combination of dynamic light scattering and single-particle optical sensing, *ACS Symp. Ser.* **1998**, *693*, 52–87.

⁵²Pelssers, E.G.M.; Stuart, M.A.C.; Fleer, G.J. Single particle optical sensing (SPOS) I. Design of an improved SPOS instrument and application to stable dispersions, *J. Colloid Interface Sci.*, **1990**, *137*, 350–361.

⁵³Pelssers, E.G.M.; Stuart, M.A.C.; Fleer, G.J. Single particle optical sensing (SPOS) II. Hydrodynamic forces and application to aggregating dispersions, *J. Colloid Interface Sci.* **1990**, *137*, 362–372.

⁵⁴Allen loc. cit.; p 328.

⁵⁵Hadlington, S. Perspectives: Solution to a sticky problem, *Chem. Brit.* December **1998**, p 16.

⁵⁶Bohren, C.F. Optics, Atmospheric, *Encycl. Appl. Phys.* **1995**, *12*, 405–434, Figure 7.

⁵⁷Hirelman, E.D.; Bohren, C.F., Eds. Optical particle sizing, *Appl. Optics* **1991**, *30*, 4673–4992.

⁵⁸Kerker, M. *The scattering of light and other electromagnetic radiation*; Academic Press: New York; 1969; p 326.

⁵⁹Phillies, G.D.J.; Billmeyer, Jr., F.W. Elastic and Quasielastic light scattering by solutions and suspensions in *Treatise on analytic chemistry*, 2nd ed.; Elving, P.J., Ed.; Wiley: New York; 1986; Part 1, Vol. 8, Chapter 8.

⁶⁰www.hach.com

⁶¹Kissa, E. *Dispersions: Characterization, testing, and measurement*; Marcel Dekker: New York; 1999; pp 291–292 and pp 448–449 which also contains a detailed description of turbidity pp 440–444.

⁶²Kourti, T.; MacGregor, J.F. Particle size determination using turbidity: Capabilities, limitations, and evaluation for on-line applications, *ACS Symp. Ser.*, **1991**, *472*, 34–63.

⁶³Hofer, M.; Schurz. J.; Glatter, O. Oil-water emulsions: Particle size distributions from elastic light scattering, *J. Colloid Interface Sci.* **1989**, *127*, 147–155.

⁶⁴Barth loc. cit.; Chapters 5 and 6.

⁶⁵Bott, S.E.; Hart, W.H. Extremely wide dynamic range, high-resolution particle sizing by light scattering, *ACS Symp. Ser.* **1991**, *472*, 106–122.

⁶⁶Koppel, D. E. Analysis of macromolecular polydispersity in intensity correlation spectroscopy: The method of cumulants, *J. Chem. Phys.* **1972**, *57*, 4814–4820.

⁶⁷Provencher, S. W.; Hendrix, J.; De Maeyer, L.; Paulussen, N. Direct determination of molecular weight distributions of polystyrene in cyclohexane with photon correlation spectroscopy, *J. Chem. Phys.* **1978**, *69*, 4273–4276.

⁶⁸Grabowski, E. F.; Morrison, I. D. Particle size distributions from analyses of quasielastic light-scattering data in *Measurement of suspended particles by quasi-elastic light scattering*; Dahneke, B., Ed.; Wiley: New York; 1983; pp 199–236.

⁶⁹Morrison, I.D.; Grabowski, E.F.; Herb, C.A. Improved techniques for particle size determination by quasi-elastic light scattering, *Langmuir* **1985**, *1*, 496–501. Herb, C.A.; Berger, E.J.; Chang, K.; Morrison, I.D.; Grabowski, E.F. Using quasielastic light scattering to study particle size distributions in submicrometer emulsion systems, *ACS Symp. Ser.* **1987**, *332*, 89–104.

[70] Cummins, P.G.; Staples, E.J. Particle size distributions determined by a multiangle analysis of photon correlation spectroscopy data, *Langmuir* **1987**, *3*, 1109–1113.

[71] Finsy, R. Particle sizing by quasi-elastic light scattering, *Adv. Colloid Interface Sci.* **1994**, *52*, 79–143.

[72] Maret, G. Diffusing-wave spectroscopy, *Current Opin. Colloid Interface Sci.* **1997**, *2*(3), 251–257.

[73] Richter, S.M.; Shinde, R.R.; Balgi, V.; Sevick-Muraca, E.M. Particle sizing using frequency domain photon migration, *Part. Part. Syst. Char.* **1998**, *15*, 9–15.

[74] Wijmans, C.M.; Horne, D.S.; Hemar, Y.; Dickinson, E. Computer simulation of diffusing-wave spectroscopy of colloidal dispersions and particle gels, *Langmuir* **2000**, *16*, 5856–5863.

[75] Shortt, D.W.; Roessner, D.; Wyatt, P.J. Absolute measurement of diameter distributions of particles using a multiangle light scattering photometer coupled with flow field-flow fractionation, *Amer. Lab.* **1996**, *28*, 21–28.

[76] Dispersion Technology, Sympatec, and Malvern Instruments.

[77] Povey loc. cit.; p 142.

[78] Holmes, A.K.; Challis, R.E.; Wedlock, D.J. A wide-bandwidth ultrasonic study of suspensions: The variation of velocity and attenuation with particle size, *J. Colloid Interface Sci.* **1994**, *168*, 339–348.

[79] Povey loc. cit.; p 7.

[80] Weiss, J.; Herrmann, N.; McClements, D.J. Ostwald ripening of hydrocarbon emulsion droplets in surfactant solutions, *Langmuir* **1999**, *15*, 6652–6657.

[81] Holmes, A.K.; Challis, R.E.; Wedlock, D.J. A wide bandwidth study of ultrasound velocity and attenuation in suspensions: Comparison of theory with experimental measurements, *J. Colloid Interface Sci.* **1993**, *156*, 261–268.

[82] Povey (1997) p 27.; Urick, R.J. A sound velocity method for determining the compressibility of finely divided substances, *J. Appl. Phys.* **1947**, *18*, 983–987.

[83] Debye, P. A method for the determination of the mass of electrolytic ions, *J. Chem. Phys.* **1933**, *1*, 611–614. Electrical conductivity depends on the number of ions per unit volume of both sign charge, the charge per ion, and the viscous drag on the ions which scales with radius. However, conclusions from conductivity measurements alone only give the average properties of the cations and the anions taken together.

[84] Oja, T.; Peterson, G.; Cannon, D.W. U.S. Patent 4,497,209, 1985.

[85] O'Brien, R.W. Electro-acoustic effects in a dilute suspension of spherical particles, *J. Fluid Mech.* **1988**, *190*, 71–86.

[86] O'Brien, R.W. The electroacoustic equations for a colloidal suspension, *J. Fluid Mech.* **1990**, *212*, 81–93.

[87] O'Brien, R.W.; Cannon, D.W.; Rowlands, W.N. Electroacoustic determination of particle size and zeta potential, *J. Colloid Interface Sci.* **1995**, *173*, 406–418.

[88] Hunter, R.J.; O'Brien, R.W. Electroacoustic characterization of colloids with unusual particle properties. *Colloids Surf. A: Physicochem. Eng. Aspects* **1997**, *126*, 123–128.

[89] James, M.; Hunter, R.J.; O'Brien, R.W. Effect of particle size distribution and aggregation on electroacoustic measurement of ζ-potential, *Langmuir* **1992**, *8*, 420–423.

[90] Ross, S.; Olivier, J.P. *On physical adsorption*; Wiley: New York; 1964.

[91] Lowell, S. *Introduction to powder surface area*, 1st ed; Wiley: New York; 1979.

[92] Lowell, S.; Shields, J.E. *Powder surface area and porosity*, 2nd ed; Chapman and Hall: New York; 1984.

[93] Brunauer, S.; Emmett P.H.; Teller, E. Adsorption of gases in multimolecular layers, *J. Am. Chem. Soc.* **1938**, *60*, 309–319.

[94] Rücker, A.W. *Presidential Address to the British Association meeting at Glasgow*, BAAS Report, 1901 September, 1901.

[95] Morrison, I.D.; Ross, S. The second and third virial coefficients of a two-dimensional gas, *Surf Sci.* **1973**, *39*, 21–36.

[96] Sacher, R.S.; Morrison, I. D. An improved CAEDMON program for the adsorption isotherms of heterogeneous substrates, *J. Colloid Interface Sci.* **1979**, *70*, 153–166.

[97] *The Micro Report*; Micromeritics Corporation: Norcross, GA; **2000**, *11*(2), 3.

[98] Rootare, H.; Powers, J.M.; Spencer, J. Particle size of abrasive powders from mercury porosimetry data, Industry & Scientific Conference Management, Inc.; 1979.

[99] Kipling, J.J. *Adsorption from solutions of nonelectrolytes*; Academic Press: New York; 1965; Chapter 17.

[100] Giles, C.H.; Nakhwa, S.N. Studies in adsorption. XVI. The measurement of specific surface areas of finely divided solids by solution adsorption, *J. Appl. Chem.* **1962**, *12*, 266–273.

[101] Giles, C.H.; D'Silva, A.P.; Trivedi, A.S. Use of dyes for specific surface measurement, *Surf. Area Determination, Proc. Int. Symp., 1969* **1970**, 317–329.

[102] The molecular areas of these dyes are based on molecular cross-sections obtained by calibration against BET measurements with nitrogen gas.

[103] Herz, A.H.; Helling, J.O. Evaluation of surface spectra in turbid silver halide dispersions, *Kolloid-Z. Z. für Polym.* **1967**, *218*, 157–158.

[104] Herz, A.H.; Danner, R.P.; Janusonis, G.A. Adsorption of dyes and their surface spectra, *Adv. Chem. Ser.* **1968**, *79*, 173–197.

[105] Patton, T.C. *Paint flow and pigment dispersion, a rheological approach to coating and ink technology*; 2nd ed.; Wiley: New York; 1979; pp 502–507.

[106] Patton loc. cit.; p 507.

[107] Beresford, J.; Smith, F.M. Dispersion of organic pigments in *Dispersion of powders in liquids with special reference to pigments*, 2nd ed.; Parfitt, G.D., Ed.; Wiley: New York; 1973; pp 267–307; especially pp 297–301.

5 Processing Methods for Making Emulsions and Suspensions

Processing methods are of two types: those that emulsify or that pull agglomerates apart by shearing forces and those that comminute aggregates by fracture. The general term is "comminution." Dry grinding is called micronizing. Often a dispersion is prepared by micronizing the dry material, mixing it with some solvent and dispersant at high solids volume fraction, milling to a fine particle size to what is called a millbase, and then diluting with more solvent and sometimes adding more dispersant in preparation for use. This chapter describes the scale-up of these processes from laboratory to manufacturing. Reviews of this subject are available.[1,2,3] An excellent source of practical information and up-to-date lists of equipment manufacturers is published each year in the June issue of *Powder and Bulk Engineering* (www.powderbulk.com).

Different equipment is used for each process. Equipment to generate shearing forces need only provide sufficient energy to attenuate an immiscible liquid within another or to separate agglomerates. Comminution requires higher energy input to break tightly bound aggregates or to shatter coherent solids.

The dry powder should be added to the liquid, taking care that the medium be of a low viscosity and a low surface tension to enhance the displacement of entrained air. The finer the powder, with smaller channels between the particles, the slower the wetting. An efficient method of mixing dry powder with a liquid is the Ejector-Mixer from Semi-Bulk Systems (www.bulk-online.com) A pipe leading from a bag of powder is inserted into the nozzle of liquid being pumped into a tank. The vacuum generated by the flow of the liquid draws the powder through the pipe as a dust cloud and combines it with the streaming liquid in the region of high shear as the two exit the nozzle. Individual particles are quickly engulfed by the flowing liquid and wetted. This efficiently eliminates the trapped air that is often encountered when dry powder is merely dumped into a tank of liquid.

Wet comminution may be processed in a single piece of equipment, but always consists of distinct steps: (1) the wetting of the particle surfaces, (2) the separation of agglomerates and aggregates into individual particles, (3) possibly the attrition or fracture of individual particles into smaller particles and (4) the adsorption of dispersant by the particle surface to obtain a stable dispersion. Each step can be considered singly to improve process efficiency. The generation of high-shear forces requires narrow gaps, or high rates of flow, or both, in comminutors. The following section gives examples of equipment in which one or the other of those modes is used to generate high-shear forces.

Full-scale manufacturing processes can include the preparation of a millbase that is composed of all the pigment but only a portion of the other ingredients. All of the size reduction is performed on this concentrate. The remaining ingredients are combined to form a "let-down" solution. After milling of the concentrate, the two are combined to make the final dispersion. Finding the most efficient combination of dry milling, millbase preparation and milling, and let-down solution is the challenge of process scale-up.

5.1 DRY GRINDING (MICRONIZING)

Many dispersion processes are more efficient if the powder is dry ground before adding the powder to the liquid. Particle-size reduction of dry materials can be accomplished by crushing, impacting, chopping, chipping, cutting, and abrasion. For example, hammer mills with free-swinging hammers to smash large particles to powders come in a wide variety of sizes and shapes, or high speed rotors with attached teeth can be used. Pairs of stationary and rotating discs with teeth arranged in concentric intermeshing rings can be used for grinding.

An air-impact pulverizer is a dry-process machine in which particles are fluidized in two opposed streams of air from high-speed jet nozzles, which project particle against particle at high kinetic energies. The nozzles are precisely aimed to hit against each other, within fractions of a second in degrees of arc, and the pressures of each stream are balanced. The mixture is size separated by being blown into a vortex. The smaller particles follow the streamlines of the air and exit with it; the larger particles are recirculated until ground small enough to escape. The equipment is well suited to break up soft solids such as carbon blacks, molybdenum disulfide, and polymers because the temperature remains low. Encapsulation is possible by injection of coating materials into the feed.

In the micronizer, compressed air, gas or steam is blown from the edges of a chamber at an acute angle to the wall, in order to impel particles into rotation. The gas escapes from an exit port at the center of the chamber. Large particles continue to circulate along the walls until they are sufficiently attrited to be carried along with the escaping gas.

The result of dry grinding is a powder more uniform in size and shape, which aids in the efficiency of subsequent mixing and milling.

5.2 HIGH-SPEED STIRRERS

5.2.a Blade Stirrers

High-speed stirrers may have a single blade, multiple blades, or intermeshing blades. The efficiency of dispersion depends on the shear rate, that is, the change of fluid velocity with distance. Velocity gradients generate the stresses that separate flocs or break up droplets. Blade stirrers are not used to break up aggregates of particles as that generally requires impact, which is not present in a stirred system. Some manufacturers call their high-speed stirrers "dissolvers" because they are also used to make liquid solutions.

The shear rate of a blade stirrer is highest near the tip of the spinning blade. Tip velocities can be up to about 4,000 ft/min. Higher velocities lead to turbulence, which reduces the dispersing action because particles or drops tend to circulate only in the eddy flow caused by the turbulence. The shape of the stirrers and their position in the tank are therefore carefully chosen to minimize turbulence. The correct speed creates a rolling vortex under the blade shaft without surging or splashing.

The shear rate is estimated by dividing the speed of the tip of the spinning blades by the distance over which the fluid velocity drops to a low value. Blade stirrers do not develop high shear rates because the distance over which the velocity of the liquid declines from its highest to its lowest value is of the order of the dimensions of the tank. For a tip velocity of 4,000 ft/min and a distance of about one foot to the edge of the tank, the shear rate would be (approximately) 70 s^{-1}, which is a relatively low shear rate.

Considerable improvement in mixing efficiency is obtained by using multiple blades. Planetary mixers are equipped with two or three vertical mixing blades moving in counter rotation. A scraping blade removes material from the walls of the tank. A high-speed blade in the center of the tank enhances the mixing. Multiple blades are essential to prevent dead zones in the mixing. The planetary mixer is particularly useful for liquids of high viscosity.

One method to increase the shear rate is to surround the spinning blades with a shield. The dispersion is pulled up to the center of rotation and thrown against the inside walls of the shield, which may have exit slots or may just let the fluid stream down the wall and out into the tank. The shear rate thus generated is much higher than attained by rotating blades in a tank without a shield.

We shall see that other milling techniques generate shear rates that are higher by orders of magnitude because the distance over which the high fluid velocities decline are much shorter. High-speed stirrers are usually used to make uniform premixes that are fed into milling equipment of higher shear rate.

Myers Engineering provides a wide variety of low-speed, high-speed, and multishaft dispersers. Gaulin's VariKinetic disperser features variable-pitch impeller vanes, the angle of which can be adjusted while the unit is operating. The tips of the vanes move at 4000–5000 ft/min. Greerco high-speed mixers are designed for applications in the pipe and for combinations with homogenizers and colloid mills. Counter-rotating paddles are useful in cosmetic and pharmaceutical applications for creams, lotions, ointments, and dispersions. Morehouse-Cowles dissolvers are blade stirrers. The single-shaft models are capable of mixing dispersions and emulsions up to 50 Pa-s viscosity. The multishaft dissolvers are capable of mixing materials with viscosities up to 2,000 Pa-s materials (see Section 5.8).

5.2.b Rotor-Stator Dispersers

Rotor-stator dispersers are continuous mixers using a high-speed rotor turning within a stator. A stator can either be simple with circumferential slots or more elaborate with tortuous channels. The rotors can be rotated from 5000 to more than 20,000 rpm to produce fluid velocities in excess of 10,000 ft/min. The liquid passes at high speed through the narrow gap between the spinning rotor and the stator and then through narrow channels in the stator producing high shears. Rotor-stator dispersers are widely used for production of paints, some printing inks, and paper coatings (Figure 5.1).

The speed of the spinning rotor is roughly twice the speed of high-speed stirrers. Liquid passing through narrow channels does not become turbulent as readily as in a tank with a high-speed stirrer. Much higher shear rates are attained in the rotor-stator dispersers because the distance over which the high fluid velocities declines is the width of the slots in the stator. This distance is easily two orders of magnitude less than high-speed stirrers sitting in containers. Hence the rotor-stators produce shear rates that are two orders of magnitude greater than high-speed stirrers. For many processes the rotor-stator disperser is adequate to obtain a good emulsion or to break up weak flocs efficiently.

A limitation of the rotor-stator design is that the flow of fluid into the spinning rotor and out through the stator is only possible for formulations of low viscosity. In some laboratory equipment catalogs rotor-stator dispersers are labeled as homogenizers, especially for biological applications. This is misleading because homogenization is a different process.

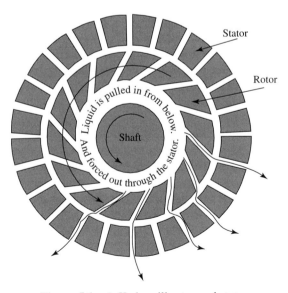

Figure 5.1 A Kady mill rotor and stator.

5.3 HIGH-SHEAR MILLS

5.3.a Colloid Mills

The operation of a colloid mill, primarily used to emulsify liquids, depends on flow through a narrow gap. The colloid mill is normally charged by feeding the material through an open funnel on the top, although pumped systems are available. The mixture should be well mixed before being fed to the mill.

In the colloid mill the rotor is a truncated cone so that the milling action is between the faces of the rotor and the stationary walls of the equipment. See Figure 5.2. The rotor is dynamically balanced to eliminate wobble and can rotate at speeds of 1,000–20,000 rpm. The gap between the rotor and stator surfaces is adjustable down to thousandths of an inch which is about an order of magnitude less than that of the Kady mill, for example. The mill is adjusted for operation by raising the rotor to create an appropriate gap, usually 1 to 5 mils (thousandths of an inch). The gap has to be readjusted to compensate for thermal expansion during use. The rotor is sometimes grooved but then care must be taken that the milling action not be turbulent. The liquids are recirculated and the stator is water cooled.

The colloid mill is best suited for emulsification. The liquid/liquid mixture as it passes through the narrow gap between the spinning rotor and the walls of the mill extends droplets into long films, which are mechanically unstable and break up into small drops. The colloid mill is less suitable to disperse pigments. The shear forces on small particles in the liquid are not large and attrition against the wall is negligible. A media mill, such as a ball mill, would have rates of attrition orders of magnitude greater.

An essential feature of any equipment in which the dispersant is blended into the mix is a rest period. The milling continuously creates new surfaces that are unstable; these require a quiescent period for adsorption of stabilizer before

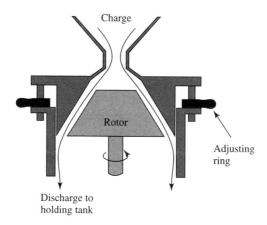

Figure 5.2 Section of a vertical colloid mill.

more surface is created, which is a relatively slow process. See Section 22.12.d, "Intermittent Milling."

5.3.b Homogenizers

Homogenizers produce emulsions by pumping the mixture at high pressure (up to 12,000 psi) with a reciprocating or piston type pump through a small orifice against a spring-loaded plunger (Figure 5.3). The gap can be as small as 1.5 mils (about 38 μm.) This is a smaller gap than can be attained in the colloid mill. Homogenization occurs in the flow around the plunger. The mixture enters the valve assembly at high pressure but low velocity. The sudden increase in velocity and decrease in pressure of the liquid as it passes through the small gap creates high-shear forces and cavitation due to vaporization. The lack of moving parts in the milling zone makes it preferable to the colloid mill in which a small malfunction of the spinning rotor at narrow gaps can lead to major damage. If the homogenizer is misused by decreasing the gap too far, the pump may stall, but a quick release of pressure prevents permanent damage. For this reason homogenizers have nearly completely replaced colloid mills. A familiar application is to reduce the size of fat globules in milk to prevent "creaming" in the bottle.

In the Microfluidizer manufactured by Microfluidics two streams of liquid constricted through microcapillaries the diameters of human hairs impact at high velocity and pressure within an interaction chamber. The pressure can be varied from 500 to 20,000 psi; the streams accelerated to velocities of 1500 ft/s. By applying 20,000 psi of pressure to a low viscosity fluid, 2×10^7 W kg^{-1} of energy are dissipated on impact. Fine droplets with a narrow size distribution are produced by a combination of shear, turbulence, impact, and cavitation forces. The equipment may be used to produce emulsions, suspensions, or foams, with applications to biology (cell rupture), pigment dispersions, and in the pharmaceutical, food, dairy, and cosmetics industries.

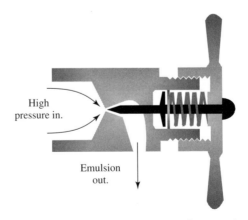

Figure 5.3 Section of a single-stage homogenizer.

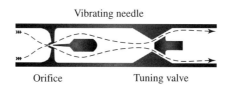

Figure 5.4 Schematic diagram of the Sonolator.

A Sonolator is a low-pressure variation of the homogenizer (Figure 5.4). Liquids are pumped through a slit-shaped orifice at pressures up to 5000 psi at high linear velocity against a stainless steel blade cantilevered in the jet stream. The dimensions of the equipment are such that the blade vibrates in the ultrasonic range causing cavitation, turbulence, and high shear in the liquid. The Sonolator uses a less expensive pump because the pressure is as much as two-thirds less than what is normally required in homogenizers. The position of the blade and the pressure drop are adjusted to produce a maximum in power, which is detected externally by a meter reading in decibels. Dual-feed systems are used to bring two different liquids together just before they pass through the equipment. This is particularly advantageous when one of the liquids is a wax and needs to be kept hot before and during emulsification.

5.3.c Ultrasonic Dispersers

Ultrasonic activators convert conventional 60 Hz ac electric current to 20,000 Hz. The high frequency is fed to an electrostrictive element, which converts the signal to mechanical vibrations in tips of various shapes called horns. The tip of the horn is immersed in the liquid and the ultrasonic vibrations cause cavitation. The higher the wattage and the larger the probe diameter, the greater the volume that can be processed. While this is satisfactory for laboratory preparation, it is obviously unsatisfactory for pilot and manufacturing scale. Hence, the best practice suggests that ultrasonic dispersers should not be used if a scale-up of the process is anticipated. For continuous use the vibrating horns must be cooled.

Sonics, Inc. provides a variety of ultrasonic dispersers. Some models are available with tips as small as 3 mm, which are useful for preparing samples for particle-size analysis directly in the cuvette. Short bursts of energy are used to avoid overheating the sample. The laboratory baths that are identified in equipment catalogs as ultrasonic cleaners provide far too little energy to disperse liquids or solids adequately.

5.4 ELECTROSTATIC DISPERSING

Electrostatic charges on the surface of a drop act to pull a liquid apart. The surface tension of the drop opposes the stretching, but if the charge is great enough, the drop will break apart.[4] A high electric potential is applied to the nozzle, which electrifies the liquid stream as it emerges. This principle can be applied to the process of dispersing one fluid into another. Electrostatic dispersion has been used extensively in many fields including electrostatic printing, paint spraying, crop spraying, and chemical processing. In providing the energy required to create fine drops or bubbles, electrostatic dispersion is more efficient than mechanical dispersion. Drops can be generated efficiently at a size far below that attainable by simple high-shear mixing.[5] Most of the applications reported, however, are limited to spraying fluids of high electrical conductivity into fluids of lower conductivity (e.g., air.) Low conductivity fluids can be dispersed into high conductivity fluids (e.g., organic liquids or air into water) only by careful design of the spray nozzle.[6,7]

5.5 IMPACT MILLS

The breaking of dispersed aggregates requires impact, which is favored when unhindered by viscous resistance. The fineness of the grind depends on the size of the grinding media: the smaller the media, the finer the grind.

5.5.a Ball and Jar Mills

A ball mill is any rolling mill in which steel or iron balls are used as the grinding medium (Figure 5.5). The containing cylinder is usually made of steel. The milling media can be steel balls, pebbles, or high-density cylinders. Steel shot is generally preferred if contamination is not too great a concern because steel has about three times the density of other common grinding media. A pebble mill uses flint pebbles* or porcelain balls as the grinding medium and the inside of the mill is lined with a nonmetallic substance. The cylinder is rotated by a shaft or on rollers. For efficient grinding, the speed, the amount and size of grinding medium, the amount of loading, and its viscosity (if a wet process) are so adjusted that the top layers of balls form a cascading, sliding stream moving faster than the lower layers, thus causing a grinding action between them. The load of balls in a ball mill should be such that the balls occupy somewhat more than one-half of the volume. Since the grinding depends on the cascading motion, the balls have to be dense and large, 1/4–2 in. diameter. Large mills are more efficient than small mills partly because the increased weight of the balls on the bottom of the cascading pile of balls contributes to faster attrition.

If the viscosity is too high or the rotation too rapid, the balls are carried all the way around the cylinder without any grinding action. The viscosity of many dispersions changes with milling so the speed of the mill needs to be adjusted during the milling. In manufacturing operations, an experienced operator can make the necessary adjustment from the sound of the cascading media. In

---

* The Paul O. Abbé company sell a wide variety of mills and milling media. They gather the best flint pebbles from a beach in Normandy where new, predominantly light-colored pebbles wash ashore with every tide.

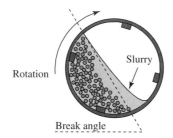

Figure 5.5 Milling action in a ball mill.

laboratories the size of the mill is sometimes so small that adjustment based on relative noise lacks precision. This leads to the danger that the speed of the mill may not be adjusted properly as the viscosity changes. Again, several samples on the same rollers or a mill located in a separate room may lead to neglect of this essential adjustment.

Jar rolling mills are small jars containing grinding media placed on motor-driven rollers. Typically, two to eight pairs of rollers that can handle up to twenty or so jars are used. These are appropriate for laboratory, pilot plant or small production facilities. Their chief advantage is that multiple small batches can be milled simultaneously. Their chief disadvantage is that the speed of the rollers is the same for all jars. This speed may not be appropriate for all the dispersions if they have different viscosities. A common experimental error is to place a series of dispersions with different dispersants or different concentrations of dispersants, looking for an optimum formulation, and mill them together. The speed appropriate for one dispersion may not be appropriate for another and a poor milling result for one dispersant or one concentration may be caused by inappropriate test conditions, not inappropriate formulation.

Grinding times can be quite long, for some materials up to a week or so. An advantage of these impact mills is that they can sometimes be safely left running unattended. Since the charging of the mill and the discharging of the final product and cleaning of the mill is labor intensive, many processes are designed to be run overnight so that the manual labor can be done without paying overtime.

Some advantages of the ball mills are that no premixing of the components is normally needed; the mill is sealed so that volatile materials can be handled, maintenance costs are low, and the ultimate particle size is low. Disadvantages include awkwardness in loading and unloading the mill, especially in separating the grinding media from the dispersion, the need for occasional adjustment of rotation speed during the process, and possible contamination of the product with media or jar components. These disadvantages combine to make ball and roller milling inefficient.

5.5.b Vibratory Mills

Vibratory mills are designed to improve the uniformity of grind in ball and jar mills by shaking the mill by mechanical action rather than relying on gravity to move the media. The method imparts a high-speed circular vibration to the mill by driving it with eccentric shafts. Vibratory mills grind about four times faster than ball and jar mills.

Small vibratory shakers are ubiquitous in retail paint stores, where new cans of paint are shaken to redisperse the pigment before being sold. A common dispersing technique in the laboratory is to use a media-filled container on the shaker. These "paint shakers" are popular in the laboratory because the milling action is much faster than ball and roller mills. The problem is that they are difficult to scale up because obtaining a similar milling action on a large scale is not practicable. A rough approximation is that one minute on a paint shaker corresponds to about one hour in a production ball mill.

5.5.c Attritors

The attritor* is a stirred-media mill in which the grinding media are moved by a series of staggered horizontal rods attached to a central shaft (Figure 5.6). Shear fields and impacts are generated by the motion of the rotating rods, leading to more efficient power consumption, finer grinding, and shorter grinding times.[8] The regularity of the stirring motion stands in contrast to the irregular cascading motion in a ball or roller miller and so allows a process to be scaled up from a laboratory or pilot sized attritor to a manufacturing size. Premixing is usually not required as the uniform milling action eliminates any dead zones in the mill. The power required is less than that required by a ball mill since the power input is

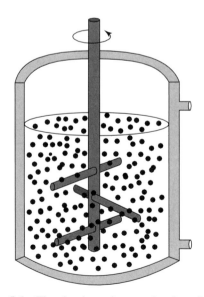

Figure 5.6 The simple attritor, a stirred media mill.

* Invented by Dr. Andrew Szegvari June 5, 1897–August 9, 1980, also founder of Union Process, Inc.

used directly to agitate the media and not used to rotate or vibrate a large, heavy vessel. Typically, the grinding media are steel shot but other types of media such as alumina, zirconia, zirconium silicate, steatite, silicon nitride, silicon carbide, tungsten carbide, mullite, and glass can be used to prevent contamination.[†]

Medium-to-high viscosity materials are attrited between 100 and 500 rpm and low-to-medium viscosity materials are attrited between 500 and 1400 rpm. Standard attritor tip speeds are 6000–10,000 ft/min; high-speed attritors are up to five times faster, resulting in a shorter milling time. The grind times are often ten times less in an attritor than in a ball mill. The grind time is related to the ball diameter and agitator speed by the empirical relation[9]

$$t = \frac{kd}{n^{1/2}} \qquad (5.1)$$

where t is the grind time to reach a certain median particle size, k is a constant that varies depending upon the slurry being processed, type of media, and the model attritor being used, d is the diameter of the media, and n is the shaft rpm.

In stirred media mills, smaller grinding media can be used than in gravity-driven ball mills, resulting in a finer grind. Recirculation of the dispersion can be provided, as can cooling for prolonged grinding. Temperature control and constant rate of stirring make stirred mills scalable.

Continuous-flow attritors are tall, narrow, jacketed vessels; the dispersion is pumped into the attritor through a bottom grid, flows up through the stirred media, and out from the top into a holding tank, typically ten times the capacity of the mill itself. This provides a convenient means to use small mills for large batches and to take advantage of intermittent milling. Wet- and dry-grinding attritors designed up for batch, circulation, or continuous mode are available from Union Process, (Figure 5.7).

A interesting variation on stirred-media mills is the Turbomill from Netzsch. The Turbomill drives a sealed rotating basket filled to between 80 and 85% by volume with grinding beads and immersed in a tank containing the dispersion. Inside the basket a fixed disc is attached to a shaft that is concentric to the basket drive shaft. This inner shaft is locked to prevent the disc from rotating. As the basket rotates the beads are forced against the disc. Shearing forces are created between the fixed disc, the moving beads, and the rotating basket. Liquid is drawn up the center of the basket, forced through the moving media and out the basket wall. This design combines the idea of the rotor-stator and the stirred media mill. Steel, glass, or ceramic beads can be used. Netzsch also produces high-speed stirrers and other stirred media mills.

5.5.d Sand or Other Small-Media Mills

In a stirred-media mill the collision of balls attrites particles in their paths. Small balls moving at high velocities can replace large, heavy balls moving more slowly.

[†] Glen Mills, Clifton, NJ is an excellent source of different types of milling media.

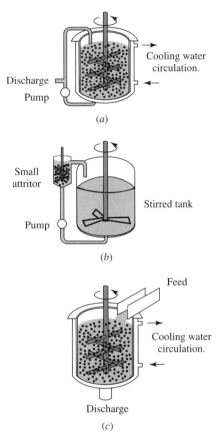

Figure 5.7 Three types of attritor (*a*) batch, (*b*) continuous, (*c*) circulating.[10]

The use of smaller balls allows more collisions because the balls are more numerous. Sand has replaced large steel shot for many fine-grinding applications.

Small-media mills consist of several impeller disks attached to a shaft rotating at speeds of 5,000–12,000 rpm, hence small-media mills are sometimes referred to as high-speed mills. The impeller disks have a width of about 70% of the tank diameter. The increased frequency of impact of the small media as they are thrown to the outside edge of the mill and the increased number of contact points because of their small size makes them more efficient. Therefore, the same throughput can be attained with smaller mills and smaller media.

The original grinding medium was Ottawa sand, 20–30 mesh (0.6–1 mm). Nowadays glass beads, ceramic beads, and steel shot of approximately that size are often used because of their greater hardness or density than sand although they are more expensive.

The dispersion is pumped into the bottom of the small-media mill, which is at least twice as tall as it is wide. It encounters intense agitation and then flows continuously from the top. The average residence time may be only a few minutes. The dispersion is sheared by media thrown outward by the rotating impeller discs. Further, the media squeeze the pigment against the chamber walls with considerable force. The temperature is controlled by cooling (or heating) water circulating in the jacket. The dwell time is controlled by the pump rate.

A limitation of the small media is that they are difficult to separate from the dispersion. The DCP mill from Draiswerke is designed to allow the separation by throwing the media to the outer wall by centrifugal force while the dispersion is discharged centrally.

Opinions vary as to whether the grinding chamber should be vertical or horizontal. Settling of the media impedes their rotation. If the viscosity of the dispersion is low, horizontal mills are often preferred, as the high speed prevents settling of the media. Similarly, vertical mills are preferred for higher viscosities because settling of the grinding media is less of a problem and these mills are simpler to construct and operate. The starting torque is less for horizontal mills.

Frequently, several mills are connected in series, each succeeding mill filled with finer grinding media.

The variables in optimizing small-media mills include dwell time in the mill and dwell time in the holding tank, rpm, media to mill volume ratio, temperature, and pigment to vehicle ratio.[11] Some advantages of the small-media mills are that they provide the finest dispersions, generally with the least color contamination and hence brightest colors, and the power consumption is small. The disadvantages are that a premix must be used, small media may block the exit screen, and shear-thickening dispersions (dilatant) do not mill well.

5.6 MILLS FOR HIGH-VISCOSITY DISPERSIONS

The equipment discussed above is limited to formulations that flow or can be pumped. The remaining option is a heavy-duty mixer, such as a three-roll mill, double-blade mixer (Banbury), kneader, screw mixer, or dough mixer. When the viscosity is high, inertial forces are quickly damped, hence particle size is not much reduced by attrition. This means that the ultimate fineness of dispersion attainable even in heavy-duty mixers is severely limited. Heavy-duty mixers produce dispersions no better than those attained by high-speed mixers for lower-viscosity formulations.

Heavy-duty mixers are provided with heaters in order to reduce the viscosity to the optimum range for effective mixing. Once it has been determined that only a heavy-duty mixer can be used, milling at the highest possible pigment loading, even if this raises the viscosity, is preferred, because the effectiveness of particle–particle attrition sharply increases with pigment concentration. Additional ingredients are added to the mill at a later stage of the processing.

5.6.a Roll Mills

A three-roll mill is a set of rolls called the feed, center, and apron rolls, rotating in opposite directions at different speeds and with a small clearance between the rolls. Heated roll mills are standard for the production of gravure and flexographic inks. Four- and five-roll mills are used to prepare lithographic printing inks. It is common practice to pass the more resistant dispersion through the unit at progressively higher pressures until an adequate degree of dispersion is achieved.

End plates are added to force the dispersion towards the center of the gap. Agglomerates are broken as they pass through the nips. High-viscosity loading is important for this type of mill; therefore, the solids content is kept as high as possible. The tremendous force required to rotate the mill with high-viscosity compositions in the nips demands unusually robust equipment. The rolls and bearings must be precisely machined to maintain accurate gaps over the entire width of the rolls. Since viscosity changes so much with temperature, the temperature of the rollers must be kept uniform.

The advantage of a roll mill is that it handles viscous materials such as printing inks. The disadvantages are that it is open and so cannot be used with volatile solvents and that the throughput is slow.

5.6.b Heavy-Duty Mixers

Heavy-duty mixers include Banbury mills and screw extruders. A Banbury mixer is used to blend carbon and other fillers into rubber or plastics. It functions at low speed with a loading of high millbase viscosity. In the Banbury mixer two kneading arms encased in a mixing chamber rotate in opposite directions and at synchronous speeds. They are so shaped that the plastic mixture is pressed against the walls of the chamber, forming a wedge during the kneading operation. The wedge is continuously formed and sheared, while the motion of the rotors ensures good mixing of the batch. The chamber is heated to mill polymers above their glass-transition temperatures.

Great care is taken that there are no dead spots in the mill, especially near the wall and in corners, since mixing is too slight to move material in these regions back into the milling region. The mixing blades are designed to pass close to the walls.

A screw extruder is the primary equipment for making alloyed, filled, reinforced, or pigmented thermoplastics. Twin-screw extruders use exchangeable corotating or counter-rotating and intermeshing or nonintermeshing modes. The rotation of the screws moves the dispersion through a cylindrical barrel while shearing the material between the blades. This feature allows the process to be continuous rather than batch as in Banbury mixers.[12,13]

5.7 CHEMICAL PROCESSING AIDS

Formulations for dispersing operations are based on determining the conditions required to produce the maximum quantity of emulsion or dispersion the most efficiently. This requires careful, but necessarily empirical, selections of dispersants, concentrations, equipment, and processing conditions.

Production of emulsions is generally easier than production of suspensions of particles. Air entrapment is a significant problem in mixing powders with liquids; this is not a problem with combining liquids. For the former, the liquid needs to penetrate powder cakes replacing air in the process. Penetration depends on the surface tension of the liquid, the contact angle of the liquid on the solid and the fineness of the powder.

The rate of penetration of the liquid in between powder particles is given by the Washburn equation [see Eq.(9.15)].

$$\frac{dL}{dt} = \frac{\sigma r \cos \theta}{4\eta L} \qquad (5.2)$$

where L is the depth of penetration, r is the "average" pore radius, σ is the surface tension, θ is the contact angle, and η is the viscosity. This expression states that higher surface tensions and lower contact angles lead to better wetting. Higher surface tensions, however, are usually associated with higher contact angles. This means that in a systematic search for the best wetting agents, it is incorrect to search for a liquid with the lowest contact angle only. It is the product of surface tension and cosine of the contact angle that needs to be maximized. The rate of penetration of liquid into the powder bed also depends inversely on the liquid viscosity so that an effective figure of merit for choosing a wetting agent is the ratio

$$\frac{\sigma \cos \theta}{\eta} \qquad (5.3)$$

A further point should be noted here. The best wetting agents are not necessarily the best dispersing agents. Aerosol OT (see Figure 13.12) is an example of an excellent wetting agent but a mediocre dispersant. A wetting agent enhances the spread of a liquid at a solid/air interface; a dispersant enhances the spread of a liquid between two solid surfaces. Many agents enhance both properties, but not all agents. In particular, polymers are excellent at dispersing solids in liquids because of their ability to keep two solid particles apart by steric interactions. Smaller molecules are more likely to be good wetting agents as wetting depends on rapid flow of liquid and rapid adsorption.

A screening test for dispersants is sometimes based on the Daniel flow point method.[14,15] Measured amounts of various dispersant solutions are added to a solid being mixed with a spatula on a glass plate until the paste just flows under the spatula. The less dispersant added before the endpoint, the better the dispersant. Some operators use the appearance of tack — the elastic snapback of a filament of the dispersion when raised with the spatula — as a measure of the endpoint. The process is repeated with various concentrations of dispersant solution. This empirical approach is most important for formulations intended to be milled at high viscosity.

The production of emulsions in a high-shear mill requires the presence of an emulsifying agent, since, if the interface is not stabilized, coalescence soon occurs. The emulsifying agent is usually added to the system before emulsification. Surface-active solutes contribute to the mechanical breakdown of agglomerated particles by promoting internal wetting. Also, by reducing the viscosity of the system and keeping the millbase fluid, surface-active solutes aid the action of ball and sand mills.

Another way in which a surface-active solute enhances milling action is an effect discovered by Rehbinder and his colleagues.[16,17,18] The Rehbinder effect is the reduction of strength of solids by the action of surface-active solutes. Originally this effect was interpreted simply as a lowering of surface energy by adsorption and hence reduction of the work required to produce new interface; but it must also include the creeping of surface-active substances along grain boundaries or dislocations while the body is under stress. Examples of the effect occur with materials as diverse as pure metals and various types of rock under the action of rock drills or during crushing and grinding.

The various effects introduced by the presence of a surface-active solute in wet milling are (a) stabilizing new interfaces by adsorption, (b) wetting of the constituent particles (primary or aggregate) of an agglomerate with elimination of air, (c) reduction of viscosity by preventing flocculation, and (d) the Rehbinder effect. All these effects operate simultaneously and hitherto have proved impossible to differentiate in practice.

5.8 MANUFACTURERS OF MILLING EQUIPMENT

Company	Telephone/FAX	Internet Address	Equipment
APV Systems 5100 River Road, 3rd floor, Schiller Park, IL 60176	(847) 678-4300 (847) 678-4407	www.apv.com	Homogenizers
Boliden Allis, Inc. P.O. Box 14888 Milwaukee, WI 53214-0888			Vibrating ball mill
Branson Ultrasonics Corp. 41 Eagle Rd. Danbury, CT. 06813-1961	(203) 796-2298 (203) 796-0320	www.bransoncleaning.com	Ultrasonic horns
Chemineer, Inc. P.O. Box 1123 Dayton, OH 45401	(937) 454-3200 (937) 454-3230	www.kenics.com	Homogenizers Colloid mills
Draiswerke, Inc. 40 Whitney Rd. Mahwah, NJ 07430	(201) 847-0600 (201) 847-0606	www.draiswerke-inc.com	Perl mills Continuous media mills Roller mills Dispersers
Farrel Corp. 25 Main St. Ansonia, CN 06401	(203) 736-5500 (203) 735-6267	www.farrel.com	Banbury mixers Extruders

MANUFACTURERS OF MILLING EQUIPMENT

Company	Telephone/FAX	Internet Address	Equipment
Five Star Technologies 21200 Aerospace Parkway Cleveland, OH 44142	(877) 513-3483 (440) 239-7015	www.fivestartech.com	High shear mixers Homogenizers
Glen Mills, Inc. 395 Allwood Rd Clifton, NJ 07012	(973) 777-0777 (973) 777-0070	www.glenmills.com	Hammer, pin, toothed, etc. dry mills Bead mills Homgenizers Ultrasonics
Kady International 127 Pleasant Hill Rd Scarborough, ME 04074	(207) 883-4141	www.kadyinternational.com	Rotor-stator mixers
Microfluidics Corp. 30 Ossipee Rd Newton, MA 02164	(617) 969-5452	www.microfluidicscorp.com	High shear homogenizers
Mikropul 20 Chatam Rd Summit, NJ 07901	(908) 598-1100 (908) 598-1455	www.mikropul.com	Cyclones Dust collection
Morehouse-Cowles 30 Ossipee Rd Newton, MA 02164	(617) 969-5452	www.morehousecowles.com	High speed dissolvers Vertical media mills Sand mills Colloid mills Horizontal media mills
Myers Engineering, Inc. 8376 Salt Lake Ave. Bell, CA 90201	(323) 560-4723 (323) 771-7789	www.myersmixer.com	High and low speed dispersers Rotor-stator mixers
Netzsch, Inc. 119 Pickering Way Exton, PA 19341-1393	(610) 363-8010 (610) 363-0971	www.netzschusa.com	Jet mills Classifiers Vacuum premix Bead mills Attrition mills Horizontal mills High speed dissolvers Three roll mills
Paul O. Abbé, Inc. 139 Center Ave. Little Falls, NJ 07424	(800) 524-2188 (973) 256-0041	www.pauloabbe.com	Pebble mills Ball mills Continuous feed mills Dry grinding Blade mixers

Continued

Company	Telephone/FAX	Internet Address	Equipment
Premier Mill One Birchmont Dr. Reading, PA 19606-3298	(610) 779-9500	www.premiermill.com	High speed dissolvers Colloid mills Media mills
Ross Engineering 710 Old Willets Path Hauppauge, NY 11788	(800) 243-7677 (631) 234-0691	www.rossmixing.com	Planetary mixers Rotor-stator mixers Three roll mills Extruders
Semi-Bulk Systems, Inc. 159 Cassens Court Fenton, MO 63026-2543	(800) 732-8769 (314) 343-2822	www.bulk-online.com	Powder injection systems
Sonic Corp. 1 Research Dr. Stratford, CT 06615	(203) 375-0063	www.sonicmixing.com	Ultrasonic emulsifiers
Sturtevant, Inc. 348 Circuit St. Hanover, MA 02339	(800) 992-0209 (781) 829-6515	www.sturtevantinc.com	Micronizers Classifiers Roll, jar, rotary crushers Cyclones
Sweco Americas 8029 US Highway 25 Florence, KY 41042	(606) 283-8400 (606) 283-8469	www.sweco.com	Vibratory separators Vibratory media mills
Thermo Haake (USA) 53 W. Century Rd. Paramus, NJ 07652	(201) 265-7865 (201) 265-1977	www.haake.de	Lab scale extruders
Union Process 1925 Akron-Peninsula Rd Akron, OH 44313	(330) 929-3333 (330) 929-3034	www.unionprocess.com	Attritors: wet and dry, continuous and batch, vertical and horizontal

REFERENCES

[1] Patton, T.C. *Paint flow and pigment dispersion, a rheological approach to coating and ink technology*, 2nd ed.; Wiley: New York; 1979; Chapters 17–24.

[2] Sheppard, I.R. Technical aspects of dispersion and dispersion equipment in *Dispersion of powders in liquids*, 2nd ed.; Parfitt, G.D., Ed.; Wiley: New York; 1973.

[3] McKay, R.B., Ed. *Technological applications of dispersions*; Marcel Dekker: New York; 1994.

[4] Crowley, J.M. *Fundamentals of applied electrostatics*; Wiley: New York; 1986 and Laplacian Press: Morgan Hill, CA; 1999; pp 29–30.

[5] Tsouris, C.; Shin, W.-T.; Yiacoumi, S. Pumping, spraying, and mixing of fluids by electric fields, *Can. J. Chem. Eng.* **1998**, *76*, 589–599.

[6] Tsouris, C. Depaoli, D.W.; Feng, J.Q.; Scott, T.C. Experimental investigation of electrostatic dispersion of nonconductive fluids into conductive fluids, *I&EC Res.* **1995**, *34*, 1394–1403.

[7] Shin, W.-T.; Yiacoumi, S.; Tsouris, C. Experiments on electrostatic dispersion of air in water, *I&EC Res.* **1997**, *36*(9), 3647–3655.

[8] Becker, J.E. Attritor grinding of refractories, *Amer. Ceram. Soc. Bull.* **1996**, *75*(5), 72–74.

[9] Johnson, J.; Szegvari, A.; Li, M. *Polym. Paint Colour J.* **1982**, *172*, 459.

[10] Doroszkowski, A. Paints, in *Technological applications of dispersions*; McKay, R.B., Ed.; Marcel Dekker: New York; 1994; p 4.

[11] www.quackco.com

[12] Wang, Y. *Compounding in co-rotating twin-screen extruders*; Rapra Technology Inc.: Shrewsbury, UK; 2000.

[13] Martin, C. In the mix: continuous compounding using twin-screw extruders, www.devicelink.com.

[14] Daniel, F.K.; Goldman, P. Evaluation of dispersion by a novel rheological method, *Ind. Eng. Chem.* (Anal. Ed.) **1946**, *18*, 26–31.

[15] Daniel, F.K., *J. Paint Tech.* **1966**, *38*, 534.

[16] Rehbinder, P. A.; Lichtman, V. Effect of surface active media on strains and rupture in solids, *Proc. Int. Congr. Surf Act., 2nd, 1957* **1957**, *3*, 563–582.

[17] Rehbinder. P. A. Formation and aggregative stability of disperse systems, *Colloid J. USSR (Eng. Trans.)* **1958**, *20*, 493–502.

[18] Shchukin, E. D.; Rehbinder, P. A. Formation of new surfaces in the deformation and destruction of a solid in a surface-active medium, *Colloid J. USSR (Eng. Trans.)* **1958**, *20*, 601–609.

6 Liquid Surfaces and Interfaces

6.1 MOLECULAR THEORY OF SURFACE AND INTERFACIAL TENSION

The surface of a liquid is in a condition of tension, the most prominent evidence of which is its tendency to contract in area. The spherical shape of a raindrop or the circular arc of a rainbow testifies to the existence of a tension at the surface of water, which makes each drop minimize its surface area. The sphere has the least surface for a given volume; hence liquid drops are spherical. The symmetry of the rainbow is a consequence of raindrops being spheres, and so, indirectly, is itself evidence of tension in the surface of water.

The surface tension of a liquid can be traced to forces of attraction between its molecules. Evidence for such intermolecular attractive forces is afforded by the very existence of the liquid state itself. Given the presence of these forces, molecules at or near the surface are subject to a net force toward the denser phase, and since they are mobile, a characteristic of the liquid phase, molecules respond to this force by moving from the region of the surface into the interior. Within a very short time, but probably not less than a few milliseconds after a new surface is created,[1] the surface region is depleted of molecules to such an extent that a lower density prevails there than in the bulk liquid. This process of depletion of the surface region by migration of molecules into the interior is soon brought to an end by a counter movement of diffusion in the opposite direction, from the higher density bulk to the lower density surface region. Thus a dynamic equilibrium is soon established in which the rate of migration out of the surface is balanced by an equal and opposite rate of diffusion into the surface, with the lower density at the surface maintained as a time average.[2]

The potential energy between two molecules as a function of the distance between them is represented in Figure 6.1. A useful mathematical model for this pair potential is the 6–12 potential, in which the attraction potential varies with the inverse sixth power of the distance of separation, and the repulsion potential varies with the inverse twelfth power of the distance of separation, that is

$$U(r) = 4\varepsilon[(s/r)^{12} - (s/r)^{6}] \tag{6.1}$$

where $U(r)$ is the net potential dependent on the separation r and s is a characteristic distance. Equation (6.1) has a minimum potential energy, $U(r_0)$, that can

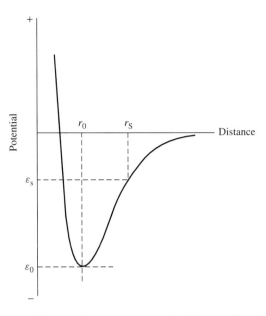

Figure 6.1 The potential energy between two molecules as a function of the distance between them.

be evaluated from $dU/dr = 0$, giving

$$r_0 = 2^{1/6}s \qquad (6.2)$$

$$U(r_0) = -\varepsilon_0 \qquad (6.3)$$

In the surface region of lower density any two molecules are on the average farther apart, r_s than a corresponding pair in the bulk, r_0; consequently the pair potential energy in the surface region ϵ_s is greater than in the bulk, ϵ_0. The excess potential energy of the molecules in the surface region is the source of the tension, for the mobility of molecules in the liquid state allows them to lose their excess potential energy by moving from the surface region into the bulk phase; this motion results in a spontaneous contraction of the surface to its minimum possible area, which, incidentally, is not always a spherical surface, as the minimizing tendency must accommodate to the action of other forces, particularly gravity. It follows that any expansion of the surface is resisted, that is, requires work, the work required being what is necessary to create the excess potential energy per unit area of additional surface. The work to create new surface is proportional to the extent of new surface created:

$$W = \sigma \Delta A \qquad (6.4)$$

where ΔA is the area of additional surface and σ is a proportionality constant, which has the units Newtons/meter (or dynes/centimeter in the CGS system).

The work W is said to be done against "the surface tension of the liquid," and the phrase provides a useful designation for the proportionality constant σ. The unit of surface tension is force per unit of length. Thus if a drop of liquid rests on a solid surface, the direction of the surface-tension force is tangential to the liquid surface and is exerted along every unit length of the line that defines the perimeter of contact of the liquid with the solid.

The potential energy of the liquid surface compared to the bulk phase is expressed as Joules per square meter, and has the same numerical value as the surface tension. The units Newtons per meter are identical to Joules per square meter. The two free-energy functions of Gibbs and of Helmholtz, designated by G and F, respectively, are essentially potential energies modified to take into account influential factors such as changes of temperature, pressure, volume, and composition as required for various processes. For most of the processes considered in this section, only surface tension and surface area are significant. Thus, for example, the Helmholtz free energy of a liquid surface is $\sigma \Delta A$. See Table 6.2.

The quantity σ varies from one liquid to another, depending on the strength of the attractive forces between their molecules. Liquids with relatively weak attractive forces create less of a density difference at their surfaces, and so less of an excess potential energy, requiring less work to extend their surfaces. Those with stronger forces would require more work and have a correspondingly larger value of σ. The relative strength of the intermolecular forces in liquids is reflected in their boiling points and vapor pressures. We should expect, therefore, that volatile liquids of low boiling point would have low values of σ and liquids of low volatility and high boiling point would have larger values of σ. The correlation of low surface tension and low boiling point is best seen in homologous series. Figure 6.2 shows the dependence of surface tension on boiling point for a variety of CH_3- and CF_3-containing materials.[3]

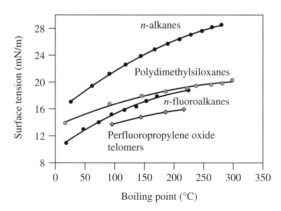

Figure 6.2 Dependence of surface tension at 20°C on boiling point for a variety of CH_3- and CF_3-containing materials.[3]

TABLE 6.1 Surface Tensions and Boiling Points of Some Pure Liquids[4]

Liquid	$T(°C)$	σ(mN/m)	Boiling Point (°C)
Acetic acid	20	27.4	117.9
Acetone	25	24	56.3
Aniline	20	42.9	184.4
Benzaldehyde	20	38.8	178.9
Benzene	20	28.9	80.1
1-Butanol	20	24.6	117.7
Bromobenzene	20	36.5	155.9
Carbon tetrachloride	25	26.4	76.7
Cellosolve	25	28.2	135.6
Chloroform	25	26.7	61.2
Cyclohexane	20	25.2	80.7
Decane	20	23.7	174.9
Diethyl ether	20	17	34.5
Dodecane	20	25.3	216.3
Ethanol	20	22.3	78.3
Ethyl acetate	20	23.7	77.1
Ethylene glycol	20	46.5	197.3
Glycerol	20	63.3	290
Heptane	20	20.1	98.4
Hexane	20	18.4	68.7
Mercury	20	484	356.6
Methanol	20	22.5	64.7
Methyl cellosolve	15	31.8	124.6
Methyl ethyl ketone	25	24	79.6
Methyl isobutyl ketone	20	23.6	116.5
Methylene chloride	20	27.8	39.7
Methylene iodide	20	50.8	(decomposes) 182
Nitrobenzene	20	43.9	210.8
Octane	20	21.2	125.7
Octanoic acid	20	29.2	239.9
1-Octanol	20	26.1	195.2
Oleic acid	20	32.8	(decomposes) 360
p-Dioxane	15	34.4	101.3
Pentane	20	16	36.1
1-Pentanol	20	25.6	137.8
Perfluoroheptane	20	11	
Perfluorohexane	20	11.9	
Perfluoropentane	20	9.9	
Polydimethylsiloxanes			
Tetramer	20	17.6	
Dodecamer	20	19.6	
1-Propanol	20	23.7	97.2
Tetrahydrofuran	25	26.4	66
Toluene	20	28.5	110.6
1,1,1-Trichloroethane	20	25.6	74
Water	20	72.9	100
	30	71.3	100
	60	67	100

This expectation is confirmed generally, with many exceptions, by reference to the data for the surface tensions and boiling points of some pure liquids collected in Table 6.1.

6.2 THERMODYNAMICS OF SURFACES AND INTERFACES

Systems in which interphase zones are significant require the introduction of terms for interfacial area and interfacial tension into their thermodynamic descriptions, analogous to the usual volume and pressure terms used for gaseous processes. If the surface tension is expressed by σ and the surface area by A, the work done on extending a surface against the force of its surface tension is $\sigma \Delta A$. The analogy to the work term for the compression of a gas, namely, pdV, is complete except for the sign of the work term: Since the spontaneous action of a gas is expansion and the spontaneous action of a surface is contraction, the signs of the two work terms are opposite. The first and second laws of thermodynamics are therefore summarized for open systems by the expression

$$dU = TdS - pdV + \sigma dA + \sum \mu_i dn_i \tag{6.5}$$

Similarly, the definition of enthalpy applied to interfacial systems requires the addition of a term σA, that is,

$$H = U + pV - \sigma A \tag{6.6}$$

A complete listing of the thermodynamic functions, differential equations, and Duhem equations are given in Table 6.2. Thermodynamic quantities are expressed

TABLE 6.2 Thermodynamic Functions for Interfacial Systems[a]

$U = TS - pV + \sigma A + \Sigma \mu_i n_i$	(6.7)
$dU = TdS - pdV + \sigma dA + \Sigma \mu_i dn_i$	(6.8)
$H = U + pV - \sigma A$	(6.9)
$dH = TdS + Vdp - Ad\sigma + \Sigma \mu_i dn_i$	(6.10)
$F = U - TS$	(6.11)
$dF = -SdT - pdV + \sigma dA + \Sigma \mu_i dn_i$	(6.12)
$G = H - TS = \Sigma \mu_i n_i$	(6.13)
$dG = -SdT + Vdp - Ad\sigma + \Sigma \mu_i dn_i$	(6.14)
$SdT - Vdp + Ad\sigma + \Sigma n_i d\mu_i = 0$	(6.15)
$SdT + pdV - \sigma dA + \Sigma n_i d\mu_i = 0$	(6.16)
$-TdS - Vdp + Ad\sigma + \Sigma n_i d\mu_i = 0$	(6.17)
$-TdS + pdV - \sigma dA + \Sigma n_i d\mu_i = 0$	(6.18)

[a] U is the internal energy, H is the enthalpy, F is the Helmholtz free energy, and G is the Gibbs free energy.

in work units: in the CGS system, ergs and in the SI system, Joules. The expression for the differential Helmholtz function, dF, leads to

$$\left(\frac{\partial F}{\partial A}\right)_{V,T,n_i} = \sigma \tag{6.19}$$

This result establishes that the surface tension, which is a physical property of a liquid surface as real as the tension in a stretched elastic band, also has the meaning of the Helmholtz function per unit area at constant volume, temperature, and composition. When expressed as a tension, the appropriate units of σ are force/unit length; when expressed as a free energy, the appropriate units are energy/unit area. The measured surface tension of water at 20°C is 73 mN/m, therefore, the Helmholtz free energy is 73 mJ/m² at the same temperature.

For a one-component system containing an interface, the entropy per unit area at constant pressure and composition is [from Eq. (6.15)]

$$\frac{S}{A} = -\left(\frac{\partial \sigma}{\partial T}\right)_p \tag{6.20}$$

Combining Eq. (6.20) with (6.7) for a closed system gives, for the surface,

$$\frac{U}{A} = \sigma - T\left(\frac{\partial \sigma}{\partial T}\right)_p \tag{6.21}$$

The internal energy per unit area, U/A, is a useful thermodynamic function to compare surface properties of members of a homologous series because it is less temperature dependent than the surface tension (or Helmholtz free energy).

We shall now use the free-energy functions listed in Table 6.2 to describe a few simple processes with closed systems (i.e., constant composition) of significant interphase areas.

6.2.a Coalescence of Droplets

The free-energy change (Helmholtz) for the process of coalescence of two droplets, Figure 6.3, at constant volume, temperature, composition, and surface tension is

$$\Delta F = F_{\text{final}} - F_{\text{initial}} = \sigma(A_{\text{final}} - A_{\text{initial}}) \tag{6.22}$$

The area decreases as drops coalesce; hence the expression for ΔA is negative and the coalescence is therefore spontaneous. The stability of aerosols and emulsions depends on the inhibition of coalescence, which requires that the thermodynamic expressions for the free energy of coalescence include additional terms of opposite sign. These terms may refer to the free energy of solvation or to the free energy of desorption of a third component (See Section 22.4).

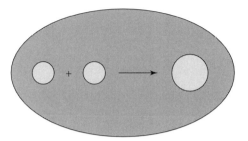

Figure 6.3 The coalescence of two droplets.

6.2.b Spreading of One Liquid on Another

The Helmholtz free-energy change *per unit area* for the process of liquid 2 spreading on the surface of liquid 1 (Figure 6.4) at constant volume, temperature, and composition is

$$\Delta F = F_{\text{final}} - F_{\text{initial}} = \sigma_2 + \sigma_{12} - \sigma_1 \quad (6.23)$$

If this expression is negative, liquid 2 spreads spontaneously; if it is positive, liquid 2 sits as a lens on the surface of liquid 1. If the drop of liquid 2 is small enough, it will still remain on the surface even though its density is greater than that of liquid 1, held there by the predominance of surface over gravitational forces.

The spreading coefficient is defined as equal to the negative of the Helmholtz free energy of spreading per unit area: $S = -\Delta F$. Therefore,

$$S_{(\text{of 2 on 1})} = \sigma_1 - \sigma_2 - \sigma_{12} \quad (6.24)$$

If the spreading coefficient is positive, liquid 2 spreads spontaneously on liquid 1; if the spreading coefficient is negative, liquid 2 sits as a lens on the surface of liquid 1. Equation (6.24) is readily tested. Harkins[5] measured values of S for 89 liquids on water, then compared the sign of S with observations of spreading or nonspreading. In all the pairs observed the sign of S agreed with the observed behavior but one. Carbon tetrachloride did not spread even though it had a calculated spreading coefficient of $+1.06$ mN/m.

6.2.c Encapsulation of One Liquid by Another

Spontaneous encapsulation of one liquid by another requires spreading of the medium around droplets of the dispersed phase. Two spreading processes are

Figure 6.4 The spreading of one liquid on another.

possible: that of liquid 2 on 1 and that of liquid 1 on 2. The corresponding coefficients are

$$S_{(of\ 2\ on\ 1)} = \sigma_1 - \sigma_2 - \sigma_{12} \qquad (6.25)$$

$$S_{(of\ 1\ on\ 2)} = \sigma_2 - \sigma_1 - \sigma_{12} \qquad (6.26)$$

Therefore,

$$S_{(of\ 2\ on\ 1)} + S_{(of\ 1\ on\ 2)} = -2\sigma_{12} \qquad (6.27)$$

This equation tells us that *both* spreading coefficients cannot be positive; one may be positive and the other negative; or both may be negative. The diagrams published by Kelvin (Figure 6.5) show two liquids for which both spreading coefficients are negative. No matter what their relative proportions may be, encapsulation of one of these liquids by the other is never achieved. If the spreading coefficient of one of these liquids on the other had been positive, it would have encapsulated the other; the reverse type of encapsulation would never occur.

Kelvin's diagrams also illustrate the constancy of the angles of contact between the two liquids, independent of their relative amounts. The relation between the angles and the surface and interfacial tensions is given by Neumann's triangle of forces (Figure 6.6),

$$\sigma_1 = \sigma_2 \cos(\pi - \theta_{12}) + \sigma_{12} \cos(\pi - \theta_2) \qquad (6.28)$$

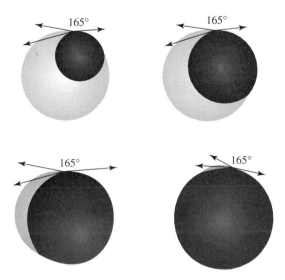

Figure 6.5 The constancy of the contact angles is independent of the relative volumes of two immiscible liquids with negative spreading coefficients on each other (W. Thomson, Lord Kelvin, 1889). The two liquids are nitrobenzene and water. The darker shade of the two represents water.

144   LIQUID SURFACES AND INTERFACES

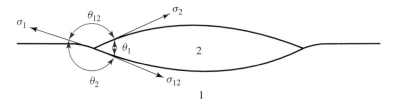

Figure 6.6 Neumann's triangle of surface forces pulling on each unit length of the line at which three fluid phases meet.

or

$$\frac{\sigma_1}{\sin \theta_1} = \frac{\sigma_2}{\sin \theta_2} = \frac{\sigma_{12}}{\sin \theta_{12}} \tag{6.29}$$

6.2.d   Works of Adhesion and Cohesion

The processes of adhesion and cohesion (Figure 6.7) may be described in terms of their opposites, namely, the processes of separation. The work of adhesion per

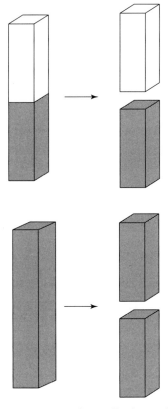

Figure 6.7   The processes of (a) adhesion and (b) cohesion.

unit area is the work done on the system when two condensed phases (1 and 2), forming an interface of unit area, are separated reversibly to form unit areas of each of the 1 and 2 surfaces.

$$W^{adh} = \sigma_1 + \sigma_2 - \sigma_{12} \tag{6.30}$$

A list of some interfacial tensions and works of adhesion, calculated from Eq. (6.30), of liquids against water is given in Table 6.3.

The work of adhesion equals the Helmholtz free energy of separation. The work of cohesion equals the work of adhesion when liquids 1 and 2 are the same. In that case σ_{12} is nonexistent and

$$W^{coh} = 2\sigma_1 \tag{6.31}$$

The work of adhesion given by Eq. (6.30) refers to the adhesion between two liquids or a liquid and a solid but an analogous expression would not measure the strength of the adhesive interface between two solid surfaces. At the dry contact, between a polymer and a metal, for example, electric charge is exchanged,

TABLE 6.3 Selected Liquid–Liquid Interfacial Tensions against Water[6]

Liquid	Temperature (°C)	σ_{12} (mN/m)	W^{adh} (mJ/m^2)
Hexane	20	51.1	40.1
Heptane	20	50.2	43.0
Octane	20	50.8	43.8
Decane	20	51.2	45.4
Carbon tetrachloride	20	45.0	54.7
Toluene	20	36.1	65.2
Benzene	20	35.0	66.6
Chloroform	20	31.6	68.3
Bromobenzene	20	38.1	71.2
Methylene chloride	20	28.3	71.0
Methylene iodide	20	48.5	75.1
Diethyl ether	20	10.7	79.1
Nitrobenzene	20	25.7	81.0
Ethyl acetate	20	6.8	89.8
Oleic acid	20	15.7	89.6
n-Octanol	20	8.5	91.8
Octanoic acid	20	8.5	91.8
n-Pentanol	20	4.4	94.0
n-Butanol	20	1.8	94.6
Benzaldehyde	20	15.5	97.3
Aniline	20	5.8	109.9
Mercury	25	427	130

Figure 6.8 The process of emersion, final and initial states.

typically 10^3-10^4 esu/cm^2 (see Section 17.5). This small charge does not contribute significantly to the interfacial free energy σ_{12}; nevertheless Derjaguin, and Toporov and Derjaguin[7,8] have shown that even such a small surface charge may contribute over a hundred times as much adhesive energy as the dispersion components when two such surfaces are separated. The reason is that the charges in the polymer film are immobile and work must be done to separate them from their countercharges in the metal. Although the electric force across the interface is weak, it extends to a greater distance than molecular forces; therefore, the work against the electric field, due to the trapped charges, is greater than the work done against the molecular forces. This and other aspects of adhesion are reviewed elsewhere.[9,10,11]

It should not be surprising that the work of cohesion of a liquid is directly measured by its surface tension. We explained above that the surface tension of a liquid arises from intermolecular forces of attraction. These, too, are responsible for its cohesion.

6.2.e Free Energy of Emersion

The process of emersion is shown in Figure 6.8. The process as measured in a laboratory is one of immersion, but for consistent thermodynamic equations, it is calculated as the negative process, emersion. The Helmholtz free energy *per unit area* of emersion at constant temperature, volume, and composition is

$$\Delta F_{\text{emersion}} = \sigma_{sv} - \sigma_{sl} \tag{6.32}$$

The connection between the spreading coefficient, the work of adhesion, and the free energy of emersion are all indexes of the interactions between two liquids and are discussed more fully in Chapter 7.

6.3 THERMODYNAMICS OF CURVED INTERFACES

At any point on a plane liquid surface the components of the tension lie within the surface, so there is none normal to the surface; on a curved liquid surface, however, the components of the tension pull normally (to the concave side) as well as tangentially to the surface. Hence, a curved surface can only be maintained by an equal and opposing force to the normal components, which force may be exerted by gas pressure or by hydrostatic pressure (Figure 6.9). For example, a spherical bubble has normal components of the surface tension pulling inward, and these are counteracted by the higher gas pressure within the bubble. For the same reason the pressure is always larger on the concave side of a curved liquid surface. Another example: The shape of a sessile drop is determined by the balance between the forces of surface tension and the counteracting hydrostatic pressure.

Consider a small element of area of a curved liquid surface of length x and width y at equilibrium. If this surface were to expand by a differential amount, dz, normal to the surface, the area expands to $(x + dx)(y + dy)$. The increase in area is, to leading order, $ydx + xdy$. The corresponding increase in volume is $xydz$. The work required to increase the area against the surface tension is balanced by the work done by the expansion against the pressure, Δp, across the curved surface, that is,

$$\Delta p (xydz) = \sigma (ydx + xdy) \tag{6.33}$$

By choosing the x and y axes to lie in the planes of the principal circles of curvature of radii, R_1 and R_2, respectively (the z axis is the line common to both planes), the properties of similar triangles give

$$\frac{x + dx}{x} = \frac{R_1 + dz}{R_1} \quad \text{and} \quad \frac{y + dy}{y} = \frac{R_2 + dz}{R_2} \tag{6.34}$$

Rearranging and substituting gives the Laplace equation

$$\Delta p = \sigma \left(\frac{1}{R_1} + \frac{1}{R_2} \right) \tag{6.35}$$

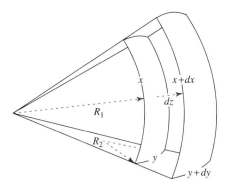

Figure 6.9 Balance of pressure drop and curvature across a curved liquid surface.

The pressure is hydrostatic in origin for liquids and is "gasic" for gases; the pressure drop across the surface is directly proportional to the curvature of the surface. The surface tension is the constant of proportionality. This remarkably simple equation describes the shapes of all free-standing liquid surfaces, no matter what their boundary conditions. Some interesting surface shapes are developed by bubbles and soap films. The spherical surface of a bubble has a curvature of $2/R$ where R is the radius of the bubble. The extra pressure of the gas inside a bubble equals $4\sigma/R$; the second factor of 2 arises because a bubble has two liquid surfaces, inside and outside, both of them equally curved (to a close approximation). A cylindrical surface has a principal radius of curvature in one direction equal to the radius of the cylinder and, in the other direction, the radius of curvature is infinite; so that the curvature of a cylindrical surface is $1/R$, that is, half that of a sphere of the same radius. A plane surface has zero curvature because the circles of curvature have infinite radii; therefore the pressure difference across a plane surface is zero. Various saddle-shaped surfaces have radii of curvature in opposite directions, which, in the case of a catenoid are also equal to each other, so that a catenoidal surface has zero curvature and no pressure drop exists across it. These results are listed in Table 6.4.

Examples of curved liquid surfaces are provided by bubbles, soap films, and the curvature induced by the contact angle of a liquid surface against a solid substrate, as, for example, the meniscus created against a glass slide positioned vertically to a water surface. The contact angle of water on clean glass is zero degrees, which requires that the water surface bend upward to meet the vertical surface of the slide; the curvature thus induced creates a lower hydrostatic pressure inside the liquid, which is balanced by the liquid held in the meniscus against the pull of gravity. Another example of curvature induced by contact angle is the approximately hemispherical meniscus of water inside a clean glass capillary tube. The pressure difference associated with this curvature is the force that causes the spontaneous rise of liquids in capillaries. The flow of liquid in response to pressure differentials created by curvature of liquid surfaces is known as capillary flow, sometimes Laplace flow.

TABLE 6.4 Pressure Drops across Curved Liquid Surfaces

Shape	Curvature	Pressure Drop
Sphere	$2/R$	$2\sigma/R$
Spherical bubble	$2/R$	$4\sigma/R$
Cylinder	$1/R$	σ/R
Cylindrical lamella	$1/R$	$2\sigma/R$
Plane	0	0
Catenoid	0	0
Spiral ramp	0	0

6.4 SESSILE AND PENDENT DROPS

A nonspreading drop supported by a solid substrate is sessile or sitting; a drop suspended from a tip is pendent or hanging. Such drops have a shape that is determined solely by a balance of the forces of gravity and of surface tension: Gravity tends to flatten sessile drops and elongate pendent drops; surface tension tends to confer spherical shape, as the sphere has minimum area for a given volume. The less dense the liquid, the more the shape approaches the sphere; the denser the liquid, the more the shape departs from the sphere. Small drops are less affected by gravity and so are almost perfectly spherical. Large drops of liquid are less affected by surface tension, hence flattened; nevertheless, their ultimate thickness is determined by contact angle and surface tension. A liquid drop in a medium in which it is insoluble forms a perfect sphere if the two liquids have the same density. A good example of a sessile drop is provided by a rain drop on a petal or leaf; another example is the shape frequently used for rural water towers, a concept developed commercially by the Chicago Bridge and Iron Works.

The balance of forces acting on a liquid drop is described mathematically by a differential equation, which gives an expression for the shape in terms of volume, surface tension, and density. Its starting point is Eq. (6.35) written in the form

$$\rho g z + C = \sigma \left(\frac{1}{R_1} + \frac{1}{R_2} \right) \tag{6.36}$$

where the pressure difference Δp is replaced by $\rho g z + C$, which is the hydrostatic pressure, $\rho g z$, at a distance z from the apex of the drop plus the pressure difference C across the apex of the drop. Equation (6.36) expressed in Cartesian coordinates is

$$\frac{d^2 z/dx^2}{[1 + (dz/dx)^2]^{3/2}} + \frac{(1/x) dz/dx}{[1 + (dz/dx)^2]^{1/2}} = \frac{\rho g z + C}{\sigma} \tag{6.37}$$

This form of the equation is a second-order nonlinear differential equation with singularities at the equatorial extremes. A numerically stable form is the expression in terms of arc lengths s and the angular inclination, Φ, of the tangent to the horizontal (Figure 6.10) as[12]

$$\sigma \left(\frac{d\Phi}{ds} + \frac{\sin \Phi}{x} \right) = \rho g z + C \tag{6.38}$$

Equation (6.38) is a transcendental differential equation. This form of the equation is the most suitable for experimental evaluation. Neither form of the equation has yet received a closed solution. A numerical solution was originally given as hand-calculated tables by Bashforth and Adams in 1883; nowadays more extensive tables giving the shape of sessile drops, pendent drops, and external menisci are available (Hartland and Hartley[12]). The inverse problem, to find the

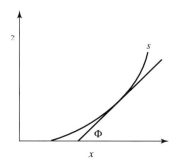

Figure 6.10 Notation for an axisymmetric surface bounding a sessile or pendant drop.

surface tension of the liquid given the shape of a sessile or pendent drop and the density of the liquid, is a more common problem, which is solved by a computer program.[13,14]

These equations describe the complete surface of revolution, but actually the drop is truncated by the plane of the substrate at a depth that determines the specific contact angle of the liquid solid system. A drop of mercury on glass, with a contact angle of 140° (see Figure 7.1), is truncated very little compared to a drop of n-octane on polytetrafluoroethane, with a contact angle of 32°.

6.5 THE KELVIN EQUATION

The effect of an increase of external pressure on a liquid is to raise its vapor pressure. An equilibrium condition obtains when the differential increase of the molar Gibbs free energy of the liquid, due to the increase in the total pressure dp, is equal to that of its vapor,

$$dG_l = dG_v \qquad (6.39)$$

At constant T

$$dG_l = V_m dp \qquad (6.40)$$

where V_m is the molar volume of the liquid and

$$dG_v = RT d \ln p \qquad (6.41)$$

where p is the vapor pressure. Assuming V_m constant with variation of p, Eq. (6.39) may be integrated between the limits p_0 and p to give

$$V_m p = RT \ln(p/p_0) \qquad (6.42)$$

Surface tension acts as a pressure on a curved surface. By combining Eq. (6.42) and (6.35), written for a sphere, setting p equal to the vapor pressure of the sphere and p_0 equal to the vapor pressure of a plane surface, gives

$$\ln \frac{p}{p_0} = \frac{2\sigma V_m}{aRT} \tag{6.43}$$

where a is the radius of the drop. Equation (6.43) is one form of the Kelvin equation where $p > p_0$ as in a convex surface. It states that small drops have higher vapor pressure than flat sheets of liquid, hence small drops evaporate more readily. Conversely, the condensation of small drops from the vapor requires so much supersaturation that the process does not occur spontaneously in the absence of solid particles to act as nucleating sites. Once small drops are nucleated, they grow rapidly as their equilibrium pressure declines.

The Kelvin equation, (6.43), also applies to the curved surface of a liquid meniscus. Where r is the radius of curvature of the meniscus,

$$\ln \frac{p}{p_0} = -\frac{2\sigma V_m}{rRT} \tag{6.44}$$

where $p < p_0$ as in a concave surface.

Equation (6.44) states that the vapor pressure of a liquid is decreased at a concave meniscus, such as occurs on a wetting liquid in a capillary tube or in the pores of a solid substrate. Experimental verification of the Kelvin equation is reported by Fisher and Israelachvili[15] for the concave meniscus of cyclohexane between crossed cylinders of molecularly smooth mica. With a clean system, where the establishment of equilibrium and the absence of significant contamination could be demonstrated, the Kelvin equation was found to hold for radii of curvature as low as 2.5 nm.[16]

6.6 OSTWALD RIPENING

In an aerosol of polydispersed drops in a closed container, the large drops grow at the expense of the smaller ones. The process is known as Ostwald ripening and is common to the coarsening of any finely dispersed phase. It applies to the growth of crystals, the digestion of precipitates and the degradation of emulsions and foams.

Equation (6.43) may be applied to the solubility of small spherical drops in liquids, replacing vapor pressures by saturation concentrations to give the Gibbs–Thomson (or Gibbs-Kelvin) equation

$$\ln \frac{c}{c_0} = \frac{2\sigma V_m}{aRT} \tag{6.45}$$

where c is the solubility of small spherical droplets of radius a and c_0 is the solubility of large drops ($r = \infty$). Equation (6.45) implies that the smaller the droplet, the greater its solubility. In a polydisperse emulsion the smaller droplets are more soluble than the larger ones so that the spontaneous process is for molecules in small droplets to transfer to larger droplets. The large droplets spontaneously grow at the expense of the smaller droplets, another instance of Ostwald ripening.

Many natural and manufactured products are emulsions, for example, foods, cosmetics, pharmaceuticals, paints, petrochemicals, explosives, and agrochemicals. The gradual increase in the droplet size and size distribution can be monitored *in situ* by ultrasonic attenuation and compared to theories for Ostwald ripening.[17]

The rate of Ostwald ripening depends on the transport mechanism for molecules from the small drop to reach the larger ones. The rate at which molecules in the disperse phase can leave surfactant-covered surfaces and the rate at which molecules penetrate the surfactant-covered surfaces of larger drop is significant and may be rate determining. The role of adsorption and desorption of molecules in micelles can also be important.[18]

Although solid particles are not usually spherical, the general conclusion that small particles have a higher solubility than large particles is well established, as is the requirement of supersaturation before nucleation occurs in clean systems, and the phenomenon of the Ostwald ripening (digestion) of precipitates.

The fierce debate about the existence of anomalous water or, as it was popularly known, "polywater," arose when investigators found that water, in capillaries of a few microns radius, had a much lower vapor pressure than predicted by the Kelvin equation. The debate ended with the discovery that the "anomalous" behavior is accounted for by the presence of various leachable impurities in solution.[19]

6.7 CAPILLARY OR LAPLACE FLOW

Capillary flow of a liquid results from differences of hydrostatic pressure within a liquid, created by local differences of curvature of the liquid surface. The pressure is less on the convex side of an interface; liquid at higher pressure then flows to the region of lower pressure. A liquid that makes a low angle of contact in a narrow vertical tube creates a meniscus that is convex to the liquid; consequently, liquid flows up the tube. Mercury in a glass tube creates a meniscus that is concave to the liquid, consequently, mercury sinks in the tube, but since the density of mercury is large, the decline of the mercury level in the tube is small; nevertheless a correction for this capillary effect has to be introduced in precise mercury manometry. The minimum pressure, Δp, required to prevent the rise of a liquid in a capillary tube of radius r, is also a result of the curvature of the liquid surface.

$$\Delta p = \frac{2\sigma \cos \theta}{r} \qquad (6.46)$$

The more acute the angle of contact, the greater the pressure required to prevent liquid rising; if the angle is 90°, no pressure is required; if the angle is obtuse, the liquid will not enter the tube spontaneously.

Another example of capillary flow is the motion of a drop of liquid in a narrow conical aperture. The capillary pressure at each end of the liquid drop is different because the radii of curvature of the two surfaces are different. In an aperture narrow enough to neglect deviations from spherical surfaces, the net pressure on the drop is given by[20]

$$\Delta p = 2\sigma \, \cos(\theta - \phi) \left(\frac{1}{R_1} + \frac{1}{R_2} \right) \quad (6.47)$$

where θ is the contact angle, 2ϕ is the angle of the cone, and R_1 and R_2 are the radii of curvature at the front and rear surfaces of the drop. For acute contact angles the drop moves toward the narrow end of the aperture; for contact angles sufficiently large, the drop moves toward the wide end.

The importance of capillary forces is measured by the Bond number,

$$B_0 = \frac{\rho g d^2}{\sigma} \quad (6.48)$$

where ρ is the density difference, g is the acceleration due to gravity, d is a characteristic dimension, and σ is the surface tension. Capillary phenomena are significant when B_0 is less than unity. The Bond number is related to the capillary constant used in the numerical solutions of the Laplace equation (6.35).[21] An interesting application of this criterion occurs in the microgravity of space, where the value of g is much reduced. Under this condition, capillary flow is significant for much larger dimensions.[22]

6.8 MARANGONI FLOW

Another kind of flow induced by surface tension is *Marangoni flow*. This flow results from local differences of surface tension. At equilibrium, no local differences of surface tension exist, but various factors can lead to nonequilibrium. Some examples of these factors are:

(a) Local thermal differences;
(b) Local differences of composition due to evaporation;
(c) Local compressions and dilatations of adsorbed films at liquid surfaces.

Each of these disturbances makes the surface tension of the liquid depart from equilibrium. Nonequilibrium surface tensions are called *dynamic* and equilibrium surface tensions are called *static*. Gradients of tension at an interface are equivalent to shear forces τ at the boundary of the two adjoining phases, α and β,

$$\text{grad}(\sigma) = \tau_\alpha + \tau_\beta \quad (6.49)$$

These shear forces cause adjoining liquids to flow.

Scriven and Sternling[23] reviewed surface movements due to Marangoni effects. They list various phenomena in which Marangoni effects play an important role, such as in crystal growth, motion of protoplasm, transport of bacteria, surface fractionation, absorption and distillation, brandy tears, foam stability, and the damping of waves by oil. The uneven drying of film coatings, creating thermal gradients, leading to cellular flow patterns at the surface, which were observed and explained correctly in 1855 by James Thomson, and are now known as Bénard cells, are also examples of Marangoni flow. Thermal gradients create both density and surface-tension gradients.

6.8.a Processes Dependent on Marangoni Flow

Many processes depend on or are plagued by Marangoni flow, caused by dynamic surface tension. James Thomson, even before Marangoni, explained the phenomenon of brandy tears. Alcohol evaporates faster than water, so the surface layer of a solution of alcohol in water is more dilute than the bulk solution. In a deep vessel, the loss of alcohol at the surface is restored by diffusion. But a thin layer of brandy adhering to the sides of a glass becomes more dilute by evaporation and remains so, with a consequent rise in its surface tension. The higher surface tension on the walls draws the surface of the brandy in the glass, until the quantity of liquid drawn up becomes so large that it falls back in the form of tears.

A household hint on how to remove grease stains from cloth was explained by Maxwell.[24] Grease has a higher surface tension than the solvent used to remove it. If the middle of the stain is wetted with solvent, the grease is driven into the clean part of the cloth. The correct method is to apply the solvent in a ring all around the stain, gradually bringing it nearer the center. The grease is thus brought to the center of the stain, from which it may be removed by a paper towel.

Marangoni flow is the explanation of a phenomenon observed by Fowkes.[25] The rate of wetting of a skein of gray unboiled cotton yarn was found to be more rapid in beakers of larger diameter. The explanation is that the solute is adsorbed from the surface of the solution on to the fiber, which depletes the surface locally, and so creates a surface tension gradient that draws the rest of the surface to the cotton. This process is much more rapid than the rate of diffusion to the surface from the bulk solution, which becomes the rate-determining step. The larger the area of the surface, the more molecules are brought into it per unit time, and then rapidly transported to the cotton.

An ingenious application of Marangoni flow was used by Leenaars, Huethorst, and van Oekel[26] to clean substrates for microelectronic chips. Surface-tension gradients are created along the meniscus during dip coating. Organic vapor is fed to the falling edge of the water meniscus. The vapor condenses on the thin water film creating a surface-tension gradient. The pull of higher surface tension drags the meniscus down toward the water-rich reservoir. The shear forces resulting from this scouring action are evidently enough to dislodge unwanted particles from the substrate.

6.8.b Thermally Induced Marangoni Flow

In the reduced-gravity environment of spacecraft, density gradients occur but do not lead to convection currents, surface tension gradients produce flow, and a type of convection called thermocapillary.[22] The term "thermocapillarity" is an unfortunate choice as it blurs the useful distinction we have made between capillary flow and Marangoni flow. Better terms are "thermally induced Marangoni flow," or "TIM flow."

A household example of TIM flow is illustrated by another effectual method to remove a grease spot. A hot iron is applied to one side of the cloth and a paper towel to the other. The thermally induced Marangoni flow drives the grease into the paper towel.

The drying of thin liquid films, such as paint, is a common process. The complexity of the mass flow, thermal gradients, and Marangoni flow makes analytic models impractical. Numerical models have provided detailed information about these processes.[27]

REFERENCES

[1] Drost-Hansen, W. Aqueous interfaces: Method of study and structural properties, *Ind. Eng. Chem.* **1965**, *57*(3), 38–44; also *Chemistry and physics of interfaces*; Ross, S., Ed.; American Chemical Society: Washington, DC; 1965; pp 13–20.

[2] Brown, R.C. The fundamental concepts concerning surface tension and capillarity, *Proc. Phys. Soc.*, London **1947**, *59*, 429–448.

[3] Owen, M.J. The surface activity of silicones: A short review, *Eng. Chem. Prod. Res. Dev.* **1980**, *19*, 97–103.

[4] Jasper, J.J. The surface tension of pure liquid compounds, *J. Phys. Chem. Ref. Data* **1972**, *1*(4), 841–1010; Riddick, J.A.; Bunger, W.B.; Sakano, K. *Organic solvents: physical properties and methods of purification*, 4th ed.; Wiley: New York; 1970.

[5] Harkins, W.D. *The physical chemistry of surface films*; Reinhold: New York; 1952; pp 44–45.

[6] Girifalco, L.A.; Good, R.J. A theory for the estimation of surface and interfacial energies. I. Derivation and application to interfacial tension, *J. Phys. Chem.* **1957**, *61*, 904–909.

[7] Derjaguin, B.V.; Toporov, Yu. P. Role of the molecular and the electrostatic forces in the adhesion of polymers, in *Physicochemical aspects of polymer surfaces*; Mittal, K. L., Ed.; Plenum: New York; 1983; Vol. 2; pp 605–612.

[8] Derjaguin, B.V.; Smilga, V.P. The present state of our knowledge about adhesion of polymers and semiconductors, *Int. Congr. Surf. Act., 3rd, 1960* **1961**, *2*, 349–367.

[9] See the *Journal of Adhesion Science and Technology*, VSP: Zeist, The Netherlands.

[10] K.L. Mittal has *edited or coedited over 50 volumes on adhesion published in ACS Symposium Series*; American Chemical Society: Washington, DC, by Kluwer: New York, by Marcel Dekker: New York; and by VSP: Zeist, The Netherlands.

[11] L.-H. Lee has *edited or coedited books on adhesion published* by Pleumum Press: New York; Kluwer: New York, and Perseus; Boulder, CO.

[12] Hartland, S.; Hartley, R.W. *Axisymmetric fluid-liquid interfaces*; Elsevier: New York; 1976; Chapter 1.

[13] Rotenberg, Y.; Boruvka, L.; Neumann, A.W. Determination of surface tension and contact angle from the shapes of axisymmetric fluid interfaces, *J. Colloid Interface Sci.* **1983**, *93*, 169–183.

[14] Butler, J.N.; Bloom, B.H. A curve-fitting method for calculating interfacial tension from the shape of a sessile drop, *Surf. Sci.* **1966**, *4*, 1–17.

[15] Fisher, L.R.; Israelachvili, J.N. Direct experimental verification of the Kelvin equation for capillary condensation, *Nature (London)* **1979**, *277*, 548–549.

[16] Fisher, L.R. Forces due to capillary-condensed liquids: limits of calculations from thermodynamics, *Adv. Colloid Interface Sci.* **1982**, *16*, 117–125.

[17] Weiss, J.; Herrmann, N.; McClements, D.J. Ostwald ripening of hydrocarbon emulsion droplets in surfactant solutions, *Langmuir* **1999**, *15*, 6652–6657.

[18] De Smet, Y.; Deriemaeker, L.; Finsy, R. Ostwald ripening of alkane emulsions in the presence of surfactant micelles, *Langmuir* **1999**, *15*, 6745–6754.

[19] Franks, F. Polywater; MIT Press: Cambridge, MA; 1981.

[20] Bickerman, J.J. *Physical surfaces*; Academic Press: New York; 1970; p 274.

[21] Boucher, E.A. Capillary phenomena: properties of systems with fluid/fluid interfaces, *Rep. Prog. Phys.* **1980**, *43*, 497–546.

[22] Ostrach, S. Motion induced by capillarity, in *Physicochemical Hydrodynamics*; Spalding, D.B., Ed.; Advance: London; 1977; Vol. 2, pp 571–589.

[23] Scriven, L.E.; Sternling, C.V. The Marangoni effects, *Nature(London)* **1960**, *187*, 186–188.

[24] Maxwell, J.C. *Theory of heat*, D. Appleton: New York; 1872; AMS Press: New York; 1972.

[25] Fowkes, F.M. Role of surface active agents in wetting, *J. Phys. Chem.* **1953**, *57*, 98–103.

[26] Leenaars, A.F.M.; Huethorst, J.A.M.; van Oekel, J.J. Marangoni drying: An extremely clean drying process, *Langmuir* **1990**, *6*, 1701–1703.

[27] Eres, M.H.; Weidner, D.E.; Schwartz, L.W. Three-dimensional direct numerical simulation of surface-tension-gradient effects on the leveling of an evaporating multicomponent fluid, *Langmuir* **1999**, *15*, 1859–1871.

7 Liquid/Solid Interfaces

7.1 THERMODYNAMICS OF LIQUIDS IN CONTACT WITH SOLIDS

The first equation of capillarity describes the thermodynamics of a liquid in contact with a solid (Figure 7.1). The equilibrium condition for a sessile (i.e., sitting) drop on a solid substrate was developed in 1805 by Thomas Young[1] by equating the horizontal components of the forces acting at each unit length of the periphery of the three-phase contact line (Figure 7.2):

$$\sigma_{sv} = \sigma_{lv} \cos\theta + \sigma_{sl} \tag{7.1}$$

Although recent literature frequently refers to this relation as the Young–Dupré equation, no support for this designation is obtained from any adequate review of the classical work in this field, such as that of Bakker (1928). If any modification is justified, it would seem more suitable to refer to the relation as the Young–Gauss equation.[2]

The contact angle is defined as the angle between the tangent to the liquid surface and the tangent to the solid surface at any point on the peripheral line of three-phase contact, as measured through the liquid. The forces themselves do not emanate from a point on the line where the three phases make contact (as the diagram in two dimensions misleadingly shows) but are measured as acting along a unit length of that line. Young's equation supposes that the contact angle is solely defined by the surface tensions of the three surfaces in contact along a line and does not vary with the size of the drop or the profile of the substrate. If the liquid spreads without limit over the surface, there is no finite angle of contact; if the liquid does not spread, the angle can vary from 0° to 180°. An angle of 180° on a smooth substrate is physically impossible: it would refer to a liquid drop that contacts a substrate at only a single point and so completely lacks any attraction or adhesion to its substrate, which is never true. The highest contact angles exist between materials of small, but still some mutual, attraction, such as mercury on glass at 140°. The contact angle is readily measured by means of a goniometer if the substrate of the solid is sufficiently plane and sufficiently extensive, which is not always the case. Young's derivation (1805) in terms of forces is not rigorous, because it assumes that the surface energy of a solid substrate or of a solid–liquid interface is associated with a tension. Only at the liquid/vapor surface or the liquid/liquid interface is the presence of a tension beyond doubt. We may avoid introducing these assumptions by discussing the equilibrium of the contact angle in terms of surface and interfacial energies, as follows.

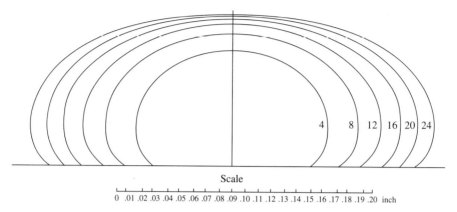

Figure 7.1 The forms of six sessile drops of mercury, weighing 4, 8, 12, 16, 2, 24 grains (1 grain equals 64.8 mg), on a clean glass substrate, as measured by Bashforth in 1882. The contact angle of 140° is the same for each drop. From Bashforth and Adams (1883).

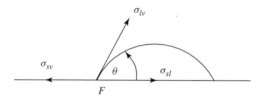

Figure 7.2 The liquid–solid–vapor contact line according to Young. The three components are acting on unit length of this line, located at F and normal to the page.

Each unit area of a liquid surface has a Helmholtz free energy numerically equal and dimensionally equivalent to its surface tension. A solid surface, because of its lack of mobility, does not develop a tension in the same way as a liquid surface; nevertheless, we may conceive that a part of the energy of a solid is proportional to its surface area and that this surface energy is thermodynamically equivalent in a solid surface or in a solid/liquid interface to the surface energy that is measured by the tension of a liquid surface. With this concept we consider the equilibrium of a liquid, in contact with its vapor, resting on a solid substrate and not subject to any other forces than those of the surface and interfacial tensions.

In Figure 7.3 let s represent the solid, l the liquid, and v the vapor, lv the liquid in contact with the vapor, sv the solid in contact with the vapor, sl the solid in contact with the liquid; ED is the surface of the solid, FG the tangent to the liquid surface at a point of contact on the periphery of the line of contact of the three phases, s, l, and v. The contact angle θ is the angle GFD in the diagram. The angle is characteristic of the nature of the three phases in contact, and if the system is in equilibrium, then its potential energy is a minimum in this

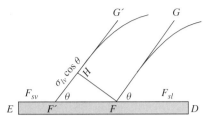

Figure 7.3 The analysis of the liquid-solid-vapor contact line according to Gauss.[3]

position. Now let the surface GF come into the position $G'F'$ parallel to GF: The angle of equilibrium is not affected by the displacement of GF to $G'F'$. This displacement of the edge of the liquid covers an differential area dA of the surface between v and s, replacing it with the same area of interface between l and s and also extends the surface area of the liquid between l and v by an amount proportional to $F'H$. Hence the total change of Helmholtz free energy is

$$dF = dA(-F_{sv} + F_{sl} + \sigma_{lv} \cos \theta) \tag{7.2}$$

At equilibrium, the free energy per unit area is a minimum; hence

$$F_{sv} - F_{sl} - \sigma_{lv} \cos \theta = 0 \tag{7.3}$$

or

$$F_{sv} = F_{sl} + \sigma_{lv} \cos \theta \tag{7.4}$$

This derivation is correct for flat, smooth surfaces; it is not a general proof. Consider, for example, the displacement of a liquid in a capillary tube: The liquid/vapor surface does not change in area. Hence $\sigma_{lv} \cos \theta dA$ in Eq. (7.2) is zero. In 1960 F. P. Buff[4] derived Eq. (7.3) by statistical thermodynamics in terms of surface free energies where

$$F_{sv} = F_{s0} - \pi_e \tag{7.5}$$

and π_e is the reduction of the free energy of the solid substrate by the adsorbed vapor of the liquid. This quantity is sometimes referred to as the equilibrium (two-dimensional) spreading pressure, by analogy with the equilibrium spreading pressure of an insoluble monolayer on a liquid surface, which is established when the spread monolayer is in equilibrium with its bulk phase. Strictly speaking, π_e is a free energy lowering per unit area, not a pressure.

For a liquid drop on a solid, the *final*, or equilibrium, spreading coefficient is

$$S = F_{sv} - F_{sl} - \sigma_{lv} \tag{7.6}$$

For a finite angle of contact, Eq. (7.4) may be introduced to give

$$S = \sigma_{lv}(\cos \theta - 1) \tag{7.7}$$

The work of adhesion of a liquid to a solid substrate is

$$W^{\text{adh}} = F_{sv} - F_{sl} + \sigma_{lv} \tag{7.8}$$

The corresponding equation for a finite contact angle is

$$W^{\text{adh}} = \sigma_{lv}(\cos\theta + 1) \tag{7.9}$$

and under the same conditions the free energy of emersion is

$$\Delta F^{\text{emersion}} = \sigma_{lv}\cos\theta \tag{7.10}$$

Equations (7.7), (7.9), and (7.10) are valuable because they have replaced two unmeasurable quantities, F_{sv} and F_{sl}, with two measurable quantities, σ_{lv} and θ. Equation (7.10) gives the Helmholtz free energy of emersion. Harkins and Jura[5] derive an expression for the enthalpy of emersion in terms of surface tension and contact angle, and their temperature variations, all of which are measurable.[6]

$$\Delta H^{\text{emersion}} = \sigma_{lv}\cos\theta - T\frac{d\sigma_{lv}\cos\theta}{dT} \tag{7.11}$$

The enthalpy of emersion can also be measured calorimetrically but depends on an accurate knowledge of the specific surface area. Given the free energy and the enthalpy changes on emersion, the entropy of emersion can be calculated. Entropy turns out to be a significant parameter in liquid/solid interactions.

7.2 DEGREES OF LIQUID/SOLID INTERACTION

The degree of interaction of a liquid with a solid substrate may be expressed by any one of a number of different but not independent quantities, each one representing the change in surface free energy per unit area for some process significant in the phenomenology of surfaces. These quantities are the spreading coefficient, the free energy of separation or work of adhesion, and the free energy of emersion, all of which have already been defined in Chapter 6.

The complete range of behavior is shown in Figure 7.4. The field is divided into three parts: (1) nonfinite contact angle, (2) contact angles less than 90°, and (3) contact angles larger than 90°. The behavior of the first region is associated with high-energy substrates and that of the third region with low-energy substrates. In the first region the liquid spreads spontaneously on the substrate, which can hold a uniform film of liquid that is stable at any thickness. In the second region the liquid "wets" but does not spread; it disproportionates to make an acute angle of contact between an adsorbed film in equilibrium with a thick lens or sessile drop. In the third region the liquid neither spreads nor wets the substrate: it forms a sessile drop, with an obtuse angle of contact, in equilibrium with the unwetted substrate. The chart of Figure 7.4 is a schematic report of the variation of the contact angle with the extent of liquid/solid interaction.

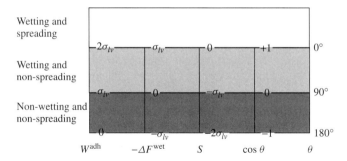

Figure 7.4 The degrees of liquid/solid interaction.

TABLE 7.1 Contact Angles, π_e and W^{adh} for Selected Systems

System	θ	π_e (mJ/m²)	W^{adh} (mJ/m²)
Water on hydrated silica	0	316	462
Water on gold	0		145.6
Water on silica	47	0	122.4
Water on graphite	85.7	15	93.3
Water on polyethylene	105	0	54.0
Water on paraffin wax	110	0	47.9
Water on Teflon®	115	0	42.0

High-energy solid surfaces interact strongly with liquids, which either spread spontaneously or give low contact angles; lower contact angles thus indicating greater surface energies of solid substrates. Table 7.1 lists the contact angles of water on various solids. Water is seen to adhere more strongly (larger values of the work of adhesion) to inorganic surfaces and to have very strong adhesion to hydrated silica (where the surface is not SiO_2 but SiOH). A bibliography of contact angle literature is available.[7]

The balance between the forces of cohesion and adhesion can be experienced when paraffin wax is splashed with water; the surface of the wax refuses to remain wet. Phenomena of this sort are complicated by kinematic effects. One problem in inkjet printing is that the liquid sometimes flows over the surface of the orifice metal. Analogous bad behavior in a teapot is for the tea to follow the underside of the spout and stain the tablecloth. Even when the spout is coated with paraffin wax the effect persists without any noticeable change. The problem, however, is not one of wetting and adhesion but one of fluid vortices in the liquid flow that press the liquid against the wall and allow it to turn corners. Some simple experiments of extreme interest are described by Reiner.[8]

7.3 FLOATING PARTICLES

Many finely divided minerals float on water, even though they may be denser than water. Gold dust, finely divided tungsten, molybdenum, silicon, sulfur, talc, arsenic trioxide, coal, graphite, galena, molybdenite, zinc blende, pyrites, and many other substances have this property. Even massive bodies, such as an oil-coated needle, will float on water. An uncut diamond weighing 0.05 g (specific gravity = 3.5) floated readily on water and remained floating for hours. The requirement for the effect is that the vertical component of the surface tension be greater than its buoyancy (see Figure 9.7). Unless the particles are very small, the contact angle must be greater than 90°. Reducing the surface tension of the water reduces the contact angle, the particle is wetted and sinks.

The foregoing principles have been used in various ways, sometimes without being understood. Dr. Matthew Hay's test for jaundice is to dust flowers of sulfur over the urine of a patient; on normal urine, sulfur floats, while the presence of bile salts causes the sulfur to sink. Again, in the East Indies the natives engaged in gold washing, who are in the habit of chewing the betel nut, spit into the washing water to prevent finely divided gold from floating away.[9] The natural protective oils on the plumage of birds, by ensuring a high contact angle, prevent air from being displaced by water, thus insulating them against the cold and permitting, for example, ducks and sea gulls to float on water. A method used to destroy a plague of starlings was to spray the flock with a solution of a powerful wetting agent; the next rain shower exposed them fatally to the cold. Water flies can walk on water, buoyed up by the high contact angle; mosquito larvae suspend themselves under the surface of stagnant water by means of a hydrophobic pseudopod penetrating the surface.

An industrial process that makes use of these principles is the separation of finely divided metal-bearing ores from the gangue by flotation. Froth flotation of minerals is possible if they can be preferentially wetted by gas rather than water. Only a small fraction of minerals is naturally hydrophobic. Hydrophobia has to be imparted to most minerals in order to float them. This is done by adding a surfactant to the aqueous suspension. This surfactant, called a *collector*, is an amphiphilic molecule which, because of primarily chemical interaction, is adsorbed on the particles with its nonpolar moiety oriented toward the bulk solution, thereby imparting hydrophobia to the particle. Collectors that are commonly used include short-chain alkyl xanthates, $ROCS_2^- M^+$, for metal sulfides; long-chain fatty acids or their alkali soaps for phosphates, hematite, and sulfates; and long-chain amines for quartz, potash, feldspars, mica, and soon. When air is introduced into a suspension of these hydrophobicized particles along with a suitable foaming agent, the particles are entrapped in the froth and so separated from the gangue, that is, the hydrophilic residue.[10]

7.4 INITIAL AND FINAL SPREADING COEFFICIENTS

High-energy substrates, unlike low-energy substrates, must be carefully protected from inadvertent contamination on contact with the atmosphere, which, especially in a laboratory, is likely to be polluted with organic vapors. The surface free energy of a substrate is reduced by adsorption; indeed, adsorption would not occur unless this were so. Some striking examples of the spontaneous reduction of surface free energy on exposure to the atmosphere were demonstrated by Quincke and by Devaux. A drop of water flashes across the surface of freshly cleaved mica, but on a mica surface exposed to the atmosphere for 20 minutes, a drop of water remains sessile. Devaux mounted a rose petal above the surface of freshly distilled mercury on which talc had been dusted to act as an indicator. Volatile odor from the petal was adsorbed by the mercury surface, as indicated by the withdrawal of the talc, leaving a bare spot. The surface energy of carefully purified water is so great that a freshly exposed surface is instantly contaminated. The surface of glass is another high-energy substrate readily contaminated on exposure to ordinary air.

For dry benzene on a clean water (or glass) surface, the initial spreading coefficient is[11]

$$S_{(b\,\text{on}\,a)} = S_{b/a} = \sigma_a - \sigma_b - \sigma_{a'b'} \tag{7.12}$$

$$= 72.8 - 28.9 - 35.0 = 8.9 \text{ mJ/m}^2 \tag{7.13}$$

That is, if the area of the spread benzene increases by one square meter and that of the clean surface of the water decreases by the same amount, there is a decrease in free energy of 8.9 mJ, which is considerable. This indicates that benzene should spread readily. However, since the benzene soon becomes saturated with water and the water ultimately becomes saturated with benzene, the surface tensions change to give the final spreading coefficient as follows:

$$S'_{(b'\,\text{on}\,a')} = S'_{b'/a'} = \sigma_{a'} - \sigma_{b'} - \sigma_{a'b'} \tag{7.14}$$

$$= 62.2 - 28.8 - 35.0 = -1.6 \text{ mJ/m}^2 \tag{7.15}$$

where the prime terms denote mutually saturated solutions. The interfacial phase is held to be mutually saturated immediately on contact.

The negative value of the final spreading coefficient indicates that benzene will not spread over the surface of water if the liquids are mutually saturated. In every case mutual saturation reduces the surface tensions, so that the final spreading coefficients are always less than initial spreading coefficients. When a drop of benzene is placed on a water surface, it spreads rapidly over the whole available area, but in a few seconds the film retracts to a lens, with a finite contact angle. This behavior conforms to the calculations of initial and final spreading coefficients given above.

Presumably had the water surface been in contact with saturated vapor of benzene its surface would have adsorbed in time sufficient benzene to reduce its surface tension to the 62.2 mJ/m^2 and the addition of a drop of benzene would not have resulted in spreading. Effects such as this, where either the presence of a saturated vapor or the contact of a liquid on the substrate, changes its surface energy by adsorption are called autophobic dewetting.

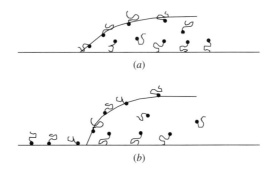

Figure 7.5 Retraction of a sessile drop caused by autophobic dewetting.

On a solid, autophobic dewetting can occur if the liquid drop contains a surface-active solute. Figure 7.5 depicts autophobic dewetting on a solid substrate in which the final contact angle is larger than the initial contact angle.

Autophobic dewetting also explains why certain binary mixtures dewet a high-energy solid, even though both pure liquid constituents wet the same solid. For example, both water and ethanol have zero contact angles on clean glass, yet a five percent solution of ethanol in water has a finite contact angle of about 40°. In Young's equation the contact angle depends on three surface energies all of which are changing with the concentration of the binary mixture.

$$\cos\theta = \frac{F_{sv} - F_{sl}}{\sigma_{lv}} \qquad (7.4)$$

Although at low concentrations of alcohol the denominator, σ_{lv}, is declining rapidly with concentration (See Figure 7.6) the solid/vapor and solid/liquid interfaces are approaching each other even faster.

Some theorists maintain that the final spreading coefficient at true equilibrium is never positive,[13] based on an argument by Gibbs[14] to the effect that the surface tension of a solution of b in a, $\sigma_{a(b)}$, can be no greater than the sum of the tensions of the surface phase, namely, that against air, $\sigma_{b(a)}$, and that against the solution, $\sigma_{b(a)/a(b)}$. The surface phase is considered by Gibbs to be a solution of a in b. It may be very thin, but cannot be as thin as a monolayer, as surface tension is a macroscopic property. A layer thick enough to have different tensions against its two adjoining phases is called a duplex film. Expressed as an equation, Gibbs' argument is

$$\sigma_{b(a)} + \sigma_{b(a)/a(b)} \geq \sigma_{a(b)} \qquad (7.16)$$

An equality in Eq. (7.16), which Gibbs anticipated would describe the usual interface between a and b, is now known as Antonoff's rule. Neither Gibbs's anticipation nor Antonoff's rule is well established in practice. The inequality itself states that the spreading coefficient is never positive. Experiments, however, consistently indicate small positive values of the spreading coefficient for some

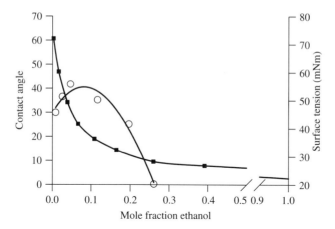

Figure 7.6 Variations of both the liquid/vapor surface tension ■ and the contact angle ○ with respect to the bulk composition of ethanol in water.[12]

systems[15] for long periods of time. The source of the contradiction may be the requirement for true equilibrium at the surface, for this may take an indefinite time. A supreme example of the persistence of a positive spreading coefficient is the spreading pressure measured on the film balance, which if enough time (eons) were allowed would always equal zero, by dissolution of the monolayer and its subsequent adsorption on the other side of the float.

The following analysis of spreading of a liquid on a solid substrate leads to the same conclusion as that of Gibbs for the fluid–fluid interface. Let the free energy of the solid substrate in a vacuum be F_{so} and in equilibrium with a vapor be F_{sv}. When the substrate is in equilibrium with the saturated vapor, the intercalation of a liquid film of macroscopic thickness (energy $F_{sl} + \sigma_{lv}$) between the solid surface and the vapor would not decrease the total surface free energy of the system since the liquid is in equilibrium with its own vapor. Therefore,[16]

$$F_{sv} = F_{sl} + \sigma_{lv} \qquad (7.17)$$

since the interfaces represented on both sides of the equation are in equilibrium with the same vapor phase, and therefore are in equilibrium with each other. This equality substituted in Young's equation,

$$F_{sv} = F_{sl} + \sigma_{lv} \cos\theta \qquad (7.18)$$

leads to $\cos\theta = 1$ or $\theta = 0°$ and a spreading coefficient of zero. Positive spreading coefficients are found experimentally because a substrate in contact with saturated vapor is in a metastable state prior to nucleation and condensation, and so has a higher energy than the same substrate covered with condensate. The metastable condition, in the absence of supersaturation, can persist indefinitely; therefore the spreading coefficient may indeed be positive to the investigator while its theoretical value of zero awaits the Greek kalends i.e., for ever. When complete spreading on a macroscopic scale is observed, the system may not be at equilibrium. If at equilibrium, $S = 0$; if not at equilibrium, $S > 0$.

The same reasoning that concludes that at equilibrium the spreading coefficient cannot be larger than zero also leads to the conclusion that at equilibrium the work of adhesion cannot be more than the work of cohesion, and that at equilibrium the free energy of wetting cannot be more than the surface energy of the wetting liquid.

These arguments assume that the liquid occupying the substrate retains the properties that it has when not in contact with the substrate. Should any interaction take place with the substrate, such as an acid–base interaction, the arguments no longer hold. The argument no longer holds because a strong interaction creates a new thermodynamic component in the system. Positive spreading is then likely to occur, even at equilibrium. A work of adhesion greater than the work of cohesion can be obtained under the same conditions, namely, an acid–base interaction between the substrate and the occupying liquid. In formulating composites, the strongest bonding between fiber and matrix is obtained by arranging for an acid–base interaction, that is, an electron transfer, between them.

For a finite contact angle, the spreading coefficient is negative. The contact angle increases and the spreading coefficient becomes more negative on approaching equilibrium. This is autophobic wetting.

If the substrate is an insoluble solid, the difference between the initial and final spreading coefficients resides entirely in the reduction of the free energy of the solid by the adsorbed vapor of the liquid, as neither the surface tension of the liquid nor the energy at the solid/liquid interface is affected by prolonged contact; hence,

$$\pi_e = F_{s0} - F_{sv} = S_{b/s} - S'_{b/s'} \qquad (7.19)$$

With some systems the energy difference between F_{s0} and F_{sv}, which is π_e of Eq. (7.19), can be quite large. With water on metallic oxides, $\pi_e \sim 300$ mJ/m^2; with organic liquids on oxides, $\pi_e \sim 60$ mJ/m^2; on the other hand some values of π_e are quite small, as shown in Table 7.1.

For substrates of low surface energy, on which vapors barely adsorb, π_e approaches a negligible quantity. Therefore such systems show hardly any difference between initial and final spreading coefficients.

Fowkes, McCarthy and Mostafa[17] introduced a laboratory technique for the direct measurement of π_e of a vapor on a low-energy substrate, using advancing contact angles of one liquid to measure the π_e generated by the vapor of a second liquid. Figure 7.7 represents the effect of cyclohexane vapor on the contact angle of water on polyethylene. In the presence of water vapor alone,

$$\sigma_{H_2O} \cos \theta_1 = F_{sv(1)} - F_{sl} \qquad (7.20)$$

When cyclohexane is added to the vapor phase, both the contact angle of the water and the free energy of the unwetted substrate are affected, giving

$$\sigma_{H_2O} \cos \theta_2 = F_{sv(2)} - F_{sl} \qquad (7.21)$$

DETERMINATION OF SPREADING PRESSURES BY ADSORPTION TECHNIQUES

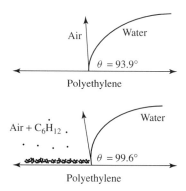

Figure 7.7 Effect of cyclohexane vapor on the contact angle of water on polypropylene.[17]

therefore,

$$F_{sv(1)} - F_{sv(2)} = \pi_e = \sigma_{H_2O}(\cos\theta_1 - \cos\theta_2) \tag{7.22}$$

assuming that the π_e due to water vapor on a low-energy substrate is negligible, that is, that $F_{sv(1)} \cong F_{s0}$. The advancing contact angle of a drop of water on polyethylene was 93.9° in air and 99.6° in air saturated with pure cyclohexane, as illustrated in Figure 7.7. The value of π_e is then 7.19 mJ/m².

7.5 DETERMINATION OF SPREADING PRESSURES BY ADSORPTION TECHNIQUES

The methods reported above to determine the equilibrium spreading pressure π_e are limited to planar substrates of large area, such as single crystal faces or polymer substrates. A method suitable for fine solid particles is provided by the technique of gas adsorption. The phenomenon of gas adsorption is owing to the spontaneous reduction of the surface free energy of the substrate; thus, for example, since water vapor is not spontaneously adsorbed by paraffin wax at room temperature, π_e is zero for this case.

The equilibrium between a gas phase and gas adsorbed on a substrate at constant temperature is expressed by equating the chemical potentials of the adsorbate in each phase. In the vapor phase

$$dG = V_g dp \tag{7.23}$$

where V_g is the molar volume of the vapor. In the adsorbed phase

$$dG = V_{ads} dp - Ad\sigma \tag{7.24}$$

where V_{ads} is the molar volume of the adsorbed gas; therefore,

$$(V_g - V_{ads})dp = -Ad\sigma = +Ad\pi_e \tag{7.25}$$

Neglecting V_{ads} by comparison with V_g and assuming the vapor is ideal gives

$$Ad\pi_e = (RT/p)dp = RTd\ln p \tag{7.26}$$

hence

$$\pi_e = RT \int_0^{p_0} \Gamma d\ln p \tag{7.27}$$

where p_0 is the saturated vapor pressure and $\Gamma = 1/A$ (moles/unit area). Equation (7.27) is the Gibbs adsorption equation applied to gas adsorption. To obtain π_e, the adsorption isotherm is plotted as the amount adsorbed, Γ, versus the natural logarithm of the equilibrium pressure: The area under the isotherm is then measured from $p = 0$ to $p = p_0$.

The major difficulty with this method is the enormous effect introduced by relatively small concentrations of high-energy sites or patches on the surface. Water is adsorbed by hydrophilic sites but not by lipophilic sites, so the measured adsorption isotherm of water on paraffin wax, for example, refers only to the hydrophilic subarea. The heterogeneity of the substrate may also be the result of large differences in the nature of the interaction of a nonpolar adsorbate with different sites on the surface. The nonspreading of drops of water on the other hand is barely affected by occasional high-energy sites. The vapor adsorption method requires great care and is time consuming. It has been used to determine π_e for adsorbed films of nitrogen, argon, and alkanes on graphite, metal oxides, and polypropylene powders[18] (See Table 8.8). For example, π_e for water on silica is approximately 450 mJ/m^2 and π_e for water on polyethylene is approximately zero.

7.6 DYNAMICS OF SPREADING

When the initial spreading coefficient is calculated to be positive, and when spreading on a macroscopic scale is observed, the liquid is expected to spread over the whole available area of substrate, and if that is sufficiently extended, to end as a monomolecular layer. Monomolecular layers of lauryl alcohol on water are so obtained. Spreading does not always occur in this way. Hardy observed that a spreading droplet is preceded and surrounded by a sensible though invisible film of liquid, which shows up ahead of the nominal contact line.[19] Certain observations of the spreading of a nonpolar liquid on steel reveal a precursor film (also called a primary film) visible in ellipsometry at the late stages of spreading, with a thickness of only a few tens of nanometers.[20] The precursor film has also been observed by measurements of electrical resistance.[21]

De Gennes has proposed a model for the spreading of a liquid film when the spreading coefficient is positive.[22] The derivation depends on the idea that the spreading liquid must flow as a thin, precursor film,

$$v \sim \frac{\sigma \theta_d^3}{\eta} \tag{7.28}$$

where v is the velocity of the spreading film front, θ_d is the dynamic contact angle, and η is the viscosity. Equation (7.28) does not depend on the spreading coefficient, which is surprising. Also surprising is that the higher the dynamic contact angle, the faster the spreading. This relation is sometimes referred to as the Tanner equation.

If volatile impurities of lower surface tension than the spreading liquid are present, they are preferentially lost by evaporation from the precursor film, the surface tension of which then increases. The higher surface tension of the precursor film draws liquid from the rest of the drop and so promotes further spreading (a Marangoni effect). In the lubrication of ball bearings, oil is transported by spreading from the grease supply to the raceway; the spreading is favored by the molecular heterogeneity of ordinary petroleum oils, whose more volatile components have lower surface tensions than the mixture. The tendency to spread can be counteracted by an additive. Undistilled squalane was made nonspreading for several days by adding 5% isopropylbiphenyl, which has a surface tension 7 mN/m higher and a boiling point 50°C lower than squalane. The same non-spreading effect can be produced by adding a liquid of less volatility than the squalane but a lower surface tension, for example, polydimethylsiloxane. Evaporation at the edge of such a film lowers the surface tension relative to the bulk liquid mixture and causes the film to retract. The polydimethylsiloxane liquids by themselves are often troublesome because of excessive spreading, but adding small amounts of a more volatile methyl phenyl silicone with a higher surface tension inhibits spreading for several weeks.[20]

7.7 FINAL SPREADING

De Gennes pointed out that the molecules in the precursor film have higher potential energies because they are separated farther than those in the bulk liquid.[16] Continued spreading is inhibited as it leads to even greater potential energies. The spreading comes to an end therefore before the drop has thinned to monolayer dimensions. What may appear to be a completely spread monolayer may actually be a film a few score nanometers thick. De Gennes estimates the final thickness of the liquid film by[23]

$$\delta \approx a \left(\frac{3\sigma_{lv}}{2S} \right)^{1/2} \quad (7.29)$$

where δ is the thickness of the film and a is the molecular dimension. When the liquid surface tension and the spreading coefficient are of the same order, the final film thickness is about a monolayer, as expected. However, when the spreading coefficient is small, the final film thickness can be quite large. If the final film thickness can be measured, say by ellipsometry, Eq. (7.29) can be used to calculate the spreading coefficient.

7.8 DETERGENCY

A cardinal principle is that a low-energy substrate cannot be covered by a higher energy fluid, as that would require potential energy to increase spontaneously. A familiar example is in soldering. The substrate is first cleaned with a flux, which is essentially hydrochloric acid, usually in the form of a paste containing zinc chloride. Once the substrate is clean it has a higher surface energy than the melted solder which then spreads on the surfaces to be joined making a strong adhesive bond. Much the same process occurs in detergency. Figure 7.8 shows a series of photographs of a clean wool fiber that had been dipped in oleic acid and then submerged in a strong detergent solution. The gradual displacement of the oil by the aqueous solution shows as a "rolling up" of the oil over a period of time. Finally the oil droplets have so little adhesion to the fiber that a little shake displaces and emulsifies them.

A technique using a quartz-crystal microbalance has been used to monitor the process of removing solid organic soils by detergents. A gold-covered quartz crystal, coated with an organic soil is suspended in a aqueous solution of a detergent, as in Figure 7.9, and the rate of removal of the soil is monitored by the change of resonance frequency of the crystal.[24,25]

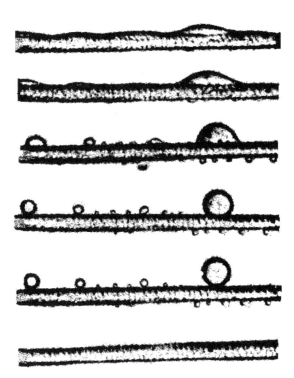

Figure 7.8 Detergent action in removing oil from a wool fiber.

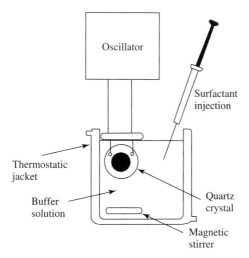

Figure 7.9 Schematic diagram of the quartz-crystal microbalance used to monitor frequency changes caused by surfactant action.[24,25]

Consider the washing of greasy chinaware. We shall assume reasonable values for the various surfaces and interfaces. Take $\sigma_{grease} = 50$ mJ/m² for the surface tension grease, $\sigma_{water} = 75$ mJ/m² for the water surface and $\sigma_{wg} = 30$ mJ/m² for the water/grease interface. Young's equation

$$50 = 75\cos\theta + 30 \tag{7.30}$$

gives $\theta = 75°$ for the water/grease contact angle; see Figure 7.10(a). That is, cold water does not spread on a greasy plate.

Next consider the effect of soapy water. Let its surface tension be 35 mJ/m² and its interfacial tension against grease be 10 mJ/m², and the interfacial tension between chinaware and grease be 25 mJ/m². Under these conditions [see Figure 7.10(b)], soapy water spreads across the grease, replacing the 50 mJ/m² grease/air interface with 35 mJ/m² soap water/air interface plus 10 mJ/m² soapy water/grease interface. The total of the surface and interfacial energies adds ups to 35+10+25, a total of 70 compared to 75 mJ/m² as in (a). But in time an even more favorable situation can be obtained. Let the soapy water/chinaware interface be 12 mJ/m². Then the soapy water will work its way onto the chinaware surface, rolling up the grease as it did on the wool fiber of Figure 7.8. The total surface energies per square meter of the surface and interfacial energies is now 47 mJ/m² instead of 70 mJ/m².

7.9 THIN-FILM COATINGS

The number of technologies dependent on thin-film coatings is almost uncountable: paints, protective coatings, publishing, lithographic printing, photographic

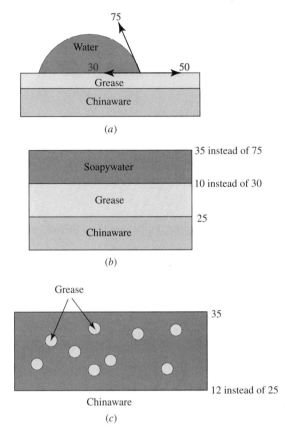

Figure 7.10 Application to detergency. (*a*) A water drop on a greasy surface. (*b*) Soapy water on a greasy surface. (*c*) Replacement of grease by soapy water.

film, magnetic tapes, computer hard drives, photoresists, optical discs, adhesives, decorative tapes, labels, electrophotography, and so on.[26,27] The simplest method to lay down a thin film is by withdrawing a solid substrate from a liquid bath or by allowing a quantity of liquid to drain down a solid surface. This is commonly called "dip" coating. In spin coating the substrate is rotated and the liquid is applied at the center of rotation; this is suitable for disks, plates, and bowls. Spray coating can be used when line-of-sight can be achieved; spray coating is used for thick coatings and falls short of dip and spin coating in uniformity. Thin films with remarkable uniformity (20 nm with thickness variations of less than 10%) can be achieved with spin coating over as much as 100 cm^2. All these techniques are referred to as free-meniscus coatings.[28] The presence of a "free" air/liquid interface means that surface forces will play a role in the formation of the thin film.

Levich modeled the hydrodynamics of dip coating to obtain an expression for the thickness of the coated film, h_0, as a function of the velocity of withdrawal of the substrate from the bulk liquid, v_0, the viscosity of the liquid, η, and the surface tension of the liquid, σ_{lv},[29]

$$h_0 = \left(\frac{\eta v_0}{\rho g}\right)^{1/2} f\left(\frac{\eta v_0}{\sigma_{lv}}\right) \qquad (7.31)$$

where the function $f(\eta v_0/\sigma_{lv})$ takes the form

$$\begin{aligned} f\left(\frac{\eta v_0}{\sigma_{lv}}\right) &\approx 0.93 \left(\frac{\eta v_0}{\sigma_{lv}}\right)^{1/6} \quad \text{for } \frac{\eta v_0}{\sigma_{lv}} \ll 1 \\ f\left(\frac{\eta v_0}{\sigma_{lv}}\right) &\approx 1 \quad \text{for } \frac{\eta v_0}{\sigma_{lv}} \gg 1 \end{aligned} \qquad (7.32)$$

The term, $\eta v_0/\sigma_{lv}$, is called the capillary number (unitless), ρ is the density of the liquid, and g the acceleration due to gravity. Equation (7.31) may be used to estimate the thickness of the liquid film on a solid of any shape, provided its radius of curvature is large compared to the capillary constant, $(\sigma_{lv}/\rho g)^{1/2}$.

Equation (7.31) also implies that the thickness of the layer is a weak function of the surface tension. However, most thin-film coatings are from volatile solvents so that the ultimate film is a dried, solid coating. The "drying line" between the wet and the dry film is a region of high curvature (from thick, wet film to thin, dry film) and, potentially, of large surface tension gradients as volatile components evaporate. The coupled processes of fluid flow, vapor flow, capillary flow, and Marangoni flow make a challenging computational problem which is only recently yielding to accurate computer calculation.[28]

The dip coating of a low-energy surface covers the surface with a liquid, but the surface is not "wetted" in the thermodynamic sense. When the film becomes thin, minute capillary waves create minute holes, especially in areas of lower energy. Once a hole is formed the negative spreading coefficient determines the contact angle. An agreement between theoretical predictions and experimental results on instabilities and various stages of dewetting of thin (less than 60 nm) films of polystyrene on low-energy substrates has been presented.[30]

7.10 ELECTRICAL CHARGES

Electrical charges at the solid/liquid interface are generally obtained by the preferential adsorption of ions from solution or by desorption of ions from the solid surface. The influence of these charges on the stability of emulsions and suspensions are the subject of many studies and theories, especially DLVO (See Section 20.3) theory for the electrostatic stabilization of suspensions.

Also, a solid surface in contact with water may be charged by application of an electric potential. The water drop may be sitting directly on an electrode or sitting on a thin film of an insulator coated on the electrode. When the solid/water interface is charged, the interfacial tension is reduced because of the greater

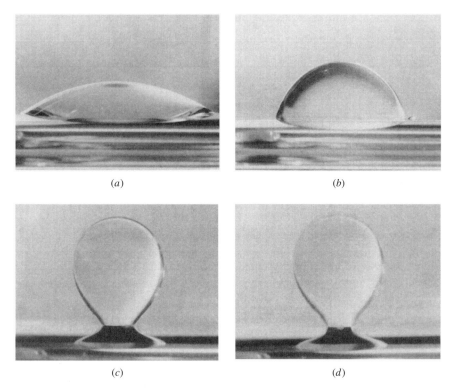

Figure 7.11 Photographs of an n-hexadecane droplet (70 μL) at a mercury–aqueous electrolyte interface taken at constant potentials of (*a*) −550 mV, (*b*) −1300 mV, (*c*) −1400 mV and (*d*) −1450 mV.[32,33]

interaction with water. The effect is much smaller with an oil drop. The lowering of the interfacial tension decreases the contact angle. A mathematical description of this effect is available.[31]

Figure 7.11 shows the effect of an electric potential on the contact angle of a drop of n-hexadecane on mercury covered by a aqueous electrolyte.[32,33] The interfacial tension of the mercury/electrolyte interface is progressively reduced by increasing the electric potential and the contact angle of the oil drop on the mercury shows a corresponding increase. In all cases displayed, adhesion counteracts buoyancy. The results on potential-controlled wetting equilibria between the mercury electrode and hydrocarbon droplets offer a rationale for electrochemical studies of disperse systems and adhesion-based electrochemical sensors for particle analysis.

REFERENCES

[1] Young, T. An essay on the cohesion of fluids, *Phil. Trans. Roy. Soc. London* **1805**, 65–87.

[2] Melrose, J.C. Evidence for solid-fluid interfacial tensions from contact angles, *Adv. Chem. Ser.* **1964**, *43*, 158–179.

[3] Gauss, C.F. *Allgemeine Grundlagen einer Theorie der Gestalt von Flüssigkeiten im Zustand der Gleichgewichts.* Translated [from the Latin original] by Rudolf H. Weber. Edited by H. Weber. Ostwald's Klassiker zur Geschichte der exakten Wissenschaften no. 135. Leipzig: Wilhelm Engelmann, 1903.

[4] Buff, F.P. The theory of capillarity, in *Encyclopedia of physics*; Flügge, S., Ed.; Springer: Berlin; 1960; pp 281–304.

[5] Harkins, W.D.; Jura, G. Surfaces of solids. XII. An absolute method for the determination of the area of a finely-divided crystalline solid, *J. Amer. Chem. Soc.* **1944**, *66*, 1362–1366.

[6] Vbranac, M.D.; Berg, J.C. The use of wetting measurements in the assessment of acid–base interactions at solid/liquid interfaces, *J. Adhesion Sci. Technol.* **1990**, *4*, 255–266.

[7] A bibliography of contact-angle use in surface science, *Technical Bulletin TB-100*; Ramé-Hart, Inc.: 43 Bloomfield Ave., Mountain Lakes, NJ 07046. www.ramehart.com

[8] Reiner, M. The Teapot effect ... a problem, *Phys. Today* September **1956**, 16–20.

[9] Edser, E. The concentration of minerals by flotation, in *Fourth report on colloid chemistry and its general and industrial applications*; HMSO: London; 1922, 263–297.

[10] Somasundaran, P. Interfacial chemistry of particulate flotation, *AIChe. Symp. Ser.* **1975**, *71*, 1–15.

[11] Harkins, W.D. *The physical chemistry of surface films*; Reinhold: New York; 1952; pp 97–99.

[12] Tronel-Peyroz, E.; Douillard, J.M.; Privat, M.; Bennis, R. Local composition fluctuations and wetting phenomena, *Langmuir* 1990, 6(3), 539–542.

[13] Rowlinson, J.S.; Widom, B. *Molecular theory of capillarity*; Clarendon Press: Oxford; 1982; p 216.

[14] Gibbs, J.W. On the equilibrium of heterogeneous substances, *Trans. Conn. Acad. 1878*, reprinted in *The scientific papers of J. Willard Gibbs, Vol. 1 Thermodynamics*; Dover: New York; 1961; pp 258–264.

[15] Ross, S.; Patterson, R.E. Innate inhibition of foaming and related capillary effects in partially miscible ternary systems, *J. Phys. Chem.* **1979**, *83*, 2226–2232.

[16] de Gennes, P.G. Wetting: Statics and dynamics, *Rev. Mod. Phys.* **1985**, *57*, 827–863.

[17] Fowkes, F.M.; McCarthy, D.C.; Mostafa, M.A. Contact angles and the equilibrium spreading pressures of liquids on hydrophobic solids, *J. Colloid Interface Sci.* **1980**, *78*, 200–206.

[18] Fowkes, F.M. Attractive forces at interfaces, *Ind. Eng. Chem.* **1964**, *56*(12), 40–52; also *Chemistry and Physics at Interfaces*; Ross, S., Ed.; ACS:Washington, DC; 1965; pp 1–12.

[19] Hardy, W.B. The spreading of fluids on glass, *Phil. Mag.* **1919** (6), *38*, 49–55.

[20] Bascom, W.D.; Cottington, R.L.; Singleterry, C.R. Dynamic surface phenomena in the spontaneous spreading of oils on solids, *Adv. Chem. Ser.* **1964**, *43*, 355–380.

[21] Ghiradella, H.; Radigan, W.; Frisch, H.L. Electrical resistivity changes in spreading liquid films, *J. Colloid Interface Sci.* **1975**, *51*, 522–526.

[22] de Gennes, P.G. *Soft interfaces, the 1994 Dirac Memorial Lecture*; Cambridge University Press: Cambridge; 1997; p 11.

[23] de Gennes, P.G. *Introduction to polymer dynamics*; Cambridge University Press: Cambridge; 1990; p 28.

[24] Caruso, F.; Serizawa, T.; Furlong, D.N.; Okahata, Y. Quartz crystal microbalance and surface plasmon resonance study of surfactant adsorption onto gold and chromium oxide surface, *Langmuir* **1995**, *11*, 1546–1552.

[25] Weerawardena, A.; Drummond, C.J.; Caruso, F.; McCormick, M. Real time monitoring of the detergency process by using a quartz crystal microbalance, *Langmuir* **1998**, *14*, 575–577.

[26] Cohen, E.D.; Gutoff, E.B., Eds. *Modern coating and drying technology*; VCH: New York; 1992.

[27] Gutoff, E.B.; Cohen, E.D. *Coating and drying defects: trouble shooting operating problems*; Wiley: New York; 1995.

[28] Schunk, P.R.; Hurd, A.J.; Brinker, C.J. Free-meniscus coating processes, in *Liquid film coating: scientific principles and their technological implications*; Kistler, S.F.; Schweizer, P.M., Eds.; Chapman and Hall: New York; 1997; Chapter 13.

[29] Levich, V.G. *Physicochemical hydrodynamics;* Prentice-Hall: NJ; 1962; pp 674–683.

[30] Sharma, A.; Reiter, G. Instability of thin polymer films on coated substrates: Rupture, dewetting, and drop formation, *J. Colloid Interface Sci.* **1996**, *178*, 383–399.

[31] Verheijen, H.J.J.; Prins, M.W.J. Reversible electrowetting and trapping of charge: Model and experiments, *Langmuir* **1999**, *15*, 6616–6620.

[32] Ivošević, N.; Žutić, V. Spreading and detachment of organic droplets at an electrified interface, *Langmuir* **1998**, *14*, 231–234.

[33] Ivošević, N.; Žutić, V.; Tomaić, J. Wetting equilibria of hydrocarbon droplets at an electrified interface, *Langmuir* **1999**, *15*, 7063–7068.

8 Theories of Surface and Interfacial Energies

The free energy of surfaces and interfaces is a consequence of intermolecular attractive and repulsive forces between molecules. Considerable progress has been made in understanding the relation between intermolecular forces and surface free energies, nevertheless, the most useful approaches are highly empirical.

The most significant intermolecular forces of attraction at interfaces are dispersion forces (sometimes called van der Waals forces) and electron donor–acceptor interactions. The intermolecular forces of repulsion are excluded volume and electrostatic repulsion. Dipole–dipole and dipole-induced dipole forces are minor by comparison. Dispersion forces are a consequence of the fluctuations of electron clouds in condensed matter and are shown to be significant for surface and interfacial energies as well as for the attractive energies between particles and droplets through liquids. Electron acceptor–donor reactions at interfaces are significant for surface and interfacial interactions but are not significant for the interactions between particles and droplets because these interactions are too short ranged.

8.1 FORCES OF ATTRACTION BETWEEN MOLECULES

Intermolecular forces are all electromagnetic in origin: (a) electrostatic, that is, Coulombic and induction forces; (b) electrodynamic, that is, dispersion forces; (c) electron donor–acceptor or proton acceptor–donor interactions, for example, hydrogen bonding; and (d) the repulsive overlapping of electron clouds, that is, excluded volume. A summary of the electrostatic and induction contributions to intermolecular forces is given in Section 26.4.

If either of the interacting molecules has a permanent electric moment, that is, a dipole moment, then the potential energy of interaction depends on their relative orientation. The average energy of interaction is found by integrating the potential as a function of orientation through all orientations multiplied by the Boltzmann probability of that orientation. The orientation-averaged forms are all temperature dependent, as at higher temperatures the rotations of the permanent electric moments reduce net intermolecular forces. The dipole–dipole contributions are often called Keesom energies. The dipole-induced dipole contributions are often called Debye energies.

Dispersion forces of attraction are a consequence of the spontaneous fluctuation of the electronic cloud in one material causing a corresponding fluctuation in neighboring materials, leading, on the average, to an attractive force. That this particular attractive force is the predominant one in determining the stability of disperse systems may be understood from the following brief summary of possible nonbonding intermolecular forces.

London, by using perturbation theory to solve the Schrödinger equation for two (hydrogen) atoms at large distances, showed that the separated atoms attract each other with an energy varying as the inverse sixth power of the distance r between them.[1] These forces are strictly quantum mechanical and are called "dispersion forces" because the equations for the attractive energy are expressed in terms of the same oscillator strengths as appear in the equations for the dispersion of light. When the fluctuations in the electric potential of a molecule are approximated as a simple harmonic oscillator and the distance between the molecules is large compared to molecular size, the dispersion energy between two such molecules is given by

$$U(r) = -\frac{3}{2}\Lambda_{ab}r^{-6} \tag{8.1}$$

where

$$\Lambda_{ab} = \left\{\frac{h\nu_a h\nu_b}{h\nu_a + h\nu_b}\right\}\alpha_a\alpha_b \tag{8.2}$$

where the ν's and the α's are the characteristic molecular frequencies and polarizabilities of the molecules. Λ is called the London constant.

The $h\nu$ terms in Eq. (8.2) are often approximated by setting them equal to the ionization potentials. Since the ionization potentials of most molecules are of the same order of magnitude, the London constant for the interaction of two different kinds of molecule, Eq. (8.2), can be approximated by the root mean square

$$\frac{h\nu_a h\nu_b}{h\nu_a + h\nu_b} \cong \left[\frac{h\nu_a}{2}\right]^{1/2}\left[\frac{h\nu_b}{2}\right]^{1/2} \tag{8.3}$$

Substituting the approximation (8.3) into Eq. (8.2) gives

$$\Lambda_{ab} \cong [\Lambda_{aa}\Lambda_{bb}]^{1/2} \tag{8.4}$$

as a good estimate. This same approximation is often used for the estimation of mixed Lennard–Jones parameters and for the calculation of second-virial coefficients of gases[2] and is called Berthelot's principle.[3,4]

The ν's and α's can also be estimated from the frequency dependence of the polarizability near the ionization frequency.[5] Alternatively, the principal frequencies can be estimated from the analysis of the refractive index in terms of a Clausius–Mossotti plot.[6] This latter method is particularly suitable for materials whose ionization potentials are not available.

Several other approximations to calculate the London coefficient Λ_{ab} have been proposed. It can be calculated in principle from the appropriate wave equations.[7,8,9]

From the analysis of observed flocculation rates of lyophobic sols, Verwey and Overbeek[10] found that when two particles are separated by distances large compared to the wavelength of the ionization potential, dispersion energies of attraction are considerably less than predicted. They postulated that the energy of the dispersion interaction is decreased at large distances because the time of propagation of electric field from one body to another and back is such that the fluctuating electric moments become out of phase. Casimir and Polder[11] analyzed the dispersion interactions, including a term for the finite speed of light, and showed that a correction factor is necessary, called retardation. The retardation correction is a monotonically decreasing function, equal to unity for small distances, and is proportional to the inverse of the separation r for long distances. The form of the retardation expression is complicated but an empirical representation is given by Overbeek.[12]

The strength of the dispersion-force attraction can be compared with the other types of molecular attractions, using the formulas and the constants of the electrical moments of some simple molecules and ions in Appendix 26.5. Potential energies are compared at the same intermolecular distance, 0.4 nm, and in units of kT (at 25°C) and are given in Table 8.1. The donor–acceptor interaction has been calculated by the Drago equation [Eq. (8.22)].

The results of these calculations show that the polar interactions, dipole–dipole (Keesom forces) and dipole-induced dipole (Debye forces), are small compared to the others. If the molecules have no net charge, then only the dispersion energy determines the long-range intermolecular attractive potential. Short-range forces are dominated by donor–acceptor interactions.[14]

8.2 THE COMPONENTAL THEORY OF INTERFACIAL TENSION

Free energies, such as surface tensions, interfacial tensions, and works of adhesion can be calculated from intermolecular forces by the methods of statistical mechanics. The state of this art is not yet so advanced as to be practicable for most interesting materials. Molecules at interfaces are often forced into preferred

TABLE 8.1 Representative Intermolecular Interactions[13]

Interaction	Species	Energy
Charge–charge	Na^+–Cl^-	$-140\ kT$
Charge–dipole	Na^+–H_2O	$-13.4\ kT$[a]
Dipole–dipole	Phenol–phenol	$-0.42\ kT$
Charge–induced dipole	Na^+–H_2O	$-1.62\ kT$
Dipole–induced dipole	Phenol–benzene	$-0.13\ kT$
Dispersion	Benzene–benzene	$-7.3\ kT$
Donor–acceptor	Phenol–benzene	$-4.3\ kT$

[a] The maximum interaction is used since the interaction energy is much greater than kT.

positions by specific interactions across the interface. This structure formation changes the entropy of the system and hence its free energy.

The empirical approximation now commonly used is to suppose that the free energy of the interface can be separated into components that are solely dependent on each of the possible intermolecular forces. That is, the free energy of a pure liquid surface can be broken into parts, each part of which depends only on dispersion forces, or only on dipole forces, or only on acid–base forces.

The second step of this approximation is to assume that the interactions across interfaces can be separated in a like manner. That is, the free energy of the interface can be broken into parts, each part of which depends only on dispersion forces, or only on dipole forces, or only on acid–base forces.

This analysis, exploited chiefly by Fowkes, is called the componential theory of capillarity. While its theoretical premise is weak, it has proved to be immensely useful for understanding interfacial phenomena.

8.2.a Dispersion Force Contributions to Interfacial Tension

Girifalco and Good[15] have indicated how interfacial tension is related to the surface tensions of the contacting liquids. Just as surface tension depends on the strength of the intermolecular forces, the interfacial tension also depends on intermolecular forces, but three different types of force come into play: homomolecular forces in each of the two liquids and heteromolecular forces between the two liquids. Instead of evaluating these forces in fundamental terms, a simpler method is to express them in terms of the surface tensions to which they give rise.

This is illustrated by Figure 8.1 in which the reduction of density at the interface is accounted for by a balance of four tendencies, two of which act to remove molecules and two of which act to restore them. Therefore the interfacial tension can be expressed in terms of surface tensions of the single and the paired tensions. The removing tendencies are expressed by the respective surface tensions of the components. The restoring tendencies are the heteromolecular interactions between the two components. The net reduction in density in the interface is both the cause and the result of the interfacial tension.

To estimate the interfacial tension from the two surface tensions, Girifalco and Good adopted a suggestion of Berthelot, who estimated the attraction between unlike gas molecules, α_{12}, as the root mean square of the attraction between like molecules (α_{11} and α_{22})[16,17].

$$\alpha_{12} = (\alpha_{11}\alpha_{22})^{1/2} \tag{8.5}$$

By analogy, Girifalco and Good[15] set up the corresponding ratio for the work of adhesion between two phases in terms of the work of cohesion of each phase, leading to

$$W_{12}^{\text{adh}} = \Phi(W_1^{\text{coh}} W_2^{\text{coh}})^{1/2} \tag{8.6}$$

THE COMPONENTAL THEORY OF INTERFACIAL TENSION

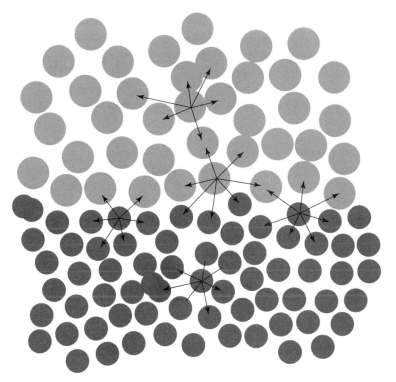

Figure 8.1 Molecular origin of interfacial tension.[18]

where
$$W_1^{coh} = 2\sigma_1, \quad W_2^{coh} = 2\sigma_2 \tag{8.7}$$

hence
$$W_{12}^{adh} = 2\Phi(\sigma_1\sigma_2)^{1/2} \tag{8.8}$$

Φ is a characteristic constant, which was introduced to confer generality and which would be evaluated for various types of systems.

The work of adhesion has been already defined in Chapter 6 by Eq. (6.30).

$$W^{adh} = \sigma_1 + \sigma_2 - \sigma_{12} \tag{8.9}$$

Therefore
$$\sigma_{12} = \sigma_1 + \sigma_2 - W^{adh} \tag{8.10}$$

Hence
$$\sigma_{12} = \sigma_1 + \sigma_2 - 2\Phi(\sigma_1\sigma_2)^{1/2} \tag{8.11}$$

Equation (8.11) shows how interfacial tension is related to the surface tensions of the contacting liquids for a simple model of homo- and heteromolecular interactions. It is referred to as the Girifalco–Good equation.

Fowkes modified this model by introducing the concept that the effects of different types of intermolecular attraction could be expressed separately in terms of their contribution to the observed surface tension.[18] Thus, for instance, in water, the intermolecular attractive forces are London dispersion and hydrogen bonding. The observed surface tension of water may be expressed in terms of the contributions made to it by each of these forces,

$$\sigma_w = \sigma_w^d + \sigma_w^H \tag{8.12}$$

where σ_w is the observed surface tension, and σ_w^d and σ_w^H are the two contributions to σ_w made by the dispersion forces and by hydrogen bonding, respectively. Hydrogen bonding is a special case of a more general type of interaction, namely, that between a Lewis base and a Lewis acid, giving σ^{ab} as an alternative designation for σ^H and also making it applicable to liquid pairs where the acid-base interaction is not one of hydrogen bonding. Liquid metals introduce another type of bonding, the metallic bond, whose contribution to the surface tension is designated σ^m. The surface tension of mercury would be written as

$$\sigma_{Hg} = \sigma_{Hg}^d + \sigma_{Hg}^m \tag{8.13}$$

Between molecules of saturated aliphatic hydrocarbons the only interaction is that of the London dispersion forces, therefore

$$\sigma_{hc} = \sigma_{hc}^d \tag{8.14}$$

London dispersion forces exist between any two molecules; consequently, the term "σ^d" enters into every expression for the observed tension. For some liquids, as we have seen, it is the only term.

When a saturated aliphatic hydrocarbon, such as octane, is superposed on water, the net density lowering of the interfacial region is the sum of that arising from the homomolecular water attraction and from the homomolecular hydrocarbon attraction, but is partially relieved by the heteromolecular attractions between water and hydrocarbon. The net difference of density is directly related to the tension that develops there, so that either the strength of the intermolecular forces or the density difference at the interface, which is the direct outcome of these forces, may be expressed in terms of surface tension. By using observed values of surface tension for this purpose, the calculation does not need to be taken all the way back to intermolecular forces, with a great gain in simplicity.

Heteromolecular forces are often evaluated as a root mean square of the homomolecular forces. Continuing to use values of surface tension in the place of homomolecular forces, and restricting the type of interaction between 1 and 2 to that of their dispersion forces only, gives

$$W_{12} = 2(\sigma_1^d \sigma_2^d)^{1/2} \tag{8.15}$$

THE COMPONENTAL THEORY OF INTERFACIAL TENSION

The factor 2 is introduced to take into account both 1-2 and 2-1 dispersion force interactions. Combining Eqs. (8.10) and (8.15) gives

$$\sigma_{12} = \sigma_1 + \sigma_2 - 2(\sigma_1^d \sigma_2^d)^{1/2} \qquad (8.16)$$

Equation (8.16), the Fowkes equation, can be applied with known values of surface and interfacial tension for a number of hydrocarbons with mercury.

$$\sigma_{hc/Hg} = \sigma_{hc} + \sigma_{Hg} - 2(\sigma_{hc}^d \sigma_{Hg}^d)^{1/2} \qquad (8.17)$$

For hydrocarbons, Eq. (8.14) applies, so there is only one unknown in Eq. (8.17), namely, σ_{Hg}^d, the dispersion-force component of the surface tension of mercury.

Table 8.2 shows the determination of σ_{Hg}^d by Eq. (8.17) based on data for surface and interfacial tensions of 10 different hydrocarbons with mercury at 20°C. The average value of σ_{Hg}^d is 200 ± 7 mN/m. Therefore, of the total measured surface tension of mercury at 20°C, 484 mN/m, about 284 mN/m is the result of a specific type of bonding between mercury atoms. The remainder of the measured surface tension is the contribution of dispersion forces, with which the dispersion forces of hydrocarbons are able to interact.

A similar use of Eq. (8.16), using the surface and interfacial tensions of hydrocarbons against water is shown Table 8.3 for the determination of σ_w^d.

$$\sigma_{hc/w} = \sigma_{hc} + \sigma_w - 2(\sigma_{hc}^d \sigma_w^d)^{1/2} \qquad (8.18)$$

The average value is 21.8 ± 0.7 mN/m.[19] Therefore, of the total measured surface tension of water at 20°C, 73 mN/m, about 51 mN/m is the result a specific type

TABLE 8.2 Determination of σ_{Hg}^d for Mercury (mN/m at 20°C)[18]

Hydrocarbon	σ_{hc}	$\sigma_{hc/Hg}$	σ_{Hg}^d
n-Hexane	18.4	378	210
n-Octane	21.8	375	199
n-Nonane	22.8	372	199
Benzene	28.85	363	194
Toluene	28.5	359	208
o-Xylene	30.1	359	200
m-Xylene	28.9	357	211
p-Xylene	28.4	361	203
n-Propyl-benzene	29.0	363	194
n-Butyl-benzene	29.2	363	193
		Average	200± 7

TABLE 8.3 Determination of σ_w^d for Water (mN/m at 20°C)[18]

Hydrocarbon	σ_{hc}	$\sigma_{hc/w}$	σ_w^d
n-Hexane	18.4	51.5	21.8
n-Heptane	20.4	50.2	22.6
n-Octane	21.8	50.8	22.0
n-Decane	23.9	51.2	21.6
n-Tetradecane	25.6	52.2	20.8
Cyclohexane	25.5	50.2	22.7
Decalin	29.9	51.4	22.0
White oil (25°)	28.9	51.3	21.3
		Average	21.8± 0.7

of bonding between water molecules. The remainder of the measured surface tension is the contribution of dispersion forces, with which the dispersion forces of hydrocarbons are able to interact.

These values of σ_w^d and σ_{Hg}^d, now determined, can be used to calculate a theoretical interfacial tension between water and mercury, assuming that the interaction between these two liquids is solely one of dispersion forces.

$$\sigma_{w/Hg} = \sigma_w + \sigma_{Hg} - 2(\sigma_w^d \sigma_{Hg}^d)^{1/2} \qquad (8.19)$$

Therefore,

$$\sigma_{w/Hg} = 72.8 + 484 - 2(21.8 \times 200)^{1/2} = 425 \text{ mN/m} \qquad (8.20)$$

The calculated value of 425 mN/m agrees with the observed values of 426–427 mN/m. This result supports the assumptions underlying the derivation. For water, it shows that the dispersion forces between the molecules are 30% of the total attractive forces, the remainder being hydrogen bonding. It also shows that the interaction between water and mercury is almost entirely dispersion force interaction, and therefore that the dipole-image forces are comparatively weak.

Equation (8.10) points out that the magnitude of an interfacial tension depends on the degree of heteromolecular interaction: If the term W^{adh} is small, the interfacial tension is large; if the term W^{adh} is large, the interfacial tension is small; if the term W^{adh} is greater than the sum of the two surface tensions, the calculated interfacial tension is negative; that is, no interface can exist and the two liquids are miscible. While surface tension is increased by the degree of homomolecular attraction, interfacial tension is diminished by the degree of heteromolecular attraction. Compare, for example, cyclohexane and benzene: Their surface tensions are about the same, but their interfacial tensions against water are very different. The former value is 50.2 mN/m and the latter value is 35 mN/m.

Clearly, on the basis of Eq. (8.10), cyclohexane has less heteromolecular interaction with water than has benzene. This conclusion may be explained by the presence of π orbitals in benzene, which have electron–donor interactions with water. This type of interaction is lacking between cyclohexane and water.

By the same token, Eq. (8.16) is inadequate to evaluate the interfacial tension between benzene and water because it lacks a term to take into account the additional heteromolecular attraction besides that of the dispersion forces. On the other hand, the interfacial tension between mercury and water, which might seem more complex, is well described by Eq. (8.16), as the capability of water to hydrogen bond is not reciprocated by the mercury and the capability of mercury to amalgamate with other metals evokes no response from water: Only dispersion forces are common to both and constitute their sole mode of heteromolecular attraction.

The underlying assumption of Fowkes' approach is that the components of the surface tension or of the work of adhesion are linearly related; that is, that the whole property can be subdivided into fractions, that each fraction can be treated separately, and that they can then be recombined to reconstitute the whole property. This would assume that each type of molecular interaction is without effect on any other type: that, for example, the simultaneous actions of dispersion forces and hydrogen bonding on any given surface site are mutually independent. Such a question cannot be decided on theoretical grounds alone; it is essentially a question of how matter behaves and can only be decided by an appeal to experimental data.

8.2.b Acid–Base Contributions to Interfacial Tension

Dispersion forces alone do not account for all the attraction between bodies across their interface. Different approaches to nondispersion forces have been proposed. The success of the treatments by Girifalco and Good[15] and by Fowkes[18] of the dispersion force interactions, using a geometric mean of the homomolecular interaction to obtain the heteromolecular, was at first merely imitated in form without reference to the nature of the molecular forces. For example, the work of adhesion between two liquids was taken by blind rote as

$$W_{12} = 2\left(\sigma_1^d \sigma_2^d\right)^{1/2} + 2\left(\sigma_1^p \sigma_2^p\right)^{1/2} \tag{8.21}$$

where the superscripts d and p refer to dispersion and polar forces, respectively. Wu[20] found better agreement by substituting the harmonic mean for the geometric mean, but still without theoretical justification.

Keesom and Debye forces, which can be approximated as the geometric mean of polar forces, are revealed by calculation to be small. The remaining polar contribution, namely, the acid–base interaction (including hydrogen bonding as a special case), is significant where it exists but should not be represented by a geometric mean. For instance, if the two immiscible liquids are both acidic or both basic, very small polar interactions take place, and the work of adhesion is given by the dispersion force interaction alone. If one material is a Lewis acid and the other a Lewis base (i.e., electron acceptor or proton donor and electron donor or proton acceptor, respectively), then the reaction between them across an interface has a profound effect on the interfacial tension. For example, the interfacial tension between a Lewis base and water is lowered appreciably by a decrease in the pH of the aqueous phase; and the interfacial tension between a Lewis acid and water is lowered appreciably by an increase in pH of the aqueous phase. Indeed, how the interfacial tension varies with the pH of the aqueous phase indicates whether the water-immiscible organic liquid is a Lewis acid or a Lewis base or neither.[21]

The major interactions of a solid surface with an adsorbate are of two kinds: the ubiquitous interactions of the London dispersion forces and the specific interactions. London dispersion forces are long range and, even though small, their sum is a significant portion of the total interaction per unit area across an interface. Most of the specific interactions are Lewis base–Lewis acid reactions (electron donor–acceptor interactions). This category includes chelation, coordination, hydrogen bonding, and the complexes between nucleophiles and electrophilic sites that are so important in catalysis.[22] The energy per mole of these interactions is large but, because of their short-range character, only those pairs in proximity contribute to the total interaction. Consequently, the magnitudes of the dispersion-force interactions and the donor–acceptor interactions are comparable. These two types of interaction have proved to be sufficient to account for the properties of interfaces.[23]

A number of empirical treatments of Lewis acid–Lewis base reactivity have been developed to predict equilibrium constants and reaction rates. These are based on empirical linear correlations between the logarithms of the equilibrium constants or of the rate constants of one series of reactions and those of another related series. An early and well-known example of a linear free-energy relation is the Hammett equation, originally developed to predict equilibrium constants for the reactions of substituted aromatic acids, but which turns out to be reasonably successful whether the substrates are attacked by electrophilic, nucleophilic, or free-radical reagents, at least within a given reaction series.

Three empirical reactivity treatments, designed particularly for acid–base reactions, are the hard soft acid–base (HSAB) principles of Pearson, the donor–acceptor numbers (DN-AN) of Gutmann, and the E&C values of Drago.[24] Of these, the E&C treatment is readily adapted to quantify the contribution of acid–base reactions at interfaces to the total work of adhesion. The DN-AN numbers bear a general resemblance to the E&C values, but are not yet in as suitable a form to calculate works of adhesion. The HSAB principles are only qualitative but provide guidelines for conditions where other approaches are unavailable.

Drago[25,26,27] proposes the following relation to calculate the enthalpy of an acid–base reaction in the gas phase:

$$-\Delta H_{ab} = E_a E_b + C_a C_b \qquad (8.22)$$

where ΔH_{ab} is the enthalpy of adduct formation per mole, E_a and C_a are empirically determined parameters for the acid, and E_b and C_b are empirically determined parameters for the base. The acid a and base b are each characterized by two independent parameters: an E value that measures its ability to participate in electrostatic bonding, and a C value that measures its ability to participate in covalent bonding. Both E and C are derived empirically to give the best agreement between calculation and experiment for the largest possible number of adducts. A self-consistent set of E and C values is now available for several dozen bases, allowing the prediction of ΔH_{ab} for over 1500 interactions.[28]

The magnitudes of the E parameters can be interpreted as measures of the susceptibility of the molecule for electrostatic interaction; the magnitudes of the C parameters can be interpreted as measures of the susceptibility for covalent interactions (similar to the concept of "hard" and "soft" acids and bases). Equation (8.22) accurately correlates over 280 enthalpies of formation. Representative values of the acid–base constants from Drago[25] are listed in Table 8.4 which, when substituted in Eq. (8.22) give enthalpies in kJ per mole.

A similar list of E and C parameters of acidic and basic polymers could be obtained from measured heats of solution, determined calorimetrically, in solvents whose E and C values are already known. The long time required to dissolve a polymer is a disadvantage of this method. A more convenient method is to measure the infrared spectral shifts of the stretching frequencies of the active groups of either polymer or solvent. The greater the enthalpy of interaction between the solvent and the polymer, the greater is the spectral shift. Both the total enthalpy of interaction and the total spectral shift are separated into contributions from the dispersion force interaction and the acid–base interaction. The dispersion-force component of the spectral shift is calculated from the dispersion-force component of the surface tension of the solvent; and the enthalpy of acid–base interaction is then calculated from the remaining component of the spectral shift. The E and C parameters for the polymer are obtained by repeating this procedure with several solvents. Ultimately, E and C parameters of various acidic and basic polymers will be available analogously to those of the solvents listed in Table 8.4.[29]

Using the enthalpy of adduct formation, Fowkes[30] proposed that the work of adhesion, were it to include acid–base interactions and other polar interactions (Keesom and Debye forces), be approximated by

$$W_{12} = W_{12}^d + W_{12}^{ab} + W_{12}^p \tag{8.23}$$

or

$$W_{12} = W_{12}^d - f N_{ab} \Delta H_{ab} + W_{12}^p \tag{8.24}$$

where N_{ab} is the moles of acid–base pairs per unit area and where the constant f therefore converts the enthalpy per unit area into surface free energy (Fowkes assumed that f is near unity since enthalpy is the major part of the free energy

TABLE 8.4 Drago C and E Parameters for a Variety of Molecular Acids and Bases[25-27]

Acids	C_a	E_a
Iodine	2.05	2.05
Iodine monochloride	1.697	10.43
Thiophenol	0.405	2.02
p-tert-Butylphenol	0.791	8.30
p-Methylphenol	0.826	8.55
Phenol	0.904	8.85
p-Chlorophenol	0.978	8.88
tert-Butyl alcohol	0.614	4.17
Trifluoroethanol	0.922	7.93
Pyrrole	0.603	5.19
Isocyanic acid	0.528	6.58
Sulfur dioxide	1.652	1.88
Antimony pentachloride	10.49	15.09
Chloroform	0.325	6.18
Water	0.675	5.01
Methylene chloride	0.02	3.40
Carbon tetrachloride	0.00	0.00

Bases	C_b	E_b
Pyridine	13.09	2.39
Ammonia	7.08	2.78
Methylamine	11.41	2.66
Dimethylamine	17.85	2.33
Trimethylamine	23.6	1.652
Ethylamine	12.31	2.80
Diethylamine	18.06	1.771
Triethylamine	22.7	2.03
Acetonitrile	2.74	1.812
p-Dioxane	4.87	2.23
Tetrahydrofuran	8.73	2.00
Dimethyl sulfoxide	5.83	2.74
Ethyl acetate	3.56	1.994
Methyl acetate	3.29	1.847
Acetone	4.76	2.018
Diethyl ether	6.65	1.969
Isopropyl ether	6.52	2.27
Benzene	1.452	1.002
p-Xylene	3.64	0.851

for these strong interactions) and the last term is usually small. For example at the benzene–water interface, $N_{ab} = 2 \times 10^{14}$ acid–base pairs per square centimeter (based on a molecular area for benzene of 0.5 nm^2); and, using Drago values from Table 8.4, Eq. (8.24) gives $W_{12}^{ab} = 20$ mJ/m^2. The total work of adhesion of benzene on water is given by

$$W_{12} = \sigma_1 + \sigma_2 - \sigma_{12} = 72.8 + 28.9 - 35.0 = 66.7 \text{ mJ/m}^2 \qquad (8.25)$$

That part of the work of adhesion due to dispersion force interaction is

$$W_{12}^d = 2(\sigma_1^d \sigma_2^d)^{1/2} = 2(22.0 \times 28.9)^{1/2} = 50.4 \text{ mJ/m}^2 \qquad (8.26)$$

The difference between W_{12} and W_{12}^d is close to the acid–base part of the total work of adhesion; that is, $66.7 - 50.4 = 16.3$ mJ/m^2, to compare with the value calculated above of 20 mJ/m^2. This result attests to the success of Fowkes' treatment of works of adhesion as a simple combination of only the two major types of molecular interaction: dispersion forces and acid–base forces.

A more sophisticated attempt to convert the enthalpy per unit area into surface free energy, that is, correct for entropy change, was given by Vrbanac and Berg[31] who calculated total works of adhesion from measured surface tensions and contact angles at various temperatures and used the following expression for f:

$$f = \left[1 - \frac{d \ln W_{12}^{ab}}{d \ln T} \right]^{-1} \qquad (8.27)$$

Their results show, at least for the systems studied, that the f factor is substantially less than unity in most cases and increases with temperature. This is not surprising since the Drago constants are derived for gas-phase reactions and not for interfaces. In a later publication, J. C. Berg has devoted more attention to the quantitative treatment of the componental theory of interfacial interactions. He concludes that the theories are only qualitatively sound.[32]

Lewis acid–Lewis base interactions have much wider applications than to their present treatment as contributory factors to interfacial tensions. A symposium dedicated to F. M. Fowkes was held in 1990 in honor of his seventy-fifth birthday.[33,38] At this symposium the application of acid–base interactions to coal and graphite surfaces, to metal oxides, to clay minerals, to carbon and graphite fibers, and to glass fibers and silicas were pointed out.

Again, the surface activity of poly(dimethylsiloxane) in a synthetic ester lubricant may be traced to an acid–base interaction between solute and solvent, which is enough to confer a degree of solubility to the poly(dimethylsiloxane). The source of the acid character of the polymer is the silicon atom because of it small size, its slight polarizability and empty d-orbitals, which can accept electrons from a Lewis base. At the same time the Lewis basicity of the polymer is caused by the unpaired electrons in the oxygen atoms. These account for the amphoteric nature of poly(dimethylsiloxane), which is demonstrated in all its interactions with solvents such as chloroform (Lewis acid) and toluene (Lewis base) in both of which it is soluble.[34]

8.3 DISPERSION ENERGIES OF SOLID SUBSTRATES

Extensive series of measurements of contact angles of various liquids on low energy polymer substrates were reported by W. A. Zisman and his coworkers at the Naval Research Laboratory, who found an empirical linear relation between the cosine of the contact angle and the surface tension of the liquid of the sessile drop. (See Figure 8.2.)

The extrapolation of the line to $\cos\theta = 1$ gives the critical surface tension σ_c of the substrate, that is, the surface tension of a liquid that just spreads on that substrate. The term "critical" is used because any liquid on the Zisman plot whose surface tension is greater than σ_c makes a finite contact angle with the substrate. The Zisman plot has no theoretical basis, although values of σ_c are used to characterize relative degrees of surface energy of polymer substrates. Table 8.5 lists Zisman's values of the critical surface tension of various polymeric substrates. Zisman's empirical prediction would fail for liquids that form hydrogen bonds or acid–base interactions with the substrate. These liquids would spread spontaneously on the substrate, but their surface tensions would not necessarily be less than σ_c.

Many years later Fowkes worked out a theoretical treatment of the interactions across interfaces, which sufficiently accounts for the contact angles of all liquids, including hydrogen-bonding liquids and liquid metals. The basic concept is to separate the net interaction into dispersion force and electron donor–acceptor contributions. His treatment follows.

The Young–Dupré equation for the contact-angle θ of the liquid l on a plane solid substrate s is

$$F_{sv} = \sigma_{lv} \cos\theta + F_{sl} \tag{8.28}$$

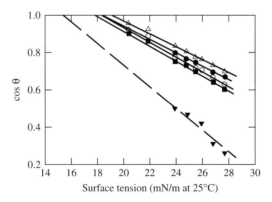

Figure 8.2 Zisman plot of $\cos\theta$ versus the surface tension of the sessile liquid. The solids are copolymers of tetrafluoroethylene (TFE) and hexafluoropropylene (HFP) containing HFP in the following concentrations: 0, 6, 11.5, 14, and 23 mole per cent; the liquids are normal alkanes.[35]

TABLE 8.5 Critical Surface Tensions of Various Polymeric Substrates[36]

Polymeric Substrate	σ_c (nN/m 20°C)
Polymethacrylic ester of perfluorooctanol	10.6
Polyhexafluoropropylene	16.2
Polytetrafluoroethylene	18.5
Polytrifluoroethylene	22
Polyvinylidene fluoride	25
Polyvinyl fluoride	28
Polyethylene	31
Polytrifluorochloroethylene	31
Polystyrene	33
Polyvinyl alcohol	37
Polymethyl methacrylate	39
Polyvinyl chloride	39
Polyvinylidene chloride	40
Polyethylene terephthalate	43
Polyhexamethylene diamide	46

where

$$F_{sv} = F_{s0} - \pi_e \tag{8.29}$$

For solid/liquid systems interacting by dispersion forces only, Eq. (8.16), originally developed to calculate the interfacial tension between two liquids, can be used to obtain F_{sl}. Rewritten in terms of the surface free energy per unit area of the substrate and the surface tension of the liquid, Eq. (8.16) becomes

$$F_{sl} = F_{s0} + \sigma_{lv} - 2(F_{s0}^d \sigma_{lv}^d)^{1/2} \tag{8.30}$$

F_{s0}, the free energy of the solid surface without adsorbed vapor, is an exact substitution for the surface tension of liquid 1 with no adsorbed vapor of liquid 2. Rearranging and introducing Eq. (8.29) gives

$$F_{sv} - F_{sl} = -\sigma_{lv} + 2(F_{s0}^d \sigma_{lv}^d)^{1/2} - \pi_e \tag{8.31}$$

Using Eq. (8.28) gives

$$\sigma_{lv} \cos \theta = -\sigma_{lv} + 2(F_{s0}^d \sigma_{lv}^d)^{1/2} - \pi_e \tag{8.32}$$

or

$$\cos \theta = -1 + \frac{2(F_{s0}^d \sigma_l^d)^{1/2}}{\sigma_{lv}} - \frac{\pi_e}{\sigma_{lv}} \tag{8.33}$$

Since the adsorption of vapors by solids of low surface energy is slight, the values of π_e are low and the term π_e/σ_l may be neglected, giving the commonly used equation

$$\cos\theta = -1 + \frac{2(F_{s0}^d \sigma_{lv}^d)^{1/2}}{\sigma_{lv}} \qquad (8.34)$$

A plot of $\cos\theta$ versus $(\sigma_{lv}^d)^{1/2}/\sigma_{lv}$ should give a straight line originating at $\cos\theta = -1$ and with a slope of $2(F_{s0}^d)^{1/2}$. The data of Zisman and his coworkers treated in this way are shown in Figure 8.3.

Figure 8.3 testifies to the merits of Fowkes' theory. The data refer to a series of liquids of known surface tension on a variety of low-energy solid substrates. Most of the liquids are hydrocarbons for which σ_{lv}^d is equal to σ_{lv}; for other liquids such as water or mercury, σ_{lv}^d can be obtained from measurements of interfacial tension as already described above. When θ, σ_{lv}^d and σ_{lv} are known, Eq. (8.34) gives F_{s0}^d. A useful test liquid to obtain F_{s0}^d is methylene iodide, CH_2I_2, for which $\sigma_{lv} = \sigma_{lv}^d = 50.8$ mN/m at 20°C. Its large surface tension leads to high contact angles, which can be measured more accurately. To obtain F_{s0}^d with this liquid, a single measurement of contact angle and the application of Eq. (8.34) is sufficient.

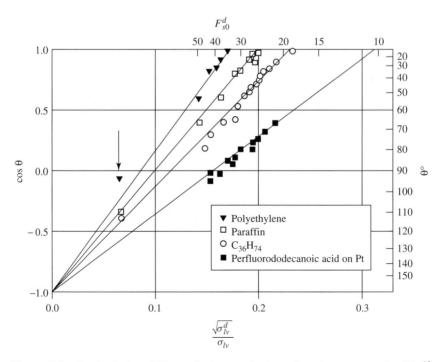

Figure 8.3 Fowkes' plot of Zisman's data to obtain surface free energy of solids.[18]

TABLE 8.6 Values of F_{s0}^d of Low Surface Energy Solids from Contact Angles at 20°C[18]

Name	F_{s0}^d (mJ/m^2)
Perfluorododecanoic acid monolayer on Pt	10.4
Perfluorodecanoic acid monolayer on Pt	13.1
Polyhexafluoropropylene (HEP)	18.9
Polytetrafluoroethylene (TFE)	19.5
n-C$_{36}$H$_{74}$ wax crystal	21.0
n-Octadecylamine monolayer on Pt	22.1
Paraffin wax	25.2
Polytrifluoromonochloroethylene (Kel-F®)	30.8
Polyethylene	35
Polystyrene	44

Table 8.6 reports values of F_{s0}^d obtained from measurement of the contact angle when the σ_{lv}^d of the liquid are known. The surface of the perfluorododecanoic-acid monolayer is composed of CF$_3$ groups; the surface of TFE is composed of CF$_2$ groups; the surface of the n-C$_{36}$H$_{74}$ wax crystal is composed of CH$_3$ groups; the surface of paraffin wax is a mixture of CH$_3$ and CH$_2$ groups; and the surface of polyethylene is composed of CH$_2$ groups. From the positions of these five materials in Table 8.6, the relative order of surface energies is

$$\text{CF}_3 < \text{CF}_2\text{H} < \text{CF}_2 < \text{CH}_3 < \text{CH}_2 \qquad (8.35)$$

The reason for these low surface energies is purely physical. The attractive force between atoms falls off as the inverse sixth power of the distance, which means that even a few angstrom units makes a profound difference in the strength of their interactions. The methylene and the methyl groups composed of carbon and hydrogen have very low densities. The fluorine atom, when combined, has a complete valence shell of eight electrons, which give it a large volume and a low density in spite of its heavier nucleus. These low-density groups have little attractive interaction and consequently low surface energies and recall the argument (see Section 6.1) leading to the statement "the quantity σ varies from one liquid to another, depending on the strength of the attractive forces between their molecules."

Another substrate composed of CF$_3$ groups is provided by vapor deposited n-perfluoroeicosane (C$_{20}$F$_{42}$) onto glass, which results in epitaxially grown crystallites with their molecular axes perpendicular to the glass surface.[37] The dynamic contact angle of water on its surface is 119° and the dynamic contact angle of methylene iodide on its surface is 107°. The dispersion-force component of the surface free energy of water is 21.8 mJ/m² and its total surface free energy is 72.8 mJ/m², respectively, and those of methylene iodide are 48.5 and 50.8 mJ/m², respectively. When these values are plotted on Figure 8.3 they agree with Fowkes' value of F_{s0}^d of 10.4 mJ/m² for CF₃.

The reference cited[37] performs a calculation that takes into account polar contributions from both substrate and sessile drop using Eq. (8.21). This gives a value of the surface free energy of 6.7 mJ/m² which these authors considered to be the lowest surface free energy of any solid, but the use of Eq. (8.21) is suspect.

Conversely, the dispersion-force contribution to the surface tension of polar organic liquids can be determined from contact-angle measurements on low-energy reference solids. Some values are reported in Table 8.7.

Equation (8.34) is applicable to obtain F_{s0}^d when the liquid makes a finite contact angle with the solid substrate, which occurs when the substrate has relatively small surface energy. On substrates of relatively large surface energy, especially when they are clean, most liquids, liquid metals excepted, spread readily with no discrete angle of contact. The values of F_{s0}^d may then be found by a different method, as follows.[18]

The final spreading coefficient in the absence of a specific interaction, is never positive at equilibrium (Section 7.4); its largest attainable value is zero. Observed examples of positive spreading are ascribed to nonequilibrium conditions. The final spreading coefficient is given by combining Eqs. (7.5) and (7.6).

$$S = F_{sv} - F_{sl} - \sigma_{lv} = F_{s0} - \pi_e - F_{sl} - \sigma_{lv} \qquad (8.36)$$

Putting $S = 0$ gives

$$\pi_e = F_{s0} - F_{sl} - \sigma_{lv} \qquad (8.37)$$

Substituting Fowkes' equation for solid/liquid systems interacting by dispersion forces only,

$$F_{sl} = F_{s0} + \sigma_{lv} - 2(F_{s0}^d \sigma_{lv}^d)^{1/2} \qquad (8.30)$$

TABLE 8.7 Values of σ_{lv}^d (mN/m at 20°C) for Polar Organic Liquids from Contact-Angle Measurements of Reference Solids[18]

Liquid	σ_{lv}	σ_{lv}^d (± std. dev.)
Tricresyl phosphate	40.9	39.2 ± 4
a-Bromonaphthalene	44.6	47 ± 7
Trichlorobiphenyl	45.3	44 ± 6
Methylene iodide	50.8	48.5 ± 9
Glycerol	63.4	37.0 ± 4
Formamide	58.2	39.5 ± 7
Polydimethylsiloxane	19.0	16.9 ± 0.5

gives

$$\pi_e = 2(F_{s0}^d \sigma_{lv}^d)^{1/2} - 2\sigma_{lv} \qquad (8.38)$$

By means of Eq. (8.38) the value of F_{s0}^d can be determined if the surface tension of the spreading liquid and its equilibrium spreading pressure π_e are both known. The value of π_e is determined from the vapor adsorption isotherm, as described in Section 7.5. Values of F_{s0}^d for metals and other high-energy solid substrates have been determined by this method and are included in Table 8.8.

The numerical value of F_{s0}^d is probably the best index of the wettability of a solid by any liquid. Substrates with high values are readily wetted and those with low values are not. High-energy substrates readily adsorb ambient impurities that lower their surface energy. A familiar example is a water-wettable glass surface that soon becomes nonwettable merely on exposure to laboratory or household air. Stainless steel when clean is water wettable but rapidly becomes water repellent by chemisorption of fatty-acid vapors or aerosols.

TABLE 8.8 Values of F_{s0}^d Obtained from π_e Measurements for Adsorbed Vapors[18]

Adsorbent	Adsorbate	Temperature	π_e (mJ/m^2)	F_{s0}^d (mJ/m^2)
Polypropylene	Nitrogen	78 K	12	26
	Argon	90 K	13	28.5
Graphite	Nitrogen	78 K	51	123
	n-Heptane	25°C	63, 56, 58	132, 115, 120
Copper	n-Heptane	25°C	29	60
Silver	n-Heptane	25°C	37	74
Lead	n-Heptane	25°C	49	99
Tin	n-Heptane	25°C	50	101
Iron	n-Heptane	25°C	53	108
	Argon	90 K	47	106
	Nitrogen	78 K	40	89
Ferric oxide	n-Heptane	25°C	54	107
Anatase (TiO$_2$)	n-Heptane	25°C	46	92
	Butane	0°C	43	89
	Nitrogen	78 K	56	141
Silica	n-Heptane	25°C	39	78
Stannic oxide	n-Heptane	25°C	54	111
Barium sulfate	n-Heptane	25°C	38	76

8.4 ACID–BASE CHARACTER OF SOLID SUBSTRATES

(See also Section 21.2.b.) Higher energy substrates contain additional potential for interactions, such as donor–acceptor interactions. These depend on both the substrate and the liquid and can be calculated from the enthalpies of adduct formation as described in Section 8.2.b. The Drago parameters (or parameters for any other treatment of donor–acceptor interactions) are obtained by microcalorimetry or Fourier-transform-infrared spectroscopy). These may also be characterized by comparing the adsorption of acid and basic polymers from solution.[23] Basic polymers are adsorbed more strongly on acid surfaces than on basic surfaces, and vice versa. Table 8.9 records some typical results.

The acid–base properties of solid substrates can be determined by a number of techniques but the results are mostly qualitative. These techniques include x-ray photoelectron spectroscopy (XPS), inverse gas chromatography, immersion calorimetry, contact angles with various solvents including thin layer wicking, and adhesion measurements.[38]

In XPS, acid–base interactions yield negative and positive chemical shifts for the acidic and basic probes, respectively. The chemical shifts arise from the transfer of electron density from the Lewis base to the Lewis acid. A typical application is to measure the binding energy shifts of an electron such as the N1s electron from adsorbed pyridine.[39]

Inverse gas chromatography measures the retention times of known gases passing through a column containing a powder to be characterized. The greater the acid–base interaction, the longer the retention time. The necessary equipment is a standard gas chromatograph. The column is packed with the powder to be characterized. The limitation of the technique is that the probe vapor passes the powder surface at high dilution so that the retention time is dominated by interactions with the highest energy sites.[40] This differs from the results of contact-angle measurements, which include interactions across the entire surface. Nevertheless the analyses are useful in that the interactions at the highest energy sites are likely to be the most significant in adhesion and catalysis.

Calorimetric methods to determine the heats of immersion are well known. The experiment is to seal the powder in a vial, suspend the vial in the test liquid, and when the system is at thermal equilibrium, break the vial. The powder is sealed in the vial along with the saturated vapor to measure the actual heat

TABLE 8.9 Acidic and Basic Solid Surfaces

Acidic	Basic
Post-chlorinated polyvinyl chloride	Polycarbonates
Asphalt components	Polymethylmethacrylate
Silica	Calcium carbonate
	Glass

of immersion. To compare the calorimetric data with contact-angle data, which are related to free energies, the contact-angle data need to be taken at several temperatures:

$$\Delta H_{\text{immer}} = \sigma_{lv} \cos\theta - T \left(\frac{\partial \sigma_{lv} \cos\theta}{\partial T}\right)_p \qquad (8.39)$$

If the powder is sealed in a vacuum, then the data have to be corrected for the heat of adsorption of the vapor on the solid, which is significant when acid–base interactions are present.

The use of immersion calorimetry to measure and distinguish between dispersive, acidic, and basic components of the surface energy of solids was further developed by Douillard and Médout-Marère.[41] These authors give some experimental examples to show that it is possible by this means to characterize the surface energy of divided solids of high surface energy.

Further indications of the acidic and basic natures of some of the above materials derive from experiments on the peeling of polymeric films cast on glass.[42] Glass has basic surface sites because strongly basic alkali-metal silicates dominate its surface. Polymethylmethacrylate films on glass are easily peeled off on application of a little water to the edge of the film. Since PMMA is itself basic it does not bond to the basic surface of glass. Cast films of post-chlorinated polyvinyl chloride, however, are strongly bonded to glass and cannot be peeled off. The acid nature of post-CPVC allows it to bond with the basic substrate. If the glass is rinsed with hydrochloric acid, its surface becomes acidic owing to a surface film of silicic acid. A cast film of PMMA is now strongly bonded and cannot be peeled off, while the adhesion of a cast film of post-CPVC is drastically reduced.

Some polymeric solids are formed by crystallization, but they are generally formed by freezing into various metastable glassy states. The surface properties of glassy materials are often studied by means of the temperature dependencies of the surface tension or of the specific heat. The temperature dependence of the surface tension has a discontinuity at the melting point but not at the temperature of the glass transition; the heat capacity has a discontinuity at the temperature of the glass transition.[43] Monographs on the characterization of polymer surfaces are available.[44]

REFERENCES

[1] London, F. The general theory of molecular forces, *Trans. Faraday Soc.* **1937**, *33*, 8–26.

[2] Hirschfelder, J.O.; Curtiss, C.F.; Bird, R.B. *Molecular theory of gases and liquids*; Wiley: New York; 1954.

[3] Berthelot, D. Sur le mélange des gaz, C.R. Hebd. Séances Acad. Sci.**1898**, *126*, 1703–1706.

[4] Berthelot loc. cit.; pp 1857–1858.

[5] Richmond, P. The theory and calculation of van der Waals forces, *Colloid Sci.* **1975**, *2*, 130–172.

[6] Gregory, J. The calculation of Hamaker constants, *Adv. Colloid Interface Sci.* **1969**, *2*, 396–417.

[7] Pitzer, K.S. Inter- and intramolecular forces and molecular polarizability, *Adv. Chem. Phys.* **1959**, *2*, 59–83.

[8] Slater, J.C.; Kirkwood, J.G. The van der Waals forces in gases, *Phys. Rev.* **1931**(2), *37*, 682–697.

[9] Gavroglu, K. *Fritz London* (1900–1954), *A scientific biography*; Cambridge University Press: London; 1995.

[10] Verwey, E.J.W.; Overbeek, J.Th.G. *Theory of the stability of lyophobic colloids*; Dover: Mineola, NY; 1999; p 266.

[11] Casimir, H.B.G.; Polder, D. The influence of retardation on the London–van der Waals forces, *Phys. Rev.* **1948** (2), *73*, 360–372.

[12] Overbeek, J.Th.G., The interaction between colloidal particles, in *Colloid science*, Kruyt, H.R., Ed.; Elsevier: New York; 1952; Vol. 1, pp 245–277.

[13] F.M. Fowkes, private communication.

[14] Marmo, M.J.; Mostafa, M.A.; Jinnai, H.; Fowkes, F.M.; Manson, J.A. Acid–base interaction in filler-matrix systems, *Ind. Eng. Chem., Prod. Res. Dev.* **1976**, *15*, 206–211.

[15] Girifalco, L.A.; Good, R.J. A theory for the estimation of surface and interfacial energies. I. Derivation and application to interfacial tension, *J. Phys. Chem.* **1957**, *61*, 904–909.

[16] Berthelot, D. Sur le mélange des gaz, *C.R. Hebd. Séances Acad. Sci.* **1898**, *126*, 1703–1706.

[17] Berthelot loc. cit.; pp. 1857–1858.

[18] Fowkes, F.M. Attractive forces at interfaces, *Ind. Eng. Chem.* 1964 *56*(12), 40–52; reprinted in *Chemistry and physics of interfaces*; Ross, S., Ed.; American Chemical Society: Washington, DC; 1965.

[19] In later work Fowkes suggests that more reliable values of σ^d for liquids and solids are to be obtained from branched or cyclic alkanes than from *n*-alkanes used as test liquids, as the latter have small anisotropic polarizabilities whereas the former have none. Fowkes, F. M. Surface effects of anisotropic London dispersion forces in n-alkanes, *J. Phys. Chem.* **1980**, *84*, 510–512.

[20] Wu, S. Surface tension of solids: Generalization and reinterpretation of critical surface tension, *Polym. Sci. Technol.* **1980**, *12A*, 53–65.

[21] Cratin, P.D.; Murray, J.M., Jr. A quantitative surface chemical characterization of pitch, *TAPPI* **1970**, *53*, 1960–1963.

[22] Fowkes, F.M. Characterization of solid surfaces by wet chemical techniques, *ACS Symp. Ser.* **1982**, *199*, 69–88.

[23] Fowkes, F.M. Donor–acceptor interactions at interfaces, *Polym. Sci. Technol.* **1980**, *12A*, 43–52.

[24] Jenson, W.B. *The Lewis acid–base concepts*; Wiley: New York; 1980.

[25] Drago, R.S.; Vogel, G.C.; Needham, T.E. A four-parameter equation for predicting enthalpies of adduct formation, *J. Am. Chem. Soc.* **1971**, *93*, 6014–6026.

[26] Drago, R.S. Quantitative evaluation and prediction of donor–acceptor interactions, *Struct. Bonding (Berlin)* **1973**, *15*, 73–139.

[27] Drago, R.S. The interpretation of reactivity in chemical and biological systems with the E and C model, *Coord. Chem. Rev.* **1980**, *33*, 251–277.

[28] Drago, R.S. Quantitative evaluation and prediction of donor-acceptor interactions, *Struct. Bonding (Berlin)* **1973**, *15*, 73–139.

[29] Fowkes, F.M.; Tischler, D.O.; Wolfe, J.A.; Lannigan, L.A.; Ademu-John, C.M.; Halliwell, M.J. Acid–base complexes of polymers, *J. Polym. Sci.* **1984**, *22*, 547–566.

[30] Fowkes, F.M. Donor–acceptor interactions at interfaces, *Polym. Sci. Technol.* **1980**, *12A*, 43–52.

[31] Vrbranac, M.D.; Berg, J.C. The use of wetting measurements in the assessment of acid–base interactions at solid/liquid interfaces., *J. Adhesion Sci. Technol.* **1990**, *4*, 255–266.

[32] Berg, J.C. Role of acid–base interactions in wetting and related phenomena, in *Wettability*; Berg, J.C., Ed.; Marcel Dekker: New York; 1993; pp 75–148.

[33] Mittal, K.L.; Anderson, H.R., Jr. Eds. *Acid–base interactions: relevance to adhesion science and technology*; VSP:Utrecht; 1991.

[34] Ross, S.; Nguyen, N. Interactions of poly(dimethylsiloxane) with Lewis bases, *Langmuir* **1988**, *4*, 1188–1193.

[35] Bernett, M.K.; Zisman, W.A. Wetting properties of tetrafluoroethylene and hexafluoropropylene copolymers, *J. Phys. Chem.* **1960**, *64*, 1292–1294.

[36] Zisman, W.A. Relation of equilibrium contact angle to liquid and solid composition, *Adv. Chem. Ser.* **1964**, *43*, 1–51.

[37] Nishino, T.; Meguro, M.; Nakamae, K.; Matsushita, M.; Ueda, Y. The lowest surface free energy based on — CF_3 alignment, *Langmuir* **1999**, *15*, 4321–4323.

[38] Mittal, K.L., Ed. *Acid–base interactions: Relevance to adhesion science and technology*; VSP: Utrecht; 2000.

[39] Chehimi, M.M.; Delamar, M.; Kurdi, J.; Arefi-Khonsari, F.; Lavaste, V.; Watts, J.F. Characterization of acid–base properties of polymer surfaces by XPS., in Mittal loc. cit.; pp 275–298.

[40] Balard, H.; Brendle, E.; Papirer, E. Determination of the acid–base properties of solid surfaces using inverse gas chromatography: advantages and limitations, in Mittal loc. cit.; pp 299–316.

[41] Douillard, J.-M.; Médout-Marère, V. Surface energy and acid–base properties of solids studies by immersion calorimetry, in Mittal loc. cit.; pp 317–347.

[42] Fowkes, F.M. The role of acid–base interfacial bonding in adhesion, *J. Adhesion Sci. Tech.* **1987**, *1*, 1–28.

[43] Lee, L.-H. Theory of the effect of phase transition on liquid surface tension, *J. Colloid Interface Sci.* **1971**, *37*, 653–658.

[44] Wu, S. *Polymer interfaces and adhesion*; Marcel Dekker: New York; 1982.

9 Experimental Methods of Capillarity

9.1 SURFACE AND INTERFACIAL TENSIONS

The Laplace equation (Section 6.3) is the basis on which all the methods of measuring surface and interfacial tensions are founded:

$$\Delta p = \sigma \left(\frac{1}{R_1} + \frac{1}{R_2} \right) \qquad (9.1)$$

where Δp is the pressure difference across the curved interface, σ is the surface tension, and R_1 and R_2 are the principal radii of curvature. In all experimental methods the pressure difference and radii of curvature are implicated in different ways.

9.1.a Capillary Height

The curved meniscus of a liquid in a narrow glass tube is created by the contact angle of the liquid against glass: If the liquid spreads on the glass, the meniscus has the form of a hemisphere; if the liquid does not wet the glass, as, for example, mercury does not, the liquid surface assumes the convex form of a sessile drop. Only if the contact angle were exactly 90° would the liquid surface be perfectly plane and coincident with the cross section of the tube. When the tube is sufficiently wide, the meniscuses along the peripheral contact with the solid do not overlap; a flat spot exists at the center of the tube, and no pressure difference is created there. However, in a narrower tube the pressure difference causes liquid to flow either up or down the tube depending on whether the curvature is concave or convex. The flow of liquid continues until the hydrostatic pressure, $\rho g h$, just balances the Laplace pressure difference, Δp.

Let a liquid of density ρ and surface tension σ make an angle of contact θ against the wall of a uniform glass tube of radius r, then

$$\Delta p = \rho g h = 2\sigma \cos \theta / r \qquad (9.2)$$

Equation (9.2) is exact when the meniscus is spherical and its weight is negligible compared to the weight of the column, which is true only for narrow tubes. When the angle is zero, Eq. (9.2) becomes

$$\sigma = \frac{\rho g r h}{2} \qquad (9.3)$$

The surface tension can be calculated from the measured height, h. The method is not recommended for finite angles of contact because the angle, if finite, is usually not known, and a finite angle reduces the capillary rise and hence the precision of the method. Harkins and Brown[1] have warned that the method has serious limitations. Capillary tubes of uniform bore must be used, readings of high accuracy require a cathetometer, and the results obtained with basic solutions may be 20% or more too low. Viscous liquids and solutions of certain organic compounds also give incorrect results; on the other hand, excellent results are obtained with water, benzene, the lower alcohols, and similar liquids. Refinements of this technique are fully described by Sugden.[2]

9.1.b Drop Weight

A stream of liquid falling slowly from the tip of a glass tube is nipped off into drops by the surface tension. A general relation is

$$V \rho g = mg = 2\pi r \sigma F \qquad (9.4)$$

where V is the volume of the drop, m is its mass, ρ is the density of the liquid, r is the radius of the tip, and F is a correction factor. The values of the factor and a discussion of the refinements of the method are given by Harkins and Brown[1] and Padday.[3]

The drop weight, mg, is found by weighing a counted number of drops or by counting the number of drops when a measured volume of liquid passes through the tip. A volumetric syringe with a motor-driven plunger is convenient to obtain an accurate volume of liquid. The nature of the method precludes the study of prolonged aging of the surface, which is an important feature of many solutions or of highly viscous liquids.

9.1.c Wilhelmy Plate

The Wilhelmy method to determine surface tension consists of measuring directly by means of a balance the weight of the meniscus formed on the perimeter of a thin plate, such as a platinum blade, partially immersed in the liquid. The Wilhelmy method has many advantages: It is absolute, simple to set up, and independent of the density of the fluid. To avoid a correction for buoyancy, the blade is positioned with its lower edge parallel to and at the same height as the level liquid surface. The measured weight is that of the meniscus created by the wetting of the plate by the liquid. The surface tension is given by the relation

$$\sigma = \frac{W}{2(l+d)\cos\theta} \qquad (9.5)$$

where W is the weight of the meniscus, $2(l + d)$ is the perimeter of the plate, and θ is the angle of contact. The method is eminently suitable for studying prolonged aging of the surface, because an electronic balance gives continuous readings while maintaining the plate at a constant height. Refinements of the method are described by Harkins and Anderson[4] and by Padday.[3] Gaines has pointed out the advantage of using plates made of filter paper, which are porous enough to give a zero contact angle and are disposable.[5]

Equation (9.5) only allows the determination of surface tension when contact angle is known or the determination of contact angle when surface tension is known. This restriction can be obviated by measuring the force on a sphere as it is pulled through a surface or interface. The curvature of the interface depends on the degree of immersion of the sphere. At one depth the curvature is zero; that is, the liquid is perfectly level all the way to contact, and from that datum the angle of contact can be obtained. At any other level the weight of the meniscus, W, can be determined, from which knowing the value of the contact angle, the surface tension can be determined. Since the method is susceptible to error because of the sensitivity of the contact angle, many points of the motion of the sphere through the interface are required.[6]

9.1.d The Du Noüy Ring

The adhesion of a liquid to a metal ring is greater than the cohesion of the liquid, if there is no finite contact angle; consequently, when a ring is detached from the surface of such a liquid, the force to be overcome is that of cohesion (2σ) rather than adhesion. This is the basis of the ring detachment method of measuring surface tensions. The surface tension is given by

$$\sigma = \frac{mg}{4\pi r} F \tag{9.6}$$

where mg is the maximum upward pull applied to the ring of radius r and F is a correction factor given by Harkins and Jordan[7] and by Padday.[3] The correction factor takes account of a small but significant volume of liquid that remains on the ring after detachment, and the discrepancy between the radius r and the actual radius of the meniscus in the plane of rupture. The method is not suited for solutions that attain surface tension equilibrium slowly or for very viscous liquids. This technique is commonly used but is more difficult and less accurate than the Wilhelmy plate.

9.1.e Sessile and Pendent Drop

The sessile and pendent drop methods for measuring surface or interfacial tensions are absolute, require only small volumes of liquid, are readily amenable to temperature control, do not require detachment, and do not depend on the contact angle. They can be brought to the highest level of accuracy by control of external vibration, high-precision measurement of drop dimensions, and computer programs that accept multiple readings of drop shape coordinates. These methods can be used for the study of aging effects and with highly viscous liquids, such as crude oils.

The shape of a sessile (or sitting) drop on a plane surface is determined by the balance between the surface tension of the liquid and hydrostatic pressures within the drop. Surface tension or interfacial tension may be determined by recording the profile of the sessile drop; and contact angle may be obtained by measuring the angle between the tangents to the liquid surface and to the solid substrate at a point of contact. The profile of the drop may be shown by a sharp silhouette thrown on a screen or by measurements of several coordinates along the drop profile. The original tables of Bashforth and Adams are now superseded by a computer program that matches observations of the drop profile to its closest theoretical counterpart.[8] The development of computer techniques has also diminished the importance of the various approximations to the drop shape that have been suggested in the past. The use of the sessile drop to measure surface or interfacial tensions is described by Butler and Bloom[9] and its use to determine contact angle is described below.

The principle of the pendent drop method is similar to that of the sessile drop method, inasmuch as the same forces of surface tension and hydrostatic pressure determine its shape. The pendent drop may be formed quickly so that changes of a newly formed surface can be monitored. The pendent drop may be vibrated to study dynamic surface or interfacial tensions, as in the pulsating-bubble technique of Lunkenheimer et al.[10] The use of the pendent drop to measure surface or interfacial tensions is described by Patterson and Ross[11] and Ambwani and Fort.[12]

Sessile and pendent bubbles are possible variants of these techniques that offer advantages under special conditions.

9.1.f Maximum Pull on a Rod

The Wilhelmy plate has the disadvantage that the contact angle often does not remain zero when measuring the surface tension of a monolayer on an aqueous substrate or with some aqueous solutions. A method that is independent of contact angle, but still makes use of essentially the same apparatus as the Wilhelmy plate, was developed by Padday,[13] and Padday, Pitt and Pashley.[14] The Wilhelmy plate is replaced by a stainless steel rod, 3 cm in length, accurately machined to a diameter of 0.500 cm. The bottom of the rod is cut to form a plane with a sharp circular edge. The plane is roughened so as to be wetted by the contacting liquid, while the cylindrical surface of the rod is highly polished. Highly polished stainless steel is not wetted by aqueous liquids, which therefore do not spread above the meniscus. The rod is hung vertically by a thread from under a standard bottom-loading balance of 1 mg sensitivity, which is supported by a motorized vertical translator that can be raised or lowered at speeds as low as 1.0 mm/min (Figure 9.1.) The axisymmetric shape of the meniscus raised by the rod above the level of the liquid (Figure 9.2) is determined only by its

204   EXPERIMENTAL METHODS OF CAPILLARITY

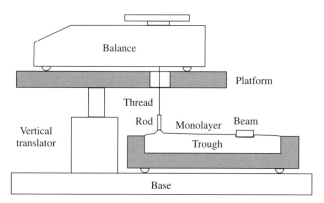

Figure 9.1 Schematic drawing of Langmuir trough and balance for measuring surface tensions.[17]

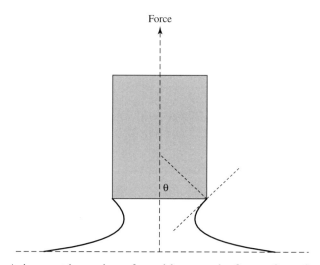

Figure 9.2 Axisymmetric meniscus formed between the free surface of a liquid and a circular rod.[14]

density, its surface tension, and the local gravitational constant and is independent of the contact angle. In this respect the method resembles that of the sessile drop or of the pendent drop. No liquid should reside on the polished side of the rod, as only the weight of the meniscus is to be measured. When the balance is raised, the reading increases to a maximum and then falls prior to detachment of the rod.

In place of the general tables given by Padday[13] a linear approximation may be used.[15] For a rod 0.500 cm diameter at 21°C,

$$\sigma_0 - \sigma = 0.54\,[171 - (W_{max} - W_{rod})] \qquad (9.7)$$

where $\sigma_0 = 72.9$ mN/m and $(W_{max} - W_{rod})$, the weight of the meniscus, is expressed in milligrams. With a balance that reads ± 1 mg or better, values of σ an be determined to at least ± 0.5 mN/m.

A variation of Padday's method in which a low surface energy (Teflon®) rod is used to contact the liquid is advantageous for the study of interfacial films that are slow to equilibrate, as such systems are likely to be disturbed by contact with a high-energy substrate. A suitable technique is described by Buboltz, Huang and Feigenson who gave results that compared well with values reported in the literature.[16] The use of a low-energy rod prevents the liquid from wetting the sides of the rod and hence less new surface is created. To minimize the creation of new surface the rod is lifted only a short distance, much less than to reach the maximum suspended weight. Analysis of the force versus distance curve gives the surface tension. This also reduces the amount of new surface created during the measurement.

9.1.g Maximum Bubble Pressure

If a bubble is blown from the bottom of a tube that dips vertically into a liquid, the pressure in the bubble increases at first as the curvature increases or as the radius of curvature decreases. A bubble small enough to be taken as spherical will grow to a hemisphere with a continuous decrease of the radius of curvature, after which farther growth increases its radius, thus decreasing its curvature and its pressure (see Figure 9.3). The highest pressure is reached at the point where the bubble is a hemisphere; at this point the pressure in the bubble is

$$P = \rho g h + 2\sigma/r \qquad (9.8)$$

where $\rho g h$ is the part of the total pressure P required to force the liquid down the tube to the level h, which is the depth of immersion of the tip of the tube, and r is the radius of the tube. If the liquid wets the tube, the radius r is its internal radius, since the liquid covers the lower edge of the tube completely. By determining the maximum pressure that is attained prior to the detachment of the bubble, the surface tension is evaluated. The method is especially suitable for aqueous solutions.

A newly developed instrument using maximum bubble pressure to monitor surface tension consists of dual capillaries of different radii using a differential pressure transducer to sense the pressure difference. The transducer output is conditioned and sent through an analog interface board to the computer where it is scaled and offset in relation to a previously computer-calculated calibration curve. The resulting value of surface tension has a resolution of 0.1 mN/m. The value is measured anew each time a new surface is formed and the bubble is released from the orifice.[18] The data are usually reported in terms of bubble frequency (Hz) rather than surface age which is indeterminate by this technique. An instrument based on this description is the SensaDyne 6000 of the Chem-Dyne Research Corporation.

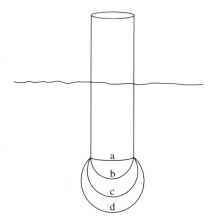

Figure 9.3 Schematic picture of the growth of a bubble showing the growth (first two interfaces) and decline (second two interfaces) of the curvature at a capillary orifice. Correspondingly, the pressure also grows and declines.[20]

Recent improvements in the maximum-bubble-pressure method make it particularly suitable to study rates of adsorption at an aqueous surface.[19] Improvements include the use of an inclined capillary with a siliconized bore, but with a hydrophilic face and outside surface. Noskov includes a detailed analysis of this method in his review article.[20]

9.1.h Spinning Drop

A drop of liquid of about 0.02 cm^3 is placed in a liquid of higher density in a straight precision-bore capillary tube, which is spun on its axis (1200–24,000 rpm). The centrifugal force throws the denser liquid to the wall of the tube and the drop moves to the axis of rotation. (See Fig 9.4) The length of the drop is determined by a balance between the centrifugal force extending it and the interfacial tension acting in the opposite direction. The lower the interfacial tension, the greater the elongation and the more precisely determined is the result; so the method is well suited for ultralow interfacial tensions. A traveling microscope is used to measure the length of the drop. For example, the interfacial tension between n-octane and water containing 0.2% Petronate TRS 10-80 (Witco Chemical Co.) and 1.0% NaCl at 27°C, measured by this technique, is 10^{-3} mN/m.[21] An analysis of the method is given by Princen, Zia and Mason.[22]

The same method was extended by Princen[23] to measure three interfacial tensions and three contact angles by means of a single experiment. Long drops of immiscible fluid phases 1 and 2, surrounded by a denser immiscible phase 3, are spun about the horizontal long axis of the tube, and the radii of the phase 1 and phase 2 drops, as well as that of the three-phase circle of contact are measured. Some stringent conditions reduce the generality of the method. Adherence of the two drops is a requirement, and this is impossible to establish if the spreading

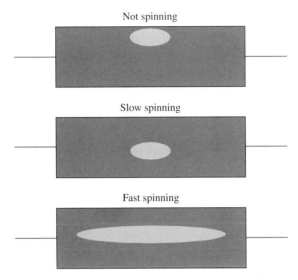

Figure 9.4 The spinning drop method for ultralow interfacial tensions.

coefficient $S_{3(1,2)}$ of phase 3 with respect to phases 1 and 2 is positive, that is, if

$$S_{3(1,2)} = \sigma_{12} - \sigma_{23} - \sigma_{13} > 0 \tag{9.9}$$

Other restrictive requirements are mentioned in the reference. But when the conditions are right, the determination of Neumann's triangle by this method is much more reliable than if the tensions and angles were measured separately, each with its own errors, in the three binary systems.

9.1.i Vibrating Jet

The vibrating jet method, originated by Lord Rayleigh, is used to determine the rate of attaining surface tension equilibrium in solutions of surface-active solutes.[24] A plate orifice is made from a thin sheet of brass (0.005-in. thick) in which an elliptical hole, approximately 1 mm by 0.8 mm, is bored. The solution is ejected through the orifice. The initial elliptical cross section of the jet is unstable because of its unequal curvatures, and oscillates around a cylindrical cross section with a frequency determined by the surface tension (see Figure 9.5). The age of the surface in seconds at the midpoint of successive waves is found from the relation $\tau = \pi r^2 L/v$, where L is the distance from the orifice, r is the radius of the jet where it has a circular cross section, and v is the flow rate. The technique is capable of measuring surface age down to 1 millisecond. Viscosity of samples for measurements by vibrating jet must be less than 5 mPa-s. Addison,[25] Burcik,[26] and Ross and Haak[27] describe the technique and report data. Noskov includes a detailed analysis of this method in his review article.[28]

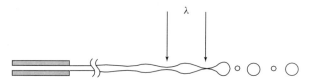

Figure 9.5 A schematic diagram of a liquid pushed through an elliptical orifice, as used to measure dynamic surface tension. Consecutive nodes become farther apart as the surface tension decreases.

Drops formed slowly from pendent liquid are larger the higher the surface tension; drops formed rapidly from a liquid jet are smaller the higher the surface tension. The former process is at equilibrium, the latter process is dynamic.

9.2 METHODS TO MEASURE CONTACT ANGLE

The following methods are exact only for perfectly uniform substrates. Nevertheless, these methods are in use to obtain average values on heterogeneous substrates. For heterogeneous substrates, see Chapter 10.

The gross geometry of the solid substrate determines the appropriate method to measure the contact angle. On flat surfaces or tubes (inside or outside), the contact angle can be measured by observation. The contact angle of a liquid on a fiber is measured by weighing the meniscus (Wilhelmy method). The contact angle of a liquid on a solid when the solid is in the form of a fine powder can obtained by measuring the pressure required to push the liquid through a porous bed of packed fines and other wetting experiments. Reviews of these techniques and the results obtained have been published.[29,30,31,32]

9.2.a On Flat Surfaces

9.2.a(i) The Contact Angle Goniometer The contact-angle goniometer is a device, mounted on an optical bench, to examine a single liquid drop resting on a smooth, planar, solid substrate. The drop is illuminated from the rear to form a silhouette, which is viewed through a telescope equipped with adjustable cross hairs, by means of which the angle of contact can be determined directly by aligning one hairline along the substrate and the other tangential to the drop at its point of contact with the substrate. The angle between the cross hairs, measured through the liquid, is the contact angle. A useful available accessory is an environmental chamber to control conditions of temperature and pressure. Elevated temperatures to 300°C are produced by integral electrical heaters, while subambient temperatures can be produced by circulating coolant through the base of the chamber. The chamber is also suitable for vacuum use and pressure of an inert gas to 0.6 atm. This method was described by Zisman and coworkers.[33] An instrument based on their design is commercially available from Ramé-Hart, Inc., which has also published a bibliography of the use of contact angle data in

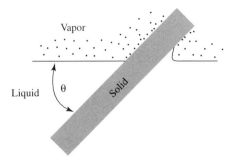

Figure 9.6 The tilting plate method for measuring contact angle on a smooth-planar solid substrate.[35]

surface science.[34] Another optical device, which also uses the outline of a drop on a plane substrate, enlarges and projects the image onto a calibrated frosted glass screen.

Another way to obtain the contact angle of a liquid on a smooth, planar, solid substrate is to tilt the substrate to the degree that the liquid surface becomes planar at the line of contact. This is called the tilting-plate method (see Figure 9.6). Langmuir suggested the use of reflected light to determine an accurate endpoint. The reflection disappears when the meniscus is flat.

9.2.a(ii) Interfacial Meniscus A direct and convenient method to measure contact angles of a liquid/vapor or a liquid/liquid interface against glass or any transparent solid substrate is by means of a photograph of the meniscus inside a cylindrical tube. Two requirements have to be satisfied: to eliminate the distortion of the image of the meniscus introduced by the tube acting as a cylindrical lens and to reduce the deviation of the meniscus from sphericity, so that the contact angle may be derived from a single dimension of the meniscus, namely, its radius. The optics of such systems were analyzed by Bock[36] who showed that an undistorted image can be obtained by immersing the tube in an external fluid of appropriate index of refraction, obtained by computation. A meniscus that departs only negligibly from a sphere is obtained by the use of tubes whose diameter is less than a limiting value that depends on the angle of contact.[37] This method is particularly applicable to measure the angle of contact as a function of temperature.

9.2.b On Fibers: The Fiber Balance

The method of weighing the meniscus is also used in the fiber balance. The instrument consists of a recording microbalance, to which the fiber is attached, and an elevator holding the liquid sample. The liquid surface is slowly moved along the fiber and the forces exerted are recorded as a function of position. The elevator is moved slowly enough to avoid effects of dynamic wetting. The downward force due to surface tension at any location of the three-phase boundary on the fiber is given by the following closed integral:

$$f(h) = \oint \sigma_{lv} \cos \theta \, ds \tag{9.10}$$

where h is the position along the fiber and ds is the element of line contact. Equation (9.10) can be integrated for a fiber of uniform perimeter and uniform contact angle.

If the fiber is not of uniform perimeter and contact angle, the balance measures only $f(h)$. The analysis is given in Section 10.8.a.

9.2.c On Powders

9.2.c(i) The Bartell Cell The wetting of a powdered solid by a liquid is often practically important. The powder can be compressed to give a flat upper surface upon which the contact angle of a sessile drop can be measured. Although this method is frequently used, the results are unreliable as porosity of the substrate affects the measured angle. A sounder method was developed by Bartell and coworkers.[38,39,40] The powder is formed into a porous plug by compression, and liquid fills the pores by dint of capillary pressure. The pressure to remove the liquid equals the Laplace pressure:

$$\Delta p = 2\sigma_{lv} \cos \theta / r \tag{9.11}$$

where r is the average pore radius, defined as the radius of the equivalent cylinder. The equivalent radius of the porous plug is obtained by measuring Δp for a liquid that spreads on the solid. Such a liquid would have a low surface tension and no contact angle. The average pore radius is then calculated by Eq. (9.11) with $\cos \theta = 1$. Care must be taken not to disturb the porous plug in any way that might affect its porosity during the time between the calibration and the test measurements. The apparatus consists of a plug holder, one end of which is connected to a pressure-measuring device and the other end connected to an indicator tube containing liquid, to observe the liquid position. Usually a simple glass tube with a fritted-glass filter makes a suitable plug holder. Metal apparatus is required only for pressures greater than one atmosphere. For acute contact angles, the pressure increase above atmospheric, Δp, required to prevent flow measures the contact angle by Eq. (9.11); for obtuse contact angles, the pressure is applied to initiate liquid flow.

If there is no movement of the meniscus in the indicator tube over a range of pressures, it is an indication of a heterogeneous substrate.[41] Advancing and receding angles can be calculated by observing the pressures at the beginning and end of the static-pressure interval. Commercially available mercury porosimeters, such as those from Micromeritics, Inc., can be adapted for this purpose.

9.2.c(ii) The Washburn Equation The flow of liquid in a capillary tube is caused by the Laplace pressure difference across the curved liquid surface. If the capillary is vertical, the liquid flow continues until the hydrostatic head created by liquid in the capillary is just balanced by the Laplace pressure. If the capillary is horizontal, the flow continues. The rate of flow of liquid in a horizontal capillary is described mathematically by combining the Laplace equation for the pressure difference across a curved surface and the Poiseuille equation for the flow of liquid in a tube.

The volume rate of flow ω of a liquid with viscosity η in a capillary of radius r and liquid length L, driven by a pressure gradient Δp, is given by the Poiseuille equation:

$$\omega = \frac{\pi \Delta p r^4}{8 \eta L} \tag{9.12}$$

The velocity of the liquid front, dL/dt, is the volume rate of flow divided by the cross-sectional area

$$\frac{dL}{dt} = \frac{\omega}{\pi r^2} \tag{9.13}$$

Hence,

$$\frac{dL}{dt} = \frac{\Delta p r^2}{8 \eta L} \tag{9.14}$$

Equation (9.14) describes the velocity of the liquid as a function of the applied pressure. If that pressure is due to the curvature of the liquid surface, then the Laplace equation (9.11) may be introduced, giving the Washburn equation[42]

$$\frac{dL}{dt} = \frac{\sigma r \cos \theta}{4 \eta L} \tag{9.15}$$

Integrating gives

$$L^2 = \frac{\sigma r \cos \theta \, t}{2 \eta} \tag{9.16}$$

where t is the time for a liquid of viscosity η and surface tension σ to penetrate a distance L into a porous material whose average pore radius is r.

The Washburn equation is used to study the penetration of liquids or wicking into fiber beds or packed powders since it relates the rate of liquid penetration to the average pore radius, surface tension, and contact angle. By measuring the rate of liquid penetration, the product $r \cos \theta$ is obtained. If the liquid has a nonfinite contact angle, the average pore radius r is obtained. Once the average pore radius for a given fiber bed or packed power is known, the contact angle of liquids with finite contact angles can be determined. Fiber beds, packed powders, or woven fabrics can be characterized either by the Bartell method or the Washburn method; the former measures the pressure to prevent liquid penetration and the latter measures the rate of liquid penetration with no applied pressure.

Considerable attention from the news media was focused on the Washburn equation when physicist Len Fisher of the University of Bristol introduced it at a press conference in London on National Biscuit Dunking Day. His presentation of the physics of biscuit dunking attracted requests for radio and TV interviews from countries as far away as Australia and South Africa. "Journalists were enthralled," he said, "to discover that there is an equation to describe biscuit dunking. Newspapers published it, TV programs showed it. More than one radio interviewer even insisted that I describe it on the air."[43]

A useful technique to obtain contact angles on powders combines the Bartell method and the Washburn method. A glass slide is coated with a suspension of fine particles and allowed to dry. This forms a thin packed layer of particles. Some care is required to preserve the porosity. A scale hung next to the slide provides both a simple and easy way to measure the rate of rise, dL/dt. When the slide is positioned vertically in contact with a wetting liquid of low surface tension, the Washburn equation, (9.16), is used with a zero contact angle to obtain the porosity, r. The whole apparatus is contained in a closed, glass vessel in which the vapor is saturated to prevent evaporation. Once the porosity, r, is ascertained, then the contact angles of liquids of higher surface tension can be determined by an application of Eq. (9.16) where the contact angle is the only unknown. The technique is called thin-layer wicking.[44]

9.2.c(iii) By Centrifugation If the contact angle of the liquid on the solid is finite and the particle size is small, then the powder will float on the surface of the liquid buoyed by capillary forces. The relevant forces are centrifugal, F_c, buoyancy, F_b, hydrostatic, F_h, and capillary, F_σ:[45]

$$F_c = -\frac{4}{3}\pi a^3 \rho_2 r \omega^2$$

$$F_b = \frac{\pi}{3} a^3 \rho_1 r \omega^2 \left(-\cos^3 \phi + 3 \cos \phi + 2\right) \quad (9.17)$$

$$F_h = \pi a^2 \sin^2 \phi h \rho_1 r \omega^2$$

$$F_\sigma = 2\pi a \sigma \sin \phi \sin(\theta - \phi)$$

where a is the particle radius, ρ_2 the particle density, R the distance from the liquid surface to the center of the centrifuge, ω, the angular velocity of the centrifuge, ρ_1, the density of the liquid, ϕ, the angle that defines the position of

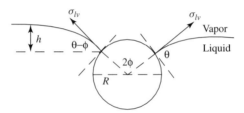

Figure 9.7 Particle floating on a liquid in a centrifugal force field.[45]

the particle in the liquid, h, the height of the hydrostatic layer (see Figure 9.7), and σ, the liquid surface tension. The minimum force required to pull the spherical particle completely into the liquid can be derived from the force balance, $F_c + F_b + F_h + F_\sigma = 0$, under the condition that $F_b + F_h + F_\sigma$ is maximal. The Laplace equation is used to obtain values for h, ϕ, and θ which describe the profile of the liquid around the particle. This is the computationally difficult part of the problem and only an iterative procedure can be used to solve the problem numerically. The contact angle is obtained from the minimum value of the angular rotation of the centrifuge that sinks a floating particle. This critical velocity is found by increasing the speed of rotation stepwise.

9.2.c(iv) By Compression on a Langmuir Balance If the contact angle of the liquid on the solid is finite and the particle size is small, then the powder will float on the surface of the liquid buoyed by capillary forces. Even a weakly hydrophobic powder with an acute angle of contact will float because any force, however slight, tending to pull it into the liquid will generate a restoring force from capillary curvature. Surface pressure (Π) versus surface area (A) can be determined for monoparticulate layers by means of the Langmuir film balance.[46,47] Upon compressing the layers, collapse occurs at a certain pressure at which the particles are forced out of the water–air interface. For hexagonally close-packed, monodisperse spheres at the point of collapse, the contact angle can be calculated by the following equation:

$$\cos\theta = \pm \left[\left(\frac{2(3)^{1/2}\Pi_c}{\pi\sigma_{lv}} \right)^{1/2} - 1 \right] \quad (9.18)$$

where Π_c is the collapse pressure. The positive sign is used when the particles are forced into the water at collapse and the negative sign when they are forced out of the water at collapse. Improvements to the technique have been published by Máté et al.[48]

9.2.c(v) By Heats of Immersion The heat of immersion is the heat released or absorbed when a clean solid is immersed into a liquid. If the solid is a powder, the heat of immersion may appear only slowly as liquid penetrates all parts of the powder. Recent improvements in microcalorimeters have made it possible to measure very small changes in heat evolved over long periods of time. The relation between contact angle and heat of immersion, ΔH_{immer}, is given by[49]

$$\Delta H_{\text{immer}} = \sigma_{lv} \cos\theta - T \left(\frac{\partial \sigma_{lv} \cos\theta}{\partial T} \right)_p \quad (9.19)$$

where T is the absolute temperature, σ_{lv} is the liquid/vapor surface tension, and θ is the contact angle. Equation (9.19) is sometimes used to calculate the temperature dependence of the contact angle from the heat of immersion by expanding the differential and collecting known terms.

Equation (9.19) has also been solved to give an approximate expression for the contact angle in terms of the measured heat of immersion.[49] The empirical derivation assumes a value for the change in the free energy of immersion, $\sigma_{lv} \cos\theta$, with temperature of -0.07 ± 0.02 mJ/m^{2}K. This is reasonable when only weak dispersion forces act across the solid/liquid interface.

$$\cos\theta \approx \frac{-0.07T + \Delta H_{\text{immer}}}{\sigma_{lv}} \qquad (9.20)$$

Equation (9.20) gives the approximate relation between heats of immersion for low-energy solid surfaces and their contact angles with the same liquid.

9.3 MANUFACTURERS OF EQUIPMENT

AST Products, Inc. 9 Linnell Circle Billerica, MA 01821-3902	(877) 667-4500 (978) 667-9778 www.astp.com	Contact angle goniometer
Brinkmann Instruments, Inc. One Cantiague Rd P.O. Box 1019 Westbury, NY 11590-0207	(800) 654-3050 (516) 334-7506 www.brinkmann.com	Drop volume Maximum bubble pressure
Camtel Ltd 5 Carrington House 37 Upper King St Royston, Herts SG8 9AZ UK	+44 (0)1763 244280 +44 (0)1763 442980 www.camtel.co.uk	Interfacial rheometer Wilhelmy plate and ring Bartell technique
CSC Scientific Company, Inc. 2810 Old Lee Highway Fairfax, VA 22031	(800) 458-2558 (703) 280-5142 www.cscscientific.com	Ring tensiometer
FIBRO system ab Box 9081, S-126 09 Hägersten, Sweden	+46 (8) 775 00 90 +46 (8) 775 00 91 www.fibro.se	Dynamic contact angle
First Ten Ångstroms 465 Dinwiddie St Portsmouth, VA 23704	(757) 393-1584 (757) 393-3708 www.firsttenangstroms.com	Dynamic contact angle
Krüss USA 9305 Monroe Rd Suite B Charlotte, NC 28270-1488	(704) 847-8933 (704) 847-9416 http://kruss.de	Wilhelmy plate and ring balances Fiber, interfacial, drop volume, maximum bubble pressure, and spinning drop tensiometers Dynamic contact angle
KSV Instruments USA P.O. Box 192 Monroe, CT 06468	(800) 280-6216 (203) 459-0437 www.ksvinc.com	Mercury porosimetry

Micromeritics Instrument Corp. One Micromeritics Dr Norcross, GA 30093-1877	(770) 662-3633 (770) 662-3696 www.mircomeritics.com	Electrozone counting Laser diffraction Porosimetry Gas adsorption Sedimentation
Ramé-Hart, Inc. 48 Morris Ave Mountain Lakes, NJ 07046	(800) 908-8934 (973) 335-2920 www.ramehart.com	Contact angle goniometer
SensaDyne Instrument Div. Chem-Dyne Research Corp. P.O. Box 30430 Mesa, AZ 85275-0430	(480) 924-1744 (480) 924-1754 www.sensadyne.com	Maximum bubble pressure tensiometer
Tantec 630 Estes Ave. Schaumburg, IL 60193	(847) 524-5506 (847) 524-6956 www.tantecusa.com	Contact angle goniometer Wilhelmy plate and ring tensiometers
Thermo Cahn 5225 Verona Rd Madison, WI 53711	(608) 276-6333 (608) 273-6827 www.cahn.com	Wilhelmy plate and ring tensiometer Dynamic contact angle

REFERENCES

[1] Harkins, W.D.; Brown, F.E. The determination of surface tension (free surface energy), and the weight of falling drops: The surface tension of water and benzene by the capillary-height method, *J. Am. Chem. Soc.* **1919**, *41*, 499–524.

[2] Sugden, S. The determination of surface tension from the rise in capillary tubes, *J. Chem. Soc.* **1921**, *119*, 1483–1492.

[3] Padday, J.F. Surface tension: Part 1. Theory of surface tension, Part 2. The measurement of surface tension, *Surf. Colloid Sci.* **1969**, *1*, 39–251.

[4] Harkins, W.D.; Anderson, T.F.I. A simple accurate film balance of the vertical type for biological and chemical work, and a theoretical and experimental comparison with the horizontal type. II. Tight packing of a monolayer by ions, *J. Am. Chem. Soc.* **1937**, *59*, 2189–2197.

[5] Gaines, G.L., Jr. On the use of filter paper Wilhelmy plates with insoluble monolayers, *J. Colloid Interface Sci.* **1977**, *62*, 191–192.

[6] Zhang, L.; Ren, L.; Hartland, S. More convenient and suitable methods for sphere tensiometry, *J. Colloid Interface Sci.* **1996**, *180*, 493–503.

[7] Harkins, W.D.; Jordan, H.F. A method for the determination of surface and interfacial tension from the maximum pull on a ring, *J. Am. Chem. Soc.* **1930**, *52*, 1751–1772.

[8] Rotenberg, Y.; Boruvka, L.; Neumann, A.W. Determination of surface tension and contact angle from the shapes of axisymmetric fluid interfaces, *J. Colloid Interface Sci.* **1983**, *93*, 169–183.

[9] Butler, J.N.; Bloom, B.H. A curve-fitting method for calculation interfacial tension from the shape of a sessile drop, *Surf. Sci.* **1966**, *4*, 1–17.

[10] Lunkenheimer, K.; Hartenstein, C.; Miller, R.; Wantke, K.-D. Investigations on the method of the radially oscillating bubble, *Colloids Surfaces* **1984**, *8*, 271–288.

[11] Patterson, R.E.; Ross, S. The pendent-drop method to determine surface or interfacial tensions, *Surf. Sci.* **1979**, *81*, 451–463.

[12] Ambwani, D.S.; Fort, T., Jr. Pendant drop technique for measuring liquid boundary tensions, *Surf. Colloid Sci.* **1979**, *11*, 93–119.

[13] Padday, J.F. Tables of the profiles of axisymmetric menisci, *J. Electroanal. Chem. Interfacial Electrochem.* **1972**, *37*, 313–316.

[14] Padday, J.F.; Pitt, A.R.; Pashley, R.M. Menisci at a free liquid surface: surface tension from the maximum pull on a rod, *J. Chem. Soc., Faraday Trans. I* **1975**, *71*, 1919–1931.

[15] D'Arrigo, J.S. *Stable gas-in-liquid emulsions, production in natural waters and artificial media*; Elsevier: New York; 1986.

[16] Buboltz, J.T.; Huang, J.; Feigenson, G.W. Surface tension determination with a Teflon rod, *Langmuir* **1999**, *15*, 5444–5447.

[17] D'Arrigo loc.cit.; p 111.

[18] Janule, V.P. Process analysis and control using surface tension measurement, Presented at 1983 Pittsburgh Conference, March 9, 1983.

[19] Mysels, K.J. Improvements in the maximum-bubble-pressure method of measuring surface tension, *Langmuir* **1986**, *2*, 428–432.

[20] Noskov, B.A. Fast adsorption at the liquid-gas interface, *Adv. Colloid Interface Sci.* **1996**, *69*, 63–129, esp. 93–100.

[21] Cayias, J.L.; Schechter, R.S.; Wade, W.H. The measurement of low interfacial tension via the spinning-drop technique, *ACS Symp. Ser.* **1975**, *8*, 234–247.

[22] Princen, H.M.; Zia, I.Y.Z.; Mason, S.G. Measurement of interfacial tension from the shape of a rotating drop, *J. Colloid Interface Sci.* **1967**, *23*, 99–107.

[23] Princen, H.M. Spinning-drop method applied to three-phase fluid equilibria, *Langmuir* **1999**, *15*, 7386–7391.

[24] Rayleigh, John William Strutt, Lord Rayleigh, On the tension of recently formed liquid surfaces, *Proc. Roy. Soc. (London)* **1890**, *47*, 281–287.

[25] Addison, C.C. The properties of freshly formed surfaces. Part I. The application of the vibrating-jet technique to surface tension measurements of mobile liquids, *Chem. Soc.* **1943**, 535–541.

[26] Burcik, E.J. The rate of surface tension lowering and its role in foaming, *J. Colloid Sci.* **1950**, *5*, 421–436.

[27] Ross, S.; Haak, R.M. Inhibition of foaming. IX. Changes in the rate of attaining surface tension equilibrium in solutions of surface-active agents on addition of foam inhibitors and foam stabilizers, *J. Phys. Chem.* **1958**, *62*, 1260–1264.

[28] Noskov, B.A. Fast adsorption at the liquid-air interface, *Adv. Colloid Interface Sci.* **1996**, *69*, 63–129, esp. 81–92.

[29] Fowkes, F.M., Ed. *Contact angle, wettability, and adhesion*; Adv. Chem. Ser. 43; ACS: Washington, DC; 1964.

[30] Johnson, R.E., Jr.; Dettre, R.H. Wettability and contact angles, *Surf. Colloid Sci.* **1969**, *2*, 85–153.

[31] Neumann, A.W.; Good, R.J. Techniques of measuring contact angles, *Surf. Colloid Sci.* **1979**, *11*, 31–91.

[32] Mittal, K.L., Ed. *Contact angle, wettability and adhesion*; VSP: Zeist: The Netherlands; 1993.

[33] Bigelow, W.C.; Pickett, D.L.; Zisman, W.A. Oleophobic monolayers. I. Films adsorbed from solution in nonpolar liquids, *J. Colloid Sci.* **1946**, *1*, 513–538.

[34] A bibliography of contact-angle use in surface science, TB-100; Ramé-Hart, Inc.: 43 Bloomfield Ave., Mountain Lakes, NJ 07046.

[35] Fowkes, F.M.; Harkins, W.D. The state of monolayers adsorbed at the interface solid–aqueous solution, *J. Amer. Chem. Soc.* **1940**, *62*, 3377–3386.

[36] Bock, E.J. Correction of astigmatic optical systems consisting of an object inside a cylinder, *J. Colloid Interface Sci.* **1984**, *99*, 399–403.

[37] Ross, S.; Kornbrekke, R.E. The wetting of the container wall as a critical-point phenomenon. I. Measurement of contact angles in cylindrical tubes: validation of a method, *J. Colloid Interface Sci.* **1984**, *98*, 223–228.

[38] Bartell, F.E.; Whitney, C.E. Adhesion tension, *J. Phys. Chem.* **1932**, *36*, 3115–3126.

[39] Bartell, F.E.; Walton, C.W., Jr. Alteration of the surface properties of stibnite as revealed by adhesion tension studies, *J. Phys. Chem.* **1934**, *38*, 503–511.

[40] Bartell, F.E.; Osterhof, H.J. The measurement of adhesion tension solid against liquid, *Colloid Symp. Monogr., 5th Nat. Symp.*, 1927 **1928**, *5*, 113–134.

[41] Lucassen-Reynders, E.H. Contact angle and adsorption on solids, *J. Phys. Chem.* **1963**, *67*, 969–972.

[42] Washburn, E.W. The dynamics of capillary flow, *Phys. Rev.* [2] **1921**, *17*, 273–283, 374–375.

[43] Fisher, L. Physics takes the biscuit, *Nature* **1999**, *397*, 469. Also *Chem. Eng. News*, March 8, **1999**, 88.

[44] Costanzo, P.M.; Giese, R.F.; van Oss, C.J. The determination of surface tension parameters of powders by thin layer wicking. in *Advances in measurement and control of colloidal processes*, Williams, R.A.; de Jaeger, N.C., Eds.; Butterworth Heinemann: Oxford; 1991; pp 223–232.

[45] Huethorst, J.A.M.; Leenaars, A.F.M. A new method for determining the contact angle of a liquid against solid spherical particles, *Colloids Surf.* **1990**, *50*, 101–111.

[46] Clint, J.H.; Taylor, S.E. Particle size and interparticle forces of overbased detergents: A Langmuir trough study, *Colloids Surf.* **1992**, *65*, 61–67.

[47] Hórvölgyi, Z.; Németh, S.; Fendler, J.H. Spreading of hydrophobic silica beads at water-air interfaces, *Colloids Surf. A: Physicochem. Eng. Asp.* **1993**, *71*, 327–335.

[48] Máté, M.; Fendler, J.H.; Ramsden, J.J.; Szalma, J.; Hórvölgyi, Z. Eliminating surface pressure gradient effects in contact angle determination of nano- and microparticles using a film balance, *Langmuir*, **1998**, *14*, 6501–6504.

[49] Spagnolo, D.A.; Maham, Y.; Chuang, K.T. Calculation of contact angle for hydrophobic powders using heat of immersion data, *J. Phys. Chem.* **1996**, *100*, 6626–6630.

10 Wetting of Irregular Surfaces

10.1 INTRODUCTION

The Young–Dupré equation is valid only for a solid substrate that is chemically homogeneous, impermeable, flat on an atomic scale, and rigid, that is, undeformable. When this ideal surface is present there should be a single and unique equilibrium contact angle. There are, however, three types of contact angles that can be measured. An advancing contact angle, θ_a, is observed when the liquid boundary moves over a dry clean surface; a retreating contact angle, θ_r, when the liquid boundary retreats over a previously wetted surface. The equilibrium contact angle has a value intermediate between θ_a and θ_r. This hysteresis of the contact angle comes from nonideality of the substrate, which can be caused by surface roughness, chemical heterogeneity, or nonequilibrium. Other causes of nonideality are deformation of the substrate or its permeability. Thermodynamics requires that the contact angle be a single-valued parameter of the system, but for most materials the observed contact angles do not have unique values.

Surface roughness makes the observed angle differ from the equilibrium angle, depending on whether the edge of the liquid is in contact with an ascending or descending slope of the substrate. Surface heterogeneity makes the contact line stick or slip as it moves. These phenomena can be seen as a drop moves along a surface, impelled by gravity, but are more readily analyzed by the Wilhelmy method of measuring contact angle.

10.2 IRREGULARITIES DUE TO ROUGHNESS: WENZEL EQUATION

Several theories have been proposed to obtain a relation between the apparent contact angle and the microscopic angle of a liquid drop on a rough solid substrate. A useful model, proposed by Wenzel,[1,2] is that the roughness of the substrate influences the apparent contact angle by providing an interfacial area larger than the apparent area. Using an argument similar to that used to derive the Young equation, the balance of free energies at the three-phase line is

$$r\sigma_{sv} = r\sigma_{sl} + \sigma_{lv} \cos \phi \qquad (10.1)$$

where r is the roughness factor, that is, the ratio of the real area to the apparent area, and ϕ is the apparent contact angle. Combining Equation (10.1) with the

Young–Dupré equation and rearranging gives

$$\cos \phi = r \cos \theta \qquad (10.2)$$

Equation (10.2) is the Wenzel equation relating the apparent contact angle to the real, microscopic contact angle and the surface roughness. When the real contact angle is acute, then surface roughness makes the apparent contact angle appear smaller; that is, the liquid appears to wet the substrate more. When the real contact angle is obtuse, surface roughness makes the apparent contact angle appear larger; that is, the liquid appears to wet the substrate less.

Equation (10.2) illustrates the effect of surface roughness on the observed contact angle but provides only a qualitative measure of surface roughness. The roughness of macroscopic surfaces can be measured quantitatively by a variety of mechanical and optical techniques.[3] These include stylus instruments with either electrical or mechanical transducers and optical techniques such as sectioning and profiling, interferometry, and the Fourier analysis of scattered light.

10.3 IRREGULARITIES DUE TO CHEMICAL HETEROGENEITY

If the substrate has patchwise heterogeneity, the advancing angle is affected more by the lower energy patches, which resist wetting and give a greater angle of contact; on receding, the liquid dewets the lower energy patches more readily and the observed angle is now the smaller one characteristic of the wetting of high-energy patches.

The modified Wilhelmy plate method is widely used to determine the wettability of solid surfaces by probe liquids. The wetting force experienced by a solid suspended from a microbalance during immersion into and emersion from the liquid is monitored. In the absence of surface roughness, chemical heterogeneity of the surface creates fluctuations in the force observed as the liquid slides over the solid and hysteresis results between the advancing and receding measurements.

If the surface heterogeneity occurs as small patches on a smooth solid substrate, the effective free energies per unit area of the solid/liquid and solid/vapor interfaces are averages around the three-phase boundary. The cosine of the observed contact angle is crudely estimated as an average of the cosines of the microscopic contact angles:

$$\cos \theta = \sum f_i \cos \theta_i \qquad (10.3)$$

where f_i is the fraction of total perimeter with contact angle θ_i. Although this derivation is not rigorous and is in general not accurate, it helps to explain some observed phenomena. Woodward and Schwartz[4] report contact angle data measured on mica surfaces covered with submonolayers of octadecylphosphoric acid.

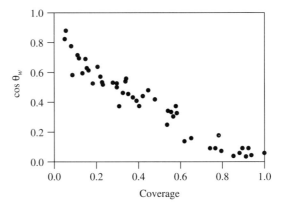

Figure 10.1 The cosine of the static contact angle of water on various subsaturated monolayers plotted versus the surface coverage measured directly using the atomic force microscope.[4]

Equation (10.3) predicts a linear dependence of the measured contact angle with the degree of coverage. The data in Figure 10.1 show a nearly linear dependence.

Mahale, Wesson and coworkers[5,6,7] proposed a model of a surface composed of equal sized patches on which the liquid has different contact angles. By means of computer simulations, the extent of hysteresis and the amplitude and frequency of force fluctuations were shown to depend upon the nature the heterogeneity and its spatial distribution. Their most important result is that hysteresis may be fully accounted for by even a small amount of surface chemical heterogeneity, as low as 5–10%. These calculations completely contradict the linearity of Eq. (10.3). The method was used to characterize plasma-treated carbon fibers.[8]

If a surface is heterogeneous either in composition or topography then the nucleation of droplets of liquid from the saturated gas phase will very from patch to patch. This variation in nucleation can be detected optically once microscopic drops have formed. The analysis can be completed in a few seconds by measuring a number of small areas at high magnification. The method is to place the surface on a temperature-controlled table within a transparent box filled with saturated water vapor. The sample is cooled and the formation of drops recorded with a CCD (charge-coupled device) camera. This measurement complements contact angle data.[9]

10.4 IRREGULARITIES DUE TO PORES

The exceedingly high contact angles, approaching 180°, observed in nature on, for example, water drops on leaves and the petals of flowers were formerly thought to be the result of extremely low surface energy; but studies reveal that composite surfaces (polymer and air) and the resulting hysteresis of the contact angle are more significant factors than the attainment of large equilibrium contact angles.[10] A fractal surface is a kind of rough surface where the coefficient r is very large. Therefore the modification of the wettability due to surface roughness can be greatly enhanced on a fractal surface. Contact angles as high as 174° have been obtained on a fractal surface made of alkylketene dimer.[11]

A drop of liquid on a felt or on a woven fabric bridges apertures containing only air. The low density of a gas means that the dispersion force of attraction is weak, which in turn justifies taking the local angle of the aperture as 180°. Equation (10.3) becomes

$$\cos\theta = f_1 \cos\theta_1 - f_2 \quad (10.4)$$

where $f_1 + f_2 = 1$ and f_1 is the fraction of the substrate occupied by the textile. Clearly, the presence of the voids has caused the contact angle to increase. That is one reason why the plumage of birds is waterproof. Once water has filled the voids, the waterproofing is destroyed. Anyone who has occupied a tent during a rainstorm is aware that rubbing the fabric brings water in. The influence of such surfaces on the apparent contact angle was analyzed by Cassie and Baxter.[12] This analysis ignores the effects of the high curvature of drops around the holes and the resulting Laplacean pressures. These become significant for small holes. A tight weave is therefore more water repellent.

Some drops with large equilibrium contact angles can remain pinned to the substrate even when the surface is tilted to a substantial angle. The real criterion for a super water-repellant surface is that water drops move freely over the surface with little or no tilting. This can sometimes be done without contact-angle hysteresis depending on the topology of the composite substrate.[10] These authors have shown that a high advancing contact angle does not always mean a low work of adhesion. Low adhesion also depends on a high receding angle of contact. The receding angle of the contact depends strongly on the nature of the topology of the substrate. Thus, for example, the same advancing angle of contact may arise on a surface containing holes (network) or its converse, a surface composed of columns, but these have very different receding contact angles. The hysteresis of the contact angles is higher in the former than the latter case. Only when the hysteresis of the angles is small do we have a genuine low adhesion.

Thorough mathematical analyses of these models are given by Johnson and Dettre[13,14] and in edited books on wetting.[15,16]

10.5 IRREGULARITIES DUE TO SCRATCHES

A surface irregularity such as a notch, produced by a scratch, causes a liquid to spread even when its contact angle θ is finite, as long as it is not greater than 90°. Figure 10.2 shows that when the angle of the notch is less than $180° - 2\theta$, the surface of the liquid in the notch is concave; when the angle of the notch equals $180° - 2\theta$, the surface of the liquid is plane; and when the angle of the notch is larger than $180° - 2\theta$, the surface of the liquid is convex. The Laplace pressure difference in the first case causes the liquid to spread along the notch; that is,

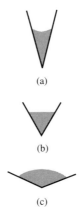

Figure 10.2 A liquid with a contact angle of 45° in (*a*) a notch of 30°, (*b*) a notch of 60°, (*c*) a notch of 135°.

there is a critical value of the notch angle, 180°−2θ, below which the liquid will spread, even though it has a negative spreading coefficient. Similarly, for a given notch angle there is a critical value of the contact angle below which the liquid will spread along the notch. This is the reason why scratches on the Teflon® coating of a frying pan reduces its efficacy in preventing sticking of fried eggs and why in general a rough surface is more readily coated than a smooth one.

Dynamic wetting on rough surfaces is important in printing and coating. In newsprint and inkjet printing the ink spreads faster in the narrow grooves between the fibers than on the smoother areas of the surface, which gives rise to a print defect known as "feathering." An attempt to study how the dynamics of liquid spreading is affected by surface roughness has been reported.[17] The enhanced velocity of spreading was demonstrated on a smooth, oxidized silicon wafer with etched, parallel, V-shaped grooves. The widths of the grooves ranged from 5 to 90 μm wide, spaced from 10 to 1000 μm apart, all with angles of 54°. Dynamic wetting was studied by following how a small silicone-oil droplet spread over the model surface. These studies allowed the observation of the point where the liquid in the grooves starts spreading, decoupled from the macroscopic droplet.

10.6 JAMIN EFFECT IN CAPILLARIES

In 1860 Jamin[18,19] noticed that an ordinary cylindrical capillary tube filled with a chain of alternate air and water bubbles is able to sustain a finite pressure. A similar effect is produced if a series of constrictions is present in the tube. Each of these two effects, which might be termed cylindrical and noncylindrical, respectively, is referred to by Jamin's name. In the former case, Jamin himself found that substituting alcohol or oil for water eliminated the effect, and subsequently[20] it was found that careful cleaning and avoidance of contamination could render a glass capillary incapable of sustaining pressure even with water. What is attained

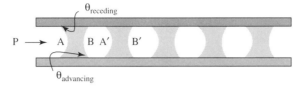

Figure 10.3 Different shapes of the meniscuses of A, A', and so on and B, B', and so on in a chain of alternating liquid drops and air bubbles in a capillary tube, shown with respect to the direction of the applied pressure, P.

with alcohol and oil is a condition conducive to spreading. When the tube is deliberately contaminated, as by a dilute solution of oleic acid in benzene, the Jamin pressure is reestablished with water.

If, as shown in Figure 10.3, the pressure P is applied from left to right, the meniscuses A, A', and so forth become more hollow and the meniscuses B, B', and so forth become flatter. The force exerted by the applied pressure is used up by distorting the meniscuses rather than by moving the chain of drops. Only when the meniscuses A, A', and so forth reach the receding contact angles and the meniscuses B, B', and so forth reach the advancing contact angles will the column containing bubbles respond to the applied pressure by moving. The magnitude of the maximum pressure that can be withstood is directly proportional to the number of bubbles and inversely proportional to the radius of the capillary. The phenomenon is the resistance to the applied pressure, and is known as the Jamin effect. It has some important consequences by obstructing the passage of liquids in narrow tubes. It is thus that the circulation of the blood can be blocked by the introduction of gas bubbles into the capillary vessels. It is the cause of the harm experienced by divers who ascend from great depths too rapidly, or by workers in underwater constructions if the vent to the outside air be opened too suddenly.

The Jamin effect for cylindrical capillaries in which the advancing and receding angles differ is analyzed as follows. Let one end of such a tube be exposed to the atmosphere and the other end acted upon by a pressure P. Then, if n is the number of bubbles, each designated by i, r the radius of the tube, σ the surface tension of the liquid, θ_a and θ_r the contact angles of the advancing and receding ends of a bubble, respectively, the following equilibrium relation holds for tubes so small that gravity effects are negligible:

$$P = \frac{2}{r} \sum_{i=1}^{n} \sigma_{lv}(\cos\theta_r - \cos\theta_a)_i + P_0 \qquad (10.5)$$

where P_0 is the atmospheric pressure. If the advancing angles and receding angles do not differ from bubble to bubble, the expression becomes

$$P = \frac{2n\sigma_{lv}(\cos\theta_r - \cos\theta_a)}{r} + P_0 \qquad (10.6)$$

The Jamin effect scales linearly with the number of bubbles, the surface tension, the difference in the cosines of the advancing and receding angles of contact, and inversely with the radius of the capillary.

10.7 TRUE CONTACT ANGLES ON ROUGH SURFACES

The above considerations have barely impinged on one aspect of rough surfaces. The importance conferred on the measurement of contact angles by the researches of Fowkes and his school makes increasingly irksome the limitation of this technique to uniform, plane, smooth, solid substrates. The advent of atomic force microscopy (AFM), with its ability to obtain a quantitative description of the roughness of a solid surface, created an opportunity to relieve the situation by correcting an observed value of a contact angle for the presence of surface roughness and so to arrive at its true value, that is, what would be obtained on an ideal smooth substrate of the same chemistry. Pioneer researchers have taken advantage of this modern instrumentation and of the well-defined substrates provided by synthetic ultrafiltration membranes to do exactly that, at least with one type of rough surface.[21]

Contact angles of water on eight different synthetic-polymer membranes, differing in pore size but all fabricated of poly(ether sulfone), were measured by the sessile-air-bubble technique. Corrections of the measured angles to eliminate the effect of surface roughness were calculated from the microgeometry of the various surfaces as revealed by AFM. The measured and corrected angles had mean and standard deviations, respectively, of $33.1 \pm 5.8°$ and $44.5 \pm 1.3°$. [See Figure 3 of Ref. 21 for data.] Clearly, by removing the effects of roughness, the mean contact angle was shown to be 34% higher and the standard error about the mean significantly reduced. Finally, for comparison, the contact angle on a smooth, relatively nonporous poly(ether sulfone) sample, which nevertheless still required a small correction for roughness, was $42.9 \pm 2.5°$. The two values agree within the error limits. [These poly(ether sulfone) membranes had been slightly hydrophilized by the manufacturer by an undisclosed process.]

10.8 TECHNIQUES

10.8.a The Wilhelmy Method (Plate or Fiber) for Measuring Contact Angles on Irregular Surfaces

The Wilhelmy plate method to obtain the contact angle on a uniform surface is described in Section 9.1.2. For such a purpose, the plate or fiber is adjusted so as to contact the surface of the liquid. The same equipment can be used to measure contact angles on irregular surfaces. The liquid surface is moved slowly along the substrate (plate or fiber) and the forces exerted on the substrate are recorded as a function of position. The elevator is moved slowly enough to avoid effects of dynamic wetting. If the substrate surface is uniform, the liquid substrate line of contact is horizontal. The downward force due to surface tension at any location of the three-phase boundary on the solid is given by the following closed integral:

$$f(h) = \oint \sigma_{lv} \cos \theta \, ds \qquad (10.7)$$

where h is the position along the fiber and ds is the element of line contact. If the surface is uniform and the perimeter constant, advancing and receding angles are obtained as follows. The liquid is first brought into contact with the substrate; the liquid level is brought up for a short distance; then the motion is reversed and the liquid is withdrawn. If the advancing and receding angles differ, the plot shows a hysteresis loop. The difference between the force on advancing the liquid and the force on its recession is

$$\Delta f = \sigma_{lv} s (\cos \theta_r - \cos \theta_a) \qquad (10.8)$$

Figure 10.4 shows a typical hysteresis loop when a plate or fiber with an advancing contact angle greater than 90° is lowered into a liquid and removed. Upon initially touching the liquid surface, point a, the force on the balance is reduced by the upward vector with magnitude $B = \sigma s \cos \theta_a$. The reduction in the force as the plate or fiber moves from b to c is its buoyancy. When the motion is reversed, the contact angle changes from θ_a to θ_r, and the reduction of the upward vector, cd is given by Eq. (10.8). From d to e the plate or fiber is

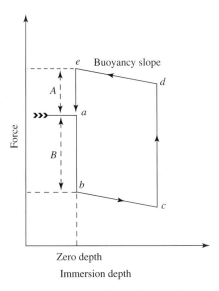

Figure 10.4 The force-depth isotherm of the fiber balance, showing a nonwetting advancing angle causing a buoyancy B, and a wetting receding angle, causing a force A. The arrows indicate the direction of motion of the fiber. While partially submerged, the fiber can be raised and lowered (recycled) to check for dynamic effects.

withdrawn from the liquid with consequent decrease in buoyancy and an increase in force. At e the plate or fiber is withdrawn from contact and the force returns to a. The change in the force, A, is $\sigma s \cos \theta_r$.

If the substrate is not of uniform perimeter or of uniform contact angle, the balance measures only $f(h)$, which cannot be farther analyzed unless the value of the perimeter is known at every height.

A variation on this technique is to pull a fiber through a small drop of liquid held within an aperture in a plate placed horizontally. The point of entry of the fiber displays the advancing contact angle and the point of exit of the fiber displays the receding angle. The force measured by the microbalance as the fiber is pulled through the drop reflects the difference in the cosines of the contact angles, Δf of Eq. (10.8). In this technique the advancing and receding angles are not measured separately. The technique is commonly used to ascertain the homogeneity of applied coatings on long sections of fibers. Differences in coating treatments leading to variations of the uniformity are readily detected.[8]

10.8.b Goniometer Methods for Measuring Contact Angle Hysteresis

The difference between the advancing and receding contact angles can also be measured by placing a sessile drop on a flat substrate and tilting the substrate until the point of incipient motion. The image of the drop at this position shows advancing and receding angles. If the advancing and receding angles are equal, the drop will slide down the substrate no matter how slight the angle. The contact-angle goniometer is well adapted to the use of this technique.

The same measurement can be made on a horizontal surface by increasing the size of the sessile drop by slow increments up to the point of incipient motion. The drop image gives the advancing contact angle. Withdrawing liquid from the drop until the contact line starts to move, causes the drop image to give the receding contact angle.

REFERENCES

[1] Wenzel, R.N. Resistance of solid surfaces to wetting by water, *Ind. Eng. Chem.* **1936**, *28*, 988–994.

[2] Wenzel, R.N. Surface roughness and contact angle, *J. Phys. Colloid Chem.* **1949**, *53*, 1466–1467.

[3] Thomas, T.R. *Rough surfaces*, 2nd ed.; Imperial College Press: London; 1999.

[4] Woodward, J.T.; Schwartz, D.K. Dewetting modes of surfactant solution as a function of the spreading coefficient, *Langmuir* **1997**, *13*, 6873–6876.

[5] Mahale, A.D.; Wesson, S.P. A computer model for wetting hysteresis. 1. A virtual wetting balance, *Colloids Surfaces A. Physicochem. Eng. Aspects* **1994**, *89*, 117–131.

[6] Wesson, S.P.; Kamath, Y.K.; Mahale, A.D. A computer model for wetting hysteresis. 2. A virtual wettability scanning balance, *Colloids Surfaces A. Physicochem. Eng. Aspects* **1994**, *89*, 133–143.

[7] Nishioka, G.M.; Wesson, S.P. Wetting behavior of spatially encoded heterogeneous surfaces, *Colloids Surfaces A: Physicochem. Eng. Aspects* **1996**, *118*, 247.

[8] Wesson, S.P.; Allred, R.E. Surface energetics of plasma-treated carbon fiber. *ACS Symp. Ser.* **1989**, *391*, Chapter 15.

[9] Hofer, R.; Textor, M.; Spencer, N.D. Imaging of surface heterogeneity by the microdroplet condensation technique, *Langmuir* **2001**, *17*, 4123–4125.

[10] Chen, W.; Fadeev, A.Y.; Hsieh, M.C.; Oner, D.; Youngblood, J.; McCarthy, T.J. Ultrahydrophobic and ultralyophobic surfaces: some comments and examples, *Langmuir* **1999**, *15*, 3395–3399.

[11] Onda, T.; Shibuichi, S.; Satoh, N.; Tsujii, K. Super-water-repellent fractal surfaces, *Langmuir* **1996**, *12*, 2125–2127.

[12] Cassie, A.B.D.; Baxter, S. Wettability of porous surfaces, *Trans. Faraday Soc.* **1944**, *40*, 546–551.

[13] Johnson, R.E., Jr.; Dettre, R.H. Contact angle hysteresis. I. Study of an idealized rough surface, *Adv. Chem. Ser.* **1964**, *43*, 112–135.

[14] Dettre, R.H.; Johnson, R.E., Jr. Contact angle hysteresis. II. Contact angle measurements on rough surfaces, *Adv. Chem. Ser.* **1964**, *43*, 136–144.

[15] Fowkes, F.M., Ed. *Contact angle, wettability, and adhesion*; Adv. Chem. Ser. 43; Amer. Chem. Soc.: Washington, DC; 1964.

[16] Berg, J.C., Ed., *Wettability*; Marcel Dekker: New York; 1993.

[17] Gerdes, S.; Cazabat, A.-M.; Ström, G. The spreading of silicone oil droplets on a surface with parallel V-shaped grooves, *Langmuir* **1997**, *13*, 7258–7264.

[18] Jamin, J.C. Mémoire sur l'équilibre et le mouvement des liquides dans les corps poreux, *C. R. Hebd. Seances Acad. Sci.* **1860**, *50*, 172–176, 311–314, 385–389.

[19] Jamin, J.C. On the equilibrium and motion of liquids in porous bodies, *Phil. Mag.* **1860** [4], *19*, 204–207.

[20] Smith, W.O.; Crane, M.D. The Jamin effect in cylindrical tubes, *J. Am. Chem. Soc.* **1930**, *52*, 1345–1349.

[21] Taniguchi, M.; Pieracci, J.P.; Belfort, G. Effect of undulations on surface energy: A quantitative assessment, *Langmuir* **2001**, *17*, 4312–4315.

11 Surface-Active Solutes

11.1 INTRODUCTION

The capillary properties of pure liquids, discussed in the previous chapters, are modified by certain solutes called surface active. Surface activity of a solute is defined as the ability to reduce the surface tension at an interface without requiring concentrations so large that the distinction between solute and solvent is blurred. The surface tension of water or the interfacial tension between water and a hydrocarbon may be reduced by 50 mN/m at concentrations of less than 0.1% of a surface-active solute. In nonpolar solutions the effects are much smaller. The quality of the solute is measured by how little of it is required for a given effect. The most marked results are obtained from a solute that combines in its molecular structure an element having a high affinity for the solvent with an element having minimal affinity for the solvent. The combination of such disparate elements produces a molecule that has its lowest potential energy at a phase boundary. This subject cannot be discussed further without a specialized vocabulary, given in Table 11.1.

Technical terms (neologisms) are formed by combinations of the words given in Table 11.1, such as the following adjectives:

amphipathic — combining both kinds (oil and water understood)
amphiphilic — with affinity for both (oil and water understood)
hydrophilic — with affinity for water
hydrophobic — lack of affinity for water
lipophilic — with affinity for oil
lyophilic — with affinity for the solvent
lyophobic — lack of affinity for the solvent

Corresponding concrete nouns are amphipaths, hydrophiles, hydrophobes, lipophiles, lipophobes, lyophiles, and lyophobes. Corresponding abstract nouns are amphipathy, amphiphilia, hydrophilia, hydrophobia, lipophilia, lipophobia, lyophilia, and lyophobia. Scholars do not consider it good form to combine Latin and Greek roots in the same word; therefore, the following terms are best avoided: hydrophilicity, hydrophobicity, oleophilic. The widely used term "surfactant" is a combination of syllables from the phrase, "surface-active agent."

The molecule of a surface-active solute may not be a combination of hydrophilic and lipophilic elements, as moderate surface activity is found in solutions

INTRODUCTION

TABLE 11.1 Glossary of Some Classical Prefixes and Suffixes[a]

English	Greek	Latin
Oil	Lipo-	Oleo-
Water	Hydro-	Aqua-
Solvent	Lyo-	
Both	Amphi-	
Flow	Rheo-	
Affinity	-philic	
Lack of affinity	-phobic	
Nature or kind	-pathic	
Science	-ology	

[a]The English words are not literal translations but interpretations of how the Greek words are understood in the vocabulary of this branch of science.

near a phase separation. Nevertheless, the more powerful surface-active solutes used as detergents, emulsifiers, or wetting agents have a molecular structure composed of hydrophilic and lipophilic elements. These elements are present in different proportions; the balance between them determines whether the solute is oil soluble or water soluble, and various degrees between determine properties useful in different applications. Every such molecule occupies a place on a scale of hydrophile-lipophile balance (HLB). (See Section 22.7). Industrial surface-active solutes are not pure material of one defined molecular weight. Figure 11.1 shows the number of major components in a widely used surfactant, Triton X100 (Union Carbide). Each of these compounds has slightly different properties including HLB, hence their HLB is based on performance rather than on molecular structure.

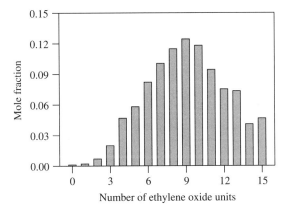

Figure 11.1 Distribution of the polyethylene oxide component in Triton X-100. The bar for $n = 15$ represents the sum of mole fractions of components $n \geq 15$.[1]

Commercial surface-active solutes are listed annually in McCutcheon's *Emulsifiers and Detergents*[2] and the *Chemcyclopedia*[3] of the American Chemical Society. A useful glossary of terms pertaining to surface-active solutes and their application as detergents is provided by the Soap and Detergent Association.[4] In January of each year *Chemical and Engineering News* issues a product report on soaps and detergents with business information on new commercial developments, new markets, company reports and sales.[5] Each quarterly issue of the *Journal of Surfactants and Detergents*[6] contains an updated list of internet addresses for professional organizations and manufacturers of surfactants.

11.2 RANGE OF SOLUTES FROM LIPOPHILIC TO HYDROPHILIC

Materials used as solutes can be classified on a scale ranging from a lipophilic extreme to a hydrophilic extreme. Starting at the lipophilic end of the scale with, for example, a long-chain hydrocarbon, materials that display various degrees of hydrophilia can be obtained by adding hydrogen-bonding substituent groups, such as hydroxyl, or amino, or carboxylate. The greatest hydrophilic character is conferred when the substituent group is ionized. In this way we proceed from stearane ($C_{18}H_{38}$), for example, to stearyl alcohol, stearyl amine, stearic acid and, finally, sodium stearate. Stearane is entirely lipophilic: It is all but completely insoluble in water and will not spread on a water surface but floats thereon as a compact liquid lens. On adding a hydrophilic group, such as hydroxyl or amino, to the stearane molecule, the solubility in water increases, although it is still practically insoluble; but the hydrogen-bonding group now enables the compound to spread spontaneously on a water surface. The spreading would continue all the way to monomolecular thickness given an appropriately small quantity of compound or a sufficiently large surface of water. This type of monomolecular layer (abbreviated as "monolayer") is known as an insoluble film. Let the hydrogen-bonding group be ionic, such as carboxylate, sulfate, sulfonate, or ammonium ions, then the corresponding compounds are much more soluble in water but are still positively adsorbed at the water/air surface or water/oil interface. By virtue of their positive adsorption these materials as solutes reduce surface or interfacial tension, the relation between adsorption and surface tension lowering being a thermodynamic consequence first made quantitatively explicit by J. Willard Gibbs (the Gibbs adsorption theorem). The surface region, that is, that which contains the excess adsorbed solute, is known as a "soluble film" to distinguish it from the insoluble monolayer already mentioned. Soluble films are usually thicker than a single layer of molecules.[7]

Figure 11.2 shows the effect of three different pure solutes in water at room temperature on the surface tension. Spectroscopically pure sucrose does not affect the surface tension. Simple electrolytes, such as NaCl, Na_2SO_4, KNO_3, increase the surface tension. And surface-active solutes, such as soap, reduce the surface tension by 40 mJ/m^2 at concentrations as low as 0.05% by weight. The data of

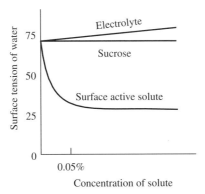

Figure 11.2 The variation of surface tension by solute types.

Figure 11.2 can be used to calculate the excess surface concentration of solute, Γ_2, in moles/m^2 by the Gibbs equation, which in its simplest form for a pure 1-1 electrolyte is

$$\Gamma_2 = -\frac{1}{2RT}\frac{d\sigma}{d\ln c_2} \qquad (11.1)$$

where σ is the surface tension of a solution of concentration c_2.

The positive slope shown for the electrolyte in solution means that it is negatively adsorbed, that is, the ions in the surface region are farther apart than they are in the bulk phase. The zero slope shown for sucrose means that it is not adsorbed; that is, the molecules in the surface region are the same distance apart as they are in the bulk phase. The negative slope shown for a surface-active solute means that it is positively adsorbed, that is, its molecules are much closer together in the surface phase than they are in the bulk phase.

As ions are the most hydrophilic groups known, increasingly hydrophilic compounds cannot be made by further substitution of hydrophilic groups; instead, the degree of lipophilia of the organic radical can be diminished by decreasing its molecular weight. Ultimately, when fewer than six carbon atoms are present in the radical, the compound, for example, sodium propionate, hardly differs from an ordinary strong electrolyte composed of a small anion and a small cation. Small ions are so extremely hydrophilic that their lowest potential energy occurs when they are fully surrounded by water molecules, which is to say, when they are in the bulk solution phase rather than in the surface region. Such a condition leads to negative adsorption, for which experimental evidence can be found in a surface tension that increases with higher concentrations of solute. Ordinary strong electrolytes show this effect.

Table 11.2 lists examples of solutes from stearane, the most lipophilic, to sodium sulfate, the most hydrophilic. Their properties and the properties they confer on the solution vary continuously. A graded series of solutes in terms of hydrophile–lipophile balance is also shown in Table 11.2. See Figure 22.5 for the HLB scale.

TABLE 11.2 Graded Series of Solutes in Terms of Hydrophile–Lipophile Balance

Lipophilic End of Scale						Hydrophilic End of Scale
Stearane	Steric Acid	Sodium stearate	Sodium laurate	Sucrose		Sodium sulfate
Soluble in oil; insoluble in water	Soluble in oil; insoluble in water	Soluble in oil and in hot water	Slightly oil-soluble; soluble in water	Insoluble in oil; soluble in water		Insoluble in oil; soluble in water
Nonspreading on water substrate	Spreads on water substrate	Spreads on water substrate	Reduces surface tension of aqueous solutions	Does not affect the surface tension in aqueous solution		Increases surface tension in aqueous solution
Does not affect interfacial tension at oil/water interface	Reduces interfacial tension at oil/water interface	Reduces interfacial tension at oil/water interface	Reduces interfacial tension at oil/water interface	Does not affect interfacial tension at oil/water interface		Increases interfacial tension at oil/water interface
Does not stabilize emulsions	Stabilizes water in oil emulsions	Stabilizes either type of emulsion	Stabilizes oil in water emulsions	Does not stabilize emulsions		Decreases the stability of emulsions
		HLB Scale				
	1			20		

11.3 TYPES: ANIONIC, CATIONIC, AND NONIONIC

A surface-active solute may be synthesized by combining in the same molecule lipophilic and hydrophilic moieties; these must also be sterically separable. For example, a block copolymer of polyethylene oxide and polypropylene oxide makes a surface-active solute, whereas the polymer produced by mixing the monomers is not surface active. Table 11.3 lists a number of hydrophilic and lipophilic moieties, which may be combined to give solutes with different hydrophile–lipophile balances.

The terms "anionic," "cationic," and "nonionic" refer in each case to the hydrophilic moiety. Of these, the anionic types account for about 75% of consumption in the United States; sodium and potassium salts of the fatty acids (soaps) are historically the first detergents known and still account for about 25% of U.S. consumption. The raw materials, animal fat (tallow) and alkali, are cheap. The chief disadvantages of the alkali-metal soaps are that they form water-insoluble salts with divalent and trivalent cations (particularly calcium, magnesium, and iron), they are insoluble in brine, and they hydrolyze to their insoluble parent fatty acids in acid solutions.

TABLE 11.3 Hydrophilic and Lipophilic Moieties

Hydrophilic		Lipophilic	
Ionic		**Hydrocarbon**	
Carboxylate		Straight-chain alkyl	(C_8–C_{18})
	—CO_2^-		
Sulfate		Branched-chain alkyl	(C_8–C_{18})
	—OSO_3^-		
Sulfonate		Alkylbenzene	(C_8–C_{16})
	—SO_3^-		
Quaternary ammonium	R_4N^+	Alkyl-naphthalene	
		Perfluoroalkyl	(C_6–C_8)
Nonionic		**Polymeric**	
Fatty acid	—CO_2H	Polypropylene oxide	$H[OCH(CH_3)CH_2]_nOH$
Primary alcohol	—CH_2OH	Polysiloxane	$H[OSi(CH_3)_2]_nOH$
Secondary alcohol	—CRHOH		
Tertiary alcohol	—CR_2OH		
Ether	—COC—		
Polyethylene oxide	H—$[OCH_2CH_2]_n$—OH		

With the advent of the petrochemical industry, stimulated by world wars, other types of anionic surface-active solutes appeared, namely, sulfonates and sulfates. At first the branched-alkyl chains were more economical but they soon introduced problems for water management because they have a poor biodegradation profile.[8] Surfactants were soon found in sewerage water and rivers and also in ground water for drinking-water supply as a result of soil infiltration. The petrochemical industry remedied the situation by converting to long chain, linear surfactants, which have a better biodegradation profile.[8] Of these, the sodium salt of the linear alkylbenzene sulfonate is the most widely used as an industrial and domestic detergent. It does not have the disadvantages listed above for the alkali-metal soaps. Examples of commonly used surface-active solutes are listed in Table 11.4.

Cationic surface-active agents are adsorbed by negatively charged surfaces by electrostatic attraction. Their first effect on such substrates is to neutralize the charge and, by virtue of their lipophile, to render the surface lipophilic. At greater concentrations the cationic is adsorbed in the usual way, that is, by the lipophile; thus placing a second layer of solute on top of the first, thereby creating a positively charged surface and converting it from lipophilic to hydrophilic. Cationic surface-active solutes have wider bactericidal activity than anionics and find their major use as germicides. Although they are detergents, they are seldom used for that purpose except where sterile conditions are important, as in the laundering of diapers or in rinsing bar glassware. Cationic solutes are usually incompatible with anionic solutes; and since latexes and dispersions are often mixed together in industrial processes, the use of cationics as emulsifying or dispersing agents is generally avoided. Special cases exist, of course, where the destabilizing of an emulsion or a latex or a foam is desired, and the addition of a cationic solute accomplishes that purpose. For example, asphalt emulsions, used in road-making, are designed to "break" and release the asphalt when they are in contact with wetted gravel. If the gravel contains flints or silicates, an emulsion stabilized by a cationic emulsifier breaks readily on contact with the negatively charged surfaces of these rocks. Gravels containing limestones, however, are positively charged on wetting and are better treated with asphalt emulsions stabilized by anionic agents. In paper-making, cationics inhibit hydrogen bonding between the cellulose fibers to produce the softer texture of tissue and toilet paper. In the same way cationics act as softeners in laundry use. Quaternary ammonium compounds, such as di-n-alkyldimethylammonium chloride, are the most commonly used cationics.

Amphoteric surface-active solutes may be either anionic or cationic in water, depending on pH. These substances have both amino and carboxylate groups: The amino group is charged positively at low pH and the carboxylate is charged negatively at high pH. An example is β-N-alkylaminopropionic acid, $RNHCH_2CH_2COOH$. Near the isoelectric point they have minimum solubility in water and greater surface activity. These are frequently used in shampoos and other personal-care products because of their mildness.

Fatty-acid esters of glycerol or sorbitol are themselves nonionic agents and soluble in oil. To confer solubility in water, the commonly used hydrophile is polyethylene oxide. This hydrophile is approved by the Food and Drug Administration for use in foodstuffs. It finds applications as a wetting agent in dehydrated milk and eggs, and in cocoa, flour, and other poorly wetted powders. Any solute containing polyethylene oxide as an adduct becomes insoluble in water at higher temperatures, as the hydrogen bonding of water to the ether oxygens is diminished

TABLE 11.4 Commonly Used Emulsifiers, Detergents, Dispersants, and Builders[2]

Chemical Class	Application
1. Anionic	
Alkyaryl sulfonates	Detergents, emulsifiers
Fatty alcohol sulfates	Detergents, emulsifiers
Lignosulfonates	Dispersants
Alkali soaps of tall oil	Anionic emulsifiers
Alkali soaps of rosin	Anionic emulsifiers
Dialkylsulfosuccinates	Wetting agents
2. Cationic	
Alkyltrimethylammonium chloride	Emulsifier, corrosion, inhibitor, textile softener, antibacterial agent, detergent
3. Nonionic	
Alkanolamides	Detergents, foam stabilizers
Glyceryl esters	Emulsifiers
Ethylene-oxide condensates of alkylphenols	Emulsifiers
Ethoxylated alkylphenols	Detergents, wetting agents, emulsifiers, dispersants
Ethoxylated fatty esters	Food emulsifiers (oil in water)
Fatty esters	Food emulsifiers (water in oil)
Polyalkylsuccinimides	Oil-soluble dispersants
Lecithins	Oil-soluble dispersants
Metal soaps	Oil-soluble dispersants
4. Builders	
Polyphosphate	Dispersing agent
Tetrasodium pyrophosphate (TSPP)	Dispersing agent
Trisodium orthophosphate	Dispersing agent
Sodium nitrilotriacetate (NTA)	Sequestering agent
Sodium ethylenediaminetetraacetate (EDTA)	Sequestering agent
Sodium carbonate, borax, silicates, citrates	Alkalis
Sodium oxydiacetate (ODA)	Sequestering agent

by heat. What was previously a hydrophile then becomes a lipophile. The cloud point of these nonionics is the temperature at which the solute becomes insoluble and the solution becomes turbid. The larger the fraction of polyethylene oxide or other hydrophiles in the molecule, the higher is the cloud point.

Builders are certain components in formulated detergents that remove metal ions, such as calcium and magnesium, either in a soluble form by sequestration or in an insoluble form by precipitation. They also maintain alkalinity, which promotes detergency. Examples of builders are listed in Table 11.4

Applications of surface-active solutes to emulsions, foams, and suspensions are treated in Chapters 22, 23, 24, and 25.

REFERENCES

[1] Gu, J.; Shelly, Z.A. Comparative phase behavior about the L_2 phase of ternary and quaternary systems of Triton X-100 and its separated *p-tert*-OPE$_n$ ($n = 5$, 7, and 9) components in cyclohexane, *Langmuir* **1997**, *13*, 4251–4244.

[2] McCutcheon's: *Emulsifiers & Detergents*, American Edition; MC Publishing: Glen Rock, NJ; An annual publication.

[3] *Chemcyclopedia; Chem. Eng. News*; Amer. Chem. Soc.: Washington DC; An annual publication.

[4] The Soap and Detergent Association. www.sdahq.org

[5] McCoy, M. Soaps and detergents, *Chem. Eng. News* January 24, **2000**, *78*(4), 37–52.

[6] AOCS Press, 2211 W. Bradley Ave., Champaign, IL 61826-3489, (217) 359-2344, www.aocs.org.

[7] Dixon, J.K.; Judson, C.M.; Salley, D.J. Study of adsorption at a solution–air interface by radiotracers, *Monomolecular Layers Symp. 1951*, **1954**, 63–106.

[8] Schwuger, M.J. *Detergents in the environment*; Marcel Dekker: New York; 1997.

12 Physical Properties of Insoluble Monolayers

12.1 OBSERVATIONS

The first reported influence of oil films on water was the observation of the damping of waves by oils such as olive oil and whale oil. This effect was known to the ancients; it is mentioned in Pliny's *Natural History* (A.D. 77), and it received widespread recognition in the eighteenth century from experiments by Benjamin Franklin. On a transatlantic voyage Franklin was informed by a sailor that a passing vessel was a whaler. On inquiring how the sailor knew, he was shown the becalmed water surrounding the ship, which was caused by a thin film of whale oil which had leaked through the wooden hull. This information inspired Franklin's classic experiment on the spreading of olive oil on a pond on Clapham Common. A surprising result of these experiments was the extent of coverage attained by a small volume of oil.

In 1891 Lord Rayleigh requested space in the British journal *Nature,* "for the accompanying translation of an interesting letter which I have received from a German lady, who with very homely [that is, domestic] appliances has arrived at valuable results respecting the behavior of contaminated water surfaces."[1] Miss Pockels' experiments demonstrated that the surface tension of a strongly contaminated water surface declines as the area of the surface is reduced. She measured the surface tension of the film by the force required to separate a small disk (6 mm diameter) from the surface. Rayleigh, repeating some of Miss Pockels' experiments, found that castor oil spreads on water to a minimum thickness of 1.3 nm, from which he concluded that the oil film attenuates to the thickness of a single molecule.

An observation made in the course of this early work is the remarkable difference in properties between bulk water and its surface. A drop of phenol or a minute crystal of stearic acid, which may a take an hour or more to dissolve when immersed in water, dissolves and spreads instantly when placed in contact with the surface.

Langmuir developed the fundamental methodology to measure the properties of insoluble monolayers as a function of their surface concentration, leading to the determination of molecular dimensions. A typical calculation goes as follows: A solution of palmitic acid (MW = 256) in benzene contains 4.24 g/L. When 0.0239 mL of this solution is placed on a water surface of 500 cm^2, the benzene evaporates and the palmitic acid forms a monomolecular film, which can be compressed to a limiting area of 0.21 nm² per molecule. The crystal structure of alkanes determined by x-ray diffraction gives a cross-sectional area of 0.186 nm² per molecule. The close agreement between these two independent results indicates that under these conditions the palmitic acid molecules are arranged with their hydrophilic acid groups in the water and with their hydrocarbon chains erected to their fullest extent and packed together in an orientation similar to that of the crystal.

Langmuir's original apparatus has been refined and automated to produce Langmuir–Blodgett layers. (see Fig 12.1).

Figure 12.2 is a schematic diagram of the possible monolayer states at three temperatures, T_1, T_2, T_3, where $T_1 < T_2 < T_3$, as typically obtained from an automated Langmuir film balance.[3] At large areas per molecule (and low spreading pressures), the film behaves as a two-dimensional analog of the ideal gaseous state, and is described by

$$\pi A = nkT \tag{12.1}$$

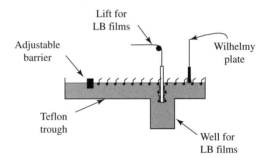

Figure 12.1 The Langmuir trough.[2]

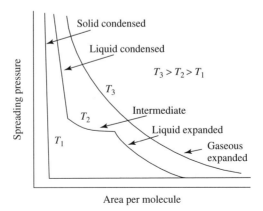

Figure 12.2 Possible monolayer states at three temperatures.[3]

where π, the lowering of the surface tension in milliNewtons per meter, is the "spreading pressure," A is the area of the water surface, and n is the number of molecules of monolayer substance. At lower temperatures and at smaller molecular areas, the monolayer undergoes a series of phase changes analogous to those in bulk matter. Using the traditional phase names, the compressed monolayer undergoes a first-order phase transition from gaseous to liquid expanded and further second-order transitions to intermediate-condensed and liquid-condensed phases (T_2) or to a solid-condensed phase (T_1). The precise behavior depends on the temperature and on the nature of both monolayer and substrate. Compression of the monolayer to smaller areas results in an increase of spreading pressure and changes the molecules from random to more organized structures, tending toward an orientation similar to that of the crystal. Ultimately, the monolayer collapses under sufficient compression. The so-called insoluble monolayer is actually a two-dimensional solution of the adsorbate in water; the spreading pressure is more closely analogous to osmotic pressure than to the pressure of a three-dimensional gas; and the monolayer states are analogous to the demixing of a two-component system.

Ries and Cook[4] compared the π-A isotherms of monolayers of stearic acid (I) and isostearic acid (II), which differ in molecular structure only by branching at the end of the hydrocarbon chain (Figure 12.3). The methyl side chain of isostearic acid should have little effect on molecular packing and film strength because it is small and at the remote end of the long molecule. Furthermore, the melting points of the isostearic and stearic acids are almost identical. Although the difference in structure is so slight, the behavior of the two monolayers is

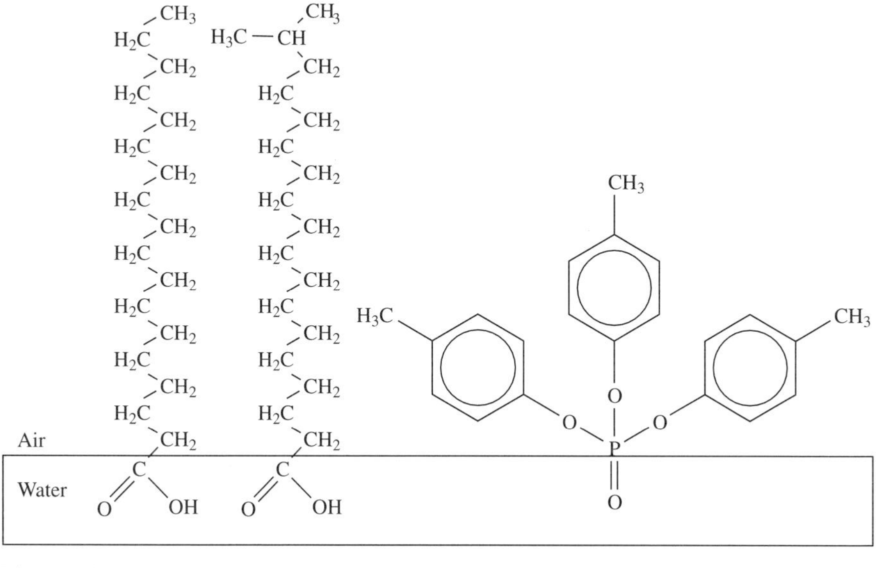

Figure 12.3 Structures of stearic acid, isostearic acid, and tri-p-cresyl phosphate at the air–water interface.[2]

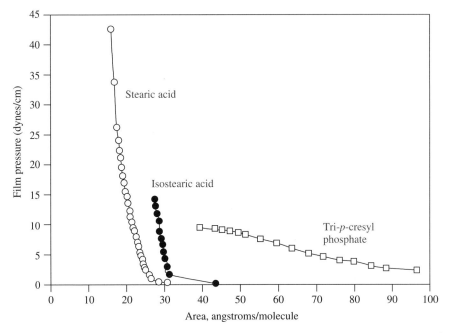

Figure 12.4 π-A isotherms of stearic acid, isostearic acid, and tri-p-cresyl phosphate.[2]

markedly different (Figure 12.4). The limiting area per molecule, extrapolated to zero pressure from the steepest part of the π-A isotherm, is 0.20 nm^2 for stearic acid but 0.32 nm^2 for isostearic acid. Evidently the tiny branching at the end of the chain in II is enough to prevent the close packing of chains achieved in I. When the chains are fully extended and oriented parallel to one another, considerable cohesion exists between molecules, just as in crystals of paraffin. Such a monolayer is very rigid and resistant to collapse: For example, the collapse pressure of I is 42 mJ/m; but the same cohesion is not reached in II: its monolayer collapses at the relatively low pressure of 15 mJ/m. A third substance, tri-p-cresyl phosphate (III) spreads out like a three-leaved clover on the surface of water. The π-A curve of the III monolayer shows great compressibility, but it collapses at 9 mJ/m and has the very extensive limiting area per molecule of 0.96 nm^2. These comparisons show the effects of differences of molecular structure.

Ries and coworkers[5,6] obtained gold-shadowed electron photomicrographs of some features of monolayers. In the discontinuous region of the first-order phase transition of n-hexatriacontanoic acid, $CH_3(CH_2)_{34}COOH$, from two-dimensional gaseous to two-dimensional liquid, at which the adsorbate should exist in two phases, namely, islands of two-dimensional liquid disks surrounded by a matrix of two-dimensional gaseous phase, the electron photomicrograph clearly shows the circular outline of the separated disklets. Successive photomicrographs, taken as the available area is reduced, show the inversion of the two-phase system to a state where the two-dimensional liquid is the continuous matrix, and finally to the state of the collapsed film, where bimolecular leaflets have been raised and then dropped on top of the underlying monolayer (Figure 12.5).

Multicomponent monolayers of biological origin constitute the lipid bilayers that are the basic structural framework of cell membranes. The inner and outer monolayers of the bilayer are oriented with their respective hydrophilic extremities toward the internal and external aqueous phases. Studies of membrane components and model compounds on aqueous substrates are intended to probe the structure and properties of lipid bilayers. Another type of biogenetic monolayer is "lung surfactant," a substance in the mammalian lung whose surface tension is greatly reduced by compression on every exhaled breath, and so regulates capillary pressure, promoting involuntary inhaling.[7] Premature infants lack lung surfactant and their breathing has to be stimulated mechanically. Biogenetic monolayers are usually extremely strong, with maximum spreading pressures at the collapse point of 68–72 mN/m, which is remarkable when compared with that of stearic acid, 40 mN/m. Ries and coworkers have published electron micrographs of collapsed films of such monolayers, revealing structures such as flat ribbons, platelets, and ridges.[8,9] Once formed these structures relax very slowly, giving rise to large hysteresis loops of the surface tension when the area is expanded and contracted. Microlayers of biological origin also occur on the

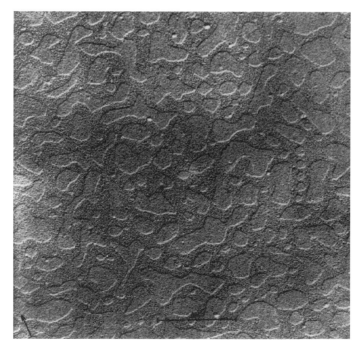

Figure 12.5 Electron photomicrograph of monolayers of n-hexatricontanoic acid.[6]

surfaces of oceans and lakes, where they may be detected by their effect on the reflection of light from the water surface.[10]

The sensitivity of the π-A curve to minute changes of molecular structure of the adsorbate makes the film-balance technique suitable for deciding fine points of difference between competing versions of structural formulas, as happened, for example, with those suggested for sterols.[11]

12.2 VISCOELASTICITY OF INSOLUBLE MONOLAYERS

A characteristic of insoluble monolayers is that their surface tension varies when the film is either expanded or compressed. Surface tension gradients lead to Marangoni flow, by which an applied strain is spontaneously opposed. This property is an elasticity caused by dilatation; liquid surfaces that are incapable of creating surface tension gradients would show no dilatational elasticity. The viscoelastic properties of insoluble monolayers can be detected and measured by means of a periodic input of stress and the recording the resulting strain, that is, by periodic contracting and expanding the monolayer attached to an automatic of the force-area curve. A technique suitable for use with the Langmuir film balance is to apply a periodic strain to the monolayer by oscillating a barrier between two points. The data are obtained as a function of frequency. The variation of the spreading pressure is measured by means of a Wilhelmy plate or a pressure transducer.[12]

Figure 12.6 shows the fluid dynamics of wave motion. When Franklin spread a monolayer of olive oil on a pond the wave motion was significantly altered. The monolayer was compressed at the crest and dilatated at the trough. The higher surface tension at the trough drew the crest down by Marangoni flow. A different mechanism operates in the damping of waves by thick layers of oil by virtue of their superior viscosity.

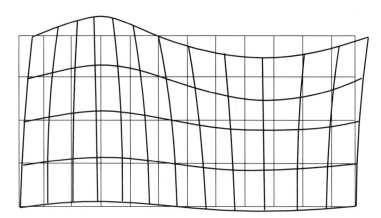

Figure 12.6 The fluid dynamics of wave motion. The volume elements at the crest are contracted and those at the trough are expanded.

12.3 BUILT-UP FILMS: LANGMUIR–BLODGETT FILMS

Langmuir[13] observed in 1919 that an insoluble monolayer could be transferred from the water substrate to a hydrophilic solid surface, such as a glass slide, by raising the slide through the liquid/gas interface. In 1934 Katharine Blodgett announced the discovery that built-up films could be formed by sequential monolayer transfer, the structures now universally referred to as Langmuir–Blodgett films.[14] On hydrophilic substrates no film is deposited during the first immersion; the first monolayer is deposited only during the first emersion. Further layers are deposited by repeated dippings of the solid through the interface. The layers are deposited hydrophile to hydrophile and lipophile to lipophile, or head to head and tail to tail, by the successive folding back and forth of the monolayer, as in Figure 12.7. The slide, in air, always has an odd number of layers on it: one on the first emersion, three on the second emersion, and so forth.

The surface is composed of the methyl groups that terminate the fatty acid chain, and these surfaces exhibit high-contact angles for water and organic liquids in accordance with the low surface energy of the methyl groups. Optical measurements by Langmuir and Blodgett[16,17] show that each double layer produces an increment of thickness nearly equal to twice the length of the fatty acid

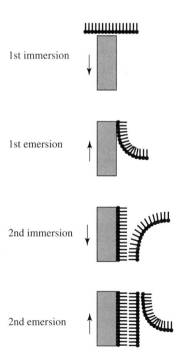

Figure 12.7 The building up of a Langmuir–Blodgett film.[15]

chain. Different arrangements of L–B films are possible depending on the initial state, surface pressure and temperature, of the insoluble monolayer.[18]

Built-up multilayers are important materials in thin-film technology, with applications to electron beam microlithography, integrated optics, electro-optics, electronic displays, photovoltaic cells, two-dimensional magnetic arrays, field effect devices, and biological membranes. Langmuir–Blodgett films are also used to build well-defined structures for fundamental studies of their electrical and optical properties. (The journal *Thin Solid Films* published by Elsevier Science, http://www.elsevier.com, frequently publishes the proceedings of symposia devoted to Langmuir–Blodgett films.)

REFERENCES

[1] Pockels, A. Surface tension (A translation by Lord Rayleigh of a letter he received), *Nature (London)* **1891**, *43*, 437–439.

[2] Ries, H.E., Jr. Monomolecular films, *Sci. Am.* **1961**, *204*(3), 152–164.

[3] Cadenhead, D.A. Monomolecular films at the air-water interface: Some practical applications, *Ind. Eng. Chem.* **1969**, *61*(4), 22–28; also *Chemistry and Physics of Interfaces II*; Ross, S., Ed.; Amer. Chem. Soc.: Washington, DC; 1971; pp 27–34.

[4] Ries, H.E., Jr.; Cook, H.D. Monomolecular films of mixtures I. Stearic acid with isostearic acid and with tri-*p*-cresyl phosphate. Comparison of components with octadecylphosphonic acid and with tri-*o*-xenyl phosphate, *J. Colloid Sci.* **1954**, *9*, 535–546.

[5] Ries, H.E., Jr.; Kimball, W.A. Structure of fatty-acid monolayers and a mechanism for collapse, *Proc. Int. Congr. Surf. Act., 2nd, 1957* **1957**, *1*, 75–84.

[6] Ries, H.E., Jr.; Kimball, W.A. Electron micrographs of monolayers of stearic acid, *Nature (London)* **1958**, *181*, 901.

[7] Exerowa, D.; Lalchev, Z.; Marinov, B.; Ognyanov, K. Method for assessment of fetal lung maturity, *Langmuir* **1986**, *2*, 664–668.

[8] Ries, H.E., Jr.; Swift, H. Monolayers of valinomycin and its equimolar mixtures with cholesterol and with stearic acid, *J. Colloid Interface Sci.* **1978**, *64*, 111–119.

[9] Ries, H.E., Jr. Interaction of cholesterol, cerebronic acid, valinomycin, and related compounds in monolayers of binary mixtures, *Colloids Surfaces* **1984**, *10*, 283–300.

[10] Cini, R.; Lombardini, P.P.; Hühnerfuss, H. Remote sensing of marine slicks utilizing their influence on wave spectra, *Int. J. Remote Sensing* **1983**, *4*, 101–110.

[11] Adam, N.K. *The physics and chemistry of surfaces*, 3rd ed.; Oxford University Press: London; 1941; pp 79–82.

[12] O'Brien, K.C.; Lando, J.B. Mechanical testing of monolayers 1. Fourier transform analysis, *Langmuir* **1985**, *1*, 301–305.

[13] Langmuir, I. The mechanism of the surface phenomena of flotation, *Trans. Faraday Soc.* **1919**, *15*, III, 62–74.

[14] Blodgett, K.B. Monomolecular films of fatty acids on glass, *J. Am. Chem. Soc.* **1934**, *56*, 495.

[15] Gaines, G.L., Jr. *Insoluble monolayers at liquid–gas interfaces*; Interscience Publishers: New York; 1966; p 337.

[16] Blodgett, K.B. Films built by depositing successive monomolecular layers on a solid surface, *J. Am. Chem. Soc.* **1935**, *57*, 1007–1022.

[17] Blodgett, K.B.; Langmuir, I. Built-up films of barium sterate and their optical properties, *Phys. Rev.* **1937**(2), *51*, 64–982.

[18] Petty, M.C.; *Langmuir–Blodgett films — an introduction*; Cambridge University Press: New York; 1996.

13 Aqueous Solutions of Surface-Active Solutes

13.1 ADSORPTION

The insolubility of water in oil and the insolubility of oil in water occur for different reasons. The cohesion of water, $W^{\text{coh}} = 146$ mJ/m², is much larger than the adhesion between water and an alkyl chain, given by Eq. (13.1), which accounts for the insolubility of water in paraffin hydrocarbons. This can be demonstrated pedagogically by gently shaking together in a Petri dish stir bars (representing polar molecules) and clear marbles (nonpolar molecules.) The stir bars coalesce and exclude the marbles. Placing the Petri dish on an overhead projector allows students to see the progress.[1] See Figure 13.1.

In contrast, while it takes work to move a water molecule into a hydrocarbon, it takes no work to move a hydrocarbon molecule into water. This statement derives from the following argument: The work of adhesion between an alkyl chain and water, due to dispersion forces, is

$$W^{\text{adh}} = 2\left(\sigma_{\text{hc}} \sigma_w^d\right)^{1/2} \cong 50 \text{ mJ/m}^2 \tag{13.1}$$

The work of adhesion between a hydrocarbon molecule and water is about the same as the work of cohesion ($= 2\sigma_{\text{hc}}$) of an alkyl hydrocarbon; that is, the enthalpy change is zero. The work is about the same to put a hydrocarbon molecule into water as it is to put it into hydrocarbon. Nevertheless, the hydrocarbon is insoluble. Therefore, the insolubility is caused by a change in entropy. This entropy change is the forced restructuring of water molecules in the vicinity of the hydrocarbon molecule, commonly called the hydrophobic effect.

One cannot claim that the alkyl chain is pushed out of the water by being repelled by water molecules, although adsorption phenomena appear to correspond to such an action. The hydrophobic effect is a key to understanding many effects in biological systems.[2]

When a certain amount of surfactant is dissolved in water at a constant temperature, a number of simultaneous equilibria are established, which are measured by their equilibrium constants. These constants are the solubility product, which determines the solubility, the distribution coefficient between dissolved and adsorbed solute, and the distribution coefficient between dissolved solute and micelles. Each of these equilibrium constants is related to a change of free energy.

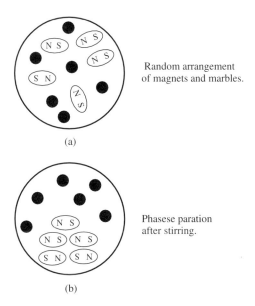

Figure 13.1 A lecture demonstration of the immiscibility of water in oil.[1]

Our present interest is in the distribution between dissolved and adsorbed solute. The hydrophile, either an ion or a nonionic group, of a surfactant containing a hydrocarbon group brings the hydrocarbon into an aqueous medium. Opposing forces come into play between the solubilizing action of the hydrophile and the opposing desolubilizing action of the hydrocarbon, which creates a dynamic equilibrium between dissolved and adsorbed solute. The distribution coefficient is influenced toward adsorption, the larger the size of the attached hydrocarbon group. (Traube's rule).

The thermodynamics of the transfer of a pure liquid hydrocarbon into water helps us to understand what takes place at the hydrocarbon/water interface of an amphiphilic solute. As this transfer does not take place spontaneously the change of free energy must be positive. That change has contributions from enthalpy and entropy changes, as described by Eq. (13.2),

$$\Delta G = \Delta H - T \Delta S \quad (13.2)$$

A collection of calorimetric data is reported by Tanford[2] for this transfer. The results for all aliphatic hydrocarbons show that the values of $T \Delta S$ far surpass those of ΔH. As an example take $C_6H_5C_3H_7$ for which ΔH is quoted as +2.3 kJ/mol and $T \Delta S$ at 298 K as −26.2 kJ/mol, giving a value of ΔG of +28.5 kJ/mol. The entropy change

$$\Delta S = S_{w/hc} - S_{hc/hc} \quad (13.3)$$

dominates the transfer, which means that a large decrease of entropy occurs at the water/hydrocarbon interface. A higher degree of local order must therefore be present in the water at that interface than in normal water. The first explanation was suggested by Frank and Evans[3] to the effect that the change in the state of the water molecules is the result of a rearrangement to regenerate the hydrogen bonds broken by the intrusion of the hydrocarbon. But new improvements in various instrumental techniques have brought into being molecular probes of the interface.[4,5,6] This more recent work definitively disproves the model of an ice-like structure at the water/hydrocarbon interface. The increase in ordered structure at the interfacial region is actually a strong orientation of water molecules brought about by dipolar interactions, accompanied by the necessary reduction of the strength of hydrogen bonding to free the water and allow the orientation to happen.

The positive free-energy change for the entry of hydrocarbon into water means a negative free-energy change for the ejection of hydrocarbon from water. Applied to the molecular hydrocarbon groups in water, thermodynamics favors their ejection, either into the adsorbed state or into micelles. Of course they are not all ejected because the hydrophile is there to favor solubility, unless the hydrocarbon group is too large.

This large entropy change is peculiar to water as a solvent for surface-active solutes. A similar feature is absent in nonpolar solvents, where spontaneous adsorption at the solvent/water/vapor interface does not occur.

Figure 13.2 depicts the initial and final states of this process showing the removal of the hydrocarbon ion from the bulk to the surface.

We have traced the adsorption behavior of hydrocarbon-based amphipathic solutes in aqueous solution to entropy changes within the water, rather than to a specific property of the lipophile. When such a solute molecule happens to arrive at a surface or an interface, it is partially withdrawn from the water; its potential energy is thereby reduced; and it lingers at that location, as it takes work to bring it back into the bulk phase. Adsorption is a state of dynamic equilibrium, which is why the adsorbed molecules are said to linger than to remain static at the interface.

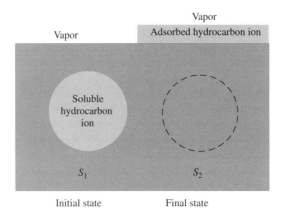

Figure 13.2 Why surfactants are spontaneously adsorbed at the water/vapor interface.

Adsorption at the water/vapor interface is not the only way for the solute molecule to withdraw from the aqueous medium. It may associate with other like molecules to form a self-assembled micelle, or it may be adsorbed at a solid/solution interface.

Amphipathic solutes other than those that are based on a hydrocarbon chain such as those that are based on a fluorocarbon chain or a polydimethylsiloxane chain, are subject to the same argument. In the complete absence of this mechanism, such as adsorption in nonpolar media, spontaneous adsorption may arise from other mechanisms such as acid–base interactions (see Section 8.4) or adsorption associated with an imminent phase change (see the Ross effect in Section 23.4.k.)

Effects of a different nature are created at the interface between water and a hydrophilic substrate such as glass or silica. Derjaguin[7] has demonstrated the presence of thick layers of liquid water of increased viscosity adjacent to the surface of glass at temperatures well below the freezing point of bulk water. This phenomenon is related to Faraday's discovery of a liquid layer of water on the surface of ice at temperatures below 0°C. At temperatures below about −10°C the liquid layer is no longer manifest.

13.2 LUNDELIUS'S RULE AND THE FERGUSON EFFECT

The lipophilic moiety limits the solubility in water of the molecule to which it is attached. The function of the hydrophile is to provide enough interaction with water to bring the insoluble lipophile into solution. A negative or a positive ion can bring a 16-carbon chain into aqueous solution at room temperature and an 18-carbon atom chain into solution in hot water. Solubility decreases with increasing chain length, and surface activity becomes more pronounced. Lundelius's rule is an explicit statement of a direct connection between the two. Its original statement is that a given solute is more adsorbed out of a solution in which it is less soluble. Clearly the competition for the solute is between the solvent and the surface or interface at which it is adsorbed: when the action of the former is weakened the action of the latter is strengthened. Lundelius, however, did not derive his rule from theory but empirically from the results of his measurements on the adsorption on blood charcoal of iodine dissolved in carbon disulfide, chloroform, and carbon tetrachloride, in each of which three solvents the iodine is of a violet color. Thus the possible question of whether the solute is the same molecular species in each solvent is forestalled. The adsorption isotherms, that is, the amount of iodine adsorbed per gram of charcoal versus the equilibrium concentration of iodine in solution, when plotted on log-log paper are linear and parallel to one another; and one can therefore compare the equilibrium concentrations, **C**, for any given amount of iodine adsorbed out of each solvent. The results in relative measure are

$$\mathbf{C}(CS_2):\mathbf{C}(CHCl_3):\mathbf{C}(CCl_4) = 4.5:2:1 \tag{13.4}$$

while the solubilities, **S**, at 14.5°C are

$$\mathbf{S}(CS_2):\mathbf{S}(CHCl_3):\mathbf{S}(CCl_4) = 69.1:14.3:8.1 = 3.8:1.8;1 \tag{13.5}$$

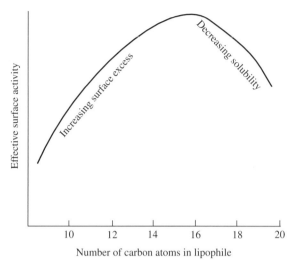

Figure 13.3 The variation of effectiveness of a surface-active solute with the number of carbon atoms in the lipophile at about 60°C. In general $C_{16} > C_{14} > C_{18} > C_{12}$.

Ferguson[8] discovered an optimum effect at a certain carbon chain length in an homologous series of surface-active solutes, when applied for various purposes, such as detergency, antibiotic action, hemolytic action, and emulsification. Lipophilic character, and therefore surface activity in an aqueous medium, is continuously increased by lengthening the hydrocarbon chain; the trend is accompanied by decreasing solubility in water, so that ultimately too few molecules are present in the system to be effective, although these few are almost all adsorbed. The balance between the two opposite trends gives a maximum in the function that relates effectiveness in a given application to the number of carbon atoms in the lipophile. The behavior is sufficiently general to be named "the Ferguson effect" (Figure 13.3).

13.3 MICELLIZATION AND SOLUBILIZATION

13.3.a Self-Assembly by Spontaneous Association

The same mechanism that causes adsorption of an amphipathic molecule also leads to spontaneous association of such molecules. The greater potential energy of water molecules in the vicinity of a hydrocarbon chain implies that any way to withdraw hydrocarbon from the surrounding water reduces the free energy and so occurs spontaneously. Adsorption of hydrocarbon at surfaces and interfaces is one such way; another is the spontaneous association of the hydrocarbon chains of amphipathic molecules to form colloidal aggregates, varying in size from dimers to complexes of 50 or more molecules. The dual equilibria of adsorption and association are depicted in Figure 13.4. Adsorption at the solution surface occurs

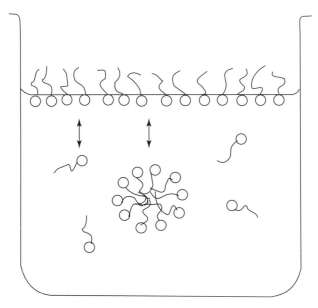

Figure 13.4 The dynamic equilibria of adsorption and micellization.

at lower concentrations than any significant association. The associated molecules form units known as "micelles" *(micellae,* L. crumbs), and the mechanism of their association is referred to as hydrophobic bonding. This infelicitous term should be construed as a shorthand reminder of a rather involved series of events: that primarily the entropic drive of the contiguous water layers, obscurely indicated by the adjective *hydrophobic,* rather than the *bonding* between hydrocarbon and hydrocarbon, is responsible for the phenomenon.

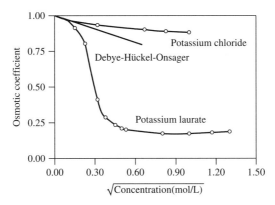

Figure 13.5 The variation of the osmotic coefficient with concentration for potassium chloride and for potassium laurate. The straight line is the Debye–Hückel–Onsager slope.[9]

Experimental evidence for the presence of micelles in aqueous solutions of amphipathic solutes is obtained from measurements of the colligative properties of such solutions, that is, the osmotic pressure, the lowering of the vapor pressure, the elevation of the boiling point, and the depression of the freezing point. These properties are linked, as the name *colligative* suggests: Each of them depends only on the concentration of osmotic units in the solution, whether these units be ions, molecules, macromolecules, or micelles. As dissolved ions or molecules of an amphipathic solute associate in solution, the concentration of osmotic units loses its proportionality to the total concentration of solute. The osmotic coefficient is defined as the ratio of the observed colligative property per mole of solute to the calculated colligative property based on the concentration of ions or molecules given by Avogadro's number and the supposition of complete electrolytic dissociation. The osmotic coefficient is equal to unity at infinite dilution and is always less than that at any finite concentration, being reduced by intermolecular or by interionic attraction as the concentration increases. The variation of the osmotic coefficient with concentration is shown in Figure 13.5 for potassium chloride and for potassium laurate, based on measurements of the lowering of the freezing point. Data for any of the other colligative properties could be used to construct a similar diagram. Using the lowering of the freezing point restricts the investigation to temperatures near 0°C. Modern techniques, using matched thermistors, make measurements of the lowering of the vapor pressure equally precise, and these measurements are not restricted to a single temperature. The abrupt decline of the osmotic coefficient at what appears to be a critical concentration of solute is clearly marked on the diagram, and certainly indicates that molecular aggregates begin to exist in the solution at and beyond a well-defined concentration, called the *critical micelle concentration,* or CMC.

At concentrations above the CMC, the micelles first formed are spherical; they become less symmetrical as concentration is increased to form cylindrical rods or lamellar structures. Figure 13.6 is a depiction of a sodium dodecylsulfate micelle

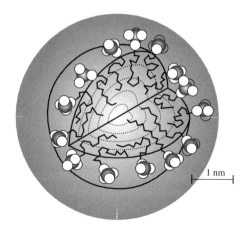

Figure 13.6 A sodium dodecylsulfate micelle.[10]

containing sixty molecules as given by Israelachvili, showing the way in which the hydrocarbon chains maintain uniform density in the interior of the micelle and the ionic groups on the outside in contact with water and their counterions. Not shown in the diagram is the ion pairing of the sulfate ions, which reduces the net charge on the micelle to about minus three. Without the ion pairing the repulsive forces of sixty anions would tear the micelle apart.

13.3.b Methods to Determine Critical Micelle Concentration

Methods to determine the CMC of amphipathic solutes can be divided into those that require no additive and those in which an additive is present in the bulk solution. The former type of method is preferable as the presence of an additive can affect the CMC.

The CMC is well defined experimentally by a number of physical properties besides the variation of the osmotic coefficient with concentration. Ionic amphipaths show a discontinuity at or near the same concentration in the variation of specific electrical conductivity with concentration. For such amphipaths, locating this discontinuity by means of conductivity measurements is the simplest experimental technique to determine CMC. Another electrical property of ionic amphipath solutions is the Hittorf transport number, which shows an interesting peculiarity at the CMC and above. After electrolysis of a solution containing micelles has proceeded for a time and a measured number of coulombs has been passed through the solution the anode and cathode compartments are analyzed for anion and cation concentrations, respectively, as is the customary procedure with this determination. The anode compartment is then found to contain more equivalents of anion than the number of electrical equivalents of current that were passed through the solution, and the cathode compartment is found to have lost rather than gained cations. The explanation is that the micelle, an aggregate of anions, must also contain cations in the form of ion pairs, which are carried to the anode with the micelle instead of to the cathode. Let us suppose, following Gonick,[11] that the micelle of the amphipath potassium laurate, KL, has a composition corresponding to the formula $K_{24}L_{27}^{-3}$, that is, an aggregate of 27 anions of which 24 are paired with cations, leaving a net ionic charge of -3; and let the transport number of the micelle be 0.227, which would give for the K^+ ion a transport number of $1 - 0.227 = 0.773$. On passing one equivalent of electricity, that is, 96,500 coulombs, 0.227 equivalents of micelles move to the anode. Now, one equivalent of micelles is one-third of a mole of micelles, and in terms of laurate ion contains one-third of the laurate content of the micelle, that is, one-third of 27 mol or 9 mol. The quantity of laurate ion in the anode compartment is therefore 0.227×9 or 2.04 mol. By analysis, therefore, the transport number of the anion appears to be 2.04, which is tantamount to saying that 204% of the current is carried by the anion. At the same time in the cation compartment, per equivalent of electricity passed through the solution, 0.773 equivalents of K^+ has passed in, but one-third of 24 mol of K^+ has moved out with the micelle, with a transport number of 0.227; that is, a net change of $0.773 - 8 \times 0.227 = -1.043$, or -104.3% of the current is transported by the cations. These transport numbers, of course, are not real. They are calculated on the wrong assumption, namely, that the ions are *not* associated; they add algebraically to 1.000, in agreement with the Hittorf mechanism, but the true transport numbers are the assumed values of $t_- = 0.227$ and $t_+ = 0.773$. The Hittorf method cannot obtain the true values; instead it yields manifestly absurd values. Below the CMC the method is valid, and the presence of micelles in the solution is marked by the deviation from reasonable to absurd values of the transport numbers.

The presence of a new species in solution can be detected by observing changes in any physical property that can be measured with high precision. Three such properties are turbidity, density, and refractive index. Solutions of amphipathic solutes (amphipaths) show a discontinuity at the CMC when these properties are plotted as a function of concentration. The most popular of these methods is the measurement of the intensity of scattered light. The intensity of the scattered light increases sharply at the CMC because the micelles scatter more light than the medium[12] (Chapter 1).

Other properties by which CMC can be measured are surface and interfacial tensions. Single molecules of an amphipath are readily adsorbed, and positive adsorption is related to the lowering of surface or interfacial tension by the Gibbs equation, which, for a two-component system, is

$$\Gamma_2 = -\frac{d\sigma}{RT d \ln a_2} \tag{13.6}$$

where Γ_2 is the excess surface concentration of solute in mol/m^2, and a_2 is the activity of the solute. Adsorption and micellization are two effects promoted by the same cause, but they are independent of each other; each has its own equilibrium constant; but once micelles form in significant quantities, the concentration of single molecules does not increase, or increases only slightly, with increase of total concentration, since all or most of the solute dissolved above the CMC goes into micelles. The adsorption equilibrium, therefore, receives no, or only slight, further contribution; and the surface or interfacial tension does not reduce any further, or reduces only slightly, below the value it reaches at CMC. The surface tension isotherm, plotted with the surface tension as ordinate versus the logarithm of concentration, shows an abrupt change of slope at the CMC. A minimum in the isotherm, should it occur, is now recognized as the consequence of contamination or the presence of hydrolysis products. The hydrolysis products are more surface active than the prime component but are removed from the surface at higher concentrations by solubilization in micelles.

Figure 13.7 shows how various properties of solutions at room temperature can be used to determine CMC, which vary within a slight range of concentration. A more highly purified solute would have less variation and would not have a minimum in the surface tension isotherm.

13.3.c Equilibrium or Phase Separation?

The major thermodynamic process taking place in micelle-forming solutions is the formation of molecular aggregates from molecularly dispersed solute. This

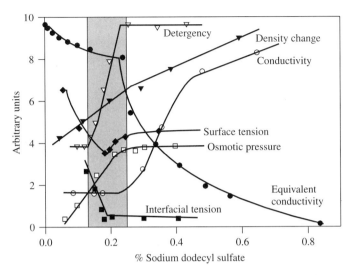

Figure 13.7 The CMC of sodium dodecyl sulfate as determined by various properties.[13]

process may be described thermodynamically either as a kinetic equilibrium between the single ions and the micellar species[14] or as a phase separation.[15] Both approaches can describe major features of micelle formation.

Micellization and adsorption are separate and independent processes, taking different paths, with different equilibrium constants and rates of reaction. Micelles are brought into the solution at a concentration of solute determined by the micellization equilibrium constant, independently of the surface concentration of adsorbate, which is determined by a different equilibrium constant. But when micelles appear at the CMC, they virtually bring adsorption to an end at whatever surface concentration has been reached, because micelles greatly inhibit further increase of concentration of the amphipathic ion or molecule. The adsorbed layer may not be fully saturated at the CMC. If the soluble amphipath is converted into an insoluble monolayer by placing it on a strong brine solution, it can be compressed further, by means of a barrier, to a greater surface concentration than it ever could reach by adsorption from solution.

Let us first consider micellization as an equilibrium process. Take as a model of a micelle the formula $K_{24}L_{27}^{-3}$ that we used before. Then

$$K = \frac{\left[K_{24}L_{27}^{-3}\right]}{[K^-]^{24}[L_{27}^-]^{27}} = \frac{[m]}{[c-24m]^{24}[c-27m]^{27}} \quad (13.7)$$

where $\left[K_{24}L_{27}^{-3}\right] = [m]$ and c is the total concentration of solute. The variation of the concentration of micelles $[m]$ with the total concentration of solute $[c]$ is calculated by Eq. (13.7) with an arbitrary value of $K = 1.000$ to illustrate the effect of large exponents in the denominator of such an expression.

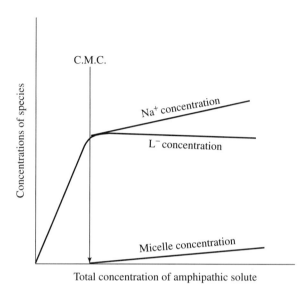

Figure 13.8 Micellization as an equilibrium process. Micelle taken as $K_{24}L_{27}^{-3}$.

Figure 13.8 shows how the concentration of micelles $[m]$ grows with an increase of total concentration $[c]$. The CMC is the point on the curve at which $[m]$ shows a sudden increase; the larger the values of the exponents in Eq. (13.7), the more abruptly does this happen and the more pronounced is the "criticality" of the "transition." It therefore becomes of interest to explore the behavior expected from a genuine transition and a genuine critical point. The model to consider is analogous to the separation of a phase when maximum solubility is reached. If micelles separate as a phase at a critical concentration, the concentration of unassociated solute does not increase beyond its saturation value reached at the CMC. Although more solute goes into solution beyond the CMC, it is taken up entirely by micelles; the concentration of micelles, therefore, continues to increase while the concentration of unassociated amphiphiles remains constant at its CMC value. If the average number of amphiphiles in a micelle is n, the molar concentration of micelles grows at a rate that is only $1/n$ times that of the moles of solute put into the solution (Figure 13.9).

Comparison of Figures 13.8 and 13.9 shows how nearly alike are the results obtained from these two models: namely, concentrations based on micellar equilibrium and concentrations based on a separation of micelles as a distinct phase. The result the two models have in common is that micelles appear in the solution abruptly enough to define a true critical or a pseudocritical micelle concentration. Experimental measurements of extreme precision on scrupulously purified solutes would be required to differentiate between the two models. Mitchell and Ninham[16] have shown that the law of mass action is an appropriate description of the process of micellization; the alternative description as a phase separation does provide a simpler basis for subsequent development of a theory of

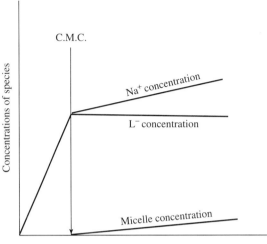

Figure 13.9 Micellization as a pseudophase separation. Micelle taken as $K_{24}L_{27}^{-3}$.

self-assembly of amphipathic molecules into micelles, but is designated "the pseudophase approximation."

13.3.d Effects of Varying Molecular Structure

A continuous variation of the hydrophile–lipophile balance of solutes is provided by a homologous series of the lipophile. The free energy of micellization, whichever mechanism of micelle formation is considered, is given by $-RT \ln [CMC]$. The logarithms of the critical micelle concentrations of a homologous series of sodium soaps (taken from Table 13.1) are plotted in Figure 13.10 against the number of carbon atoms in the alkyl chain. The straight line obtained on this plot shows that each additional carbon atom makes the same contribution to the change of free energy on micellization. The slope of the straight line gives a value of 1.64 kJ/mol per carbon atom. This value may be compared with the free energy of adsorption of a homologous series of fatty acids at a water/air surface, which is 2.72 kJ/mol (see Figure 15.5). The two values are of the same order, indicating the common source of the two effects; but complete agreement is not to be expected because of the different hydrophiles and the different structures of micelles and adsorbed films.

The same relation that holds for the variation of the CMC with the chain length of an homologous series also holds for the concentrations of successive members of an homologous series of sodium soaps required to give the same lowering of surface tension, where it is known as Traube's rule. This rule can be interpreted, by the same thermodynamic reasoning, as an indication of the increase in the surface activity for each additional CH_2 group and in the work done when a molecule passes from the bulk phase to the surface layer.

TABLE 13.1 Critical Micelle Concentrations of Aqueous Solutions of Sodium Soaps without Added Electrolyte

Name	Temperature (°C)	CMC (M)
Sodium pentanoate	20	2.35
Sodium hexanoate	20	1.6
Sodium heptanoate	20	9.5×10^{-1}
Sodium octanoate	25	3.45×10^{-1}
Sodium nonanoate	20	2×10^{-1}
Sodium decanoate	25	9.5×10^{-2}
Sodium dodecanoate (laurate)	25	2.4×10^{-2}
Sodium tetradecanoate (myristate)	25	6.9×10^{-3}
Sodium hexadecanoate (palmitate)	50	3.2×10^{-3}
Sodium octadecanoate (stearate)	50	1.8×10^{-3}

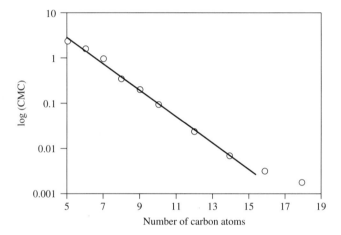

Figure 13.10 Critical micelle concentration of aqueous solutions of sodium soaps without added electrolyte as a function of chain length.

Surface tension and interfacial tension isotherms show that the initial steep decline of the tensions is arrested at the CMC. Only the single ions are surface active; the micelles are surface inactive as they are symmetrically hydrophilic. Micelle formation is one possible way to remove the lipophile from the water; it is a parallel property to adsorption and competitive with it. If micelle formation can be inhibited, therefore, the lowering of surface and interfacial tension might persist to higher concentrations, so that much lower tensions would be attained. The ease with which micelles are formed (which is reflected in a lower CMC) is reduced by using branched, asymmetric, or subdivided alkyl chains as the lipophile. The potassium salts of the sulfonate of hexadecyl phenyl ether and of the sulfonate of resorcinol dioctyl ether are powerful surface-active solutes. Their behavior provides a comparison, as lipophiles, between a single straight-chain paraffin and the same chain divided in two. Interfacial tension isotherms at room temperature, against cyclohexane, were measured by Hartley[17] and are shown in Figure 13.11. At low concentrations the hexadecyl ether is the more effective surface-active solute as shown by its greater adsorption at the interface; but it forms micelles at a lower concentration than does the dioctyl ether. Its ability to lower interfacial tension further essentially terminates at its CMC. The dioctyl compound continues to lower the interfacial tension to values well below those obtainable with the hexadecyl analogue. A disadvantage of the double-chain or branched-chain compounds is their comparatively low solubility, which is also related to their difficulty in forming micelles.

Hartley observed that the dioctyl solution at a concentration of 0.12%, the most concentrated clear solution of this compound obtainable at room temperature, emulsified various oils easily. Gentle handshaking in an open test tube produced emulsions stable for at least several weeks. In terms of Rosen's "efficiency" of surface-active solutes (the lower the concentration at which a given surface tension is reached), and "effectiveness" (the lower the surface tension at the CMC), the hexadecyl ether is more efficient and less effective.[18]

Oil/water interfaces with ultralow tensions are used to penetrate porous systems, as in the removal of oil from oily soil, and in promoting spontaneous emulsification. Interfacial tensions as low as 10^{-5} mN/m have been measured in connection with tertiary oil recovery. Compounds in which the hydrophile is located near the middle of the carbon chain rather than at the end are among the most powerful wetting agents and penetrants. A well-known example is sodium di-2-ethylhexylsulfosuccinate (Aerosol OT, Cytec Corporation). (See Figure 13.12.)

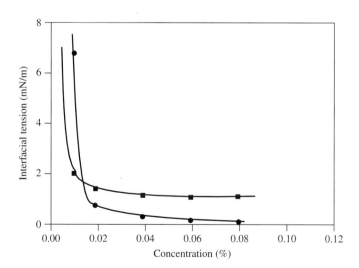

Figure 13.11 Interfacial tensions of water against cyclohexane as a function of the concentration of the sulfonates of hexadecyl phenyl ether (■) and resorcinol dioctyl ether (●).[17] Without any surfactant, the interfacial tension between water and cyclohexane is 50.2 mN/m.

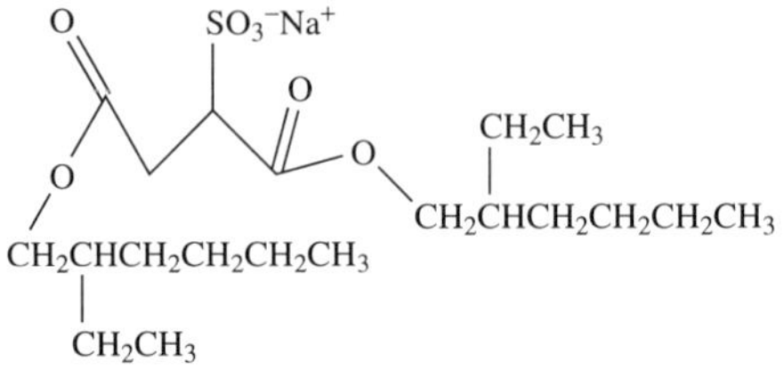

260 AQUEOUS SOLUTIONS OF SURFACE-ACTIVE SOLUTES

Figure 13.12 Sodium di-2-ethylhexylsulfosuccinate.

Traube's rule for a homologous series, originally found for a series of soaps, also holds for the behavior of the salts of the secondary alcohol sulfates, but does not hold for salts of primary alcohol sulfates.[19] All such attempts to obtain simple relations for molecular and atomic contributions to surface activity share the same defect of limited application. Davies and Rideal have worked out a quantitative evaluation of the HLB of a surface-active solute by means of molecular and atomic contributions (see Section 22.7), but the HLB scale as a measure of surface activity also has many exceptions. While certain variations in molecular structure are clearly influential, constitutive effects overshadow any simple rule. Since the CMC of a surface-active solute, even of commercial agents that are far from being single-component systems, is readily measured by any of a number of techniques, the relation between the CMC and HLB can be established. A linear relation obtains between the logarithm of the CMC of members of a homologous series and their HLB number (Figure 13.13).

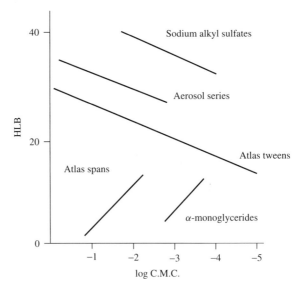

Figure 13.13 The linear relation between the logarithm of the CMC and HLB for a number of homologous series. Aqueous solutions have negative slopes and nonpolar solutions have positive slopes.[20]

13.3.e Mixtures of Homologs

Commercial surfactants rarely consist of a single pure component because the cost of purification is high and the demand for pure materials is low. Commercial agents whose lipophiles are hydrocarbon chains or whose hydrophiles are polyethylene-oxide chains are composed of a distribution of homologs. These mixtures actually offer advantages over any single component of the distribution. All the homologs comicellize in aqueous solution. Those of higher molecular weight, which are better adsorbates, are available when the micelles dissociate. Commercial mixtures are therefore more effective in applications.

If research on surfactants is to be relevant to applications, it has to include the study of properties of mixtures of known composition. Such systems are more complex and researches on such systems are naturally subsequent to the investigation of single components. For example, commercial alcohol ethoxylates are composed of a distribution of ethoxymers. As a first step to understanding how the distribution affects performance, the cloth-wetting time was used as an indicator of performance by D. L. Smith.[21] Binary mixtures of octaethylene glycol monodecyl ether (C12-8) and hexaethylene glycol monotetradecyl ether (C14-6) were studied. The wetting time was found to be inversely proportional to the average diffusion constant of various mixtures, as calculated from the diffusion constant of their two components and their composition. Therefore it is possible to predict the relative cloth-wetting time for any alcohol ethoxylate distribution by knowing the diffusion constant of each ethoxymer in the distribution and the composition of the distribution. Diffusion constants can be calculated from the time dependence of surface tension.

13.3.f Solubilization by Micelles (Microemulsions)

Solubilization was discovered when it was found that a concentrated aqueous soap solution absorbs liquid benzene to form a clear isotropic liquid. Other organic compounds, normally insoluble in water, also dissolve spontaneously in aqueous solutions of surface-active solutes at concentrations above the CMC. The solubility of 2-nitrodiphenylamine in dilute potassium laurate solutions is shown in Figure 13.14. The solubility remains constant and is the same as in pure water until the concentration of potassium laurate reaches its CMC; the solubility of the amine then increases with concentration of the soap, due to solubilization in the micelles. That micelles are responsible for the solubilizing action is also shown clearly by the variation of the Henry's law constant for the solubility of a gas as a function of concentration of surface-active solute.

The absorption isotherms at 25.0°C of butadiene in water and in aqueous solutions of a cationic surface-active solute, p-diisobutylphenoxyethoxy-ethyl dimethylbenzyl ammonium chloride (Hyamine 1622, Rohm & Haas Co.) are shown in Figure 13.15. The regular decrease of the Henry's law constant is linear with the logarithm of the concentration of Hyamine 1622, as shown in Figure 13.16. The Henry's law constant of butadiene in pure water is 3.95×10^8 Pa/mol gas/mol solution. Extrapolating the experimental straight line to the

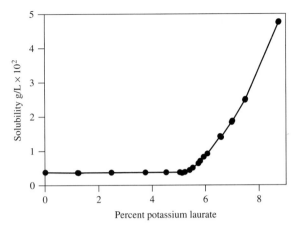

Figure 13.14 Solubility of 2-nitrodiphenylamine in aqueous solutions of potassium laurate.[22]

concentration of solute corresponding to the value of the Henry's law constant of 3.95×10^8 would presumably yield a minimum concentration of solute below which no additional solubility of butadiene occurs. The value for the minimum concentration obtained from Figure 13.16 is 0.19% Hyamine 1622. The CMC of Hyamine 1622 from conductivity measurements is 0.16%. The role of micelles in solubilization is demonstrated by this agreement.

X-ray diffraction of micellar solutions shows that the hydrocarbon chains of the micelle interior are essentially liquid-like (liquoid). The photographs have a broad, diffuse halo corresponding to a spacing of 0.45 nm, similar to that given by liquid alkanes.[24] The application of NMR techniques also indicate that the hydrocarbon is in a liquoidal state.[25]

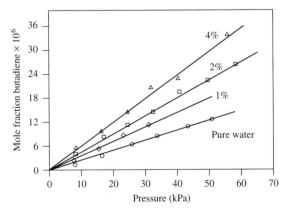

Figure 13.15 Mole fraction of butadiene dissolved in pure water and in solutions of Hyamine 1622 as a function of the equilibrium partial pressure of butadiene at 25.0°C.[23]

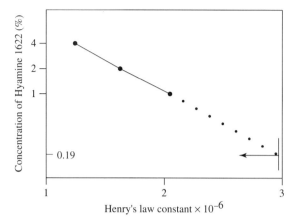

Figure 13.16 Variation of the Henry's law constant for the solubility of butadiene in aqueous solutions of Hyamine 622 with the concentration of solute. The vertical dotted line is the Henry's law constant for pure water. The left arrow is the concentration of Hyamine 1622 below which the solubility is not enhanced.[23]

As more solubilizate is brought into the micelles, they change gradually to swollen micelles (microemulsion) and then to miniemulsion droplets. At this stage the system certainly contains two phases. The interior of the micelle is essentially hydrocarbon in a liquid state; the solubilizate is a second component in the micelle and accordingly has all the thermodynamic properties of a solute, including osmotic pressure. These droplets have also capillary (or Laplace) pressure due to their small radius of curvature. The capillary pressure causes solubilizate to diffuse from the smaller droplets to the larger ones (Ostwald ripening); at the same time the loss of solubilizate reduces the osmotic pressure in the smaller droplets with respect to the larger ones. These two opposing effects of diffusion balance at equilibrium, which occurs at a particular droplet size. Uniform microemulsion droplets in the range 10–100 nm may be stabilized in this way at an equilibrium point.[26]

The same principle is applied in emulsion polymerization, by including a water-insoluble nonvolatile paraffin along with the monomer in the micelles. The droplets of paraffin act similarly to the hydrocarbon core of the micelle. Capillary pressure promotes diffusion of monomer from smaller to larger droplets, but in doing so the osmotic pressure inside the smaller droplets is reduced and the osmotic pressure of styrene in paraffin droplets is increased. This difference in osmotic pressure reverses the outward flow of styrene from micelles, and so terminates the Ostwald ripening. When, for example, styrene is homogenized with water containing emulsifier, the resulting emulsion is unstable due to Ostwald ripening. A small amount of water-insoluble component strongly inhibits the Ostwald ripening because, by dissolving styrene, it creates an osmotic pressure. Since only the styrene diffuses, the osmotic pressure in the larger droplets increases, compared to that in the smaller droplets. The gradient of osmotic pressure tends to send the styrene back to the smaller droplets. Thus the Ostwald ripening sets up a counteraction that brings about a state of equilibrium. Ugelstad, El-Aasser and Vanderhoff[27] prepared finely dispersed styrene microemulsions stabilized by the presence of a coemulsifier that had a low solubility in water and a low molecular weight. Examples of such a coemulsifier are long-chain fatty alcohols and alkanes, like hexadecanol and hexadecane. When polymerization takes place, the resulting latex has uniform particles of a controlled size.

An ingenious use of solubilization for the purpose of detecting extremely small concentrations, for example, 9–18 ng/g, of a herbicide in lake water is to concentrate hydrophobic compounds such as steroids or quinine into micelles of sodium laurate sulfate. When a voltage is applied during electrokinetic chromatography, electrophoresis brings the micelles into thin, highly concentrated zones, which are then readily analyzed by chromatography and a UV-absorption detector.[28]

13.4 THE KRAFFT POINT

In 1895 Krafft and Wiglow noticed[29] that the solubility of a homologous series of sodium soaps in water shows an abrupt increase with temperature at what has since been called the Krafft point. They ascribed the effect to the melting of the hydrocarbon chains of the soaps, based on the coincidence of the Krafft points of the soaps with the melting points of their parent fatty acids. The modern interpretation is that the Krafft point marks a transition between the dissolution of the soap to form ions and its dissolution to form micelles. Krafft's notion of partial melting is supported, however, by modern evidence that the hydrocarbon core of the micelle is in the liquoidal state. The solubility of soap increases sharply with temperature once micelles have appeared, by a process that can only be described as self-solubilization, which packs added solute into micelles. Mixed micelles are also capable of being formed with numerous additives above the Krafft point, further evidence of the liquid character of the micellar core.

The solubility relations of a surface-active solute are shown in Figure 13.17 in the form of a phase diagram. The CMC does not vary much with temperature, as micellization is chiefly an entropic effect. Solubility increases with temperature to a greater degree than does micellization: Consequently at some temperature the solubility curve intersects the CMC curve. The Krafft point can be seen from the diagram to be the temperature at which the CMC and the solubility coincide. At temperatures below the Krafft point, the CMC occurs at a higher concentration than the solubility, that is, micelles are formed only in supersaturated solutions. The CMC can still be determined quite readily, however, as these solutions can be supercooled for many days without crystallizing. At temperatures above the Krafft point, the solute forms micelles in solutions that are unsaturated. Surface-active solutes do not function well as detergents or emulsifying agents below their Krafft point, either because of too low a solubility or because micelles are lacking to act as solubilizers and as reservoirs of the surface-active molecules.

At 60°C the detergent power of the sodium alkyl sulfates is in the order C_{16} > C_{14} > C_{18} > C_{12} (see Figure 13.3). Measurements of foam number at 60°C by

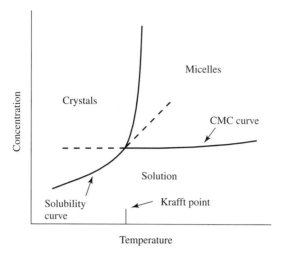

Figure 13.17 The pseudophase diagram of sodium dodecyl sulfonate and water.[30]

the Stiepel method show the compounds to lie in the same order.[31] But at 40°C the C_{14} compound gives maximum foam, while at 20°C the C_{12} is the most powerful profoamer.[32] These results confirm the rule that surface activity at higher temperatures is promoted by using a longer alkyl chain in the lipophile, that is, an amphipath with a higher Krafft point.

The Krafft points of some surface-active solutes in aqueous solution are listed by Rosen.[33] These are for purified compounds. Mixtures generally have Krafft points considerably lower than those of single components.

13.5 ELASTICITY OF SURFACE FILMS

The elasticity of a surface film under any given conditions is defined as the ratio of any small increase in spreading pressure π to the areal compression. The areal compression is the relative contraction of the area, that is, the ratio of the change of area to the area:

$$E = -\frac{d\pi}{dA/A} = -\frac{d\pi}{d \ln A} \tag{13.8}$$

Since the areal compression is dimensionless, surface elasticity has the same units as spreading pressure. The elasticity at the point P on the π-A curve in Figure 13.18 is obtained by extending the tangent at point P to intersect the π axis at point G.

Let F be the value of π at P; then

$$E = -\frac{A d\pi}{dA} = \frac{FP \cdot FG}{FP} = FG \tag{13.9}$$

Hence, if the relation between the spreading pressure and the area per molecule under certain conditions, as for instance at a given temperature, is represented

266   AQUEOUS SOLUTIONS OF SURFACE-ACTIVE SOLUTES

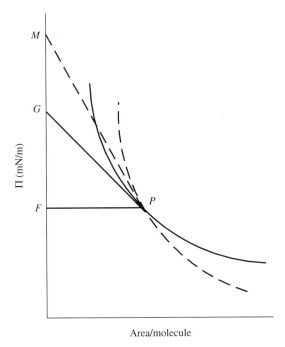

Figure 13.18 How to obtain elasticity of a surface film from a π-A curve. The solid line represents the equilibrium isotherm; the doted line represents a nonequilibrium curve. The equilibrium elasticity is *FG*; the dynamic elasticity is *FM*.

by the locus of *P*, the elasticity of the film when in the state represented by *P* may be found by drawing *PG* a tangent to the curve at *P*, and *PF* a horizontal line. The portion *FG* of the π axis represents, on the scale of π, the elasticity of the film.

In most bodies, compression introduces an increase of temperature, the effect of which is to increase the elasticity. Elasticity can be measured isothermally or adiabatically. The former elasticity pertains when stresses and strains occur slowly enough to remain at thermal equilibrium. The latter pertains in the case of rapidly changing forces, where there is not time for the temperature to be equalized. The adiabatic elasticity is larger than the isothermal elasticity.

π-A curves can be drawn that do not represent equilibrium. If the surface of the solution has recently undergone rapid expansion or contraction, a nonequilibrium state of dynamic surface tension exists. On compression the film becomes more concentrated and the value of π increases; on expansion the film is more dilute and the value of π decreases. In Figure 13.18 the dynamic π-A curve at point *P* lies above the equilibrium curve on compression and below it on expansion. If enough time is allowed, these changes in π disappear, but if measurements are made before equilibrium is reached, the condition is as shown in the diagram. The tangent to the dynamic curve at point *P* is represented by *PM*, and the dynamic elasticity is represented by *FM*.

If the surface of a solution containing a surface-active solute is expanded suddenly, then, before adsorption from the bulk phase to the newly created surface has had time to take place, the same number of adsorbed molecules remain on the surface, so the differential increase in area per molecule is

$$da = d\left(\frac{A}{n_s}\right) = \frac{dA}{n_s} \tag{13.10}$$

where a is the area per adsorbed molecule, A is the total area, and n_s is the number of molecules of adsorbed solute on the total area. The suddenly expanded surface is not a stable state, but subsequent changes are so slow on a molecular scale that one can assume it to be at thermal equilibrium. Thermal equilibrium implies that thermodynamics is applicable even though the states of the system are not in mechanical equilibrium. We may, therefore, use equilibrium equations of state for the surface film at each stage of its path to final equilibrium. Let us suppose initially that the adsorbed surface film is described by the two-dimensional ideal equation of state:

$$\pi a = (\sigma_0 - \sigma)a = kT \tag{13.11}$$

or

$$(\sigma_0 - \sigma) = RT\Gamma_2^G \tag{13.12}$$

where $\Gamma_2^G = n_s/N_0 A$ mol/m².

Elasticity is defined as the ratio of the increase in the surface tension resulting from an infinitesimal increase in area and the relative increment of the area. For a lamella with adsorbed films on both sides, the elasticity E is given by

$$E = \frac{2d\sigma}{d \ln A} = -\frac{2d\sigma}{d \ln \Gamma_2^G} \tag{13.13}$$

Differentiating Equation (13.12) and substituting in Equation (13.13) gives

$$E = 2RT\Gamma_2^G \tag{13.14}$$

Equation (13.14) says that at a given temperature the elasticity is proportional to the Gibbs excess surface concentration in the case of a two-dimensional ideal equation of state. For one-sided and more compressed films, the elasticity is given by a few terms of a power series in Γ_2^G:

$$E = RT\Gamma_2^G + b\left(\Gamma_2^G\right)^2 + c\left(\Gamma_2^G\right)^3 \tag{13.15}$$

Elasticity determined isothermally from the equilibrium π-A curve is called Gibbs elasticity, and nonequilibrium elasticities measured by any dynamic method are known as Marangoni elasticities.[34] Marangoni elasticity is larger in value than the Gibbs elasticity that could be obtained in the same system.

While Gibbs elasticity is a thermodynamic property of a surface film, Marangoni elasticities depend on the nature of the stresses and strains applied to the surface, as do the dynamic surface tensions. Neither Marangoni elasticities nor dynamic surface tensions have characteristic values. Some investigators measure dynamic surface tensions occurring at rather low frequencies of dilatation-compression cycles, from 1 per minute to 1 every 30 minutes; others[35] have used frequencies as high as 15–135 Hz (cycles/second,) although such disturbances are far from corresponding to the extension-contraction cycles occurring in foam. The measurement must be made coincidentally with measurements of changes of surface area.

The modulus of surface elasticity of a solution can be determined by measuring the damping of transverse ripples as a function of their frequency. The surface of a wave is contracted at the crest and extended at the trough. In the absence of any surface-active material, this contraction and extension does not change the surface tension; but if an adsorbed layer is present, it is compressed on the contraction and dilatated on the extension, causing the surface tension to decline at the crest and to increase at the trough. These local differences of tension alter the pattern of subsurface flow, giving rise to a greater rate of energy dissipation by viscous friction, and consequently to a greater damping of the waves than would otherwise occur. The distance-damping coefficient β may be measured by the logarithmic decrease of the wave amplitude, A, with distance x from the source:

$$\beta = \frac{-d \ln A}{dx} \quad (13.16)$$

The damping coefficient β is constant for Newtonian liquid surfaces: The surface of a solution contains adsorbed solute if the solute is surface active, and such a surface may not be Newtonian. The plot of $\ln A$ versus x would then not be linear. The testing of Eq. (13.16) therefore is informative about the presence of a non-Newtonian shear viscosity at the surface of a solution. If, however, the surface should be Newtonian, the data have a further use: The change of the damping coefficient compared to pure solvent ($\beta = \beta_0$), as a function of wave frequency, is directly related to the surface elasticity,[36] assuming that diffusional interchange between bulk and surface is negligible during dilatation and compression of the adsorbed layer.[37]

13.6 DYNAMIC SURFACE TENSIONS

Rapidly expanding liquid surfaces occur in several processes, such as spraying, painting, and coating, and in the generation of foams. The surface tension of a pure liquid is established almost immediately on creation of the surface. Solutes with low surface activity, such as short-chain alcohols, reach equilibrium after formation of a fresh interface in about a millisecond; for some high-molecular weight surfactants, this time approaches a day. To reach equilibrium the molecules have to rearrange to a preferred orientation at the created surface. Figure 13.19

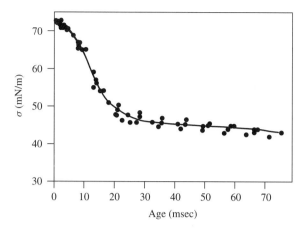

Figure 13.19 Dynamic surface tension of Aerosol OT in water at a concentration of 500 mg/l as measured by the vibrating jet.[38]

shows the dynamic surface tension of Aerosol OT in water at a concentration of 500 mg/L as measured by the vibrating jet.

In solutions of surface-active solutes the rate at which equilibrium is reached depends on the rate of diffusion of solute molecules to the surface. The rate of diffusion, in turn, is determined by the size of the molecule and its concentration. The rate of solute adsorbed at a newly created surface in the absence of stirring or of an energy barrier to adsorption, increases with concentration, according to an equation given by Ward and Tordai:[39]

$$n = 2000 c N_0 \left(\frac{Dt}{\pi} \right)^{1/2} \quad (13.17)$$

where n is the number of molecules/m² after time t in seconds, D is the bulk diffusion constant in m²/s, c is the bulk concentration in mol/L, and N_0 is Avogadro's number. The time needed to replace solute at a new surface becomes progressively less with increasing concentration.

Diffusion brings molecules to the surface rapidly; yet experience shows that some solutions take hours or even days to reach constant (static) surface tension. Such long periods of aging are ascribed to traces of polyvalent ion impurities, if the water had not been distilled in quartz or treated with ion exchange resins,[40] or to highly surface-active impurities present in trace amounts.[41] These impurities, because of their low concentration, may require hours to reach the surface and produce their full effect.

Dynamic surface tensions are defined as any nonequilibrium values of surface tension that arise when the surface of a solution is extended or contracted. As the surface moves toward equilibrium, either by adsorption or desorption of solute, the surface tension changes toward its static value. If the time scale is in milliseconds, the techniques available are the vibrating jet,[42,43] the maximum bubble-pressure method, and the falling meniscus method[44] (see Section 9.1). For time scales larger than 30 s, the ordinary static methods of measuring surface tension, such as the Wilhelmy plate, are applicable. All these methods give the variation of surface tension with time. To measure the change of surface *concentration* in jets, ellipsometry offers a method.[45] To convert the coefficient of ellipsometry into surface concentration a calibration is required against a direct measurement of surface excess concentration. Other methods give indirect evidence of the presence of a surface-active solute by the effect of dynamic surface tension on the damping of capillary waves or the reduction of the velocity of a rising bubble, but most readily by the stability of bubbles. These methods are useful in nonpolar solutions where the surface-tension lowering is small and measurements of surface tension are not sufficiently precise to detect dynamic surface tensions directly.

A major difference between aqueous and nonpolar solutions lies in the magnitude of the effects produced by surface-active solutes. The surface tension of water is reduced from 73 to 25 mN/m quite readily by amphipathic organic solutes; but the surface tension of most organic solvents is already in the low range of 25–30 mN/m, so that only a small reduction can be achieved by an organic solute. Thus, although Marangoni effects may arise in nonpolar solutions, they are usually much less pronounced than in aqueous solutions of soaps or detergents. Oil lamellae, therefore, have a relatively low resistance to mechanical shock; consequently oil foams are transient or evanescent, resembling the foam produced from very dilute aqueous solutions of detergents or more concentrated solutions of weakly surface-active solutes. Special solutes have been developed to stabilize nonpolar foams for application in the field of cellular plastics. These solutes incorporate polyalkylsiloxane or perfluoroalkyl moieties in the molecule, which are able to reduce the surface tension of organic liquid monomers by 12–15 mN/m to boost surface-tension gradients, hence Marangoni effects, and the result is obvious in an increased stability of the foam.

13.7 ULTRALOW DYNAMIC SURFACE TENSION

Some processes that use solutions of surface-active solutes depend on the creation of regions where the interfacial tension is lower by a few orders of magnitude than its equilibrium value. Contraction of a surface that contains adsorbed solute compresses the layer and causes the surface tension to decline. If subsequent relaxation is rapid, the ultralow tension is fleeting, but slow relaxation processes are known that allow such ultralow dynamic surface tensions to be retained long enough to be useful. A prime example is a monolayer of dipalmitoyl lecithin, which has an equilibrium surface tension of more than 70 mN/m but can be compressed to a long-persisting tension of 1 mN/m. This compound is physiologically important in reduction of the muscular effort to expand the lungs and in preventing the collapse of lung alveoli. No synthetic surface-active solute has yet been found to compete in effectiveness with this product of millennia of evolution (or Divine creation!).[46,47]

13.8 PHASE BEHAVIOR OF SURFACE-ACTIVE SOLUTES

A thorough treatment of the phase rule applied to surface-active solutes is provided in Laughlin's book, which provides a comprehensive review of the fundamentals of phase science of surface-active systems.[48] The author remarks "There presently exists within academic physical chemistry an intense preoccupation with its theoretical and quantum mechanical aspects, and a comparatively low-level interest in experimental phase science and thermodynamics. Viewed from the perspective of an industrial research chemist, this situation is difficult to understand and impossible to justify."

Micellar solutions are only a part of the total phase behavior of binary systems containing water and a surface-active solute. Many such solutes form elaborate structures at higher concentrations. The milkiness of concentrated soap solutions is not due to micelles (they are too small to scatter a significant quantity of light) but is due to larger drops of liquid crystals. The structures of liquid crystals are determined by putting them in small (approximately 1 mm) capillaries and by examining them through a microscope with crossed polarizers. Figure 13.20 represents idealized structures that may occur as concentration of solute increases in a binary aqueous system. Mesomorphic phases, or mesophases, are structures in which one or two dimensions are highly extended. Such a structure cannot but be anisotropic. A mesophase, that is, an in-between or intermediate phase, has some properties characteristic of crystals, such as birefringence, and yet flows

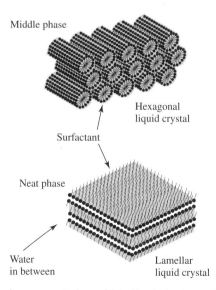

Figure 13.20 Schematic representation of idealized structures formed by water and a surface-active solute as concentration of solute increases. The terms "middle" and "neat" are derived from nineteenth-century soap-boilers' practice.[51]

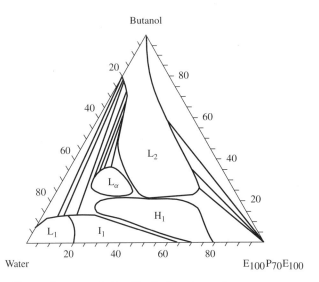

Figure 13.21 Phase diagram of the $E_{100}P_{70}E_{100}$-butanol-water system at 25°C. The tie-lines are represented by full straight lines. L_1 denotes the water-rich (micellar) solutions region, I_1 the normal ("oil"-in-water) micellar cubic liquid crystalline region, H_1 the normal hexagonal liquid crystalline region, L_α the lamellar liquid crystalline region, and L_2 the alkanol-rich solution region.

like a liquid. Lehmann called it a liquid-crystal (LC) phase as it combines long-range order with fluidity.[49] The application in liquid-crystal displays depends on the ability to change orientation in an electric field: an ability that requires both structure and fluidity.[50]

Phase diagrams are used to specify the temperature and concentrations at which various structures exist at equilibrium. In ternary systems in which the third component is an alkane derivative, such as an acid, alcohol, or amide, various structures can be identified. The following system is of interest because it displays several self-assembled structures, that is, micellar solutions, as well as cubic, hexagonal (middle phase), and lamellar (neat phase) lyotropic liquid crystals, depending on the polymer concentration and the temperature. Figure 13.21 describes the phase behavior of a ternary system consisting of an amphiphilic block copolymer, water, and butanol at 20°C.[52] The surfactant is a nonionic block copolymer of poly(ethylene oxide) (E) and poly(propylene oxide) (P), designated $E_{100}P_{70}E_{100}$. It has a molecular weight of 13.3 kDa and corresponds to the commercially available Pluronic F127.®

The diagram is not complete; some three-phase systems are certainly present in the heterogeneous regions; but the main interest is in the location and nature of the one-phase regions. First, we see that the polymer surfactant is water soluble, forming micelles at low concentrations (CMC ≈ 1% by weight). The isotropic solution region, designated L_1, is a micellar solution and exists along the polymer-water axis up to 20% by weight $E_{100}P_{70}E_{100}$. Butanol up to 10% by weight is solubilized in this range of concentration. The L_1 solutions are optically transparent and fluid; their viscosity increases as concentrations approach the phase boundary with the liquid-crystalline phase (I_1). An isotropic liquid-crystalline region of cubic structure (I_1) replaces the L_1 solution region along the water-polymer axis and extends up to 65% by weight $E_{100}P_{70}E_{100}$. Its ability to incorporate butanol decreases as concentrations approach the upper boundary. No two-phase region was found between the L_1 and I_1 phases, though one would normally be expected. The accuracy of the determination of the phase boundary is 2–3% by weight, so if the expected two-phase region exists it had to be within a very narrow range of concentration. A birefringent (anisotropic) region (H_1) exists along the polymer-water axis from 70 to 80% by weight $E_{100}P_{70}E_{100}$. In this range of concentrations the H_1 phase incorporates butanol and extends between the cubic (I_1) and the alkanol-rich solution region (L_2) down to approximately 20% by weight copolymer. The structure of this region, established by SAXS (small-angle x-ray scattering) measurements, is hexagonal. The birefringent liquid-crystalline region stable in the 20–30% by weight $E_{100}P_{70}E_{100}$ and 25–30% by weight butanol ranges had a lamellar (smectic) structure (L_α). A large one-phase isotropic solution region (L_2) extends from the butanol-rich corner down to just 20% by weight butanol. Solutions in the L_2 region are similar in fluidity to butanol at high concentrations of butanol, but are more viscous at butanol concentrations below 50% by weight. The L_2 phase is capable of solubilizing water because it contains inverse micelles. The difference between the two micellar solutions L_1 and L_2 is that in the L_1 phase the copolymer is bent in the form of a U with the ethylene oxide in the aqueous medium; while in the L_2 phase the U is inverted with the propylene oxide in the nonpolar medium. Hence the term "inverted micelle."

To obtain a stable foam from an oil solution, a combination of the oil solution and liquid crystal is required. Figure 13.22 is the generalized phase diagram for the ternary system of hydrocarbon, water, and surface-active solute showing an isotropic oil solution with inverse micelles, A, and a liquid-crystal phase, B. Water is not miscible with the hydrocarbon; the surface-active solute is soluble in oil but not in water. The solution phase alone, or the liquid-crystal phase alone, does not foam. Only compositions in the two-phase area, T, between the liquid crystal and the oil solution produce stable foams, which have a lifetime of several hours. Compositions outside this two-phase area have a foam life of seconds only.

13.9 THE EFFECTS OF GEOMETRIC PACKING

The shape of surfactant molecules also determines the nature of the surfactant phases. What shapes best match the volume required by the lyophobic groups as modified by the packing of the lyophilic heads?[54] For example, spherical micelles form when the hydrocarbon chains fit within a volume defined by the optimum spacing of the head groups. Large structures have lower entropy than smaller ones

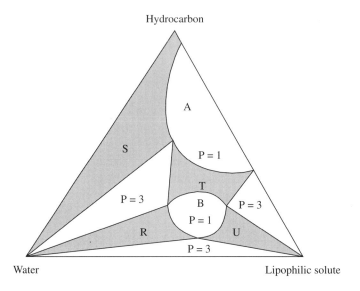

Region	Number of phases	Composition
A	1	Isotropic solution with inverse micelles
B	1	Lamellar liquid crystals
R	2	Water phase plus liquid crystal phase
S	2	Water phase plus solution containing inverse micelles
T	2	Solution containing inverse micelles plus lamellar liquid crystal phase
U	2	Excess emulsifier plus lamellar liquid crystals

Figure 13.22 Ternary phase diagram of hydrocarbon, water, and surface-active solute. P is the number of phases within each region.[53]

so another tendency is for the surfactants to form the smallest shape consistent with good geometric packing. The argument for spheres leads to the following requirements for surfactants to form spherical micelles:[55]

$$\frac{v}{a_0 l_c} < \frac{1}{3} \tag{13.18}$$

and

$$M = \frac{4\pi l_c^3}{3v} \tag{13.19}$$

where v is the volume of the micelle, a_0 is the optimum area for the head groups and l_c is the critical chain length, a length approximately the extension of the hydrocarbon before it loses a liquid-like state, and M is the mean aggregation number. The expression $v/a_0 l_c$ is called the critical-packing parameter.

Surfactants that do not match the requirement of the critical-packing parameter, Equation (13.18), will form shapes other than spheres. Derivations based

TABLE 13.2 Mean (Dynamic) Packing Shapes of Lipids and the Structures They Form[56]

Surfactant	Critical-packing parameter, $v/a_0 l_c$	Critical-packing shape	Structures formed
Single-chained with large head-group areas: SDS in low salt.	<1/3	Cone	Spherical micelles
Single-chained with small head-group areas: SDS and CTAB in high salt and nonionics.	1/3–1/2	Truncated cone	Cylindrical, rod-like micelles
Double-chained with large head-group areas: lecithin, dihexadecyl phosphate.	1/2–1	Truncated cone	Flexible bilayers, vesicles
Double-chained with small head-group areas: anionics in high salt, saturated frozen chains.	≈1	Cylinder	Planar bilayers
Double-chained with small head group areas: nonionics.	>1	Inverted truncated Cone	Inverse micelles

on geometry lead to a general relation between the critical-packing parameter and the shape of the amphiphilic structure formed. These are given in Table 13.2.

Extensions of these geometric factors include the consideration of the effect of cosurfactants and, especially, biological structures.

REFERENCES

[1] Pravia, K.; Maynard, D.F. Why don't water and oil mix? *J. Chem. Ed.* **1996**, *73*, 497.

[2] Tanford, C. *The hydrophobic effect: formation of micelles and biological membranes*; Wiley: New York; 1980.

[3] Frank, H.S.; Evans, M.G. Free volume and entropy of condensed systems, *J. Chem. Phys.* **1945**, *13*, 507–532.

[4] Du, Q.; Freysz, E.; Shen, Y.R. Surface vibrational spectroscopic studies of hydrogen bonding and hydrophobicity, *Science* **1994**, *264*, 826–828.

[5] Chang, T.-M.; Dang, L.X. Molecular dynamics simulations of $CCl_4 - H_2O$ liquid–liquid interface with polarizable potential models, *J. Chem Phys.* **1996**, *104*, 6772–6783.

[6] Scatena, L.F.: Brown, M.G.; Richmond, G.L. Water at hydrophobic surfaces: Hydrogen bonding and strong orientation effects, *Science* **2001**, *292*, 908–912.

[7] Derjaguin, B.V. The world of neglected thicknesses and its place and role in nature and technology, *Colloids Surface* **1993**, *79*, 1–9.

[8] Ferguson, J. The use of chemical potentials as indices of toxicity, *Proc. Roy. Soc. London* **1939**, *127B*, 387–404.

[9] McBain, J.W. *Colloid science*; D.C. Heath: Boston; 1950; p 248.

[10] Israelachvili, J. *Intermolecular and surface forces*; Academic Press: New York; 1st ed., 1985; 2nd ed., 1991.

[11] Gonick, E. Stokes' law and the limiting conductance of organic ions, *J. Phys. Chem.* **1946**, *50*, 291–300.

[12] Zimm, B.H. The scattering of light and the radial distribution function of high polymer solutions, *J. Chem. Phys.* **1948**, *16*, 1093–1099.

[13] McBain, J.W. Solutions of soaps and detergents as colloidal electrolytes in *Colloidal chemistry, theory and applications*; Alexander, J., Ed.; Reinhold Publishing: New York; 1944; Vol. 5; pp 102–120.

[14] Jones, E.R.; Bury, C.R. The freezing-points of concentrated solutions. II. Solutions of formic, acetic, propionic, and butyric acids, *Phil. Mag.* **1927**, *4*(7), 841–848.

[15] Stainsby, G.; Alexander, A.E. Studies of soap solutions. II. Factors influencing aggregation in soap solutions, *Trans. Faraday Soc.* **1950**, *46*, 587–597.

[16] Mitchell, D.J.; Ninham, B.W. Micelles, vesicles and microemulsions, *J. Chem. Soc., Faraday Trans. 2* **1981**, *77*, 601–629.

[17] Hartley, G.S. Interfacial activity of branched-paraffin-chain salts, *Trans. Faraday Soc.* **1941**, *37*, 130–133.

[18] Rosen, M.J. *Surfactants and interfacial phenomena*; Wiley: New York; 1st ed., 1978, pp. 153–159; 2nd ed., 1989; pp. 212–214.

[19] Dreger, E.E.; Keim, G.I.; Miles, G.D.; Shedlovsky, L.; Ross, J. Sodium alcohol sulfates–properties involving surface activity, *Ind. Eng. Chem.* **1944**, *36*, 610–617.

[20] Little, R.C. The physical chemistry of nonionic surface-active agents, Ph.D. dissertation, Rensselaer Polytechnic Institute, 1960; University Microfilm Abstract 60-2689.

[21] Smith, D.L. Prediction of cloth-wetting times from diffusion constants of alcohol ethoxylate mixtures, *J. Surfactants Detergent* **2000**, *3*(4), 483–490.

[22] Harkins, W.D. *The physical chemistry of surface films*; Reinhold Publishing: New York; 1952; p 323.

[23] Ross, S.; Hudson, J.B. Henry's law constants of butadiene in aqueous solutions of a cationic surfactant, *J. Colloid Sci.* **1957**, *12*, 523–525.

[24] Luzzati, V.; Mustacchi, H.; Skoulios, A. The structure of the liquid-crystal phases of some soap + water systems, *Discuss. Faraday Soc.* **1958**, *25*, 43–50.

[25] Lawson, K.D.; Flautt, T.J. Nuclear magnetic resonance studies of surfactant mesophases, *Mol. Cryst.* **1966**, *1*, 241–262.

[26] Adamson, A.W. A model for micellar emulsions, *J. Colloid Interface Sci.* **1969**, *29*, 261–267.

[27] Ugelstad, J.; El-Aasser, M.S.; Vanderhoff, J.W. Emulsion polymerization: Initiation of polymerization in monomer droplets, *J. Polym. Sci., Polym. Lett. Ed.* **1973**, *11*, 503–513.

[28] Quirino, J.P.; Terabe, J. Exceeding 5000-fold concentration of dilute analytes in micellar electrokinetic chromatography, *Science* **1998**, *282*(5388), 465–468.

[29] Kraft, F.; Wiglow, H. Über das verhalten der fettsauren alkalien und der seifen in gegenwart von wasser III. Die seifen als kristalloïde (On the behavior of soaps in the presence of water III. Soaps as crystalloids), *Ber. Dtsch. Chem. Ges.* **1895**, *28*, 2566–2573.

[30] Shinoda, K.; Nakagawa, T.; Tamamushi, B-I; Isemura, T. *Colloidal surfactants, some physicochemical properties*; Academic Press: New York; 1963; p 7.

[31] Stiepel, C. Die schaumzahl der seifen (The foaminess of soaps), *Seifensieder Ztg.* **1914**, *41*, 347.

[32] Götte, E. Ein beitrag zur kenntnis der waschwirkung (A contribution to detergency), *Kolloid Z.* **1933**, *64*, 222–227.

[33] Rosen loc. cit.; pp 216–217.

[34] Rusanov, A.I.; Krotov, V.V. Gibbs elasticity of liquid films, threads, and foams, *Prog. Surf Membr. Sci.* **1979**, *13*, 415–524.

[35] Malysa, K.; Lunkenheimer, K.; Miller, R.; Hartenstein, C. Surface elasticity and frothability of n-octanol and n-octanoic acid solutions, *Colloids Surfaces* **1981**, *3*, 329–338.

[36] Cini, R.; Lombardini, P.P. Experimental evidence of a maximum in the frequency domain of the ratio of ripple attenuation in monolayered water to that in pure water, *J. Colloid Interface Sci.* **1981**, *81*, 125–131.

[37] Lucassen-Reynders, E.H.; Lucassen, J. Properties of capillary waves, *Adv. Colloid Interface Sci.* **1969**, *2*, 347–395.

[38] Thomas, W.D.E.; Potter, L. Solution–Air Interfaces I. An oscillating jet relative method for determining dynamic surface tension, *J. Colloid Interface Sci.* **1975**, *50*, 397–412.

[39] Ward, A.F.H.; Tordai, L. Time-dependence of boundary tensions of solutions I. The role of diffusion in time-effects, *J. Chem. Phys.* **1946**, *14*, 453–461.

[40] Davies, J.T.; Rideal, E.K. *Interfacial phenomena*, 2nd ed.; Academic Press: New York; 1963; p 167.

[41] Mysels, K.J. Surface tension of solutions of pure sodium dodecyl sulfate, *Langmuir* **1986**, *2*, 423–428.

[42] Noskov, B.A. Fast adsorption at the liquid-gas interface, *Adv. Colloid Interface Sci.* **1996**, *69*, 63–129 (234 references).

[43] Miller, R.; Dukhin, S.S.; Kretzschmar, G. *Dynamics of adsorption at liquid interfaces*; Elsevier: Amsterdam, 1995.

[44] Delay, R.; Pétré, G. Dynamic surface tension, *Surf. Colloid Sci.* **1971**, *3*, 27–81.

[45] Hutchison, J.; Klenerman, D.; Manning-Benson, S.; Bain, C. Measurements of the adsorption kinetics of a cationic surfactant in a liquid jet by ellipsometry, *Langmuir* **1999**, *15*, 7530–7533.

[46] Exerowa, D.; Lalchev, Z.; Marinov, B.; Ognyanov, K. Method for assessment of fetal lung maturity, *Langmuir* **1986**, *2*, 664–668.

[47] Park, S.Y.; Hannemann, R.E.; Frances, E.I. Dynamic tension and adsorption behavior of aqueous lung surfactants, *Colloids Surf. B.* **1999**, *15*, 325–338.

[48] Laughlin, R.G. *The aqueous phase behavior of surfactants*; Academic Press: London; 1994.

[49] Otto Lehmann (1855–1922) in a note to Friedrich Reinitzer (1857–1927) "... my new results confirm your [previously] declared view, that the [substance] consists of very soft crystals ... It is absolutely homogeneous, and another liquid — as you assumed formerly — is not present ... It is of a high interest for the physicist that crystals exist which are of such a considerable softness that one could almost call them liquid."

[50] Drzaic, P.S. *Liquid crystal dispersions*; World Scientific: New Jersey; 1995.

[51] Corkill, J.M.; Goodman, J.F. The interaction of nonionic surface-active agents with water, *Adv. Colloid Interface Sci.* **1969**, *2*, 297–330.

[52] Holmqvist, P.; Alexandridis, P.; Lindman, B. Phase behavior and structure of ternary amphiphilic block copolymer-alcohol-water systems, *Langmuir* **1997**, *13*, 2471–2479.

[53] Friberg, S. Liquid crystals and foams, *Adv. Liq. Cryst.* **1978**, *3*, 149–165.

[54] Laughlin loc. cit.; pp 229–232.

[55] Israelachvili, J. *Intermolecular and surface forces*, 2nd ed.; Academic Press: New York; 1991; 380–382.

[56] Israelachivili loc. cit.; p 381.

14 Surface Activity in Nonpolar Media

Surface activity in nonpolar media is not usually observed at the liquid/air surface as the surface tension of the medium is already low, so that adsorption of solute does not occur appreciably, unless the solute is a material of exceptionally low surface energy, such as a siloxane polymer or a compound that contains perfluoroalkyl substituents. But given an opportunity to react with either a Lewis acid or base, the hydrophilic group of the oil-soluble amphiphile can be brought to a liquid/liquid interface or to a liquid/solid interface or to the interior of an inverse micelle. Such groups interact readily with water, which accounts for the formation of inverse micelles and for the low interfacial tension at the oil/water interface. Indeed an oil-soluble amphiphile in an organic medium will extract dissolved water from the solvent and transfer it into the center of a micelle.

Most amphiphilic substances in nonpolar organic solvents have groups of ionic character, such as carboxylates or sulfonates, which, with a certain amount of water, bond together to form inverse micelles. Values of CMC of various solutes in benzene and in carbon tetrachloride are listed by Rosen.[1] In general, the CMCs in these solvents are lower than in aqueous solutions.

14.1 THE INVERSE MICELLE

The interior of the nonpolar micelle is an environment of exceptionally high ionic strength. Fowkes[2] points out that the concentration of water can be as low as one-tenth of a molecule per ion, corresponding to a 550 M solution! This is like the ionic environment of crystals, but the small size of the micelle and its dynamic character allow ready access to reactants and solvents. The micellar core can also be strongly acidic or basic, and hence a medium for catalysis of organic reactants.[3]

The inverse or nonpolar micelle is capable of carrying a net charge conferred by an excess ion in the core. This property is important in the application of such structures to provide conductivity to carry away electrostatic charges and to prevent buildup of electric fields to dangerous levels in the handling of low conductivity fluids, as in the filling or draining of fuel tanks or supertankers.[4]

14.2 ADSORPTION FROM NONPOLAR SOLUTION

Dispersions of finely divided solids in nonpolar media are stabilized by the same two mechanisms that are found to operate in aqueous systems, namely, electrostatic and steric stabilization, but salient differences exist that have to be taken into account. Particularly with respect to electrostatic stabilization, popular misconceptions are rife. Some authors have dismissed this stabilizing factor in nonpolar media as unimportant, but industrial scientists in the oil industry are aware that it is actually equally as important as in aqueous systems. The mechanism of charging in media of low conductivity leads to an opposite result from that which occurs in aqueous media. In aqueous solutions an acid surfactant, such as lauryl sulfonic acid, ionizes and the anion is adsorbed by interfaces to give *negatively charged surfaces*. In organic media, however, acidic dispersing agents are not ionized and are adsorbed as neutral molecules onto basic surface sites where they release protons to charge the surfaces positively. If the concentration of dispersing agent dissolved in the oil is sufficiently high that dynamic adsorption and desorption occur, some negatively charged dispersant molecules are desorbed from the surface into the solution, providing the counterions to give *positively charged surfaces*.

Whereas adsorption of hydrocarbon in water is promoted by an entropy change taking place in the water, no analogous phenomena exist in oil media. Here adsorption has to rely on the enthalpic drive of Lewis acid–Lewis base interaction between the dispersant and the surface.

Figure 14.1 shows the essential difference. In water a hydrocarbon-based ion is spontaneously adsorbed and reduces the surface tension as described by Gibbs' adsorption-isotherm theorem. But most oil-soluble detergents do not reduce the surface tension of the oil in which they are dissolved. Adsorption occurs at the oil/water interface because the orientation of the dispersant finds a lower potential energy at the oil/water interface. The same is true at the hydrophilic interface of a solid particle.

Adsorption of solute by an inorganic interface in a nonpolar medium differs from adsorption out of an aqueous medium in that the former process lacks the entropy difference arising from the change of structure of the water (see Section 13.1). Here the significant difference of entropy resides in the withdrawal of solute from the bulk organic phase to the interface, where it has fewer degrees of freedom, which means that the change of entropy of the solute on adsorption is negative. Since the change of entropy is negative, the Helmholtz free energy, that is, the condition for spontaneous adsorption to occur, would be negative only if the enthalpy change had a large negative value. To obtain a large negative value for the enthalpy change, a specific chemical interaction between the solute and the substrate is required. The most significant such interaction is that between a Lewis acid and a Lewis base. Lacking such an interaction, the interface cannot compete with the solvent for the solute.

This adsorption mechanism is the one commonly found in nonpolar systems. Dispersions of carbon black in aliphatic hydrocarbons, for example, are stabilized

ADSORPTION FROM NONPOLAR SOLUTION   **281**

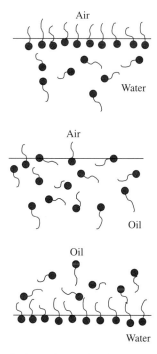

Figure 14.1 A comparison of adsorption at the air/water, air/oil, and the water/oil interfaces.

by aromatic hydrocarbons with long aliphatic side chains.[5] The acidic surface of the carbon black interacts with the basic aromatic group of the stabilizer. Again, amine bases are effective reagents for the deflocculation of carbon black in cellulose esters and oleoresinous vehicles,[6] which is another example of an acid–base interaction.

A solute that is an electron acceptor (or Lewis acid) will interact with an electron donor (or Lewis base,) whether it finds that base on the substrate or in the solvent. Fowkes and Mostafa[7] measured the extent of adsorption of a basic polymer (polymethyl methacrylate or PMMA) on an acidic substrate (silica, acidity due to surface OH groups), as a function of the Lewis acidity and basicity of six solvents used to dissolve the polymer (see Figure 14.2). Carbon tetrachloride was the most neutral solvent, benzene a very weak base, and dioxane and tetrahydrofuran stronger bases. The acid solvents were dichloromethane and chloroform. The basicity of the basic solvents was calculated by their heat of mixing with *tert*-butyl alcohol from Drago's table (see Table 8.4) using *tert*-butyl alcohol as a model for the acid OH groups on silica; the acidity of the acid solvents was measured by their heat of mixing with ethyl acetate, an ester comparable to the methacrylate groups of the polymer. In the PMMA–SiO_2 system, polymer adsorption decreases as solvent basicity increases because the solvent now successfully competes with the polymer for the acidic surface sites. Likewise,

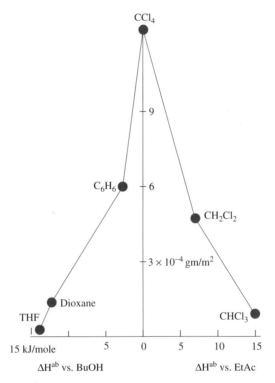

Figure 14.2 Adsorption of PMMA (basic) on silica (acidic) as a function of the acidity or basicity of the solvent. The basicity is measured as the heat of acid–base interaction with *tert*-butyl alcohol and the acidity is measured as the heat of acid–base interaction with ethyl acetate. Poor adsorption occurs at the right-hand side because the acidic solvent dissolves the polymer too well for it to be taken out of solution by adsorption; poor adsorption occurs on the left-hand side because the basic solvent preempts the acidic surface of the silica so successfully that the basic polymer is excluded.[7]

polymer adsorption decreases as solvent acidity increases because the solvent successfully competes with the acidic surface sites for the basic PMMA. These authors correlated the observed degree of polymer adsorption with the acidity and basicity of the solvent, thus showing the overriding importance of acid–base interactions in determining the outcome of the competition. Similar results were obtained for the PVC–$CaCO_3$ system (Figure 14.3). The same two polymers were used to determine the ratio of acidic to basic surface sites in a series of iron-containing pigments (iron oxides tend to have both kinds of site).[8]

14.3 ELECTRICAL CHARGES IN NONPOLAR MEDIA

The inability of electrolyte to ionize in solvents of low conductivity has led to a mistaken belief that electrostatic stabilization of particles in a nonpolar

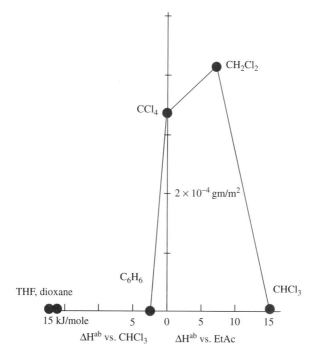

Figure 14.3 Adsorption of post-chlorinated PVC (acidic) on calcium carbonate (basic) as a function of the basicity or acidity of the solvent. The basicity is measured as the heat of acid–base interaction with chloroform and acidity is measured as the heat of acid–base interaction with ethyl acetate. Poor adsorption occurs at the left-hand side because the basic solvent dissolves the acid polymer too well for it to be taken out of solution by adsorption; poor adsorption occurs on the right-hand side because the acid solvent preempts the basic surface of the calcium carbonate so successfully that the acid polymer is excluded.[7]

medium is nonexistent. Dispersed carbon in benzene develops appreciable zeta potentials and stability in the presence of calcium alkylsalicylate, which gives positive zeta potentials, or quaternary ammonium picrates, which give negative zeta potentials.[9,10] These two suspensions, when mixed together, became unstable. Such mutual antagonism is evidence of an electrostatic mechanism. The sign of the charge on the particles in these benzene suspensions is opposite to what would be expected if the charging mechanism were the adsorption of the larger ion, as it is in aqueous solution. It follows that electrostatic stabilization *is* important in stabilizing nonpolar suspensions[11] and that the charging mechanism is *not* the same as in aqueous suspensions.[12]

Fowkes et al.[13] have shown by many examples that in organic media the mechanism of particle charging is the formation of ions in adsorbed films on particle surfaces, where acid–base (or donor–acceptor) interactions occur between the particle surface and the suspending agent (suspendant).

Potentials develop when adsorbed suspendant ions are desorbed into the organic medium, where they form the diffuse electrical double layer. Zeta potentials well over 100 mV result from the stronger acid–base interactions. DLVO (Derjaguin–Landou — Verwey–Overbeek) energy barriers often exceed 25 kT, leading to stability ratios of 10^8 or more (see Section 20.5).

Basic suspendants, for example, are adsorbed as neutral molecules on acidic sites on the surface of the particle, where proton transfer confers on them a positive charge. During the dynamic equilibrium of adsorption and desorption, some of the charged suspendant is desorbed, leaving the particle negatively charged. The low dielectric constant of nonpolar solvents causes electrostatic forces, both repulsion and attraction, to be about 40 times greater than in water. In order to free the counterion from the surface it has to be held at a distance from the surface, approximately tens of nanometers. The counterion is often thereafter incorporated into an inverse micelle, where its electrical charge finds a hydrophilic environment in the form of solubilized water. This feature of the charging process is important, for a bulky micelle keeps the charge of the counterion at a distance from the oppositely charged surface, thus reducing the force of attraction sufficiently to prevent the counterion being drawn into the Stern layer.[12] Acidic or basic polymers, therefore, are the most effective suspendants of particles in nonpolar media. Adsorption of a basic suspendant is depicted in the first transition in Figure 14.4. Desorption of the basic suspendant, now carrying a bound proton, is depicted in the second transition in Figure 14.4. The particle is now negatively charged. Carbon particles that can acquire either a negative or a positive charge

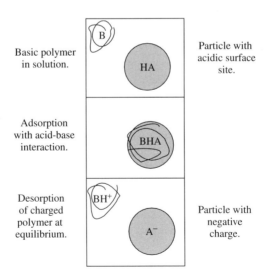

Figure 14.4 Mechanism of electrostatic charging of dispersed acidic particles (with acidic sites HA) by basic dispersants (with basic sites B) in solvents of low dielectric constant.[13]

from basic or acidic suspendants, respectively, have both acidic and basic sites on their surfaces (amphoteric).

Electrically charged particles in nonpolar media figure in a number of technological applications, such as the electrophoretic development of latent images (electrophotography), electrophoretic displays, electrodeposition of special coatings, and removal of particulate contaminants from nonpolar media by an imposed electric field. Liquid development of an electrostatic image on a photoreceptor surface, such as a selenium drum, depends on the electrophoresis of charged particles in a nonpolar medium. Electrophoretic displays are based on the migration of highly reflecting, charged pigment particles to a viewing electrode. Filtration efficiency is improved with positively charged microporous membranes that combine submicron sieving with electrostatic adsorption.

Harmful electrical effects in nonpolar media are found in automotive applications in which electrified surfaces attract dispersed particles from crankcase oils, thus blocking flow through small orifices. Electric fields develop at metal surfaces by an exchange of charge with acid or basic components in the oil. These electric fields become large when tanks are emptied, as counterions are carried away with the oil.

The cause of a number of explosions and fires in oil refineries, depots, and tankers, although never established with absolute certainty, is probably traced to the discharge of electrostatic fields. Splash filling is a frequent cause of electrification inside a tank. Other sources of electrification are mixing, blending, and agitating. These operations are frequently accompanied by emulsification of the tank contents and the stirring up of water bottoms. Explosions may even occur inside grounded containers. Klinkenberg and van der Minne[4] report that a large tank in Shell's refinery at Pernis exploded 40 minutes after the start of a blending operation in which a tops-naphtha mixture was being pumped into a straight-run naphtha. On the following day a second attempt was made to blend these materials and again an explosion occurred 40 minutes after starting the pumps. This striking and unusual coincidence could only be explained by assuming that both explosions were caused by static electricity.

A large electrical potential can develop on removing mobile countercharges from the vicinity of an interface such as a pipe wall or a sedimenting water droplet. In aqueous solutions, the high conductivity prevents the growth of any large electric potentials; but in poorly conducting media, electric potentials can become quite large. These large potentials spark on discharge and can ignite explosive vapor mixtures. A method to reduce electric potential is to increase the conductivity of the liquid by the addition of antistatic additives. The conductivity cannot be increased by means of electrolytes in media of low dielectric constant as ionization does not take place, but conductivity can be increased by the addition of a combination of chemicals that form charged micelles.

The half-value time of an organic liquid is the time taken for the charge in a liquid, completely filling a metal container, to decrease to half its original value. The half-value time is inversely proportional to the specific conductivity and directly proportional to the dielectric constant. Safety of refinery operations requires a conductivity of at least $5 \times 10^{-3} \Omega^{-1}$ cm^{-1}. Table 14.1 lists the conductivities and half-value times of several liquids. The half-value times vary by nine orders of magnitude from microseconds to hours. Electric fields in aqueous media dissipate instantaneously, even in distilled water, while taking all day to dissipate in petroleum distillates. High voltages are generated because of the slow dissipation if, in the meantime, the countercharges are carried far away from the charged interface.

Because of the ease with which countercharges are removed from the vicinity of a charged interface in a low-conductivity medium, the charge density can be determined directly by allowing the liquid with its entrained countercharges to flow out of a container into a receiving vessel connected to an electrometer (Figure 14.5).

The same principle can be applied to a suspension in a liquid of low conductivity. The separation of the mobile charge from the dispersed particles is

TABLE 14.1 Half-Value Times of Various Liquids[14]

Liquid	Conductivity (Ω^{-1} cm^{-1})	Half-Value Time (s)
Highly purified hydrocarbons	10^{-17}	12,000
Light distillates	10^{-16}–10^{-13}	1,200–1.2
Crude oil	10^{-11}–10^{-9}	0.012–0.00012
Distilled water	10^{-6}	4.8×10^{-6}

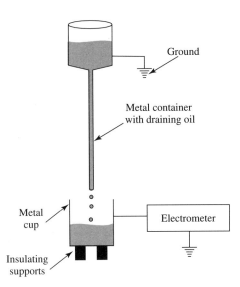

Figure 14.5 Sketch of apparatus to measure the electric charging in a metal capillary.[15]

brought about by passing the system through a microporous filter, which retains the charged particles while the countercharges are collected in an isolated metal cup connected to an electrometer. The charge-to-mass ratio is calculated from the net charge collected per gram of particles. The charge per particle is calculated from the charge per gram and the particle size. Charge per particle is the end determination of most electrophoresis measurements; by this method, it is obtained directly. (see Section 17.4.e).

Electrophoretic mobilities are difficult to measure in nonpolar systems because low conductivities and irreversible electrodes conduce to nonuniform, time-varying electric fields. As the charges move, their distribution becomes nonuniform, generating an internal electric field. This phenomenon is not significant in aqueous media as the large number of charges per unit volume neutralizes any such field, but in media of low conductivity, these internal electric fields are superimposed on the applied electric field. Initially, the electric field is only the applied field, but during the time required to come to a steady state (Table 14.1), the field decreases and is no longer known. Since the magnitude of the electric field varies in an unknown manner throughout the cell, the electrophoretic mobility cannot be calculated as the ratio of the particle velocity to the applied electric field.

The varying current flow through the system can be monitored, however, along with the velocity of the particles, and these data can be combined to yield the electrophoretic mobility under some conditions and with some assumptions.[16,17] The usefulness of these electrical transients for the determination of electrophoretic mobility depends on the fact that in low-conductivity media the particles carry a substantial part of the total current (in aqueous media the particles carry an insignificant part of the total current).

14.4 THEORIES OF BUBBLE RISE IN VISCOUS MEDIA

The presence of a surface-active solute in a lubricating oil may cause problems of lubrication due to emulsified water, air entrainment, or foam. Such striking effects, when they occur, clearly imply a condition of surface activity as their precursor, but surface activity in a solution does not always manifest itself by conspicuous phenomena: it may be subtly present. An awareness of the condition is desirable in order to guard against the development of problems, which may arise either by further buildup of the concentration of the surface-active solute or by its synergistic combination with additives. The detection of the condition, in the absence of clear evidence afforded by foaming or emulsification, is not as straightforward as might appear. Lubricating oils, like most organic liquids, have low surface tensions ($\sigma < 30$ mN/m) and surface activity in such solvents may be a matter of lowering the surface tension by one 1 mN/m or less. The measurement of the lowering of the surface tension of the solvent, which is the direct method to detect the presence of a surface-active solute, is not a feasible procedure with solvents that initially are not chemically pure. Even if the oil were a single chemical component, this direct approach is not required, as methods to detect dynamic surface tension are better suited to discover whether a surface-active solute is present. Of these the easiest to perform is the measurement of the rate of ascent of a bubble through the solution. One limitation, however, common to all methods, must be remarked: they can all definitively demonstrate the *presence* of surface activity in a solution, but the absence of a positive response, in whatever method used, is a necessary but not sufficient reason to conclude that a surface-active solute is *not present*. The disequilibrium, on which the phenomenon depends, may be over before it is detected.

The rate of rise of single air bubbles in a solution is determined by the behavior of newly formed surfaces and how they respond to forces of dilatation and compression. In a pure liquid, under conditions of laminar flow, a rising bubble moves faster than predicted by Stokes' law, since the mobility of its interface allows lower velocity gradients in the liquid than those that develop at an immobile, or rigid, interface. When a surface-active solute is adsorbed at the interface, however, the movement of the interface and of air inside the bubble is restricted; the velocity gradient in the outer fluid is increased, until at the limit the bubble acquires the property of a rigid sphere, and its rate of rise is reduced to that given by Stokes' law.

14.4.a Rise in Pure Liquids: Hadamard Regime

Stokes' theory for the terminal velocity of a solid sphere in a viscous medium was extended by Rybczynski[18] and Hadamard[19] to fluid spheres. For a liquid drop or a gas bubble of radius a, density ρ_1, and viscosity η_1, moving through an infinite volume of a medium of density ρ_2, and viscosity η_2 the terminal velocity is given by

$$v_s = \frac{2(\rho_1 - \rho_2) g a^2}{9\kappa \eta_2} \tag{14.1}$$

where κ, the Rybczynski–Hadamard correction factor, has the value

$$\kappa = \frac{3\eta_1 + 2\eta_2}{3\eta_1 + 3\eta_2} \tag{14.2}$$

The derivation of Eq. (14.2) postulates that the medium exerts a viscous drag on the surface of the bubble or liquid drop and so sets up a circulation of the fluid contained inside, whether gas or liquid. According to the theory, a bubble containing a circulating gas, with $\eta_1 \ll \eta_2$, would move 50% faster ($\kappa = 2/3$) than one in which the gas, for any reason, does not circulate; for in the latter case the Rybczynski–Hadamard factor is unity, and the velocity of the bubble is given by the unmodified form of the Stokes' law.

Garner and Hale[20] and Garner and Hammerton[21] demonstrated experimentally the existence of the circulation inside air bubbles and examined the effects of bubble size and shape. They showed that the validity of the above equations is limited to the range of Reynolds number less than 1, and to the same conditions as for Stokes' law to hold, including the requirement that the gas bubbles be spheres. Ryskin and Leal[22] derived a numerical solution of an ascending bubble in a liquid by integrating the forces in the surface for Reynolds numbers in the range $0.5 < R_e < 200$ and for Weber numbers up to 20. The experimental data of Furler and Ross[23] extended to these higher Reynolds numbers and conformed to the theoretical description of a fluid or a rigid interface, whichever happened to occur in the systems observed.

14.4.b Effect of Surface-Active Solutes: Stokes Regime

Levich[24] was the first to provide a satisfactory explanation of the retardation of the velocity of a rising bubble caused by surface-active solutes in the medium. He postulated that adsorbed solute is not uniformly distributed on the surface of a moving bubble. The surface concentration on the upstream part of the bubble is less than the equilibrium concentration, while that on the downstream part is greater than equilibrium. This disequilibration of the concentrations is brought about by the viscous drag of the medium acting on the interface, which in turn creates a disequilibrium of surface tensions, with the lower tension where the concentration of adsorbate is greater. The liquid interface then flows (Marangoni flow) from the region of lower tension to that of higher tension, and the direction of this flow offsets the flow induced by the shear stress in the outer fluid acting on the interface. The increasing rigidity of the interface inhibits the circulation of internal gas to a greater or lesser degree; when completely inhibited, the interface is effectively completely rigid and the terminal velocity of rise is reduced to that given by Stokes' law.

Figures 14.6(b) and 14.6(c) show that the ratio of the observed velocity of ascent of a bubble to the calculated Stokes' velocity varies between the limits $0.99 < \kappa < 1.52$ for the solutions in mineral oil, and the limits $0.95 < \kappa < 1.47$ for the solutions in trimethylolpropane heptanoate. The value of $\kappa = 1.52$ obtained for the mineral oil and 1.47 for the trimethylolpropane heptanoate demonstrates the virtual absence of any surface-active contaminant in these solvents. No significance is attached to the difference of 0.05 units from the theoretical value at the lower limit, $\kappa = 1.00$, as fluctuations of that magnitude at that limit are within the experimental error. Figure 14.6a and 14.6b also show that the range in which κ is concentration dependent is $0.1\ \mu g/g < c < 20\ \mu g/g$ for solutions of polydimethylsiloxane in mineral oil and in trimethylolpropane heptanoate. For solutions of Span 20 in mineral oil, the lowest concentration for any measurable variation in the value of κ is 500 times greater, but the range of concentration through which the variation of κ takes place is relatively narrower, that is, $50\ \mu g/g < c < 500\ \mu g/g$. The solute N-phenyl-1-naphthylamine in trimethylolpropane heptanoate requires a still greater concentration before its effect is found [Figure 14.6(c)], showing that it is even less surface active in this solvent.

The inhibition of movement of the surface and of air circulation inside the rising bubble is accepted as the explanation of the remarkable slowing of its rate

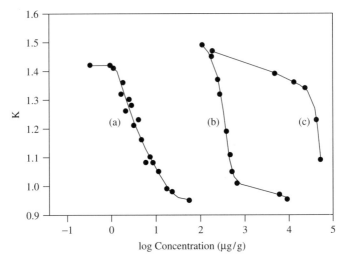

Figure 14.6 Ratio of the observed velocity of ascent of a bubble to the calculated Stokes' velocity in solutions of various concentrations. (*a*) Polydimethylsiloxane in trimethylolpropane-heptanoate, (*b*) Polydimethylsiloxane in mineral oil, (*c*) N-phenyl–1–1napthylamine in trimethylolpropane-heptanoate.[25] Each figure shows the transition from the Hadamard to the Stokes regime.

of ascent. Okazaki, Hayashi and Sasaki[26] found that a concentration of sodium dodecyl sulfate (SDS) in water as low as 10^{-5} M retards the rate of ascent of a bubble in that solution and also confers stability on a single bubble at its surface, even though the static properties of the liquid, that is, the density, the bulk compressibility, the viscosity, and the surface tension, all remain unchanged from those of pure water. A time dependence of the surface tension is also absent in 10^{-5} M SDS in water; and no measurable increase of the shear viscosity of the surface of a solution so dilute was observed.

Oil solutions behave in many respects as do the very dilute aqueous solutions of SDS in water at concentrations of 10^{-6}–10^{-4} M, investigated by Okazaki, Hayashi and Sasaki. In oil systems, as in these aqueous systems, certain static and dynamic properties of the solution that might appear to be pertinent to surface activity show no change from those of the solvent, yet the rate of bubble rise is greatly affected by the presence of the solute.

REFERENCES

[1] Rosen, M.J. *Surfactants and interfacial phenomena*, 2nd ed., Wiley: New York; 1989; p 149.

[2] Fowkes, F.M. The interactions of polar molecules, micelles, and polymers in nonaqueous media, in *Solvent properties of surfactant solutions*; Shinoda, K., Ed.; Marcel Dekker: New York; 1967; pp 65–115.

[3] Fendler, J.H.; Fendler, E.J. *Catalysis in micellar and macromolecular systems*; Academic Press: New York; 1975.

[4] Klinkenberg, A.; van der Minne, J.L. *Electrostatics in the petroleum industry: the prevention of explosion hazards*; Elsevier: New York; 1958.

[5] van der Waarden, M. Stabilization of carbon-black dispersions in hydrocarbons, *J. Colloid Sci.* **1950**, *5*, 317–325.

[6] Fischer, E.K. *Colloidal dispersions*; Wiley: New York; 1950; p 245.

[7] Fowkes, F.M.; Mostafa, M.A. Acid–base interactions in polymer adsorption, *Ind. Eng. Chem., Prod. Res. Dev.* **1978**, *17*, 3–7.

[8] Fowkes, F.M. Characterization of solid surfaces by wet chemical techniques, *ACS Symp. Ser.* **1982**, *199*, 69–88.

[9] van der Minne, J.L.; Hermanie, P.H.J. Electrophoresis measurements in benzene: Correlation with stability. I. Development of method, *J. Colloid Sci.* **1952**, *7*, 600–615.

[10] van der Minne, J.L.; Hermanie, P.H.J. Electrophoresis measurements in benzene: Correlation with stability. II. Results of electrophoresis, stability and adsorption, *J. Colloid Sci.* **1953**, *8*, 38–52.

[11] Morrison, I.D. Criterion for electrostatic stability of dispersions at low ionic strength, *Langmuir* **1991**, *7*, 1920–1922.

[12] Morrison, I.D. Electrical charges in nonaqueous media, *Colloids Surf.* **1993**, *71*, 1–37.

[13] Fowkes, F.M.; Jinnai, H.; Mostafa, M.A.; Anderson, F.W.; Moore, R.J. Mechanism of electric charging of particles in nonaqueous liquids, *ACS Symp. Ser.* **1982**, *200*, 307–324.

[14] Klinkenberg loc. cit.; p 28.

[15] Klinkenberg loc. cit.; p 50.

[16] Novotny, V.J. Physics of nonaqueous colloids, *ACS Symp. Ser.* **1982**, *200*, 281–306.

[17] Kornbrekke, R.E.; Morrison, I.D.; Oja, T. Electrophoretic mobility measurements in low conductivity media, *Langmuir* **1992**, *8*, 1211–1217.

[18] Rybczynski, W. Über die fortschreitende bewegung einer flüssigen kugel in einem zähen medium, *Bull. Int. Acad. Pol. Sci. Lett., Cl. Sci. Math. Nat., Ser. A* **1911**, 40–46.

[19] Hadamard, J. Mouvement permanent lent d'une sphère liquide et visqueuse dans un liquide visqueux, *C. R. Hebd. Seances Acad. Sci.* **1911**, *152*, 1735–1738.

[20] Garner, F.H.; Hale, A.R. The effect of surface active agents in liquid extraction processes, *Chem. Eng. Sci.* **1953**, *2*, 157–163.

[21] Garner, F.H.; Hammerton, D. Circulation inside gas bubbles, *Chem. Eng. Sci.* **1954**, *3*, 1–11.

[22] Ryskin, G.; Leal, L.G. Numerical solution of free-boundary problems in fluid mechanics, *J. Fluid Mech.* **1984**, *148*, 1–17.

[23] Furler, G.; Ross, S. Experimental observations of the rate of ascent of bubbles in lubricating oils, *Langmuir* **1986**, *2*, 68–72.

[24] Levich, V.G. *Physicochemical hydrodynamics*; Prentice Hall: Englewood Cliffs, NJ; 1962; pp 395–452.

[25] Suzin, Y.; Ross, S. Retardation of the ascent of gas bubbles by surface-active solutes in nonaqueous solutions, *J. Colloid Interface Sci.* **1985**, *103*, 578–585.

[26] Okazaki, S.; Hayashi, K.; Sasaki, T. Mechanism of antifoaming according to the classification of antifoamers, *Chem., Phys. Appl. Surf. Act. Subs. Proc. Int. Congr., 4th, 1964* **1967**, *3*, 67–73.

15 Thermodynamics of Adsorption from Solution

The stability of interfaces depends on adsorption, inasmuch as the presence of the spontaneously adsorbed species confers an electrical charge or a steric barrier, or both, to retard coalescence of particles, drops, or bubbles. A gas adsorbed at a solid surface constitutes (or may be treated as) a one-component system, with applications to catalysis; a single solute adsorbed at a solvent surface constitutes a two-component system, with applications to foam, wetting, and spreading; a single solute distributed between two immiscible liquids and adsorbed at the liquid/liquid interface constitutes a three-component system, with applications to emulsions and solvent extraction. In what follows we discuss two- and three-component systems, the discussion of which can be simplified for one-component systems.

15.1 THE GIBBS ADSORPTION FROM SOLUTION

Consider the interfacial phase σ between two bulk liquid phases α and β (Figure 15.1). The σ phase, however, is not an autonomous phase because it is subject to an external force field, measured by the interfacial tension, caused by the difference in nature of the two adjacent bulk phases. The external field creates heterogeneity of composition, just as the presence of a gravitational or electric field might act analogously. The σ phase, consequently, has a continuously varying composition and density.

By the first and second laws of thermodynamics, the energy of the entire three-phase system is (see Section 6.2)

$$dU = TdS - pdV + \sigma dA + \sum \mu_i dn_i \tag{15.1}$$

where U, S, V, A, and n_i are the extensive variables and T, p, σ, and μ_i are the intensive variables. The two work terms, pdV and σdA have opposite signs because the spontaneous tendency of a gas is to expand ($dV > 0$) and that of a liquid surface is to contract ($dA < 0$).

Integrating Eq. (15.1) gives

$$U = TS - pV + \sigma A + \sum \mu_i n_i \tag{15.2}$$

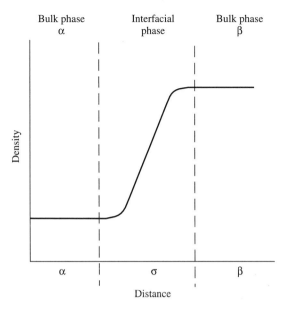

Figure 15.1 Schematic of interfacial phase between two bulk phases.

Taking the total differential of Eq. (15.2) and comparing with Eq. (15.1) gives a Gibbs–Duhem equation,

$$SdT - Vdp + Ad\sigma + \sum n_i d\mu_i = 0 \qquad (15.3)$$

Equation (15.3) can be written for each of the bulk phases separately:

$$S^\alpha dT - V^\alpha dp + \sum n_i^\alpha d\mu_i^\alpha = 0 \qquad (15.4)$$

$$S^\beta dT - V^\beta dp + \sum n_i^\beta d\mu_i^\beta = 0 \qquad (15.5)$$

Subtracting Eqs. (15.4) and (15.5) from (15.3), and using the condition for equilibrium that $d\mu_i = d\mu_i^\alpha = d\mu_i^\beta$ gives

$$(S - S^\alpha - S^\beta) dT - (V - V^\alpha - V^\beta) dp + Ad\sigma + \sum \left(n_i - n_i^\alpha - n_i^\beta\right) d\mu_i = 0 \qquad (15.6)$$

The surface excess quantities are defined as

$$S^\sigma = S - S^\alpha - S^\beta \qquad (15.7)$$

$$V^\sigma = V - V^\alpha - V^\beta \qquad (15.8)$$

$$n_i^\sigma = n_i - n_i^\alpha - n_i^\beta \qquad (15.9)$$

THERMODYNAMICS OF ADSORPTION FROM SOLUTION

Hence

$$A d\sigma + S^\sigma dT - V^\sigma dp + \sum n_i^\sigma d\mu_i = 0 \qquad (15.10)$$

At constant temperature and pressure,

$$-d\sigma = \sum \frac{n_i^\sigma}{A} d\mu_i = \sum \Gamma_i d\mu_i \qquad (15.11)$$

where

$$\Gamma_i = \frac{n_i^\sigma}{A} \quad \text{mol m}^{-2} \qquad (15.12)$$

Equation (15.11) is the Gibbs adsorption isotherm. The surface concentrations defined by Γ_i are *excess* surface concentrations.

A qualitative understanding of the Gibbs adsorption theorem may be obtained by recalling that any spontaneous process at a liquid surface must be accompanied by a loss of potential energy. Now since the surface tension measures the potential energy difference between molecules at the surface and those in the bulk liquid, we should expect spontaneous adsorption of solute to be linked to a reduction of the surface tension. The quantitative relation is the Gibbs adsorption isotherm, which shows that the rate of change of the surface tension with the chemical potential (or, in dilute solutions, the logarithm of the concentration of the solute) is related to the excess surface concentration of solute due to adsorption. If the adsorption is positive, the change in the surface tension is negative, and vice versa.

15.1.a Two-Component Systems

For a two-component system, Eq. (15.11) is

$$-d\sigma = \Gamma_1 d\mu_1 + \Gamma_2 d\mu_2 \qquad (15.13)$$

The Gibbs–Duhem equation applied to the bulk phase at constant temperature and pressure is

$$X_1 d\mu_1 + X_2 d\mu_2 = 0 \qquad (15.14)$$

where X_1 and X_2 are the mole fractions of the two components. Therefore,

$$d\mu_1 = -\frac{X_2}{X_1} d\mu_2 \qquad (15.15)$$

Substituting in Eq. (15.13) gives

$$-d\sigma = \left(\Gamma_2 - \Gamma_1 \frac{X_2}{X_1}\right) d\mu_2 \qquad (15.16)$$

The whole coefficient of the term in $d\mu_2$ has the units of mole fraction per unit area; it is therefore some kind of surface concentration of the solute, which is convenient to designate by a separate symbol, Γ_2^G, or

$$\Gamma_2^G = \Gamma_2 - \frac{X_2}{X_1}\Gamma_1 \tag{15.17}$$

Although variations in the conventions for fixing the plane boundary between the bulk and surface phases cause variations in the separate values of Γ_1 and Γ_2, they cannot cause any variation in $d\sigma$. Therefore the expression given in Eq. (15.17) is invariant regardless of how one regulates the convention. The invariant quantity given by Eq. (15.17) is the same as the quantity defined by Gibbs, denoted by him with the symbol $\Gamma_{2(1)}$ and usually called the Gibbs surface-excess concentration.

Using the definition of Γ_2^G for a two-component system in Eq. (15.16) and using activities instead of chemical potentials gives

$$d\sigma = -RT\Gamma_2^G d\ln a_2 \tag{15.18}$$

where Γ_2^G is the Gibbs surface-excess in moles per square meter and a_2 is the activity of the surface-active solute in the bulk solution. Since we are using the Gibbs convention to define the position of the boundary between bulk and surface phases, only a term for the solute appears in Eq. (15.18).

For a nonionizing single solute, in dilute aqueous solution where it may be taken as ideal in behavior,

$$d\mu_2 = RT d\ln X_2 \tag{15.19}$$

where X_2 is the mole fraction of the surface-active solute in the bulk solution. The Gibbs equation then becomes

$$-d\sigma = RT\Gamma_2^G d\ln X_2 \tag{15.20}$$

An ionic solute introduces the possibility of having both ions in the surface phase, as well as H^+ if surface hydrolysis occurs. Taking sodium lauryl sulfate, NaLS, as an example,

$$-d\sigma = RT\,(\Gamma_{Na^+} d\ln X_{Na^+} + \Gamma_{LS^-} d\ln X_{LS^-} + \Gamma_{H^+} d\ln X_{H^+}) \tag{15.21}$$

The pH of the bulk solution does not change appreciably as the result of adsorption, even if surface hydrolysis does occur, hence $d\ln X_{H^+} = 0$. Furthermore, in the absence of other electrolytes that might contribute Na^+ to the solution,

$$d\ln X_{Na^+} = d\ln X_{LS^-} \tag{15.22}$$

and the Gibbs equation is

$$-d\sigma = 2RT\Gamma_2^G d\ln X_2 \qquad (15.23)$$

In general, for dilute aqueous solutions of ionic surface-active solutes, in the absence of other electrolytes, the Gibbs equation takes the form

$$-d\sigma = \nu RT\Gamma_2^G \alpha\, d\ln X_2 \qquad (15.24)$$

where ν is the number of ions per surface-active solute and α is the degree of dissociation. For a 1:1 strong electrolyte, ν is 2 and α is 1.

For dilute solutions of a completely dissociated 1:1 electrolyte in the presence of a swamping amount of an electrolyte containing a common (but not surface-active) ion to the surface-active solute, the dissociation is suppressed, and the Gibbs equation has the same form as that for a nonionic surface-active solute, namely, Eq. (15.20). That situation exists, for example, if H^+ is adsorbed rather than Na^+, which is the result of surface hydrolysis. The literature contains many examples where surface hydrolysis is not assumed and still Eq. (15.20) is used, whereas Eq. (15.24) would be the correct form.

The wide use and fundamental importance of the Gibbs adsorption theorem required that its validity should receive direct experimental confirmation, especially as anomalous surface-tension isotherms of "pure" surface-active solutes showed minima, which, according to Gibbs' theorem, are inexplicable for two-component systems. McBain and Humphreys[1] devised the microtome method to determine the absolute amount of adsorption at *static* air/water interfaces. The experiment was a heroic effort, beset with experimental problems, to skim off the surface of a solution 50–500 μm thick by means of a small microtome blade, propelled by catapult action, moving at a speed of about 10.7 m/s. The solution so obtained was analyzed and its concentration compared with that of the bulk of the solution. The results successfully confirmed Gibbs' theory for both positive and negative adsorption, using solutions of hydrocinnamic acid (β-phenyl propionic acid), phenol, and sodium chloride.[2]

15.1.b Three-Component Systems

We shall use subscripts 1, 2, and 3 to designate oil, solute, and water, respectively; and L and W to designate the oil (lipoid) and water phases. In applying Eq. (15.11), two geometrical plane surfaces have to be placed to separate the interfacial phase from the bulk oil and water phases. More than one convention is possible in placing these surfaces because Eq. (15.11) remains true no matter where the imaginary surfaces are drawn, although the individual values of Γ_i are affected. A convenient and natural choice would be to select the limits that separate a well-defined adsorbed layer from the solution, which, at low concentrations, is likely to be no more than a monolayer. This convention corresponds to the U-convention of Hutchinson.[3,4,5] The argument that follows is not, however, restricted to that case. Consider the same solute distributed between two

THE GIBBS ADSORPTION FROM SOLUTION   297

bulk phases, oil and water, making a three-component system. Equation (15.11) for three components reads

$$-d\sigma = \Gamma_1 d\mu_1 + \Gamma_2 d\mu_2 + \Gamma_3 d\mu_3 \tag{15.25}$$

The Gibbs–Duhem equation applied to the oil phase (L) and to the water phase (W) gives us two more relations, which, if we suppose the oil and water phases to have negligible mutual solubility, are as follows:

$$X_1^L d\mu_1 + X_2^L d\mu_2 = 0 \tag{15.26}$$

$$X_2^W d\mu_2 + X_3^W d\mu_3 = 0 \tag{15.27}$$

We can use Eqs. (15.26) and (15.27) to obtain the following:

$$d\mu_1 = -\frac{X_2^L}{X_1^L} d\mu_2 \tag{15.28}$$

$$d\mu_3 = -\frac{X_2^W}{X_3^W} d\mu_2 \tag{15.29}$$

Substituting Eqs. (15.28) and (15.29) in Eq. (15.25) gives

$$-d\sigma = \left[\Gamma_2 - \frac{X_2^L}{X_1^L}\Gamma_1 - \frac{X_2^W}{X_3^W}\Gamma_3\right] d\mu_2 \tag{15.30}$$

The whole coefficient of the term in $d\mu_2$ in Eq. (15.30) has the units of moles of component 2 per unit area; it is, therefore, some kind of surface concentration of the solute, which it is convenient to designate by a separate symbol, Γ_2^G, or

$$\Gamma_2^G = \left[\Gamma_2 - \frac{X_2^L}{X_1^L}\Gamma_1 - \frac{X_2^W}{X_3^W}\Gamma_3\right] \tag{15.31}$$

Although different conventions for fixing the plane boundaries between the bulk and the interfacial phases cause variations in the separate values of Γ_1, Γ_2, and Γ_3, they cannot cause any variation in $d\sigma$. Therefore, the expression for Γ_2^G given by Eq. (15.31) is invariant regardless of what convention is adopted. The invariant quantity given by Equation [15.31] is essentially the same as the quantity defined by Gibbs, denoted by him by the symbol $\Gamma_{2(1,3)}$ [Eq. (15.37)]. When both liquid phases contribute the same solute to the interface, the quantity adsorbed is the sum of the two quantities from each immiscible phase.

A clearer understanding of the meaning of the U-convention (or any other convention for defining Γ_i) can be gained by using small whole numbers to designate relative concentrations in the bulk and in the interfacial phases. Suppose the oil and aqueous phases are in equilibrium with the solute distributed in each phase, having a mole fraction of 1/5 and 1/7, respectively. Let the solute be positively adsorbed at the interface from each bulk phase according to the numbers in the diagram (Figure 15.2). Applying Eq. (15.31) gives

$$\Gamma_2^G = 51 - \tfrac{1}{4}(84) - \tfrac{1}{6}(102) = 13 \tag{15.32}$$

The variation of concentration of component 2 across the interface for the numerical example shown in Figure 15.2 is represented in Figure 15.3, showing the heterogeneity of composition in the σ phase and the homogeneity of the L and W phases.

An equivalent expression for Eq. (15.31) can be obtained by multiplying Eq. (15.26) by a multiplier x and (15.27) by a multiplier y and subtracting from

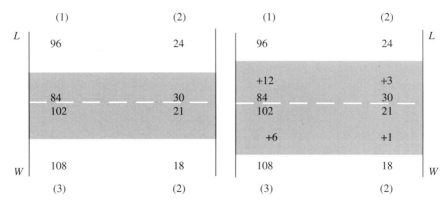

Figure 15.2 A numerical example of the variation of concentration of component 2 across an interface.[6]

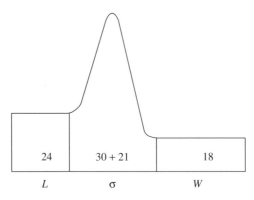

Figure 15.3 The variation of concentration of component 2 in the σ phase and in the homogeneous L and W phases.

Eq. (15.11), giving

$$-d\sigma = \left(\Gamma_1 - x\Gamma_1^L\right) d\mu_1 + \left(\Gamma_2 - x\Gamma_2^L - y\Gamma_2^W\right) d\mu_2$$
$$+ \left(\Gamma_3 - y\Gamma_3^W\right) d\mu_3 = \sum \Gamma_i(x, y) d\mu_i \quad (15.33)$$

where $\Gamma_i(x, y)$ is a surface concentration dependent on the values selected for x and y. Equation (15.33) is valid for any numerical values assigned to x and y; in particular, it is convenient to choose the multipliers in such a way as to make vanish two of the terms in Eq. (15.33). Let $\Gamma_{1(2,3)}$ represent $\Gamma_1(x, y)$ when $\Gamma_2(x, y)$ and $\Gamma_3(x, y)$ are made to equal zero by an appropriate choice of x and y (say, x_1, y_1); similarly, let $\Gamma_{2(1,3)}$ represent $\Gamma_2(x, y)$ when $\Gamma_1(x, y)$ and $\Gamma_3(x, y)$ are made to equal zero by a second appropriate choice (say, x_2, y_2); and let $\Gamma_{3(1,2)}$ represent $\Gamma_3(x, y)$ when $\Gamma_1(x, y)$ and $\Gamma_2(x, y)$ are made to equal zero by a third appropriate choice of x and y (say, x_3, y_3). Then by Eq. (15.33),

$$-d\sigma = \Gamma_{1(2,3)} d\mu_1 = \Gamma_{2(1,3)} d\mu_2 = \Gamma_{3(1,2)} d\mu_3 \quad (15.34)$$

For our example of a three-component system, the terms in μ_1 and μ_3 are selected to vanish, that is,

$$\left(\Gamma_1 - x\Gamma_1^L\right) = 0 \quad \text{and} \quad \left(\Gamma_3 - y\Gamma_3^W\right) = 0 \quad (15.35)$$

These operations are equivalent to bringing the concentrations of component 1 in the L phase and component 3 in the W phase up to the interface unchanged, and so bring Eq. (15.33) to the following:

$$-d\sigma = \left(\Gamma_2 - x\Gamma_2^L - y\Gamma_2^W\right) d\mu_2 = \Gamma_{2(1,3)} d\mu_2 \quad (15.36)$$

Comparing Eqs. (15.30) and (15.36) yields

$$\Gamma_2^G = \Gamma_{2(1,3)} \quad (15.37)$$

Thus the two derivations [of Eqs. (15.31) and (15.36)] are equivalent. In Eq. (15.36), $\Gamma_{2(1,3)}$ is expressed as the surface concentration of component 2 reduced by two terms that are functions of how much of component 2 is present in the lipoid and the water phase; that is the reason for describing $\Gamma_{2(1,3)}$ as an *excess* concentration.

A numerical representation of $\Gamma_{2(1,3)}$ is shown in Figure 15.2, with the planes of separation drawn so as to make the surface excess of solvent in each phase equal zero. To do so 12 units of solvent 1 is added to the σ phase to bring its total up to 96 and also 6 units of solvent 3 is added to the σ phase to bring its total up to 108. These additions are required to meet conditions expressed by Eq. (15.31). The number of solute molecules in the α phase has, meanwhile, been increased proportionately by 3 units from the L phase and 1 unit from the W phase. The total excess surface concentration of solute is $33 - 24 = 9$ units with respect to

300 THERMODYNAMICS OF ADSORPTION FROM SOLUTION

the L phase plus $22 - 18 = 4$ units with respect to the W phase, or 13 units in all. This is the same answer already obtained by means of the U-convention.

The numerical values quoted in Figure 15.2 correspond to a weakly surface-active solute, such as a lower alcohol. Soaps and detergents are much more strongly adsorbed, so that the terms other than Γ_2 in Eq. (15.30) may be neglected. The Gibbs excess concentration Γ_2^G may then be taken as equal to the actual surface concentration without significant error. The Gibbs concentration Γ_2^G, can be evaluated from a plot of the surface tension against the logarithm of the *concentration* (for a dilute solution) or the logarithm of the *activity* for a more concentrated solution. The reciprocal of Γ_2^G is the area per mole of component 2 in the surface, that is, $A = 1/\Gamma_2^G$, where A is the area per mole.

15.2 THE SURFACE TENSION ISOTHERM

At sufficiently low concentrations of a surface-active solute, the surface tension isotherm is linear at constant temperature:

$$\sigma_0 - \sigma = mX_2 \qquad (15.38)$$

where m is the slope of the isotherm and X_2 is the mole fraction of solute. The lowering of the surface tension of solvent by a solute, $\sigma_0 - \sigma$, is called the spreading pressure and is designated by π: that is,

$$\pi = \sigma_0 - \sigma \qquad (15.39)$$

Differentiating Eq. (15.39) gives

$$d\pi = -d\sigma \qquad (15.40)$$

Substituting Eq. (15.39) into Eq. (15.38), taking logarithms, and differentiating, gives

$$d \ln \pi = d \ln X_2 \qquad (15.41)$$

Substituting Eqs. (15.39) and (15.41) into Eq. (15.20) gives

$$d\pi = RT\Gamma_2^G d \ln \pi$$

or
$$\pi = \Gamma_2^G RT \qquad (15.42)$$

or
$$\pi A = RT \qquad (15.43)$$

Equation (15.43) is the ideal equation of state of an adsorbed film; it appears therefore that a linear surface tension isotherm signifies an ideal solution in the interface.

The linear relation between surface tension and concentration, which is the behavior by which we recognize an ideal interfacial solution, is observed only with dilute solutions. With solutes of pronounced surface activity, such as soaps or other detergents, the dilution required to get into the range of ideality of the interfacial solution is so extreme that such minute concentrations are rarely the subject of measurements. Solutes of low or moderate surface activity do, however, produce ideal solutions in the interfacial phase at equilibrium bulk concentrations that are several orders of magnitude greater.

At concentrations greater than the range of application of Eq. (15.38), the surface tension isotherm is no longer linear. An empirical equation due to von Szyszkowski[7] is often used:

$$\pi = \sigma_0 - \sigma = RT\Gamma_m \ln\left(\frac{X_2}{a} + 1\right) \qquad (15.44)$$

where Γ_m is the number of moles of component 2 per unit area at saturation in the interface and a is an empirical constant.

15.2.a Adsorption Isotherm: Langmuir

The Szyszkowski equation, when combined with Eq. (15.20) gives the adsorption isotherm equation, as follows. Differentiating Eq. (15.44) gives

$$-\frac{d\sigma}{dX_2} = \frac{\Gamma_m RT}{X_2 + a} \qquad (15.45)$$

Rewriting the Gibbs equation (15.20) as

$$-\frac{d\sigma}{dX_2} = \frac{RT\Gamma_2^G}{X_2} \qquad (15.46)$$

Equating Eqs. (15.45) and (15.46), and rearranging, gives

$$\Gamma_2^G = \frac{\Gamma_m X_2}{a + X_2} \quad \text{or} \quad \theta = \frac{X_2}{a + X_2} \qquad (15.47)$$

Equation (15.47) is the adsorption isotherm equation of Langmuir.

15.2.b Equation of State: Frumkin

By eliminating the concentration of solute X_2 between Eqs. (15.44) and (15.47), the equation of state of Frumkin is obtained.[8]

$$\sigma_0 - \sigma = -RT\Gamma_m \ln\left(1 - \frac{\Gamma_2^G}{\Gamma_m}\right) \qquad (15.48)$$

302 THERMODYNAMICS OF ADSORPTION FROM SOLUTION

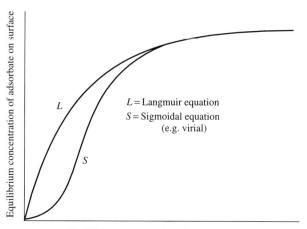

Figure 15.4 The extrapolation to low concentrations of the description of data at high concentrations by a Langmuir equation.[9]

The derivations given demonstrate that the surface tension isotherm of von Szyszkowski, the adsorption isotherm of Langmuir, and the equation of state of Frumkin are equivalent descriptions. Unfortunately, these equations do not apply exactly to most experimental systems although they fit data well in the region of practical interest, that of close-to-maximum surface tension lowering for nearly surface-saturated solutes in solution. The danger lies in extrapolating the fit found in the region of large surface tension lowering to the region of small surface tension lowering at very low concentrations of solute. The region of low concentration is the region of ideal behavior. When data at these low concentrations are considered, the surface tension isotherm often has a sigmoidal shape, which these equations entirely overlook. Figure 15.4 shows the nature of the error that may be introduced by assuming that the Langmuir equation or its congeners, when extrapolated to zero concentration gives a good description of behavior in situations where a sigmoidal representation would be a more accurate description.

15.3 STANDARD-STATE FREE ENERGIES

The constants found for the fit to experimental data of these congeneric equations at higher concentrations may not be extrapolated to calculate changes in standard-state thermodynamic functions where the standard states are defined by the Henry's law slope. The limiting initial slope of the surface tension isotherm, $-d\sigma/dX_2$, if determined by measurements in the Henry's law region, may have the form $RT\Gamma_m/a$; but if determined by extrapolation from data at higher concentrations fitted to Eq. (15.44) equals $RT\Gamma_m/\mathbf{a}$ where **a** may or may not equal a.

The Henry's law region is the linear portion of the adsorption isotherm, corresponding to the linear surface tension isotherm, Eq. (15.38), and refers to an ideal surface solution described by $\pi A = RT$. If the surface solution is ideal, the concentration of the bulk solution in equilibrium with it is also low enough to be ideal. When both solutions are ideal their activity coefficients are unity, and the ratio of their concentrations is equal to the ratio of their activities. Let K be the equilibrium constant for the process of adsorption; therefore,

$$K = \frac{\gamma_2^\sigma \theta}{\gamma_2^\alpha X_2} = \frac{\theta}{X_2} = \frac{1}{a} \tag{15.49}$$

where

$$\theta = \frac{\Gamma_2^G}{\Gamma_m} \tag{15.50}$$

where γ_2^σ is the activity coefficient of the solute in the interfacial or σ phase and γ_2^α is the activity coefficient of the solute in the bulk or α phase.

From the expression for the standard change of the Gibbs function on adsorption

$$\Delta G^0 = -RT \ln K = RT \ln a \tag{15.51}$$

An alternative and more convenient standard state can be chosen. The slopes of Eqs. (15.44), (15.47), and (15.48) applied to data at high concentrations, when extrapolated to infinite dilution, may be substituted into the equation

$$\Delta G^0 = RT \ln \mathbf{a} \tag{15.52}$$

This purely mathematical standard is not the real behavior of the solute but rather an imaginary Henry's law based on its behavior near saturation. This standard state of the solute (pure solute) still derives from a Henry's law slope but on an extrapolated Henry's law slope based on $\mathbf{a}$ rather than the actual Henry's law slope based on a. The practical consequence of using this new standard state is that the new standard change of the Gibbs function, given by Eq. (15.52), can be calculated directly from the constants of the best fit of the Szyszkowski or Langmuir equations to experimental data.

The usual forms of the Szyszkowski and Langmuir equations are

$$\Delta = \frac{\sigma_0 - \sigma}{\sigma_0} = \mathbf{b} \ln \frac{c_2}{\mathbf{c} + 1} \tag{15.53}$$

$$\Gamma_2 = \frac{\Gamma_m c_2}{c_2 + \mathbf{c}} \tag{15.54}$$

where c_2 is the molality and $\mathbf{b}$ and $\mathbf{c}$ are constants. The empirical constants have the following significance:

$$\mathbf{b} = \frac{RT\Gamma_m}{\sigma_0} \quad \text{and} \quad \mathbf{c} = \frac{1000\mathbf{a}}{18(1-\mathbf{a})} \tag{15.55}$$

from which Γ_m and **a** can be evaluated from data. The standard free energy of adsorption is then calculated by means of Eq. (15.52).

The meaning of this new standard state is that the solute is considered to behave at infinite dilution as it would in the nearly saturated surface layer. This is thermodynamically equivalent to the standard state suggested by Rosen and Aronson[10] and Ross and Morrison,[9] who selected an imaginary condition based on the extrapolation of the Szyszkowski–Langmuir equations, valid at high concentrations, to establish an apparent Henry's law constant. It is evaluated by the extrapolation, $\pi \rightarrow 0$, of the portion of the surface tension isotherm in which in X_2 is linear with π. At the high concentrations where this linearity is found, Eq. (15.44), the Szyszkowski equation, reduces to

$$\pi = RT\Gamma_m \ln\left(\frac{X_2}{\mathbf{a}}\right) \qquad (15.56)$$

so the intercept $X_2 = \mathbf{a}$ occurs at $\pi = 0$. This method provides a ready manner to evaluate **a** without having to determine the Szyszkowski constants for the surface tension isotherm.

The surface tension isotherms for a homologous series of carboxylic acids were described by the Szyszkowski equation by Freundlich[11] From the values of these constants, Γ_m and ΔG^0 were calculated by means of Eqs. (15.44) and (15.52) (Table 15.1). The values of ΔG^0 of fatty acids in aqueous solutions bear a linear relation to the number of carbon atoms in the chain, as would be expected from Traube's empirical rule[12,13] as well as from the theoretical concept that the free energy of adsorption is due to a decrease in entropy of the aqueous medium caused by the presence of hydrocarbon in the bulk phase.[14] The slope of the straight line (Figure 15.5) gives the change of the standard Gibbs free energy of adsorption per carbon atom as -2.72 kJ/mol (-649 cal/mol), which is close to the values for the standard free energy of adsorption per carbon atom for homologous series of organic alcohols and ethers.

TABLE 15.1 Values of the Constants, Γ_m and $-\Delta G^0$, for a Homologous Series of Carboxylic Acids from Surface Tension Isotherms

Acid	T(°C)	$RT\Gamma_m/\sigma_0$	c (mol/L)	$1/\Gamma_m$ (nm²/molecule)	$-\Delta G^0$ (kJ/mol)
Formic	15	0.1252	1.38	0.433	8.85
Acetic	15	0.1252	0.352	0.433	12.12
Propionic	15	0.1319	0.112	0.411	14.86
Butyric	18.5	0.1792	0.052	0.302	16.95
Valeric	17.5	0.1792	0.0146	0.302	19.91
Hexanoic	19	0.1792	0.0043	0.302	22.98
Heptanoic	18	0.2575	0.0018	0.210	25.01
Octanoic	18	0.3489	0.00045	0.155	28.36
Nonanoic	18	0.2389	0.00014	0.227	31.19

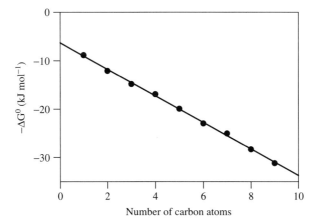

Figure 15.5 The variation of $-\Delta G^0$ for a homologous series of fatty acids as a function of the number of carbon atoms.[9]

For complete transfer of a hydrocarbon chain from water to a hydrocarbon medium, the ΔG^0 per carbon atom also remains constant with a value of -3.70 ± 54 kJ/mol (-884 ± 13 cal/mol), for chains from 3 to 8 carbons. Comparison of these two values suggests that the adsorbed state of the solute at the surface of an aqueous solution is not wholly a hydrocarbon environment. More significant, perhaps, is the further deduction that the hydrophobic character of a chain is hardly likely to vary linearly from 3 to 8 carbons if the chain is curled up in water. Mukherjee[15] suggests that coiling of a hydrocarbon chain in water barely commences at 16 carbon atoms and that shorter chains are fully extended.

15.4 TRAUBE'S RULE

Molecules that combine hydrophilic and hydrophobic moieties in their structure are spontaneously adsorbed by hydrophobic substrates out of aqueous solution because they can orient themselves at the interface with their hydrophilic moiety towards the medium and their hydrophobic moiety towards the substrate. Among molecules of this type we find all surface-active solutes; the free energy of their adsorption measures the strength of their adsorption and their consequent effectiveness as suspending agents.

By comparing curves of σ versus $\log c_2$ for members of a homologous series, Traube[12,13] saw that the surface activity of each member of the series increased regularly with the number of carbon atoms in its molecule. In the series of normal fatty acids in aqueous solution, the surface activity, as measured by the concentration required to reduce the surface tension of water by a constant amount, say 20 mN/m, approximately triples for each additional —CH_2— in the molecule:

THERMODYNAMICS OF ADSORPTION FROM SOLUTION

that is, the surface activity increases geometrically as the number of carbon atoms in the solute molecule increases arithmetically. A similar relation obtains between the geometrically decreasing solubility of a hydrocarbon in water and the arithmetically increasing number of carbon atoms in the chain.

The lowering of surface tension produced by a homologous series of normal aliphatic alcohols, from 6 to 10 carbon atoms, in aqueous solution at 20°C is reported in Figure 15.6 as a function of the logarithm of the concentration. The slopes of these curves at the lower end are almost linear and are the same for each solute, showing, by Eq. (15.20), that the same number of molecules of each is adsorbed per unit area of surface. This result implies that hydrocarbon chains in the adsorbed film are normal to the surface. The quasi-linearity shows, again by the same equation, that Γ_2^G barely increases as the solution concentration is increased, from which we deduce that the adsorbed molecules are quite closely packed together. The area per molecule, calculated by Eq. (15.12), lies between 0.27 and 0.28 nm^2 for each alcohol. This molecular area is 25–30% larger than the limiting area per molecule of a hydrocarbon chain that is measured with insoluble monolayers of long-chain alcohols and acids on a water substrate, which implies that spontaneous adsorption of molecules out of solution does not create as great a limiting compression in the soluble monolayer as can be produced by mechanically pushing together the molecules in an insoluble monolayer.

Traube's rule follows from Eq. (15.56) if Γ_m is independent of the chain length of the members of a homologous series. If Γ_m is such a constant, the same ratio of $X_2/\mathbf{a}$ for each member gives the same lowering of the surface tension. For a homologous series the values of $\mathbf{a}$ are found to decrease in a geometric progression (Table 15.1), and therefore the values of X_2 do so as well. These conditions appear to hold.

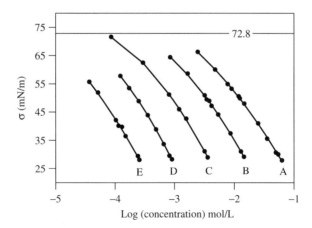

Figure 15.6 The lowering of surface tension produced by a homologous series of normal aliphatic alcohols as a function of the number of carbon atoms.[16]

REFERENCES

[1] McBain, J.W.; Humphreys, C.W. The microtome method of the determination of the absolute amount of adsorption, *J. Phys. Chem.* **1932**, *36*, 300–311.

[2] McBain, J.W.; Swain, R.C. Measurements of adsorption at the air–water interface by the microtome method, *Proc. Roy. Soc. (London)* **1936**, *A154*, 608–623. Professor McBain was awarded the Humphry Davy Medal in **1939** by the Royal Society for his astute selection of an experimental method.

[3] Hutchinson, E. Films at oil–water interfaces. I, *J. Colloid Sci.* **1948**, *3*, 219–234.

[4] Hutchinson, E. Films at oil–water interfaces. II, *J. Colloid Sci.* **1948**, *3*, 235–250.

[5] Hutchinson, E. Films at oil–water interfaces. III, *J. Colloid Sci.* **1948**, *3*, 531–537.

[6] Ross, S.; Chen. E.S. Adsorption and thermodynamics at the liquid–liquid interface, *Ind. Eng. Chem.* **1965**, *57*(7), 40–52.

[7] von Szyszkowski, B. Experimental studies of the capillary properties of aqueous solutions of fatty acids, *Z. Phys. Chem., Stoechiom. Verwandschaftsl.* **1908**, *64*, 385–414.

[8] Frumkin, A. Die kapillarkurve der höheren fettsäuren und die zustandsgleichung der oberflächenschicht, *Z. Phys. Chem., Stoechiom. Verwandschaftsl.* **1925**, *116*, 466–484.

[9] Ross, S.; Morrison, I.D. Thermodynamics of adsorbed solutes, *Colloids Surfaces* **1983**, *7*, 121–134.

[10] Rosen, M.J.; Aronson, S. Standard free energies of adsorption of surfactants at the aqueous solution/air interface from surface tension data in the vicinity of the critical micelle concentration, *Colloids Surfaces* **1981**, *3*, 201–208.

[11] Freundlich, H. *Colloid and capillary chemistry*; Hatfield, H.S., Trans.; Methuen: London; 1926.

[12] Traube, J. Capillaritätserscheinungen in beziehung zur constitution und zum molekulargewicht, *Bet. Dtsch. Chem. Ges.* **1884**, *17*, 2294–2316.

[13] Traube, J. Über die capillaritätsconstanten organischer stoffe in wässerigen lösungen, *Justus Liebig's Ann. Chem.* **1891**, *265*, 27–55.

[14] Tanford, C. *The hydrophobic effect: formation of micelles and biological membranes*; Wiley: New York; 1980.

[15] Mukerjee, P. The nature of the association equilibria and hydrophobic bonding in aqueous solutions of association colloids, *Adv. Colloid Interface Sci.* **1967**, *1*, 241–275.

[16] Defay, R.; Prigogine, I.; Bellemans, A. *Surface tension and adsorption*; Everett, D.H., Trans.; Longmans, Green: London; 1966; p 95.

16 The Relation of Capillarity to Phase Diagrams

16.1 INTRODUCTION

The term "capillarity" is meant to include all those phenomena in which surface tension plays a part. The word "capillarity" is derived from the first such effect to attract the attention of eighteenth-century natural philosophers, namely the rise of liquids in a capillary tube. Other capillary phenomena are coalescence of drops, wetting and spreading, encapsulation, flotation, adsorption, Marangoni flows, the geometry of bubbles and drops, the stability of emulsions and foams, the action of antifoams, and so on. Ross and Nishioka discovered a fundamental principle that capillary phenomena occur near certain defined locations within phase diagrams, namely, the phase boundaries.[1] The lines on a phase diagram represent compositions, temperatures, and pressures, where the number or kind of phases in a system change abruptly. Examples are freezing and boiling point curves, solubility limits, and so on. When a system begins to exhibit some capillary phenomenon, some component or components are near a phase boundary. Phase diagrams represent equilibria, hence minimum chemical potentials, whereas fields of investigation such as calorimetry or intermolecular forces only give direct information about enthalpy, excluding entropy. The distinction is extremely important when interactions at surfaces are considered since entropic contributions are almost always significant.

Surface chemists do not need to be reminded that the surface of a liquid has properties different from its interior or bulk phase. For example, it has been known for a long time that water-insoluble amphipathic substances, such as myristic acid, can dissolve at the surface of water, where they form monolayers, although they are insoluble in bulk water. McBain pointed out that a drop of phenol or a minute crystal that may take a hour or more to dissolve when immersed in water spreads instantly and dissolves in a few seconds when it happens to touch the surface.[2] The "anomalous water" that was so publicized a few years ago[3] though perhaps not a factual discovery, had nevertheless enough basis in theory and in practice to lend credence to the claim that an allotropic form of water exists at an interface. D. W. Thompson wrote[4] "the free surface of every liquid... is no mere limit or simple boundary; it becomes a region of great importance and peculiar activity... It is a morphological field with a molecular structure of its own, and a dynamical field with energetics of its own."

… INTRODUCTION 309

The higher potential energy of a liquid surface compared to the bulk liquid is, of course, the cause of the proclivity of liquids to minimize their surface area and is also the cause of spontaneous adsorption at liquid surfaces. These processes require work to be done, equivalent to the reduced potential energy of the system. Fortunately, the resulting phenomena are readily observable: the former by the characteristic shape of drops or of other minimum surfaces, and the latter by the reduction of surface tension concomitant with adsorption, as formulated by the Gibbs adsorption theorem. Nor does the reduction of surface tension require the use of an instrument to be detected, at least qualitatively, as the stability of bubbles or of froth is enough to allow one to infer the presence of that effect.

A surface-active solute can only be so called with respect to a particular solvent. Such a solute contains in its molecular structure some moiety that interacts strongly with the solvent, whether it be by solvation, hydrogen bonding, or acid–base interaction (these may be merely different names for the same effect) and another moiety in which interaction with the solvent is insufficient to overcome the cohesional energy of the solute. The former moiety is termed "lyophilic" and the latter "lyophobic," meaning "affinity for the solvent" and "lack of affinity for the solvent," respectively. The lyophilic moiety confers solubility, and the lyophobic moiety ensures that the solubility is limited. An example of an amphipathic solute for a hydrocarbon solvent is a molecule containing a hydrocarbon moiety, which is soluble in the hydrocarbon, and a hydrogen-bonding moiety, which, because of its cohesional energy, has limited solubility in hydrocarbon, but which segregates by adsorption at the oil/water interface. Such a solute would be recognized as a typical oil-soluble emulsifying agent, for example, the esterified sorbitols (Spans).

The moieties need not even be chemically linked in one molecule: A combination of two solutes, incompatible when together, may be brought into solution by a third component, as, for example, a mixture of water and benzene may be dissolved by a cosolvent such as ethanol. Such a combination is likely to include partial miscibility at some part of its (ternary) phase diagram. The position of a solution of a given concentration on the phase diagram indicates degrees of heteromolecular interaction: compositions near a phase boundary have a weaker solute–solvent interaction than those farther from a phase boundary. Adsorption is therefore the precursor of imminent phase separation, as the surface offers a region for partial segregation of molecules prior to their more complete separation as a bulk phase. As compositions approach those of a phase boundary, adsorption increases and other interfacial phenomena associated with surface activity begin to occur. A propensity towards phase separation is therefore a general guide to surface-active behavior. A prescient but passing and incidental remark to this effect was made by Langmuir nearly a century ago: "In mutually saturated liquids, especially near the critical temperature, the conditions are favorable for orientation and segregation of the molecules in the liquid."[5] Adsorption of solute is usually accompanied by micellization (segregation of solute molecules).

The surface activity of a solute is not primarily due to its amphipathic molecular structure but to its weak interaction with the solvent. This interaction must

THE RELATION OF CAPILLARITY TO PHASE DIAGRAMS

of course still be sufficient to dissolve the solute but need be no greater than the least degree required to do so. A tendency towards phase separation is a general indicator of surface activity. This statement is akin to Lundelius's rule[6] that the least soluble materials are the most readily adsorbed. Thus, for example, the foam stability of gelatin solutions is greatest at the isoelectric point where the gelatin is least soluble. Similarly, the foaminess of polymer solutions is maximum in poorer solvents, declining again, however, when insolubility supervenes.[7] See also the Ferguson effect (Section 13.2). Not surprisingly, therefore, the surface activity displayed by solutions or mixtures can be related to the phase diagram.

The fundamental index of the surface activity of a solute in any solvent is positive adsorption, which is measured quantitatively by the excess surface concentration of solute. For a binary solution, this quantity is given by Γ_2^G in the adsorption isotherm equation of Gibbs.

$$-d\sigma = RT\Gamma_2^G d \ln X_2 \tag{16.1}$$

where T is the surface tension and X_2 is the mole fraction.

If Eq. (16.1) is applied to treat experimental data of a binary system at concentrations greater than about 0.01 mM, activities rather than concentrations must be used. Nishioka, Lacey and Facemire[8] used available data on surface tensions and activity coefficients as functions of composition and temperature to calculate Gibbs excess concentrations of solute for the binary system diethylene glycol and ethyl salicylate. The Gibbs excess concentrations were then plotted as cosorption contours superimposed on the phase diagram for the system (Figure 16.1).

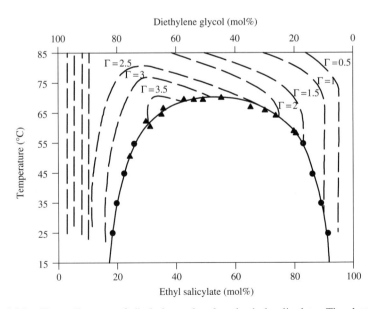

Figure 16.1 Phase diagram of diethylene glycol and ethyl salicylate. The dotted lines are the Gibbs excess concentrations ($\mu m/m^2$) of ethyl salicylate.[8] The dotted lines are cosorption contours.

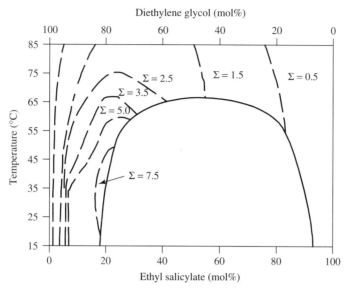

Figure 16.2 Phase diagram and interpolated isaphroic lines of the two-component system diethylene glycol and ethyl salicylate. The average lifetime of a bubble, Σ, is measured in seconds.[9]

The thesis that surface activity is the precursor of phase separation is amply confirmed by these observations of increasing surface activity of the unsaturated solutions as they approach saturation, with maximum surface activity near the critical-solution point. In these solutions ethyl salicylate, the component of lower surface tension, is the surface-active component. The maximum surface excess has a focal point that does not correspond exactly to the consolute point, but is shifted towards the component of higher surface tension (diethylene glycol).

Foam stabilities measured at different points on the same phase diagram are shown on Figure 16.2. Σ is the average lifetime of a bubble in seconds and the dotted lines are contours of equal foam stability or isaphroic contours (from *aphros*, Greek for foam). The maximum foam stability ($\Sigma > 7.5$ s) does not occur at the same temperature and composition as the maximum surface activity shown in Figure 16.1. The explanation is that foam stability does not depend solely on surface activity but is also strongly influenced by viscosity and the viscosity of these solutions is greater at lower temperatures.

16.2 REGULAR SOLUTION THEORY

Nishioka et al.[8] compared these experimental results with a theoretical model of the same kind of system, using the "two-surface-layer" regular solution model of Defay and Prigogine.[10] In this model only the top two molecular layers are considered to differ in composition from the bulk phase, arising from molecular coordination numbers less than those of molecules in the bulk phase. Consider a molecule in the bulk phase: it has z nearest neighbors, of which lz are in the same lattice plane, where l is the fraction of nearest neighbors in that plane (for example, 6/12 for a close-packed lattice, 4/6 for a cubic lattice); and mz are in either contiguous plane, where m is the fraction of nearest neighbors in that plane (for example, 3/12 for a close-packed lattice, 1/6 for a cubic lattice). Then

$$l + 2m = 1 \tag{16.2}$$

A molecule in the surface layer has a smaller number z' of nearest neighbors given by

$$z' = (l+m)z = (1-m)z \tag{16.3}$$

Using the Bragg–Williams approximation and assuming that the molecular surface areas of the two components are the same, the compositions in the upper two layers are given by the following two equations:

$$\log \frac{x_1 x_2''}{x_1'' x_2} - \frac{2\alpha (x_2'' - x_2)}{RT} - \frac{2\alpha m (x_2 + x_2' - 2x_2'')}{RT} = 0 \tag{16.4}$$

$$\frac{RT}{\alpha} \log \frac{x_1 x_2'}{x_2 x_1'} + \frac{a(\sigma_2 - \sigma_1)}{\alpha} + 2l(x_2 - x_2') + m(x_2 - x_1) + 2m(x_2 - x_2'') = 0 \tag{16.5}$$

where x_1, x_2 = mole fraction of component 1 or 2 in the bulk phase
x_1', x_2' = mole fraction of 1 or 2 in the top monolayer
x_1'', x_2'' = mole fraction of 1 or 2 in the intermediate layer
α = interaction constant ($2RT_c$ for a regular solution)
σ_1, σ_2 = surface tension of pure component 1 or 2
a = area per molecule

Eq. (16.4) and (16.5) were solved numerically for x_2' and x_2'' as a function of the composition of the solution (x_1). A close-packed lattice was assumed and the term $a(\sigma_2 - \sigma_1)/\alpha$ in Eq. (16.5) was assumed to be constant and equal to -0.5. This assumption was based on reasonable values of $a = 0.30$ nm²/molecule, $T_c = 300$ K, and $(\sigma_2 - \sigma_1) = -6.9$ mN/m. The dimensionless surface-excess concentration of component 2 is then

$$\omega \Gamma_2^G = \frac{x_2' - x_2 + x_2'' - x_2}{x_1} \tag{16.6}$$

where ω is the surface occupancy in square meters per mole.

The dimensionless surface excess, $\omega \Gamma_2^G$, calculated for a regular solution is shown in Figure 16.3 as a series of contours of equal surface concentration, called cosorption lines. In this hypothetical system the component of lower surface tension is surface active at all points in the phase diagram. The maximum surface activity, which is called the focal point, occurs at a point on the solubility curve,

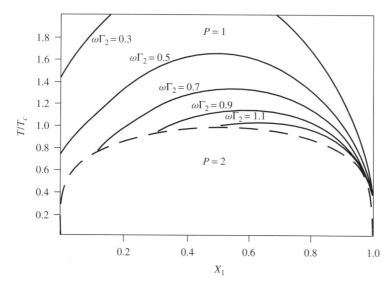

Figure 16.3 Dimensionless surface-excess concentration of component 2, $\omega\Gamma_2^G$, calculated for a two-layer regular solution model. Component 2 was assumed to have an appreciably lower surface tension than component 1.[8]

thus agreeing with Lundelius's rule, but it does not coincide with the critical point, being biased towards a lower concentration of the component of lower surface tension. Calculations based on the theory disclose that the greater the difference in surface tension between the two components, the greater is the bias towards that side of the composition scale and also that the greater that bias, the greater is the surface activity.

Figure 16.3 shows that at higher temperatures a regular solution tends towards having maximum surface activity at a mole fraction of one-half, but experimental data do not confirm this prediction: They show the maximum shifting towards still lower concentrations of the component of lower surface tension. This behavior is also typically observed for solutions at temperatures far above a critical point, such as water–ethanol solutions at room temperature, which have a maximum surface activity at 5–10% alcohol. Nevertheless, Figure 16.3 bears a marked resemblance to published diagrams reporting observed foam stabilities of two-component systems as functions of composition and temperature.

16.3 CRITICAL POINT WETTING

Wetting phenomena, close to the critical point of a binary liquid mixture, present an interesting situation, where a layer of a heavier liquid can wet the walls of the vessel and can reside over a lighter one.[11] The theory of this effect was expounded by Cahn[12] who stated: "It is shown [by an argument based on theory alone] that in any two-phase mixtures of fluids near their critical points, contact angles against any third phase become zero in that one of the critical phases completely wets the third phase and excludes contact with the other critical phase." At first glance the above prediction seems to be contradicted by reasoning that, at the limit, T tending towards T_c, the two conjugate phases are identical and so ultimately would be expected to make the same angle of contact on any solid substrate; that is, the value of the contact angle is expected to tend towards 90° as T tends towards T_c.

A simple but effective way to study the wetting of the third phase is to measure the contact angle made by the liquid/liquid meniscus against the wall of a transparent container, with due precautions to eliminate sources of error.[13]

The experimental results of Ross and Kornbrekke[14,15] do not confirm Cahn's prediction but reveal that the contact angle reaches a limiting value of 90° at the critical temperature in both binary and ternary systems. Also, again in both binary and in ternary systems, a region of perfect wetting of the substrate by one of the phases precedes the approach of the contact angle to 90° at the consolute point. A revision of the theory by Chatterjee and Gopal[16] pointed out that a condition for the wetting layer to be stable is that there exists a difference of density between the two conjugate solutions. A density difference exists only below the critical temperature, consequently the thickness of the wetting layer goes to zero as T tends towards T_c.

Conditions are more complex and less predictable when particles of a dispersed solid are present, as the nature of the solid substrate determines the nature of the contact angle and hence the adsorption. The late E. A. Boucher drew attention to the relevance of the results of Ross and Kornbrekke to the partitioning of inorganic colloids in three-component, two-phase systems and provided a number of pertinent references.[17,18,19,20] A cognate research by Rubinovich et al.[21] reports measurements at room temperature of the sticking force between filaments of silica (both hydrophilic and hydrophobic) in butanol–water mixtures. A considerable increase in the sticking force occurs in the range of relative concentrations of water ($0.8 \leq c/c_s < 1$, where c_s is the saturated concentration of water in butanol). The effect is ascribed to the formation by adsorption of a water annulus in the contact zones between the filaments. The mechanism is precisely that postulated for flocculation of a dispersion by the formation of bridges of a liquid immiscible with the medium, as reported years ago by Fischer.[22] When the composition of the solution was altered so as to enter the two-phase region, the silica transferred from the butanol-rich to the water-rich congugate. These results take their place within the growing body of data that support the relation between capillarity and the phase diagram.

REFERENCES

[1] Ross, S.; Nishioka, G. Foaming behavior of partially miscible liquids as related to their phase diagrams in *Foams*; Akers, R.J., Ed.; Academic Press: New York; 1976; pp 17–32.

[2]McBain, J.W.; Woo, T.M. *Proc. Roy. Soc.* **1937**, *163A*, 182–188.

[3]Franks, F. *Polywater*; MIT Press: Cambridge, MA; 1981.

[4]Thompson, D.W. *On growth and form*; University Press: Cambridge; 1942; p 464.

[5]Langmuir, I. The distribution and orientation of molecules, *Colloid Symp. Monogr., 3rd Nat. Symp., 1925* **1925**, *3*, 48–75.

[6]Lundelius, E.F. Adsorption und löslichkeit, *Kolloid Z.* **1920**, *26*, 145–151.

[7]Ross, S.; Nishioka, G. The relation of foam behavior to phase separations in polymer solutions, *Colloid Polym. Sci.* **1977**, *255*, 560–565.

[8]Nishioka, G.M.; Lacy, L.L.; Facemire, B.R. The Gibbs surface excess in binary miscibility-gap systems, *J. Colloid Interface Sci.* **1981**, *80*, 197–207.

[9]Ross, S.; Townsend, D.F. Foam behavior in partially miscible binary systems, *Chem. Eng. Commun.* **1981**, *11*, 347–353.

[10]Defay, R.; Prigogine, I. Surface tension of regular solutions, *Trans. Faraday Soc.* **1950**, *46*, 199–204.

[11]Rowlinson, J.S.; Widom, B. *Molecular theory of capillarity*; Clarendon Press: Oxford; 1982; pp 227–232.

[12]Cahn, J.W. Critical point wetting, *J. Chem. Phys.* **1977**, *66*, 3667–3672.

[13]Ross, S.; Kornbrekke, R.E. Wetting of the container wall as a critical point phenomena I. Measurement of contact angles in cylindrical tubes: Validation of a method, *J. Colloid Interface Sci.* **1984**, *98*, 223–228.

[14]Ross, S.; Kornbrekke, R.E. Wetting of the container wall as a critical point phenomena II. Wetting by contacting congugate solutions of binary systems, *J. Colloid Interface Sci.* **1984**, *99*, 446–454.

[15]Ross, S.; Kornbrekke, R.E. Wetting of the container wall as a critical point phenomena III, Wetting by contacting congugate solutions of ternary systems, *J. Colloid Interface Sci.* **1984**, *100*, 423–432.

[16]Chatterjee, S.; Gopal, E.S.R. Effects of capillary waves on the thickness of wetting layers, *J. Phys. France* **1988**, *49*, 675–680.

[17]Boucher, E.A. Capillary phenomena: Properties of systems with fluid-fluid interfaces, *Rep. Prog. Phys.* **1980**, *43*, 497–546.

[18]Boucher, E.A. Separation of small-particle dispersions by the preferential accumulation in one of two liquid phases, or by static flotation at their interface, *J. Chem. Soc., Faraday Trans. I* **1989**, *85*, 2963–2972.

[19]Boucher, E.A. Critical wetting, flocculation of silica particles in near–critical lutidine–water mixtures and related phenomena, *J. Chem. Soc., Faraday Trans.* **1990**, *86*, 2263–2266.

[20]Boucher, E.A. Surface activity in three-component systems near the Plait point and the implications for the partitioning of inorganic hydrosols and bioparticles in phase-separated solutions, *J. Colloid Interface Sci.* **1990**, *137*, 593–596.

[21]Rabinovich, Ya. I.; Movchan, T.G.; Churaev, N.V.; Ten, P.G. Phase separation of binary mixtures of polar liquids close to solid surfaces, *Langmuir* **1991**, *7*, 817–820.

[22]Fischer, E.K. *Colloidal dispersions*; Wiley: New York; 1950; pp 128–135.

17 Electrical Charges in Dispersions

17.1 IONS AND CHARGED PARTICLES IN DISPERSIONS

"The study of ions in gases is a part of physics, while the study of ions in solution is classed as chemistry."[1] The ions in a plasma are kept apart by high kinetic energy at high temperatures or by a high electric field or both. These processes are in the realm of physics.

What energy keeps oppositely charged ions apart in solution? That free ions are plentiful in aqueous solution and rare in organic liquids provides a clue. The shell of associated water molecules around an ion provides a "steric" barrier to close approach. When this hydrate shell around the ion is thick enough that oppositely charged ions on closest approach attract each other with less than their average kinetic energy, that is, approximately the Boltzmann factor of kT, they remain dispersed. Molecules of nonpolar solvents do not usually form tightly adsorbed shells around ions so that oppositely charged ions quickly recombine if they are ever separated. This is the realm of chemistry.

Most dispersions contain electrically charged particles or droplets. The whole system must be electrically neutral so that an equivalent number of oppositely charged ions, called counterions, are present in the solvent surrounding the particles. What energy keeps the oppositely charged ions from neutralizing the charged particles? The answer is that the charges on the surface of the particle and the ions in solution are hydrated with layers of water molecules and they do not neutralize each other for the same reason that ions do not recombine in solution.

Even though the counterions and the charged surface of a particle or drop are not in close combination, the counterions are held near to the surface of the particle by electrical attraction. The layer of counterions surrounding a charged particle is called the diffuse double layer. The structure of that layer, that is, the concentration of counterions as a function of distance from the surface of the particle, is described by the Poisson–Boltzmann equation.[2]

Particles in suspension with the same surface chemistry have an electrical charge of the same sign, therefore they repel one another. This electrical repulsion acts to counteract dispersion forces of attraction. An essential consequence of DLVO theory is that the stability of a dispersion of charged particles or droplets depends on the electrostatic repulsion generated as the surrounding electrical double layers interpenetrate.

17.1.a Origin of Charges at Surfaces in Aqueous Media

Quincke[3,4,5] and later investigators showed that electrokinetic phenomena and, accordingly, the presence of surface charges at the solid–liquid interface exist in nearly all systems, especially when the liquid is water. Generally an interface in distilled water is negatively charged. On contact with other liquids, however (e.g., oil of turpentine), the surface is frequently positive; still other liquids (e.g., ether, petroleum ether, or oil of hartshorn) produce no surface charge. The surface charge may be acquired by surface ionization or by preferential adsorption of anions or cations. In water, mineral oxides and sulfides (e.g., silica, iron oxide, arsenic sulfide) become positively charged at low pH and negatively charged at high pH, with a zero charge at an intermediate pH, which is known as the point-of-zero charge (PZC) or isoelectric point. The PZCs of acidic oxides occur at a low pH and for a basic oxide, such as Al_2O_3, at a high pH.

All dispersed particles in water spontaneously acquire electric charges by two principal mechanisms: dissociation of ionogenic groups or preferential adsorption of ions, usually anions, from solution. Examples of the first mechanism are shown by the pH dependence of the charge on metal oxides and the dependence of the charge of the silver halides on the concentration of silver ions in solution. These are represented by the following equilibria:

Dissociation of ionogenic groups:

$$\text{lattice–SiOH} \Leftrightarrow \text{lattice–SiO}^- + H^+$$
$$\text{lattice–AlOH} \Leftrightarrow \text{lattice–Al}^+ + OH^-$$
$$\text{lattice–AgCl} \Leftrightarrow \text{lattice–Cl}^- + Ag^+$$
$$\text{lattice–AgCl} \Leftrightarrow \text{lattice–Ag}^+ + Cl^-$$

Preferential adsorption:

$$\text{metal lattice–OH} + H_2O \Leftrightarrow \text{metal lattice–O}^- + H_3O^+$$
$$\text{metal lattice–OH} + H_3O^+ \Leftrightarrow \text{metal lattice–OH}_2^+ + H_2O$$
$$\text{metal lattice–OH} + OH^- \Leftrightarrow \text{metal lattice–O}^- + H_2O$$

17.1.b Point of Zero Charge in Aqueous Media

The electrical properties of sparingly soluble oxides in contact with aqueous solutions are determined by the pH value in the absence of preferential adsorption of soluble ions in solution. The particles are charged positively at pH values below the PZC and negatively above the PZC.

The addition of anionic surfactants to a dispersion makes a particle more negative. Addition of cationic surfactants makes the particles more positive. If the particle is below its PZC (positively charged) then the adsorption of an anionic surfactant destabilizes the dispersion by charge neutralization. Increased adsorption of anionic surfactant reverses the charge on the particle and restabilizes the dispersion. Continued adsorption of anionic surfactant by a particle above its PZC increases the total negative charge on the particle. Therefore careful attention to the pH of an aqueous dispersion and knowledge of the PZCs of the oxides present is critical to understanding dispersion stability. The PZC of charged particles can be determined by measuring the electrophoretic mobility of the particle as a function of pH.

The PZC is defined as the negative of the Briggsian logarithm of the activity of the potential-determining ion at which the net charge on the dispersed particle is zero. At the PZC the electrostatic repulsion between the particles is lost. Typical values of the PZC for various materials are given in Tables 17.1 and 17.2. At values of pAg less than PZC, the silver halide particles are charged positively.

The PZC for many oxides have been determined, but the results are somewhat variable. The essential reasons are that the oxides occur in various crystallographic forms, often mixed in a single sample, impurity levels vary, especially at the surface which is most influential on particle charge, and the hydration of the surface often depends on the heat pretreatment. Extensive reviews of the data have been published.[6,7,8]

17.1.c Origin of Charges at Surfaces in Nonpolar Media

For free electrical charges to be present in nonpolar media, oppositely charged ions need to be held apart sufficiently far by some chemical structure that the attraction is less than thermal energy. In water this structure is provided by hydration layers. Most nonpolar solvents do not associate strongly with ions to form a solvation layer yet solutions of oil soluble detergents are conductive. The inverse micelles formed in such solutions have an aqueous interior that dissolves any ions present in the system. Both the surfactant molecules and the ions are mobile in dynamic equilibria. Most micelles are electrically neutral but

TABLE 17.1 Points of Zero Charge for Some Oxides in Water[9]

Oxide	PZC	Oxide	PZC	Oxide	PZC
Ag_2O	11.2	HgO	7.3	SnO_2	5.6
Al_2O_3	9.1	La_2O_3	10.1	Ta_2O_5	2.8
BeO	10.2	MgO	12.4	ThO_2	9.2
CdO	11.6	MnO_2	5.3	TiO_2(rutile)	5.7
CeO_2	8.1	MoO_3	2.0	TiO_2(anatase)	6.2
CoO	10.2	Nb_2O_5	2.8	V_2O_3	8.4
Co_3O_4	7.4	NiO	10.2	WO_3	0.4
Cr_2O_3	7.1	PuO_2	9.0	Y_2O_3	8.9
CuO	9.3	RuO_2	5.3	ZnO	9.2
Fe_2O_3	8.2	Sb_2O_5	1.9	ZrO_2	7.6
Fe_3O_4	6.6	SiO_2	2.0		

TABLE 17.2 Points of Zero Charge of Some Ionic Solids in Water[10]

Substance	PZC
Fluorapatite, $Ca_5(PO_4)_3(F, OH)$	pH = 6
Hydroxyapatite, $Ca_5(PO_4)_3(OH)$	pH = 7
Calcite, $CaCO_3$	pH = 9.5
Fluorite, CaF_2	pCa = 3
Barite (synthetic), $BaSO_4$	pBa = 6.7
Silver iodide, AgI	pAg = 5.6
Silver chloride, AgCl	pAg = 4
Silver sulphide, Ag_2S	pAg = 10.2

a few have excess anions that are compensated by excess cations in others. In that way charged species, both positive and negative, are present even in hydrocarbon solutions.[11] Charge exchange between micelles and mobile ions on a particle surface leads to a charged particle.[12] Besides micelles, any other structure in solution such as a large polymer or copolymer capable of strongly associating with an ion would also be an agent to charge suspended particles. This situation is illustrated in Figure 17.1, showing the transfer of a cation from one micelle to another and thus accounting for the conductivity of oil-soluble electrolytes such as Aerosol OT, lecithin and OLOA 1200 (now sold as OLOA 371). Two micelles of AOT, containing three molecules each, exchange a single sodium cation to form two oppositely charged micelles, each still containing three AOT molecules. Of all the micelles in solution, a small fraction, for example, about 1% of OLOA 1200 micelles, are charged. Doubly charged micelles are probably extremely rare because of Coulombic repulsion in the interior of the micelle.

Figure 17.2 shows the conductivity of OLOA 1200 in (*a*) dodecane and (*b*) dodecane containing dispersed carbon black. The conductivity of OLOA 1200 in free solution increases linearly with concentration. In the presence of the carbon black, the conductivity does not change until the carbon black is about half saturated with adsorbed OLOA 1200 after which the free OLOA 1200 plus the charged carbon particles increase the conductivity of the dodecane. Unlike charged particles in water, charged particles in nonpolar media often contribute significantly to the electrical conductivity.

Figure 17.3 shows the zeta potential in mV of the dispersed carbon particles in the same dispersions. At low concentrations of the agent, all of it is adsorbed and the particles have little or no charge. Until free OLOA 1200 is in solution there is no dynamic equilibrium to charge the particles and therefore no appreciable increase in zeta potential occurs until there is free OLOA 1200 in solution. A comparison of Figure 17.2 and Figure 17.3 shows that a region of low conductivity is also a region of low zeta potential. Once the particles are fully saturated the zeta potential levels off but the total conductivity continues to rise because the concentration of micelles is increasing.

320 ELECTRICAL CHARGES IN DISPERSIONS

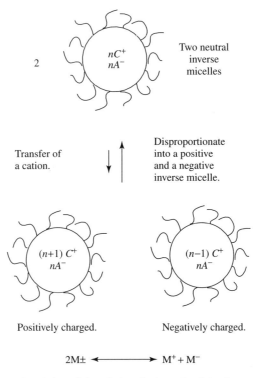

Figure 17.1 Schematic of the origin of electrical conductivity in nonpolar media where a cation is transferred from the core of one inverse micelle to another core. The micellar core contains cations and anions, which dynamically rearrange to give micelles with excess cations and micelles with excess anions.

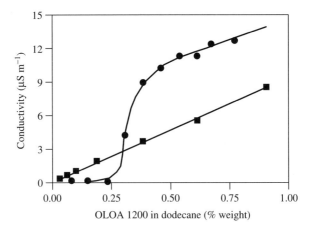

Figure 17.2 The conductivity of OLOA 1200 in dodecane and dodecane containing dispersed carbon black.

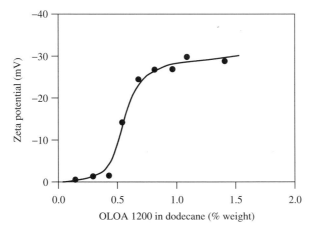

Figure 17.3 The variation in zeta potential of carbon black dispersed in dodecane as a function of added OLOA 1200.

17.1.d Preferential Adsorption

Preferential adsorption is the adsorption of more anions than cations or vice versa. The adsorption of hydrogen ions or hydroxyl ions just discussed is an example of preferential adsorption. The anion, alkylcarboxylate, formed when soap dissociates in water, is preferentially adsorbed at a solid or liquid surface, resulting in a net negative charge at the interface.

A more complex example of preferential adsorption of a potential-determining ion is the deflocculation of kaolin by tetrasodium pyrophosphate (TSPP), $Na_4P_2O_7$. Kaolin particles have the form of a stack of platelets. Each platelet has a silica surface on one side and an alumina surface on the other. The platelets are stacked together by virtue of a weak negative charge on the silica side and a weak positive charge on the alumina side. The edges of the stacked platelets are positively charged. Every particle of kaolin is such a booklet. Before being dispersed, the particles are flocculated by a face-to-edge electrostatic bonding as shown in Figure 17.4.

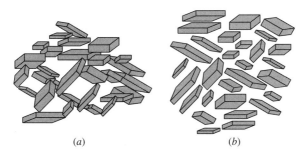

Figure 17.4 An example of (a) flocculated and (b) deflocculated dispersed particles.

The change of charge on a particle on adding TSPP is reflected by a change in various properties of the suspension. A graphical correlation of the measured properties of kaolin suspensions as a function of percent TSPP by weight of kaolin is shown in Figure 17.5. The electrophoretic mobility of the particles increases rapidly with the first small additions of TSPP. The rapid increase in particle mobility at TSPP additions up to about 0.05% is interpreted as the result of specific adsorption of anion on the positively charged edges of the kaolin. The electrical conductivity, the flocculation value and the pH, all indicate that TSPP continues to be adsorbed up to about 0.10%, but the electrophoretic mobility does not increase as rapidly as at first. This is interpreted as evidence of continued adsorption of phosphate on the negatively charged faces of the kaolin particles, but with less influence on the net charge and on the mobility.

Another example of preferential adsorption of surface-active anions by positively charged particles of silver iodide is shown in Figure 17.6, which reports the zeta potential as a function of the equilibrium concentration of various solutes in water. The sign and magnitude of the zeta potential reflect the sign and magnitude of the charge on the particle. The adsorption of the anions causes reversal of the sign of the charge on the particles at concentrations far below the CMC values of the solutes.

Hydrolysable metal cations in solution form different ions at different pH and these also can act as potential-determining ions. The adsorption of these ions changes the nature of the particle surface. Figure 17.7 shows the change in zeta

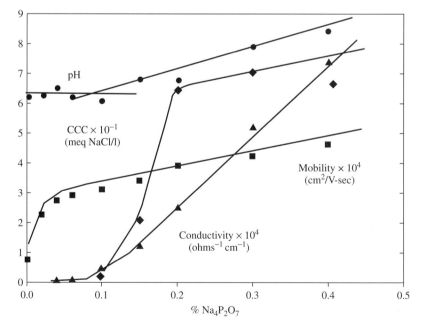

Figure 17.5 A graphical correlation of properties with weight percent TSPP added.[13]

IONS AND CHARGED PARTICLES IN DISPERSIONS 323

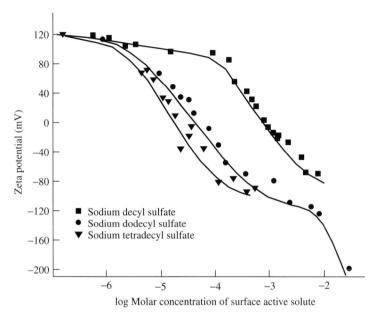

Figure 17.6 Zeta potential as a function of concentration of various surface-active solutes in water (AgI–water interface.) ■ Sodium decyl sulfate; • sodium dodecyl sulfate; ▼ sodium tetradecyl sulfate.[14]

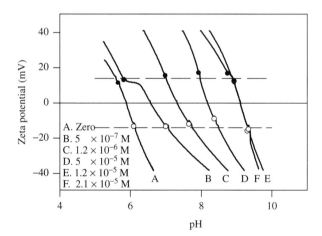

Figure 17.7 The zeta potential of titania dispersions as a function of pH and concentration of aluminum nitrate. The transition from a stable positive dispersion to a stable negative dispersion occurs through a range of instability marked on the diagram by horizontal dotted lines.[15]

potential of titania particles as a function of pH and concentration of aluminum nitrate.

For each concentration of aluminum nitrate, the transition from stable, positively charged particles to stable, negatively charged particles occurs at a definite pH. However that pH varies with the concentration of aluminum nitrate, the higher the concentration of the aluminum nitrate; the higher the pH for neutrality. This makes sense since the aluminum nitrate forms positive ions that, when adsorbed, make the surface more positive and require a higher pH to make it neutral.

Continuing to consider Figure 17.7, the curves show that at a fixed pH the charge on the particle varies from positive to negative as the aluminum nitrate concentration is removed. This change in charge also passes through the region of charge neutrality and unstable dispersion. Therefore if aluminum nitrate were added to a dispersion the stability could vary from stable at low concentrations, to unstable at intermediate concentrations, to stable again at higher concentrations.

Knowing the composition of the solution phase in dispersions is the key to understanding the stability of the system. In particular, one needs to be aware of any potential determining ions, like the aluminum nitrate example just considered.

17.1.e Ions Near Charged Particles; the Electrical Double Layer

17.1.e(i) Gouy–Chapman theory   Adsorbed ions leave their counterions in the solution, and these excess charges cluster near the surface forming an electrical double layer. The double layer consists of two regions: an inner region composed of counterions strongly attached to the surface and is carried with it when the particle moves (the Stern layer) and a diffuse region containing the remainder of the excess counterions. In the diffuse region (Gouy–Chapman layer) the counterions are distributed according to a balance between their thermal motion and the forces of electrical attraction. They can be perturbed from their equilibrium distribution by the motion of the particle.

The model used to describe the ion distribution functions and the potential of the region near the charged interface was developed independently by Gouy[16] and Chapman[17] by combining the Poisson equation for the second derivative of the electric potential as a function of charge density with the Boltzmann equation for the charge density as a function of potential. The problem is formulated as: Given a surface charge, find the spatial distribution of its counter ions. The assumptions are:

The surface charge is continuous and uniform.
The ions in the solution are point charges.

Despite the dubiety of these assumptions, the Poisson–Boltzmann equation has been found to agree with observation over a wide range of experimental conditions.[18,19,20]

IONS AND CHARGED PARTICLES IN DISPERSIONS   325

Let the electrical potential be Φ_0 at the Stern layer and $\Phi(x)$ at a distance x from the Stern layer in the electrolyte solution. Taking the surface to be positively charged for argument's sake, the Boltzmann equation gives

$$n_i = n_{i0} \exp\left(-\frac{z_i e \Phi(x)}{kT}\right) \qquad (17.1)$$

where n_i is the concentration of ions of kind i at a point where the potential is $\Phi(x)$; n_{i0} is the concentration in the bulk of the solution; z_i is the valency including the sign of the charge: positive for cations, negative for anions.

The charge density ρ in the solution is the algebraic sum of the ionic charges per unit volume:

$$\rho = \sum z_i e n_i = \sum z_i e n_{i0} \exp\left(-\frac{z_i e \Phi}{kT}\right) \qquad (17.2)$$

The Poisson equation relates the charge density to the Laplacean of the electric potential (see Section 26.3)

$$\nabla^2 \Phi = -\frac{\rho}{D\varepsilon_0} \qquad (17.3)$$

where D is the dielectric constant and ε_0 is the permittivity of free space, namely, 8.854×10^{-12} C/V-m. The negative sign in Eq. (17.3) results from the charge density in the solution, ρ being negative for a positively charged surface.

Combining Eqs. (17.2) and (17.3) gives the differential equation (the Poisson–Boltzmann equation) for the potential Φ as a function of the coordinates,

$$\nabla^2 \Phi = -\frac{1}{D\varepsilon_0} \sum z_i e n_{i0} \exp\left(-\frac{z_i e \Phi}{kT}\right) \qquad (17.4)$$

No exact analytical solution has been found for Eq. (17.4) that is valid for all values of the parameters. The common approximations used to simplify Eq. (17.4) are the special cases of (i) a small surface potential or (ii) a symmetrical electrolyte.

(i) *Exact Solutions of the Poisson–Boltzmann Equation for Small Potentials and Any Electrolyte.* Equation (17.4) can be solved approximately when the surface potential is small, that is, for example, when $|z_i e \Phi / kT| \ll 1$, when $z_i \Phi$ is less than 25 mV at room temperature. The procedure is to expand the exponential: the first term, $\sum z_i e n_{i0}$, is zero because the solution has a negligible net charge, and only the second term of the series is retained:

$$\nabla^2 \Phi = \kappa^2 \Phi \qquad (17.5)$$

where

$$\kappa^2 = \frac{e^2}{D\varepsilon_0 kT} \sum n_{i0} z_i^2 \qquad (17.6)$$

$$\kappa = 2.328 \left(\sum n_{i0} z_i^2\right)^{1/2} nm^{-1} \qquad (17.7)$$

for water at 25°C when the n_{i0} are in mol/L and z_i is the charge per ion. Note that κ has units of reciprocal length, so $1/\kappa$ is called the Debye length or the double-layer thickness. In most systems $1/\kappa$ varies from about 1 to 100 nm.

The summation in Eq. (17.6) is proportional to the ionic strength, $I = \frac{1}{2}\sum c_i z_i^2$, where the c_i are the concentrations of each ion in mol/L. For a 1-1 electrolyte the ionic strength equals the concentration of salt added.

Equation (17.5) can be solved for different geometries. For the planar interface

$$\Phi = \Phi_0 \exp(-\kappa x) \qquad (17.8)$$

and for the sphere of radius a

$$\Phi = \frac{a\Phi_0}{r} \exp[-\kappa(r-a)] \qquad (17.9)$$

where r is the distance from the center of the sphere. In these approximations the potential declines exponentially with distance. At a distance from the surface equal to $1/\kappa$, the potential has fallen by a factor of $1/e$, that is, 36.8%, and the number of counterions in the diffuse double layer has decreased by an even larger fraction; which perhaps justifies the designation of $1/\kappa$ as the "thickness" of the double layer.

The effect of increasing ionic strength on $1/\kappa$ is described by Eq. (17.7) and is illustrated in Figure 17.8.

The surface charge density σ is provided by adsorbed potential-determining ions. It has the same sign as the surface potential; its units are Cm^{-2} and it is

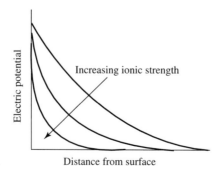

Figure 17.8 The variation of the Debye length, $1/\kappa$, with ionic strength.

given by Eq. (17.10). (The net charge of the diffuse layer, which is the charge contained by the volume subtended by unit area extending from the surface ($x = 0$ or $r = a$) to infinity, has the same numerical value but the opposite sign.)

$$\sigma = -\int_a^\infty \rho\, dr = \int_a^\infty D\varepsilon_0 \nabla^2 \Phi\, dr \tag{17.10}$$

The surface charge density on a plane with low surface potential is found by using Eq. (17.8), the Poisson equation (17.5) in planar coordinates (see Section 26.3), and Eq. (17.10).

$$\sigma = D\varepsilon_0 \kappa \Phi_0 \tag{17.11}$$

The surface charge density on a sphere is found by using Eq. (17.9), Poisson's Eq. (17.3) in spherical coordinates (see Appendix C), and Eq. (17.10).

$$\sigma = \int_a^\infty D\varepsilon_0 \left(\frac{d^2\Phi}{dr^2} + \frac{2}{r}\frac{d\Phi}{dr} \right) dr = -D\varepsilon_0 \left(\frac{d\Phi}{dr} \right)_{r=a} \tag{17.12}$$

The surface charge density is found by differentiating Eq. (17.9) with respect to r and evaluating at $r = a$, to give

$$\sigma = \frac{D\varepsilon_0 \Phi_0 (1 + \kappa a)}{a} \tag{17.13}$$

The charge on a sphere, Q, is $4\pi a^2$ times the surface charge density σ; therefore, the charge on a sphere with low surface potential is

$$Q = 4\pi a D\varepsilon_0 \Phi_0 (1 + \kappa a) \tag{17.14}$$

Equation (17.14) was used by Debye and Hückel to calculate the potential around an ion. Equations (17.11) and (17.13) show that the surface potential Φ_0, the surface charge density σ, and the ionic composition of the solution through its influence on κ are interrelated. The surface charge density σ is affected only by the adsorption of potential-determining ions. If inert electrolyte is added, κ increases; hence the surface potential decreases without any change of surface charge density.

(ii) *Solutions of the Poisson–Boltzmann Equation for Symmetrical z-z Electrolytes* (e.g., *1-1, 2-2, 3-3*) For the special case of a symmetrical electrolyte, Eq. (17.4) reduces to

$$\nabla^2 \Phi = -\frac{2zen_0}{D\varepsilon_0} \sinh\left(-\frac{ze\Phi}{kT} \right) \tag{17.15}$$

with no restriction on the magnitude of the surface potential. Equation (17.15) has been solved analytically only for the planar interface. For a spherical particle the Poisson–Boltzmann equation has no tractable analytical solution, so must be solved by numerical integration. A detailed description of various approximate solutions of the Poisson–Boltzmann equation for spherical symmetry is reported by Dukhin and Derjaguin.[21] O'Brian and White reexamined the equations that govern the ion distributions and velocities, the electrostatic potential and hydrodynamic flow field around a solid colloidal particle in an applied electric field. They provided a numerical solution to the case of a spherical colloidal particle in a general electrolyte solution.[22]

For the planar interface and symmetrical electrolyte, the solution of the Poisson–Boltzmann equation (17.15), is[23]

$$\Phi = \frac{2kT}{ze} \ln \frac{1 + \gamma \exp(-\kappa x)}{1 - \gamma \exp(-\kappa x)} \tag{17.16}$$

for which

$$\sigma = (8n_0 D\varepsilon_0 kT)^{1/2} \sinh \frac{ze\Phi_0}{2kT} \tag{17.17}$$

where

$$\kappa^2 = \frac{2n_0 z^2 e^2}{D\varepsilon_0 kT} \quad \text{for a symmetric electrolyte} \tag{17.18}$$

$$\kappa = 3.288 z c^{1/2} \text{ nm}^{-1} \text{ at } 25°C$$

where c is the bulk electrolyte concentration (mol/L) in water and

$$\gamma = \tanh \frac{ze\Phi_0}{4kT} \tag{17.19}$$

A useful approximation for γ for small surface potentials is

$$\gamma \cong \frac{ze\Phi_0}{4kT} \tag{17.20}$$

These same equations are used to approximate the solution of the Poisson–Boltzmann equation for asymmetrical electrolytes. This simplification causes little error in the description of colloidal phenomena because the valency of the ion of the same charge as the particle is unimportant and its value of z can be set equal to the valency of the counterions. For example, if $CaCl_2$ is used as a flocculant for a negatively charged sol, z is taken as 2.

Added inert electrolyte increases the value of κ [Eq. (17.18)], thus decreasing the value of $1/\kappa$, the double layer thickness, and causing the double layer to contract. At the same time the added electrolyte decreases the surface potential Φ_0 [Eq. (17.17)], as it does not affect the surface charge density. Thus, for example, at low values of Φ_0 increasing n_0 by 10% decreases the surface potential by about 5%. Hence, the effect of added electrolyte on the potential curve is twofold.

A useful approximation for Eq. (17.16) is the solution for the potential at a large distance from a particle carrying any charge, large or small. Equation (17.16) in the limit of large x reduces to

$$\Phi = \frac{4\gamma kT}{ze} \exp(-\kappa x) \quad (17.21)$$

Adding the requirement that the surface potential be small, then introducing Eq. (17.20), gives

$$\Phi = \Phi_0 \exp(-\kappa x) \quad (17.22)$$

17.1.e(ii) Stern Theory The inadequacies of the Gouy–Chapman theory become evident when observations are made of the properties of electrodes where surface charge density, surface potential, and capacitance can be measured. The electrical double layer at the mercury/solution interface is the one most thoroughly studied. This system has a well-defined interfacial geometry, a wide range of potentials, and the possibility of working with a series of new surfaces to minimize contamination. Other materials can be dispersed as stable sols with an electrode reversible to the dispersed phase. An example is an aqueous suspension of silver iodide, for which the potential-determining mechanism is known and stable AgI electrodes exist.[24]

The capacitance C (per unit area) of the double layer at a planar interface is[25]

$$C = \frac{d\sigma}{d\Phi_0} \quad (17.23)$$

By means of the Gouy–Chapman theory, the capacitance of a planar interface can be evaluated, using Eq. (17.17), by

$$C = \left(\frac{2n_0 D\varepsilon_0}{kT}\right)^{1/2} \cosh \frac{ze\Phi_0}{2kT} \quad (17.24)$$

The observed capacitances are always much less than the theoretical capacitance obtained from Eq. (17.24). The source of this difference is ascribed to the neglect of the finite size of ions in the Gouy–Chapman model, especially for the counterions near the charged surface.

Stern[26] introduced ionic dimensions into the model of the double layer for its inner localized portion (the Stern layer). The constitution of a very small area of the particle, and the Stern layer, along with the diffuse double layer associated with the surface of the particle, is depicted diagrammatically in Figure 17.9 for (a) a negatively charged sol of silver iodide and (b) a positively charged sol of silver iodide.[27] The negatively charged sol is stabilized with a slight excess of hydrogen iodide in solution, and the positively charged sol is stabilized with a slight excess of silver nitrate in solution. Consider first the negatively charged particle: roman letters Ag and I represent the two ions in the lattice, while I$^-$ represents the potential-determining ion adsorbed out of the solution. The Stern layer contains a few counterions (H$^+$) and the remainder of the counterions are in the diffuse double layer. The diagram depicts electrical neutrality: The surface of the negatively charged particle has a net represented charge of 8I$^-$ adsorbed on it, with 6H$^+$ in the Stern layer, and 7H$^+$ plus 5I$^-$ in the diffuse double layer, to give no total excess of charge. The positively charged particle is represented as charged by adsorbed silver ions, Ag$^+$. Again the positively charged particle has a net represented charge of 8Ag$^+$ adsorbed on it, with 6NO$_3^-$ in the Stern layer, and 6NO$_3^-$ plus 4Ag$^+$ in the diffuse double layer, to no total excess of charge. The depicted Stern layer is separated from the depicted diffuse double layer by a dotted line. The Stern layer moves with the particle when it moves with respect to its medium. The Stern layer is within the plane of shear. The excess ions in the diffuse double layer beyond the plane of shear are not trapped by the electrostatic attraction of the surface, and do not move with the particle.

The total surface potential Φ_0 is divided into a potential drop Φ_s over the diffuse part of the double layer and a potential drop, $\Phi_0 - \Phi_s$ over the Stern layer. The Gouy–Chapman theory applies without change to the diffuse part of

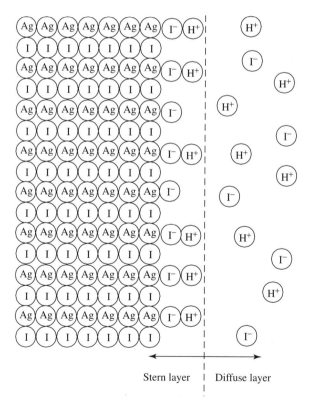

Figure 17.9 Diagrammatic representation of the constitution of a portion of a particle of silver iodide. (*a*) Negatively charged by adsorbed iodide ions. (*b*) Positively charged by adsorbed silver ions.

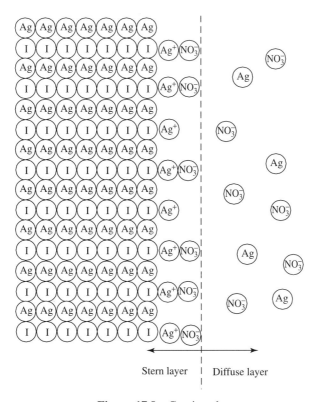

Figure 17.9 Continued.

the double layer, which originates at a short distance from the charged surface. The theoretical capacity of the total double layer, C_t, is the capacity of the diffuse layer, C_d, and that of the Stern layer, C_s, in series.

$$C_t = \frac{C_d C_s}{C_d + C_s} \quad (17.25)$$

Equation (17.25) explains the discrepancy between the observed capacitance and the impossibly high values of the Gouy–Chapman model; because if C_d, is large, the total capacitance given by Eq. (17.25) reduces to C_s, the capacitance of the Stern layer. The measured capacitance is then the smaller capacitance of the Stern layer. Sparnaay[28] gives a detailed analysis of the ion size corrections to the Poisson–Boltzmann equation.

17.2 ELECTROKINETIC PHENOMENA

Volta's invention of the voltaic pile, published in 1800, created great excitement in the scientific community. Immediately afterwards voltaic piles were constructed

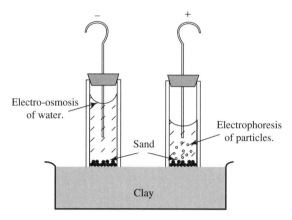

Figure 17.10 Reuss's discovery of the rise of the water level in the left tube (electro-osmotic flow) and the rise of particles in the right tube (electrophoresis.)

in laboratories throughout Europe and one discovery followed another within a few years. Among them was the electrolysis of water and the electrodeposition of new elements. In 1809 F. F. Reuss published a paper entitled *Sur un nouvel effet de l'électricité galvanique*.[29] Reuss reported the electrokinetic effects of electrophoresis and electro-osmotic flow in one experiment. He forced two open glass tubes into a mass of wet clay. At the bottom of the tubes he placed a layer of sand to prevent a layer of particles from being stirred up, placed water in the tubes, and put in two electrodes. On applying a potential difference, the water level rose in the tube containing the negative electrode; at the same time clay particles migrated to the positive electrode. See Figure 17.10.

A similar experiment is to substitute a potato for the mass of wet clay and observe the electro-osmotic flow in both tubes and the electrophoresis of starch granules. See Figure 17.11.

Methods of measuring zeta potential use the electrokinetic phenomena that arise from the mutual effects of electric field and of tangential motion of two phases with respect to each other. An applied electric field directed along the phase boundary causes relative motion of one phase with respect to another, and, conversely, motion along the charged phase boundary creates an electric potential. Small charged particles suspended in a medium move (electrophoresis) when a potential gradient is applied, or if the solid is stationary, the effect of moving the medium past the interface generates a potential difference (streaming potential). A value of the zeta potential is obtained from the electrophoretic mobility in the former case or from the streaming potential in the latter case. Conversely, the suspended particles can be moved, either by sedimentation or centrifugation, and the potential generated can be measured, or a potential gradient can be applied to a stationary interface and the motion of the medium measured. Four distinctive phenomena are possible, as tabulated in Table 17.3.

A excellent source of information on electrokinetic phenomena is Hunter.[30,31]

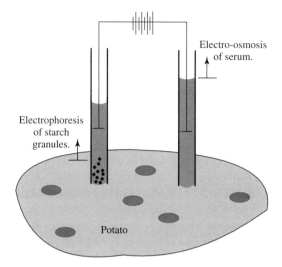

Figure 17.11 Electro-osmosis and electrophoresis in the potato.

TABLE 17.3 Electrokinetic Phenomena

Name	Solid phase	Applied stress	Measured response
Electrophoresis	Dispersed particles	Electric field	Motion of dispersed phase
Sedimentation potential	Dispersed particles	Gravitational field	Potential gradient
Electro-osmosis	Packed bed	Electric field	Motion of medium
Streaming potential	Packed bed	Motion of medium	Potential gradient

17.3 ZETA POTENTIAL

In all the electrokinetic phenomena one phase moves with respect to another. In electrophoresis, the particle moves with respect to the liquid. Between the two phases is a plane of shear. The electric potential at the plane of shear is called the zeta potential.[32] This is the experimentally measured quantity. The zeta potential, ζ, is not exactly the surface potential, Φ_0, but is the value used for surface potential in calculations of electrostatic stabilization in DLVO theory and is one reason why DLVO theory is approximate. (See Chapter 20.)

A more rigorous analysis of the electric potential within the plane of shear shows that the electric potential can vary widely right up to the surface of the particle. The adsorption of positive or negative ions within the Stern layer can increase the charge or reverse its sign. Therefore it is not accurate to say that the zeta potential measures the charge of the particle. The zeta potential is the potential at the plane of shear. This, however, is the potential of practical interest in colloid stability, because it determines the net interparticle forces. The zeta potential is not as useful in understanding the chemistry at the particle surface. For this the electric potential at distances closer than the shear plane are needed. For instance, measurements of the dropping mercury electrode give the potential at the metal surface, not at the plane of shear.

Zeta potentials are measured for various purposes in industrial research. As a minimum inquiry, the measurement answers the question: Is the electrical charge on the dispersed particle positive or negative? This information is often sufficient to suggest further steps in processing. The next higher level of inquiry has to do with quality control: Does this batch have the same electrostatic property as the previous batch? A final level of inquiry has to do with stability: Has the product sufficient electrostatic repulsion to maintain its stability or should more dispersing agent be provided? An aqueous clay slurry, containing tetrasodium pyrophosphate, is stabilized by electrostatic repulsion. Hence, electrophoretic measurement is useful as a quality control test in the clay-mining and paper-making industries. In R&D, electrophoretic measurements are used to screen potential suspending agents. Another question that often arises is how various processing steps affect the stability of the suspension; these changes, too, can be monitored by means of zeta potentials. In systems that are stabilized by a combination of electrostatic and steric repulsions, the contribution of the electrostatic component is determined by measuring the zeta potential. Finally, quantitative calculations of the stability ratio, to determine the magnitude of the repulsive barrier and hence the lifetime of the dispersion, are based on measurements of zeta potential.

17.4 MEASUREMENTS OF ZETA POTENTIAL

A critical analysis of recent advances in electrokinetic measurements is available.[33]

17.4.a Electrophoresis

For suspensions, emulsions, or macromolecules in solution, electrophoresis is the most common technique by which to measure zeta potential. The electric force acting on the double layer causes the particle to move at a constant velocity. The ratio of the velocity to the electric field is the electrophoretic mobility, measured as (meters/second) per (volts/meter). For aqueous systems, the electrophoretic mobility is generally independent of the electric field.

Smoluchowski derived an equation for the zeta potential,[34]

$$\zeta = \frac{\eta v}{E D \varepsilon_0} \qquad (17.26)$$

where η is the viscosity, v/E is the electrophoretic mobility and ζ is the zeta potential. The same equation describes the relation between the applied average linear velocity of the medium, v, and the consequent potential field, E, in the other electrokinetic effects. Equation (17.26) applies to the electrophoresis of nonconducting particles aqueous media of high ionic strength. Smoluchowski's derivation essentially assumes that the applied electric field lines conform to the shape of the particle, which in turn implies that the field lines are bent by the ionic double layer around each particle.

Hückel, fresh from his triumph with the theory of interionic attraction in electrolyte solutions, also derived a relation between zeta potential and electrophoretic mobility. He assumed that the applied electric field lines run straight between the electrodes, without conforming to the shape of the particle. His derivation, therefore, applies to point particles, or, in practice, to very small particles in water or to particles in oil, because those are examples of dilute double layers. According to Hückel,[35]

$$\zeta = \frac{3\eta v}{2ED\varepsilon_0} \quad (17.27)$$

Equation (17.27) gives a higher value of the zeta potential by a factor of three halves than Eq. (17.26). The transition between these two limiting cases for small zeta potentials was quantified by Henry.[36] The shape of the field depends on the electrical conductivities of the particle and the surrounding liquid, the size of the particle, and its electrical double layer.

A more general solution of the relation between electrophoretic mobility and zeta potential is provided by O'Brien and White in the form of graphs calculated by a computer program.[37] Figure 17.12 and Figure 17.13 show the reduced electrophoretic mobility, U,

$$U = \frac{3\eta e}{2D\varepsilon_0 kT} \frac{v}{E} \quad (17.28)$$

as a function of the reduced zeta potential, $e\zeta/kT$, for various values of κa.[38] Figure 17.12 shows the Hückel limit at $\kappa a = 0$ and Figure 17.13 shows the Smoluchowski limit at $\kappa a = \infty$.

Table 17.4 shows the limits of particle size for various concentrations of electrolyte for which either the Hückel or the Smoluchowski formulation is applicable within 10% or so.[39]

17.4.a(i) Microelectrophoresis Electrophoresis refers to the motion of particles caused by an applied electric field. The dispersion is held either in a tube or a narrow glass cell. When the motion of the particles is measured through a microscope, the technique is called microelectrophoresis. The motion of particles in a microelectrophoretic cell is influenced by the electroosmotic flow arising from the charge on the capillary walls. A stationary layer exists at about 28% of the distance between the walls where the flow and counterflow neutralize each other. The electrophoretic mobility of particles is measured by focusing on the particles in the stationary layer. Several commercial instruments are available to measure electrophoretic mobility. Instruments marketed by Zeta-Meter and Rank

336 ELECTRICAL CHARGES IN DISPERSIONS

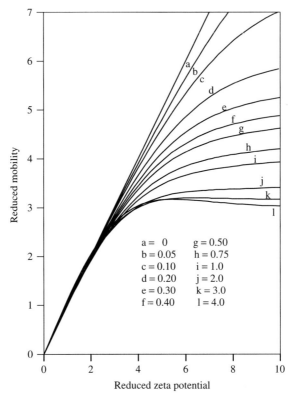

Figure 17.12 Reduced electrophoretic mobility of a spherical particle in a 1-1 electrolyte as a function of the reduced zeta potential, for $\kappa a < 2.75$.[37]

TABLE 17.4 Range of Applicability of Smoluchowski (S) and the Hückel (H) Equations for $\zeta = -50$ mV

Electrolyte	Concentration (M)	κ at 20°C (μm^{-1})	$1/\kappa$ (μm)	Diameters (μm) S	H
Pure water		1.0	1.0	>600	<1.0
Uni-univalent	10^{-5}	10	0.1	>17	<0.32
	10^{-3}	100	0.01	>1.7	<0.032
	10^{-1}	1000	0.001	>0.17	<0.003
Uni-divalent	10^{-5}	18	0.056	>9.4	<0.03
	10^{-1}	1800	0.00056	>0.094	<0.0003
Di-divalent	10^{-5}	210	0.048	>8.1	<0.048
	10^{-1}	2100	0.00048	>0.081	<0.0005

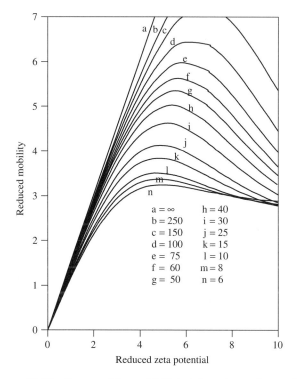

Figure 17.13 Same as Figure 17.12 but for values of $\kappa a > 3$.[37]

Brothers measure electrophoretic mobility by timing the motion of individual particles viewed through a microscope. A more automated version is available from Laval Lab, where images of the moving particles are captured and analyzed electronically. A spinning prism to visually arrest the motion of a large number of independently moving particles can be used to obtain an average velocity much more easily than measuring the velocity by eye. Several companies sell a rotating prism system as an attachment to their microelectrophoresis apparatus. (See Section 17.6).

In order to focus on the motion of single particles, dispersions have to be diluted, preferably with a serum of the same composition as the sample. A suitable serum can be obtained by a freeze-thaw cycling of the dispersion or cross-flow filtration. Concentrated dispersions are measured by moving boundary electrophoresis, the Mass Transport Analyzer of Micromeritics Corporation, or electroacoustics techniques (see Section 17.4.d.)

17.4.a(ii) Laser Doppler Microelectrophoresis Laser Doppler velocimetry (LDV) is a well-known technique to measure the velocity of particles.[40,41,42] Aerodynamic and hydrodynamic flows can also be measured by seeding the gas or water with a few particles and using LDV to determine the local velocity. The usual experimental method is to split a laser beam into two beams, cross the two beams at a point by means of two appropriately directed mirrors, and measure the light scattered as a particle traverses the region defined by the crossed beams. The crossed beams form an optical grating and a particle moving through the grating produces a beat signal proportional to its velocity across the grating.

Laser Doppler microelectrophoresis replaces microscope imaging at the stationary layer of the electrophoresis cell. The beam crossing of the LDV apparatus is adjusted to a focus at the stationary layer.[43] The velocity of the particles is measured with the LDV optics, the electric field is known by the voltage applied and the distance between the electrodes, and the electrophoretic mobility is the ratio of the measured velocity to the known electric field. This microelectrophoretic method is capable of measuring a large number or particles and can be fully automated. Instruments that use laser Doppler velocimetry are sold commercially by Brookhaven Instruments, Coulter Instruments and Malvern Instruments.

The electrophoretic mobility in a microelectrophoretic cell has to be measured at the stationary layer inside the cell. Electroosmotic flow, however, does not arise instantly, but after a time, on the order of a second.[44] On the other hand, the electrophoretic response of the particles is nearly instantaneous (on the order of microseconds).[43] If the electrophoretic motion were measured quickly (say in tens of milliseconds) then electro-osmosis has not started and the electrophoretic motion could be measured anywhere in the cell. LDV can indeed measure the velocity of moving particles in this short time so the commercial equipment makes the measurement with approximately a 50 Hz, square-wave signal. Using this ac signal also minimizes electrode polarization.

Electrophoretic mobilities of particles in oils are generally lower by an order of magnitude than those in water. Higher electric fields, say greater than 10^4 V/m, are required to get sufficient motion to be observed. These high fields are obtained by having the electrodes close together, often only a few millimeters apart, but they introduce electrohydrodynamic instabilities. Novotny and Hair were the first to show that the LDV technique avoids the disturbing electrohydrodynamic instabilities by taking measurements quickly and at various positions within the gap.[45]

17.4.a(iii) ***Phase-Angle Light Scattering*** The measurement of low electrophoretic mobilities, especially those often found in nonpolar dispersions or liquids of high viscosity, offers a challenge to experiment. The reason is that the lower the mobility, the lower the velocity to be measured, requiring the electric field to be correspondingly increased. Quite often high electric fields cannot be used because they cause experimental difficulties, particularly electrode reactions and dielectrophoretic motion (i.e., the motion of uncharged particles in an electric field gradient). LDV measurements depend upon particles moving across the optical grating and, if the motion is slow, the frequency of the signal is too low to measure. An improvement is to measure the phase shift of the signal, a technique named phase-angle light scattering or PALS.[46,47] Mobilities as low as 10^{-10} m^{2}V^{-1}s^{-1} have been measured with commercially available equipment, the Brookhaven ZetaPALS.

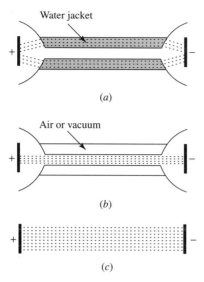

Figure 17.14 Schematic drawings of capillary-tube electrophoretic cells and a parallel-plate electrophoretic cell containing oil dispersions. The electric field lines are indicated by dotted lines. (*a*) and (*b*) are diagrams of the capillary tube cell when filled with a dispersion in a low conductivity liquid with and without a surrounding water bath. (*c*) is a diagram of the parallel plate cell. (*b*) and (*c*) are not suitable for aqueous dispersions because they lack temperature control.[48]

17.4.a(iv) **Electrophoresis at Low Conductivity** Determining the sign and magnitude of the electric charge of colloidal particles suspended in low-conductivity liquids (10^{-12} $\Omega^{-1}\text{m}^{-1}$ range) by electrophoresis continues to be experimentally difficult and controversial. One reason is that when the dielectric constant of the liquid is less than 10, especially less than 5, and when one is working with the type of apparatus commonly used, irregular movements, rather than electrophoretic motion, are often observed. The usual apparatus is the microelectrophoresis cell with electrodes a known distance apart at each end of a long capillary tube. This cell is usually placed inside a cooling jacket to minimize Joule heating. If a dispersion in a low conductivity medium is placed inside such an apparatus the particles often move erratically, because the electric field lines terminate at the cell walls, since the water jacket is a direct path to ground (see Figure 17.14). Removing the water bath restores the electric field.

A set of parallel plate electrodes placed in the dispersion works well for dispersions of low conductivity where Joule heating is negligible; such a setup for aqueous dispersions would result in too much thermal motion.[48]

17.4.b Electro-osmosis

Electro-osmosis is used to measure the zeta potential of particles, or fibers, or chips, or any other shape that is too large to remain suspended for electrophoretic

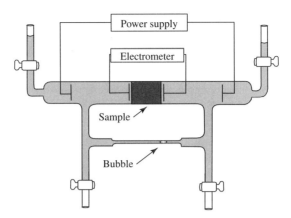

Figure 17.15 An electro-osmotic apparatus for measuring zeta potential.

measurements. The method is to apply an electric field to a packed bed and measure the rate of flow of liquid, which is measured by the velocity of an air bubble in a capillary (see Figure 17.15). The zeta potential in terms of the measured velocity of the bubble, v(m/s), is given by

$$\zeta = \frac{\eta v}{E D \varepsilon_0} \frac{r^2}{R^2} \quad (17.29)$$

where r is the radius of the capillary and R is the radius of the packed bed. When the packed bed is rather porous the pressure drop across the capillary itself cannot be neglected and a correction term is needed.[49]

Rather than measuring the velocity of the liquid, the electro-osmotic flow can be offset by applied pressure across the packed bed, and the current measured. The advantage of this technique is that the velocity of the liquid is no longer a factor.[50]

$$\zeta = -\frac{\eta I}{D \varepsilon_0 P} \quad (17.30)$$

where I is the current per unit area between the electrodes and P is the pressure gradient across the packed bed (N/m^3).

17.4.c Streaming Current and Streaming Potential Techniques

For particles greater than about 50 mm in diameter, the zeta potential is measured more easily by pumping the liquid while the particles are held stationary. The moving liquid carries with it ions from the electrical double layers. The motion of the ions is an electric current, which is measured with an ammeter. The zeta potential is calculated by[51]

$$\zeta = -\frac{\eta I}{D \varepsilon_0 P} \quad (17.31)$$

where I is the measured current per unit area, P is the measured pressure gradient (change of pressure with length). Streaming current equipment is commercially available as the BI-EKA from Brookhaven Instruments, as the Streaming Current Monitor and Titrator from Chemtrac, and as the Charge Analyser from Rank Brothers.

Instead of measuring current, the voltage drop across the packed bed, the electric field can be measured and the zeta potential for nonconducting particles given by[52]

$$\zeta = \frac{\eta E \lambda_0}{D \varepsilon_0 P} \qquad (17.32)$$

where E is the measured electric field across the sample and λ_0 is the conductivity of the medium. Laval Lab sells the Zetacad, which determines zeta potentials by streaming potential.

When the particles are conductive, the solution conductivity, λ_0, is replaced in Eq. (17.32) with the term $\lambda_0 + 2\lambda_s a^{-1}$ where λ_s is the surface conductivity of the particles with radius a. Unfortunately, the surface conductivity of particles is not generally known.

17.4.d Electroacoustic Techniques

When a high-frequency electric field is applied to a dispersion, the particles vibrate electrophoretically in the applied field. If there is a density difference between the particle and the liquid, this vibration generates an acoustic signal. The acoustic signal can be detected with an ultrasonic acoustic sensor. The magnitude of the acoustic signal depends on the electrophoretic mobility of the particles, the density difference, and the volume fraction of the particles as related by

$$ESA = A(\omega) \Phi \frac{\Delta \rho}{\rho} Z \mu_d \qquad (17.33)$$

where ESA (electrokinetic sonic amplitude) is the acoustic signal, $A(\omega)$ is a calibration constant, Φ is the volume fraction of particles, and $\Delta \rho$ is the difference between the density of the particles and the liquid, which has a density ρ, Z is an instrumental factor dependent on the acoustic impedance of the suspension and is automatically measured by the commercial equipment, and μ_d is the dynamic electrophoretic mobility. Equation (17.33) holds at essentially any particle concentration, so this technique is well suited to determine the zeta potential of particles at high concentration, 0.5% to over 40% by volume. The dynamic electrophoretic mobility is related to the zeta potential by

$$\zeta = \frac{3 \mu_d \eta}{2 D \varepsilon_0 \, G(a, \omega)} \frac{1}{[1 + f]} \qquad (17.34)$$

where D is the dielectric constant of the liquid, ε_0 is the permittivity of free space, and η is the viscosity of the liquid. Both G and f are functions calculated by the designers of the instruments. G depends on the particle size and the measuring frequency.

The application of the electric field and the resulting acoustic wave provide a truly nonintrusive measurement which does not alter the properties of the sample. Electroacoustic equipment is available from Colloidal Dynamics, Dispersion Technology, and Matec Applied Sciences.

An analogous technique is to apply a high-frequency acoustic signal to the dispersion. (See Table 17.5.) If there is a density difference between the particle and the liquid, the particles respond to the pressure, the electric double layer is polarized back and forth, generating an electric field in phase with the applied acoustic signal. (See Figure 17.16.) By placing electrodes at one-half wavelength apart, which amounts to a few millimeters distance, the magnitude of the effect can be measured as a voltage called the colloid vibration potential, CVP. An instrument is available from Dispersion Technology. Both techniques can be carried out at high-volume fractions.

Electroacoustic instruments move particles at high frequencies, on the order of a MHz. At higher frequencies the particles have too much inertia to keep up with the signal and so the amplitude of their motion decreases. The attenuation of signal with frequency provides a method to measure particle size distribution in the dispersion. Electroacoustic measurement is therefore a means to obtain both size and zeta potential at high particle concentrations.

Electroacoustic measurements have been made on a wide variety of complex samples, all without dilution: cadmium adsorption on kaolin, cobalt adsorption on silica, in the synthesis of metal oxide dispersions, the surface modifications of titania with alumina, and concentrated dispersions of titania, silica, silicon nitride, yttrium oxide, inorganic phosphors, coal, carbon black (water and oil), barium titanate, and polymethylmethacrylate. The dispersions were not transparent nor dilute; some of the samples were flowing.[53]

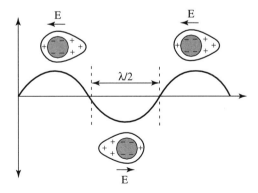

Figure 17.16 A schematic showing the alternating polarization of the electric double layer created by an acoustic wave and the consequent CVP.

TABLE 17.5 Electroacoustic Techniques

Technique	Applied Stress	Measured Response
ESA	AC electric field	Acoustic signal
CVP	Acoustic signal	AC electric field

17.4.e Charge-to-Mass Ratio in Low-Conductivity Media

In low-conductivity media, countercharges can be physically separated from their substrates as was shown by Klinkenberg and van der Minne in their study of explosion hazards in the petroleum industry (see Section 14.3). This cannot be done in aqueous solutions because the ionic relaxation is so rapid. An apparatus to take advantage of this phenomenon retains charged particles on a porous support while the countercharges are carried away, collected, and measured in the isolated Faraday cup. Solvent is pumped onto the top of a filter upon which has been placed a small sample of the dispersion. The charged particles from the dispersion are held on the filter. The flow of solvent carries the countercharges into a Faraday cup, where the charge collected is measured by an electrometer.[54] All the excess ions in the dispersion come in electrically neutral pairs so that they do not contribute to the net charge measured.

The charge per unit mass, measured as described above, the electrophoretic mobility, and the size of particles in low conductivity dispersion can be measured independently. These, however, are not independent characterizations but can each be used to check internal consistency.[55] The three properties are related by a function that depends on the fluid viscosity and the particle density. For a monodispersed dispersion of nonconducting spheres this function is

$$\frac{(Q/M)d^2}{\zeta} = \frac{12D\varepsilon_0}{\rho_2} \qquad (17.35)$$

where Q/M is the average charge to mass, d is the diameter, ζ is the zeta potential, D is the dielectric constant of the medium, ε_0 is the permittivity of free space, and ρ_2 is the density of the particle. If the results are not self-consistent, then one looks for unsuspected factors such as a different degrees of dispersion in the different experiments.

17.5 ELECTRICAL EFFECTS IN DRY POWDERS

If two solids are brought into contact and then separated, they are generally found to be electrically charged; this is the phenomenon of contact electrification. The phenomenon has also been called triboelectrification since the charging of the two solids is often brought about by rubbing, but the rubbing is not necessary. Contact electrification is a key element in such processes as self-clinging wrapping materials, spray painting, adhesion, and xerography. It is also responsible

344 ELECTRICAL CHARGES IN DISPERSIONS

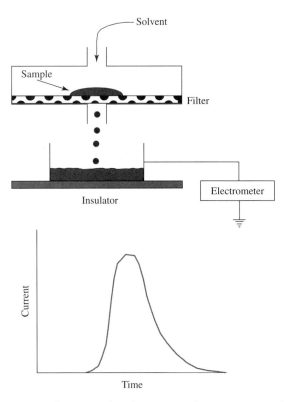

Figure 17.17 Apparatus for measuring the average charge-to-mass ratio of particles in nonpolar dispersions.

for explosions in mines and flour mills, where the electric charges accumulated in moving dust clouds are eventually discharged by an electric spark. The distinguishing characteristic of all these phenomena is that charges and countercharges can be separated because of low conductivity of the medium. We have discussed the same effects in suspensions of solids in media of low conductivity (see Section 14.3).

Contact electrification was first announced by Volta in 1797 while studying the potential difference generated by dry metals in contact. From measurements of these electric potentials, Volta developed a work-function series by means of his law of successive contacts. He attributed the electric potentials to an attraction of electricity by matter, varying with the difference in the nature of the substance, which produces unequal potentials and then sets up an opposition to their equalizing. In modern terms the electrons in the conduction band, also called the Fermi level, of one metal spontaneously transfer to the other metal, that is, from a low-work function solid to a high-work function solid. Further separation of charges is brought to an end by the electric potential it has itself created. This potential is called the Volta potential.

For interfaces between a solid and a liquid, that is, the wet contact, charge exchange is almost always due to ionic motion, either by the adsorption of charged surface-active solutes or by the desorption of ionic components of the solid surface. The determination of the distribution of ionic charges at a solid–liquid interface is the study of the electrical double layer. The electric potential measured across a wet contact is called the Nernst potential.

A Voltaic pile is constructed by replacing every second dry contact of a pile of plates of two alternating metals, with a conducting wet contact, for example,

$$Zn|wet|Cu|dry|Zn|wet|Cu|dry|Zn|wet|Cu \qquad (17.36)$$

The change of the Gibbs free energy that determines the electromotive force of the cell can be indifferently ascribed either to the difference in chemical potential of the electrons in the metals, or to the cell reaction, whatever it might be, by means of which the electrons are transferred. The cell reaction is simply the mechanism by which electrons are transported from one electrode to the other.[56]

Contact electrification is most often analyzed with the model of dry contacts. For every dry contact there is a conceptually equivalent wet contact; an analysis based on dry contacts does not exclude an alternative analysis based on wet contacts. In practice, in the presence of dirt and moisture, the actual state of the dry contact is difficult to determine.

Quantitative measures of the nature and extent of charge exchange at metal insulator and insulator–insulator interfaces are difficult to obtain. Useful experimental techniques are:

(a) Dry-contact charging with various metals, with or without an applied electric field, in a vacuum or not;
(b) Vibrating capacitor method;[57]
(c) Cascading metallic beads on an insulator surface;[58]
(d) Spectroscopic measurement of electrons emitted during separation of surfaces;[59]
(e) Field-effect capacitor;[60,61]
(f) Contact-charge spectrograph.[62]

A general review of contact electrification is given by Lowell and Rose-Innes.[63]

Several investigators[64,65,66,67] have reported that the charge transferred into a polymer by a contacting metal depends on the work function of the metal. This observation led to the hypothesis that polymers could be arranged into a "triboelectric series" analogous to the metallic work-function series. The work function of a polymer is assigned the value of the work function of the metal with which it does not exchange charge. The order of the materials in the triboelectric series is likely to be the same as the order of ionization potentials, electron affinities, acid or base strengths, reduction potentials, or Hammett sigma values.[68,69,70] The charging is believed to be governed by the Fermi level of the metal and the highest occupied molecular orbital (HOMO) for electron transfer to the metal or the lowest unoccupied molecular orbital (LUMO) for electron transfer to the insulator. Typical proposed triboelectric series are listed in Table 17.6 and Table 17.7.

The absolute value of the work function for insulators and even the order of the materials in a triboelectric series are uncertain, partly because of ambient contamination, but also depending on the nature and number of contacts. Even two pieces of the same material exchange charge; for instance, one rubber rod rubbed across another. When the charge exchange is not great, an insulator does not always retain the same place in a triboelectric series.

The Duke–Fabish model[72,73,74] of the nature of electron states in insulators attempts to account for the deficiencies in the simple idea of a triboelectric series. They propose (a) each metal interacts with only a narrow range of electron states

TABLE 17.6 Triboelectric Series for Various Insulators[71]

Positive	Negative
Wool	Teflon®
Nylon	
Cellulose	
Cotton	
Natural silk	
Cellulose acetate	
Polymethyl methacrylate	
Polyvinyl alcohol	
Polyethylene glycol-co-terephthalic acid	
Polyvinyl chloride	
Polyacrylonitrile-co-vinyl chloride	
Polyethylene	

TABLE 17.7 Triboelectric Series for Various Polymers According to Davies[65]

Polymer	Work function (eV)
Polyvinyl chloride	4.85 ± 0.2
Polyimide	4.36 ± 0.06
Polytetrafluoroethylene	4.26 ± 0.05
Polycarbonate	4.26 ± 0.05
Polyethylene terephthalate	4.25 ± 0.10
Polystyrene	4.22 ± 0.07
Nylon 66	4.08 ± 0.06

in the insulator; the amount of charge exchanged depending on the distribution of acceptor or donor energies; (b) injected charge is trapped to a depth of 1–2 μm; (c) the trap states are not electronic levels of the molecule but some lower energy states stabilized by intermolecular relaxation; and (d) polymer films can have both acceptor and donor states and the number of these states accessible to the interface is limited.

17.6 MANUFACTURERS OF EQUIPMENT TO MEASURE PARTICLE CHARGE

Company	Telephone/FAX	Internet	Technique
Beckman-Coulter, Inc. 4300 N. Harbor Blvd. P.O. Box 3100 Fullerton, CA 92834-3100	(714) 871-4848 (714) 773-8283	www.beckmancoulter.com	Electrophoresis by laser velocimetry
Brookhaven Instruments Corporation 750 Blue Point Rd. Holtsville, NY 11742-1896	(631) 758-3200 (631) 758-3255	www.bic.com	Electrophoresis by laser velocimetry Steaming potential
Chemical ElectroPhysics Corp. 705 Yorklyn Rd. Hockessin, DE 19707	(302) 239-4677	www.cep-corp.com	In-line dielectric analyzer
ChemTrac Systems, Inc. 6991 Peachtree Industrial Blvd., 600 Norcross, GA 30092	(770) 449-6233 (770) 447-0889	www.chemtrac.com	Streaming current monitors Electrokinetic charge titrators
Colloidal Dynamics, Inc. 11 Knight St. Building E18 Warwick, RI 02886	(401) 738-5515 (401) 738-5542	www.colloidal-dynamics.com	Electroacoustics
Dispersion Technology, Inc. 3 Hillside Ave. Mount Kisco, NY 10549	(914) 241-4791 (914) 241-4842	www.dispersion.com dispersi@dispersion.com	Electroacoustics
Laval Lab, Inc. 2567, Boul. Chomedey Laval, Québec Canada, H7T 2R2	(888) 667-7077 (450) 681-9939	www.lavallab.com	Microelectrophoresis Streaming potential
Malvern Instruments 10 Southville Rd Southborough, MA 01772	(508) 480-0200 (508) 460-9692	www.malvern.de	Electrophoresis by laser velocimetry
Matec Applied Sciences 56 Hudson St. Northboro, MA 01532	(508) 393-0155 (508) 393-5476	www.matec.com	Electroacoustics

Company	Telephone/FAX	Internet	Technique
Micromeritics Instrument Corp. One Micromeritics Dr Norcross, GA 30093-1877	(770) 662-3633 (770) 662-3696	www.mircomeritics.com	Mass transport electrophoresis
Particle Sizing Systems 75 Aero Camino Suite B Santa Barbara, CA 93117	(805) 968-1497 (805) 969-0361	www.pssnicomp.com	Electrophoresis by laser velocimetry
Rank Brothers, Ltd. 56 High St. Bottisham Cambridge CB5 9DA UK	+44(0) 122-381-1369 +44(0) 122-381-1441	www.rankbrothers.co.uk	Microelectrophoresis Rotating prism electrophoresis Streaming current titration
TSI, Inc. P.O. Box 64394 St. Paul, MN 55164	(651) 483-0900 (651) 490-2748	www.tsi.com	Laser Doppler velocimetry
Zeta-Meter, Inc. 765 Middlebrook Ave. PO Box 3008 Staunton, VA 24402-3008	(800) 333-0229 (540) 886-3728	www.zeta-meter.com	Microelectrophoresis

REFERENCES

[1] Gurney, R.W. *Preface to Ionic processes in solution*; Dover: New York; 1962; McGraw-Hill: New York; 1953.

[2] Verwey, E.J.W.; Overbeek, J. Th. G. *Theory of the stability of lyophobic colloids*; Elsevier: Amsterdam; 1948; Dover: Mineola, NY; 1999; Part I.

[3] Quincke, G. Über eine neue art elektrischer ströme, *Ann. Phys. Chem. (Poggendorf's)* **1859**, *107*, 1–47.

[4] Quincke, G. Über die fortführung materieller theilchen durch strömende ecektricität, *Ann. Phys. Chem. (Poggendorf's)* **1861**, *113*, 513–598.

[5] Quincke, G. Über capillaritäts-erscheinungen an der gemeinschaftlichen oberfläche zweier flüssigkeiten, *Ann. Phys. Chem. (Poggendorf's)* **1870**, *139*, 1–89.

[6] Lyklema, J. *Fundamentals of interface and colloid science*, Vol. 2 Solid–liquid interfaces. Academic Press: New York, 1995, Appendix 3.

[7] Ardizzone, S.; Trasatti, S. Interfacial properties of oxides with technological impact in electrochemistry, *Adv. Colloid Interface Sci.* **1996**, *64*, 173–251.

[8] Kosmulski, M. Attempt to determine pristine points of zero charge (PPZC) of Nb_2O_5, TaO_5, and HfO_2, *Langmuir* **1997**, *13*, 6315–6320.

[9] Averages of PPZC values of oxides from different sources in Ref. 8.

[10] Fuerstenau, D.W. The adsorption of surfactants at solid–water interfaces, *Chem. Biosurfaces* **1971**, *1*, 143–176.

[11] Morrison, I.D. Electrical charges in nonaqueous media, *Colloids Surf. A. Physicochemical Eng. Aspects* **1993**, *71*, 1–37.

[12] Morrison, I.D. The influence of electric charges in nonaqueous dispersions in *Dispersion and Aggregation: Fundamentals and Applications*; Moudgil, B.M.; Somasundaran, P. Eds.; Engineering Foundation: New York; 1994.

[13] Olivier, J.P.; Sennett, P. Electrokinetic effects in kaolin-water systems. I. The measurement of electrophoretic mobility, *Clays Clay Miner.* **1967**, *15*, 345–356.

[14] Ottewill, R.H.; Watanabe, A. Studies on the mechanism of coagulation. Part 2. The electrophoretic behavior of positive silver-iodide sols in the presence of anionic surface-active agents, *Kolloid Z.* **1960**, *171*, 132–139.

[15] James, R.O.; Wiese, G.R.; Healy, T.W. Charge reversal coagulation of colloidal dispersions by hydrolysable metal ions, *J. Colloid Interface Sci.* **1977**, *59*, 381–385.

[16] Gouy, G. Sur la constitution de la charge électrique a la surface d'un électrolyte, *J. Phys. Theor. Appl.* **1910** (4), *9*, 457–467.

[17] Chapman, D.L. A contribution to the theory of electrocapillarity, *Phil. Mag.* **1913**, (6), *25*, 475–481.

[18] Grahame, D.C. The electrical double layer and the theory of electrocapillarity, *Chem. Rev.* **1947**, *11*, 441–501.

[19] Lau, A.; McLaughlin, A; McLaughlin, S. The adsorption of divalent cations to phosphatidylglycerol bilayer membranes, *Biochim. Biophys. Acta* **1981**, *645*, 279–292.

[20] Pashley, R.M. Hydration forces between mica surfaces in electrolyte solutions, *Adv. Colloid Interface Sci.* **1982**, *16*, 57–62.

[21] Dukhin, S.S.; Derjaguin, B. V. Equilibrium double layer and electrokinetic phenomena, *Surf Colloid Sci.* **1974**, *7*, 49–272.

[22] O'Brien, R.W.; White, L.R. Electrophoretic mobility of a spherical colloidal particle, *J. Chem. Soc., Faraday Trans. 2* **1978**, *74*, 1607–1626.

[23] Verwey, E.J.W.; Overbeek, J. Th. G. *Theory of the stability of lyophobic colloids*; Elsevier: Amsterdam; 1948; Dover: Mineola, NY; 1999.

[24] Lyklema, J.; van Leeuwen, H.P. Dynamic properties of the AgI–solution interface: implications for colloid stability, *Adv. Colloid Interface Sci.* **1982**, *16*, 127–137.

[25] Hunter, R.J. *Zeta potential in colloid science, principles and applications*; Academic Press: New York; 1981; p 33.

[26] Stern, O. Zur Theorie der elektrolytischen doppelschicht, *Z. Elektrochem. Angew. Phys. Chem.* **1924**, *30*, 508–516.

[27] Weiser, H.B. A *textbook of colloid chemistry*, 2nd ed.; Wiley: New York; 1949; p 215.

[28] Sparnaay, M.J. Ion-size corrections of the Poisson–Boltzmann equation, *Electroanal. Chem. Interfacial Electrochem.* **1972**, *37*, 65–70.

[29] Reuss, F.F. Sur un nouvel effet de l'électricité galvanique, *Mémoires de la Société Imperiale de Naturalistes de Moskou* **1809**, *2*, 327.

[30] Hunter, R.J. *Zeta potential in colloid science*; Academic Press: New York; 1981.

[31] Hunter, R.J. *Foundations of colloid science*, 2nd ed.; Oxford University Press: New York; 2001; Chapters 7 and 8.

[32] Hunter (1981), pp 4–6; Hunter (2001) Sections 8.4 and 10.2.3.

[33] Hidalgo-Álvarez, R.; Martín, A.; Fernández, A.; Bastos, D.; Martínez, F.; de las Nieves, F.J. Electrokinetic properties, colloidal stability and aggregation kinetics of polymer colloids, *Adv. Colloid Interface Sci.* **1996**, *67*, 1–118.

[34] Hunter (1981) Section 3.5; Hunter (2001) Section 8.2.3.

[35] Hunter (1981) p. 69; Hunter (2001) p. 381.

[36] Henry, D.C. The cataphoresis of suspended particles. Part 1. The equation of cataphoresis, *Proc. Roy. Soc. London* **1931**, *A133*, 104–129.

[37] O'Brien, R.W.; White, L.R. Electrophoretic mobility of a spherical colloidal particle, *J. Chem. Soc., Faraday Trans. 2*, **1978**, *74*, 1607–1626. A copy of the program as a self-contained Fortran subroutine is available from the authors.

[38] Hunter, R.J. *Introduction to modern colloid science*; Oxford University Press: New York; 1993; pp 238–241.

[39] Wiersema, P.H. *On the theory of electrophoresis*; Drukkerij Pasmans: The Hague; 1964.

[40] Menon, R. Laser Doppler velocimetry: Performance and applications, *Amer. Lab.* February **1982**, 122–142.

[41] Cummins, H.Z.; Pike, E.R., Eds. *Photon correlation spectroscopy and velocimetry*; Plenum Press: New York; 1977.

[42] Thompson, H.D.; Stevenson, W.H., Eds. *Laser velocimetry and particle sizing*; Hemisphere Publishing: Washington; 1978.

[43] Xu, R. Methods to resolve mobility from electrophoretic laser light scattering measurements, *Langmuir* **1993**, *9*, 2955–2962.

[44] Minor, M.; van der Linde, A.J.; van Leeuwen, H.P.; Lyklema, J. Dynamic aspects of electrophoresis and electro-osmosis: A new fast method for measuring particle mobilities, *J. Colloid Interface Sci.* **1997**, *189*, 370–375.

[45] Novotny, V.; Hair, M.L. Charges and dynamics of colloidal particles by quasielastic light scattering, in *Polymer Colloids*; Fitch, R.M., Ed.; Plenum: New York; 1971; 37–50.

[46] Miller, J.F.; Schätzel, K.; Vincent, B. The determination of very small electrophoretic mobilities in polar and nonpolar colloidal dispersions using phase angle light scattering, *J. Colloid Interface Sci.* **1991**, *143*, 532–554.

[47] Tscharnuter, W.W.; McNeil-Watson, F.; Fairhurst, D. A new instrument for the measurement of very small electrophoretic mobilities using phase analysis light scattering (PALS), *Colloid Surf. A. Physicochemical Eng. Aspects* **1998**, *140*, 53–57.

[48] Kornbrekke, R.E.; Morrison, I.D.; Oja, T. Electrophoretic mobility measurements in low conductivity media, *Langmuir* **1992**, *8*, 1211–1217.

[49] Ryde, N.P.; Matijević, E. Electroosmosis in packed beds, *J. Colloid Interface Sci.* **1996**, *177*, 675–676.

[50] Hunter (1981) p. 62; Hunter (2001) Section 8.2.1.

[51] Hunter (1981) p. 68; Hunter (2001) Section 8.5.1.

[52] Hunter (1981) p. 66; Hunter (2001) Section 8.2.2.

[53] *Electroacoustics for characterization of particulates*. Workshop held at the National Institute of Standards and Technology, Gaithersburg, MD, February 2–4, 1993.

[54] Morrison, I.D.; Thomas, A.G.; Tarnawskyj, C.J. A method to measure the average charge to mass ratio of particles in low-conductivity media, *Langmuir* **1991**, *7*, 2847–2852.

[55] Morrison, I.D.; Tarnawskyj, C.J. Toward self-consistent characterizations of low conductivity dispersions, *Langmuir*, **1991**, *7*, 2358–2361.

[56] Ross, S. The story of the Volta potential, in *Nineteenth-century attitudes: Men of science*; Kluwer: Dordrecht; 1991.

[57] Zisman, W.A. A new method of measuring contact potential differences in metals, *Rev. Sci. Instrum.* **1933**, *3*, 367–370.

[58] Gibson, H.W.; Pochan, J.M.; Bailey, F.C. Surface analyses by a triboelectric charging technique, *Anal. Chem.* **1979**, *51*, 483–487.

[59] Derjaguin, B.V.; Smilga, V.P. The present state of our knowledge about adhesion of polymers and semiconductors, *Proc. Int. Congr. Surf. Act., 3rd, 1960* **1961**, *2*, 349–367.

[60] Fowkes, F.M. Interface acid-base–charge-transfer properties, *Surf. Interfacial Aspects Biomed. Polym.* **1985**, *1*, 337–372.

[61] Fowkes, F.M.; Hielscher, F.H. Electron injection from water into hydrocarbons and polymers, *Org. Coat. Plast. Chem.* **1980**, *42*, 169–174.

[62] Fabish, T.J.; Saltsburg, H.M.; Hair, M.L. Charge transfer in metal/atactic polystyrene contacts, *J. Appl. Phys.* **1976**, *47*, 930–939.

[63] Lowell, J.; Rose-Innes, A.C. Contact electrification, *Adv. Phys.* **1980**, *29*, 947–1023.

[64] Arridge, R.G.C. The static electrification of nylon 66, *Br. J. Appl. Phys.* **1967**, *18*, 1311–1316.

[65] Davies, D.K. Charge generation on dielectric surfaces, *Br. J. Appl. Phys.* **1969**, *2(2)*, 1533–1537.

[66] Davies, D.K. Charge generation on solids, *Int. Congr. Static Electr., 1st, 1970* **1970**, *1*, 10–21.

[67] Nordhage, F.; Bäckström, G. Electrification in an electric field as a test of the theory of contact charging, *Inst. Phys. Conf. Set.* **1975**, *27 (Static Electrif.)*, 84–94.

[68] Skinner, S.M.; Savage, R.L.; Rutzler, J.E., Jr. Electrical phenomena in adhesion 1. Electron atmospheres in dielectrics, *J. Appl. Phys.* **1953**, *24*, 438–450.

[69] Webers, V.J. Measurement of triboelectric position, *J. Appl. Polym. Sci.* **1963**, *7*, 1317–1323.

[70] Gibson, H.W. Control of electrical properties of polymers by chemical modification, *Polymer* **1984**, *25*, 3–27.

[71] Harper, W.R. *Contact and frictional electrification*; Oxford University Press: London; 1967; p 352.

[72] Duke, C.B.; Fabish, T.J. Contact electrification of polymers: A quantitative model, *J. Appl. Phys.* **1978**, *49*, 315–321.

[73] Duke, C.B.; Fabish, T.J. Charge-induced relaxation in polymers, *Phys. Rev. Lett.* **1976**, *37*, 1075–1078.

[74] Fabish, T.J.; Duke, C.B., Molecular charge states and contact charge exchange in polymers, *J. Appl. Phys.* **1977**, *48*, 4256–4266.

18 Forces of Attraction Between Particles

Flying south of the city of New Orleans, one can see the dark-brown, silt-laden Mississippi River emptying into the Gulf of Mexico. The locals call the river "the big muddy." The dark-brown plume abruptly disappears not far from shore leaving just the deep blue of the open sea. At the interface between the fresh river and the salty sea, the suspended silt flocculates and settles quickly out of sight.[1]

Dispersions flocculate because attractive forces hold particles together when they collide. Dispersions remain deflocculated if particles do not stick when they collide. The forces between dispersed particles can be altered drastically: The addition of electrolyte to a stable dispersion can eliminate the repulsive interactions and cause the dispersion to flocculate rapidly; addition of small concentrations of stabilizing agents can produce large repulsive interactions.

The phenomenological and mathematical descriptions of dispersion stability depend on understanding the nature of the forces of attraction between particles, called dispersion-force attraction,* and repulsions between particles, either electrostatic or steric. Electrostatic repulsion is caused by particles having electric charges of the same sign. Steric repulsion is by caused the compression of adsorbed polymer layers. This chapter treats the attractive forces between particles by first considering the attractive forces between molecules.

18.1 THE FORCES OF ATTRACTION BETWEEN MOLECULES

Attractive forces between molecules are well understood. Chemical bonds can form when the molecules are within atomic distances. Lewis acid–base attractions, weaker than chemical bonds, can also form when molecules are within atomic distances. Molecules still interact even when they are beyond bond-forming distances. Ionic molecules will interact by Coulomb forces. Molecules

---

* An unfortunate coincidence brings two unrelated meanings of the term "dispersion" into juxtaposition in this particular subject. In one usage, a dispersion is a stable suspension of a finely divided phase in a phase in which it is immiscible. London's theory of intermolecular forces, based on the interaction of electric fields, makes use of "the dispersion relation" between wavelength and frequency. Because of this feature of his model, London referred to intermolecular forces as "dispersion forces." Another example of two unrelated meanings occurs with the term "phase," which denotes both a physical state of matter and a property of a wave.

with electric dipoles also attract each other. And all molecules attract each other by London dispersion forces.

The London dispersion-force attraction between molecules first appeared as an empirical constant, a, a measure of the cohesion between molecules, in van der Waals' equation of state for nonideal gases[2]

$$\left(p + \frac{a}{V_m^2}\right)(V_m - b) = RT \tag{18.1}$$

where p is the pressure, V_m is the molar volume, b is the excluded volume of the molecules, R is the gas constant, and T is the absolute temperature. The ratio, a/V_m^2, implies that the intermolecular attractive energy varies inversely as the sixth power of the intermolecular distance. This is a longer range interaction than molecular bonds or acid–base interactions.

The empirical Lennard–Jones potential, $\phi(r)$, between two molecules also includes this long-range attraction (the negative inverse sixth term) and a shorter range repulsion (the positive inverse twelfth term) due to the repulsion of interpenetrating electronic clouds at short distances[3]

$$\phi(r) = 4\varepsilon\left[\left(\frac{\sigma}{r}\right)^{12} - \left(\frac{\sigma}{r}\right)^{6}\right] \tag{18.2}$$

where ε is the potential energy of maximum attraction, and σ is the distance of neutral approach, both empirical constants. Equations (18.1) and (18.2) are empirical equations.

A theoretical explanation for this long-range attractive force between molecules was given by Fritz London (in English[4]). The full derivation relies on quantum mechanics; but London also provided a simplified model. The electrons and nuclei of all molecules are in constant motion. At any instant a molecule has an uneven distribution of electric charges, a fluctuating dipole, which creates an electric field. The motion of the electrons and nuclei of an adjacent molecule will be influenced by this electric field so as to create another fluctuating dipole. The interaction between two induced dipoles creates a net attraction. The magnitude of the interaction depends on three factors: the distance between the molecules, their polarizabilities, and their ionization potential. The London dispersion attraction between two molecules is[5]

$$\phi_{12}^d(r_{12}) = -\frac{3}{2}\left(\frac{h\nu_1 h\nu_2}{h\nu_1 + h\nu_2}\right)\frac{\alpha_1\alpha_2}{r_{12}^6} = -\frac{\Lambda_{12}}{r_{12}^6} \tag{18.3}$$

where $h\nu$ is the characteristic energy of a molecule approximately equal to the ionization energy, α is the polarizability, and r_{12} is the intermolecular distance. The London constant, Λ_{12}, refers to the interaction of two different materials as indicated by the subscripts $_1$ and $_2$. This is called the heterointeraction. When the two materials are the same, the London constant is Λ_{11}. This is called the homointeraction.

The London constant for the heterointeraction can be approximated from the homointeractions by the relation

$$\Lambda_{12} \simeq (\Lambda_{11}\Lambda_{22})^{1/2} \qquad (18.4)$$

This approximation is best when the characteristic energies are not too different.

The polarizability of a molecule is a familiar quantity. The more polarizable the molecule, the more loosely are electrons held to the nucleus. The electrons in hydrocarbons are held tightly in localized orbitals between the carbon and hydrogen atoms, hence hydrocarbons have small polarizabilities. Consequently, they have small dispersion forces of attraction. Aromatic molecules have delocalized electron clouds and so are more polarizable and have stronger dispersion forces of attraction. Metals and graphite have free electrons and so are highly polarizable and have the largest forces of attraction. Molecular energies and polarizabilities are available in tables.[6,7]

18.2 DISPERSION FORCES OF ATTRACTION BETWEEN PARTICLES

The possible forces of attraction between particles are similar to those discussed above between molecules: electrostatic, dipolar, and dispersion.

Suspended particles and droplets generally have a net electrical charge, especially when dispersed in water, where the charge is caused either by the dissociation of surface ions or by the adsorption of ions from solution. Electric charges on particles give rise to electrostatic forces between particles. The magnitude of the electrostatic forces cannot be simply calculated by Coulomb's equation. Each charged particle is surrounded by its own counterions and multibody interactions are important. The electrostatic forces between charged particles are discussed in Chapter 19.

Dipole–dipole forces between particles are not significant. A particle may be composed of molecules having dipoles, but in the formation of the particle, these molecules would line up in such a way that the total dipole moment is zero.

The second significant force between particles arises from the same spontaneous electron fluctuations (induced dipoles) that give rise to long-range attractive forces between molecules. Even though a particle does not have significant permanent dipole moment, it can have significant induced-dipole interactions with another particle. The fluctuating electron clouds in a particle are able to induce an instantaneous dipole in a particle close by. These instantaneous dipoles create an attractive force.

The third significant force between particles arises from the interaction of adsorbed polymer layers. This interaction is called "steric" and is discussed in Chapter 21.

18.3 THEORETICAL APPROACHES

Long ago, Clerk Maxwell pointed out two approaches to any scientific problem.[8] He wrote that mathematicians usually begin by considering a single molecule, or unit charge, and then conceive its relation to another molecule or unit charge. But the conception of an elemental unit is an abstraction because our perceptions are related to extended bodies. Is there not, he asked, an alternative and less abstract method in which we begin with the whole rather than building it up from the parts? He went on to illustrate the application of a holistic, or molar, approach to electromagnetism by means of field theory.*

Similarly, our understanding of the dispersion-force attraction between two particles has taken two paths. In 1937 Hamaker[9] proposed a molecular model to calculate the dispersion force attraction between particles. In 1956 Lifshitz[10] proposed a molar approach to the attractive forces between particles. The Hamaker model is more familiar and simpler. The Lifshitz model is more complex but more reliable.

In Hamaker's molecular model the dispersion-force attraction between particles is calculated by summing the dispersion-force attraction between all pairs of molecules in the separate particles. Corrections were introduced to account for such factors as third-body perturbations, the effects of intervening material, and attenuation (called retardation) of induced-dipole interactions at longer distances as electronic fluctuations fall out of phase. The data required for the calculation are the ionization potentials and polarizabilities of the molecules. This molecular approximation is roughly equivalent to predicting the spectrum of a particle as the sum of the molecular spectra of its constituents.

In Lifshitz's molar model the dispersion-force attraction between particles is calculated from the coordination of the instantaneous electron fluctuations of a particle with those of a nearby particle by means of quantum electrodynamics. The data required for calculations are the dielectric constants of the particles as a function of frequency. Parsegian and Ninham[11] developed a numerical method that uses readily measured optical spectra to make this theory calculable.

18.3.a Molecular Approach: Hamaker Theory

Hamaker[9] took London's expression for the dispersion attraction between two isolated molecules, Eq. (18.3), and summed it for all combinations of molecules between two discrete particles. The work done to bring the two particles from infinity to a given separation distance is the Gibbs free energy. When the distance of separation is more than a few molecular diameters, the summation can be

---

* Here is Francis Bacon, even earlier (1620), on the same theme: "For that school [Leucippus, Democritus] is so busied with the particles [i.e., molecules] that it hardly attends to the structure; while the others are so lost in admiration [i.e., wonder] of the structure that they do not penetrate to the simplicity of nature. These kinds of contemplation should therefore be alternated and taken by turns; that so the understanding be rendered at once penetrating and comprehensive." *Novum Organum*, Aphorism LVII.

replaced by an integral. The dispersion-force contribution to the free energy between particles 1 and 2 is

$$\Delta G_{12}^d = -\pi^2 \rho_1 \rho_2 \Lambda_{12} \oint_{V_1} dV_1 \oint_{V_2} \frac{dV_2}{\pi^2 r_{12}^6} \tag{18.5}$$

where dV_1 and dV_2 are differential volume elements of V_1 and V_2, and ρ_1 and ρ_2 are the molecular number densities (molecules per unit volume.) By convention, the material-dependent term in front of the integral is called the Hamaker constant, A_{12},

$$A_{12} = \pi^2 \rho_1 \rho_2 \Lambda_{12} \tag{18.6}$$

The integral depends on the shape of the particles and their distance apart. Therefore Hamaker theory provides equations that depend on three terms: (1) the material-dependent Hamaker constant, A_{12}, (2) the interparticle distance, and (3) a geometric shape factor. These constants for many industrially important materials can be found in the literature. A internet search with the key words "Hamaker constant" and the name of the material is often enough to locate the necessary quantity.

18.3.a(i)   The Effects of Intervening Substances   We are primarily interested in the dispersion-force attraction between two particles or drops dispersed in a liquid. The fundamental Hamaker expression, Eq. (18.6), does not provide for the effects of this intervening medium, since the London constants, Λ_{12}, are calculated for the interaction of two molecules *in a vacuum*.

The effect of an intervening substance, 2, between two bodies of composition 1 and 3, is calculated by the principle of Archimedean buoyancy, for which the Hamaker constant is

$$A_{123} = A_{13} + A_{22} - A_{12} - A_{23} \tag{18.7}$$

The case of most interest is the interaction of two particles of material 1 dispersed in a liquid of material 2. For this the Hamaker constant is given by

$$A_{121} = A_{11} + A_{22} - 2A_{12} \tag{18.8}$$

If the root-mean-square approximation for London coefficients,

$$\Lambda_{12} \simeq (\Lambda_{11} \Lambda_{22})^{1/2} \tag{18.4}$$

is used in the expression for the Hamaker constant and the density of the two materials is about the same then

$$A_{12} \simeq (A_{11} A_{22})^{1/2} \tag{18.9}$$

Substituting Eq. (18.9) into Eq. (18.8) gives

$$A_{121} \simeq \left(A_{11}^{1/2} - A_{22}^{1/2}\right)^2 \qquad (18.10)$$

as a useful approximation. The Hamaker constant, A_{121}, is always positive. In Eq. (18.5) if the material-dependent term is positive, then the free-energy expression is negative; hence, particles composed of the same substance always attract each other regardless of the dispersion medium. This result is called the "De Boer-Hamaker" theorem.

Analogously, the interaction for two different materials across a liquid is obtained from Eq. (18.7) and (18.9),

$$A_{123} \simeq \left(A_{11}^{1/2} - A_{22}^{1/2}\right)\left(A_{33}^{1/2} - A_{22}^{1/2}\right) \qquad (18.11)$$

If the value of A_{22} is intermediate between the values of A_{11} and A_{33}, then the Hamaker constant is negative; hence, the free energy of interaction is positive and the particles of substances 1 and 3 repel each other through the medium 2. The occurrence of a negative A_{123} is usually accompanied by sensible physical phenomena, for example, a detachment or elution of particles, cells, or macromolecules from a substrate.[12]

18.3.a(ii) Values of Hamaker Constants for Some Common Materials

Table 18.1 gives values of the Hamaker constants for some common materials interacting through a vacuum or air. Note that conductors have the strongest interactions and insulators have the least. Conductors have free electrons and hence higher polarizabilities than insulators, which have no free electrons and low polarizabilities. Small metal particles have free electrons, so are highly polarizable and hence have strong dispersion forces of attraction.

Table 18.2 gives values of the Hamaker constants, A_{121}, for some common materials where the medium is water. These values are useful for understanding the stability of aqueous dispersions.

Table 18.3 gives values of the Hamaker constants, A_{123}, for dissimilar particles dispersed in water. The negative values in Table 18.3 indicate that the two particles repel each other. The repulsive interaction implies that the two phases will not be drawn into intimate contact; a thickness of medium is maintained between them. When one of the phases is air, a negative Lifshitz–Hamaker constant implies that the medium spreads on the dispersed phase,[13] for example, the system octane-water-air has a negative Hamaker constant of -0.24×10^{-20} J, which implies that a film of water intervenes between air and octane. Clearly there is an intimate connection between a negative Hamaker constant and a positive spreading coefficient, but only when the interactions are limited to dispersion forces.

18.3.a(iii) The Hamaker Equation for Different Geometries

The Hamaker integral, Eq. (18.5), has been evaluated for many geometries.[19] We report here

TABLE 18.1 Lifshitz and Hamaker Constants for Interaction Across a Vacuum or Air

Substance	Reference	A_{11} (10^{-20} J)
Graphite	17	47.0
Gold	14,17,15	45.3, 45.5, 37.6
Silicon carbide	16	44
Rutile (TiO$_2$)	16	43
Silver	14,17	39.8, 40.0
Germanium	14,17	29.9, 30.0
Chromiun	15	29.2
Copper	17	28.4
Diamond	17	28.4
Zirconia (n-ZrO$_2$)	16	27
Silicon	14,17	25.5, 25.6
Metals (Au, Ag, Cu)	16	25–40
Iron oxide (Fe$_3$O$_4$)	16	21
Selenium	14,17	16.2, 16.2
Aluminum	14,16,17	15.4, 14, 15.5
Cadmium sulfide	17	15.3
Tellurium	17	14.0
Polyvinyl chloride	18	10.82
Magnesia	14,17	10.5, 10.6
Polyisobutylene	18	10.10
Mica	16,15	10, 10.8
Polyethylene	17	10.0
Polystyrene	18,14,16	9.80, 6.57, 6.5, 6.4
	17,15	7.81
Polyvinyl acetate	18	8.91
Polyvinyl alcohol	17	8.84
Natural rubber	18	8.58
Polybutadiene	18	8.20
Polybutene-1	18	8.03
Quartz	15	7.93
Polyethylene oxide	18	7.51
Polyvinyl chloride	16	7.5
Hydrocarbon (crystal)	16	7.1
CaF$_2$	16	7
Potassium bromide	17	6.7
Hexadecane	18	6.31
Fused quartz	16	6.3
Polymethylmethacrylate	17	6.3
Polydimethylsiloxane	18	6.27
Potassium chloride	17	6.2
Chlorobenzene	18	5.89
Dodecane	18,16	5.84, 5.0
Decane	18	5.45
Toluene	18	5.40

TABLE 18.1 (*Continued*)

Substance	Reference	A_{11} (10^{-20} J)
1,4-Dioxane	18	5.26
n-Hexadecane	16	5.1
Octane	18,16	5.02, 4.5
Benzene	16	5.0
n-Tetradecane	16	5.0
Cyclohexane	18,16	4.82, 5.2
Carbon tetrachloride	18,16	4.78, 5.5
Methyl ethyl ketone	18	4.53
Water	14,16,17	4.35, 3.7, 4.38
Hexane	18	4.32
Diethyl ether	18	4.30
Acetone	18,16	4.20, 4.1
Ethanol	16	4.2
Ethyl acetate	18	4.17
Polypropylene oxide	18	3.95
Pentane	18,16	3.94, 3.8
PTFE	16	3.8
Liquid He	16	0.057

the three of most general interest: parallel plates, equal-sized spheres, and a sphere with a slab. Each geometry produces a different algebraic form for the energy but all depend explicitly on the Hamaker constant and the interparticle distance.

18.3.a(iii)1 Planar Parallel Slabs The Hamaker equation for two thick, planar, parallel slabs separated by a distance H is

$$\Delta G^d = -\frac{A}{12\pi H^2} \quad (18.12)$$

where ΔG^d is the *free energy per unit area* and A is the Hamaker constant which could be A_{11}, A_{12}, A_{121}, or A_{123} depending on conditions. The dependence of the dispersion force on the materials is reflected by the choice of Hamaker constant and on the distance between the plates, H.

18.3.a(iii)2 Equal Spheres The Hamaker equation for two equal spheres of radii a, with centers a distance R apart is

$$\Delta G^d = -\frac{A}{6}\left[\frac{2a^2}{R^2 - 4a^2} + \frac{2a^2}{R^2} + \ln\left(1 - \frac{4a^2}{R^2}\right)\right] \quad (18.13)$$

where ΔG^d is *the total free energy (not* per unit area) and the Hamaker constant could be A_{11}, A_{12}, A_{121}, or A_{123} depending on conditions. When two equal

360 FORCES OF ATTRACTION BETWEEN PARTICLES

TABLE 18.2 Lifshitz and Hamaker Constants for Interaction Across Water

Substance	Reference	A_{121} ($\times 10^{-20}$ J)
Gold	14,17,15	33.5, 33.4, 27
Silver	14,17	26.6, 28.2
Titania (Rutile)	16	26
Chromium	15	18.8
Copper	17	17.5
Germanium	14,17	16.0, 17.7
Diamond	17	14.0
Zirconia	16	13
Silicon	14,17	11.9, 13.4
Tellurium	17	5.38
Sapphire	32	5.32
Cadmium sulfide	17	4.85
Selenium	14,17	4.77, 5.82
Alumina	14,17	4.12, 4.17
Air	16	3.70
Calcite	32	2.23
Mica	16,15	2.0, 2.33
Crystalline quartz	32	1.70
Magnesia	14,17	1.60, 1.76
Polyvinyl chloride	32	1.30
Quartz	15	1.08
Polymethyl methacrylate	32	1.05
Calcium fluoride	32	1.04
Polystyrene	32,14,17,15	0.931, 0.27, 0.35, 1.01
Fused silica	32	0.849
Fused quartz	32,16	0.833, 0.63
Polyisoprene	32	0.743
Potassium bromide	17	0.54
Polyvinyl alcohol	17	0.54
Hexadecane	32	0.540
Pentadecane	32	0.526
Tetradecane	32	0.514
Tridecane	32	0.504
Dodecane	32,16	0.502, 0.50
Hexadecane	16	0.50
Undecane	32	0.471
Decane	32	0.462
Nonane	32	0.435
Octane	32,16	0.410, 0.41
Polyethylene	17	0.40
Heptane	32	0.386
"Teflon FEP"	32	0.381
Hexane	32	0.360
Pentane	32	0.336
Polytetrafluoroethylene	32,16	0.333, 0.29
Potassium chloride	17	0.31

TABLE 18.3 Lifshitz and Hamaker Constants for Interaction of Two Different Materials Across a Third

Substance	Reference	A_{123} (10^{-20} J)
Gold : Water : Chromium	15	21.0
Selenium : Water : Gold	14	12.2
Selenium : Water : Silicon	14	7.72
Gold : Water : Quartz	15	3.2
Selenium : Water : Magnesia	14	2.83
Polystyrene : Water : Gold	14,15	2.98, 3.2
Polystyrene : Water : Silver	14	2.67
Polystyrene : Water : Selenium	14	1.15
Water : Octane : Air	16	0.51
Water : Pentane : Air	16	0.08
Octane : Water : Air	16	−0.24
Fused quartz : Tetradecane : Air	16	−0.4
Fused quartz : Octane : Air	16	−0.7
Fused quartz : Water : Air	16	−0.87
Quartz : Water : Vacuum	15	−1.85
Polystyrene : Water : Vacuum	15	−1.85
Mica : Water : Vacuum	15	−2.76
Gold : Water : Vacuum	15	−9.19

spheres approach each other closely, Eq. (18.13) reduces to

$$\Delta G^d = -\frac{Aa}{12H} \quad (18.14)$$

$$H = R - 2a \quad (18.15)$$

where H is the distance between the surfaces of the spheres. Equation (18.14) is the most commonly used equation to estimate the strength of the dispersion-force attraction between suspended particles. Equation (18.13) is used for more accuracy.

For two equal spheres far apart, Eq. (18.13) reduces to

$$\Delta G^d = -\frac{16Aa^6}{9R^6} \quad (18.16)$$

One important implication of these equations is that the dispersion attraction between particles is shown to be significant at distances of the order of their own dimensions.*

---

* The attractive interactions between molecules were known to decrease quickly with intermolecular distances. The Lennard–Jones potential used an inverse sixth power of distance. Data from the rates of flocculation of particles indicated that interparticle attractive forces did not decrease so rapidly with distance and therefore might be a consequence of a new kind of interaction. Hamaker's contribution was to explain the long-range interaction between particles in suspension as just a consequence of molecular interactions and not a consequence of any new kind of interaction.

18.3.a(iii)3 A Sphere and a Thick Slab The Hamaker equation for a sphere of radius a whose surface is a distance H from a slab is

$$\Delta G^d = -\frac{A}{6}\left[\frac{a}{H} + \frac{a}{2a+H} + \ln\frac{H}{2a+H}\right] \qquad (18.17)$$

where ΔG^d is the total free energy change and A has the same options as before. If the sphere is very close to the slab, then $H \ll a$ and

$$\Delta G^d = -\frac{Aa}{6H} \qquad (18.18)$$

which is twice the sphere–sphere interaction given by Eq. (18.14)

These equations imply that in general small particles are attracted to walls and to larger particles more than to each other in the absence of dispersing agents.

18.3.a(iii)4 Other Geometries These integrations are readily extended to multilayer structures.[20] The interaction energies for other shapes are orientation dependent and hence less general. Of notable interest, however, is the work of Marjorie Vold,[21] in which she shows that the preferred orientation of two prolate spheroids is end to end, a particularly useful result to describe the flocculation of certain lubricating greases (and possibly the interactions of micelles). She also derived the expression for the Hamaker constant for adsorbed layers on colloidal particles[22] and for the interaction of various anisometric particles.[23] Also useful is the analysis of the interaction of a cone and a flat plate (a model of surface roughness) by Sparnaay,[24] who concludes that the effects of irregularities on a flat surface begin to become significant when their linear dimensions are on the order of 10–20% of the interparticle distance.

18.3.a(iv) Dispersion-Force Contributions to Interfacial Tensions The work required to divide a material into two separate free surfaces from an initial intermolecular distance of r_1 to an infinite separation, if the only attraction is the dispersion force, is

$$W^d_{11} = -\Delta G^d_{11} = \frac{A_{11}}{12\pi r_1^2} \qquad (18.19)$$

The work so characterized is the dispersion work of cohesion, equal to twice the surface tension. That is,

$$W^d_{11} = 2\sigma^d \qquad (18.20)$$

where σ^d is the theoretical surface tension due to dispersion energies alone. Hence from Eqs. (18.19) and (18.20),

$$\sigma^d = \frac{A_{11}}{24\pi r_1^2} \qquad (18.21)$$

Equation (18.21) was originally derived by Fowkes[25] and used to calculate the dispersion-force contributions to surface tension from Hamaker constants and vice versa. Similarly, if the only force of attraction between two materials is dispersion force, then the work to separate the two phases is given by

$$W_{12}^d = -\Delta G_{12}^d = \frac{A_{12}}{12\pi r_{12}^2} \qquad (18.22)$$

If the London coefficient Λ_{12} is approximated by the root mean square of the individual London coefficients, Λ_1 and Λ_2(Berthelot's principle), and the intermolecular distance r_{12} is approximated by the root-mean-square of the individual intermolecular distances,

$$r_{12} = (r_1 r_2)^{1/2} \quad \text{and} \quad A_{12} = (A_{11} A_{22})^{1/2} \qquad (18.23)$$

then

$$W_{12}^d = \frac{(A_{11} A_{22})^{1/2}}{12\pi r_1 r_2} \qquad (18.24)$$

and

$$W_{12}^d = 2 \left(\sigma_1^d \sigma_2^d\right)^{1/2} \qquad (18.25)$$

Equation (18.25) is a key element in Fowkes' treatment of dispersion force contributions to the work of adhesion and shows the rationale for the use of the root-mean-square approximation to calculate interfacial tensions (see Section 8.2).

18.3.a(v) The Effect of Retardation Dispersion-force attractions arise from the fluctuations of the electrons in one particle creating an electric field that induces electron fluctuations in a second particle. The electron fluctuations are attractive when they are in phase. To be in phase, the length of time for the electric field to propagate from one particle to another must be small compared to the period of the fluctuation.[26] The propagation of electric induction between two particles moves at the speed of light. A typical frequency of electronic fluctuation is about 3×10^{15} Hz. At the speed of light, this corresponds to a wavelength of about 25 nm. Therefore at distances greater than about a quarter of this, 5 nm, the fluctuations are correlated less and the energy of attraction reduced.

The retardation correction of the dispersion-force interaction for two molecules[27] was integrated numerically by Overbeek to obtain the interaction of two flat plates[28] and presented in the form of a table. In the limit of a fully retarded interaction between two flat plates,

$$U^r(r) = \frac{0.49c}{\pi v_0 H} U(r) \qquad (18.26)$$

where $U(r)$ is the unretarded energy, $U^r(r)$ is the retarded energy, c is the speed of light, v_0 is the natural frequency of interaction, and H is the distance of separation. Equation (18.26) shows that the dispersion force attractions fall more rapidly with distance than predicted by the calculations that neglect retardation.

The correction factor for the interaction of two spheres is readily obtained from a diagram given by Fowkes and Pugh,[29] which was calculated from the exact analytic expression derived by Clayfield, Lumb and Mackey.[30]

18.3.a(vi)   A Criticism of the Molecular Approach   The essence of the molecular approach to calculate the energy of interaction between two particles is to add up all possible molecular interactions. The molecular interactions are based on the coordinated motions of electrons in molecules in the separate particles. A strong criticism is that the interactions of molecules in separate particles must be far less that the interaction of molecules in the same particle, which is ignored. Therefore, the entire Hamaker calculation is suspect. Fortunately, the calculations agree reasonably with experiments so the approach is used to this day, but clearly a theory that takes into account all the molecular interactions simultaneously is preferred. This is called the molar, as opposed to molecular, approach.

18.3.b   Molar Approach: Lifshitz Theory

The molecular approach starts from molecular-pair interactions and works up to the attraction between particles, making various simplifying assumptions on the way. An alternative approach is to start with the electrical properties of a particle and derive attractive potentials directly. The frequency dependence of the dielectric constant of a solid or liquid is related to the fluctuation of the electronic clouds of the whole material. Starting from that premise, Lifshitz and coworkers[10,31] calculated the dispersion attractions between particles through a medium using Feynman diagrams and quantum electrodynamics. The details of that derivation are beyond the scope of this book, but the final formulas for the free energies of interaction for a few important configurations are worth quoting. By means of these equations, the attraction between particles may be calculated from the dielectric responses of the materials at frequencies close to the peaks of their absorption spectra.

In the Hamaker model the free energy of interaction separates into a material-dependent constant (called the Hamaker constant) and a geometry-dependent integral. Lifshitz theory, on the other hand, gives such a separation of terms only for the special case of the interaction between two half-spaces (parallel plates) when retardation is not significant. The reliability of Lifshitz theory is so much better than Hamaker theory that the Lifshitz constant for parallel plates is usually substituted for the Hamaker constants for that and all other geometries [Eqs. (18.12) to (18.24)].

The mathematics of Lifshitz theory will be daunting to most readers (as it is to the present authors) but if only a few materials are of great interest, then an investment of time to do the calculations for those few materials is time well spent. The necessary calculations for many common materials have been published. The data needed for the calculation are the optical spectra of the materials, particularly in the UV and visible ranges, usually only including the strongest absorption peak or two. The calculation itself is a double summation [Eq. (18.28)].

18.3.b(i) Two Half-Spaces, Nonretarded Interaction

Lifshitz theory depends on knowing dielectric spectra. Optical spectra are more familiar to chemists. Optical spectra are absorption coefficients at different wavelengths; dielectric spectra, which may be obtained from the optical spectra by the Krammers–Kronnig relation, are dielectric "constants" at different frequencies, ω. The free energy of interaction per unit area for two half-spaces with dielectric responses $\varepsilon_1(\omega)$ and $\varepsilon_3(\omega)$ separated by a material of dielectric response $\varepsilon_2(\omega)$ of thickness H (for the nonretarded approximation) is[32]

$$\Delta G_{123}(H) = -\frac{A_{123}}{12\pi H^2} \tag{18.27}$$

where

$$A_{123} = \frac{3kT}{2} \sum_{n=0}^{\infty}{}' \sum_{m=1}^{\infty} \frac{(\Delta_{12}\Delta_{32})^m}{m^3} \tag{18.28}$$

$$\Delta_{12} = \frac{\varepsilon_1(i\xi_n) - \varepsilon_2(i\xi_n)}{\varepsilon_1(i\xi_n) + \varepsilon_2(i\xi_n)} \quad \text{and} \quad \Delta_{32} = \frac{\varepsilon_3(i\xi_n) - \varepsilon_2(i\xi_n)}{\varepsilon_3(i\xi_n) + \varepsilon_2(i\xi_n)} \tag{18.29}$$

$$\xi_n = n\frac{4\pi^2 kT}{h} \tag{18.30}$$

where k is the Boltzmann constant, T is the absolute temperature, h is Planck's constant, and the prime on the summation indicates that the $n = 0$ term is given half-weight. The quantity ξ_n is a frequency. At 21°C, ξ_1 is 2.4×10^{14} rad/s, a frequency corresponding to a wavelength of light of about 1.2 µm. As n increases, the value of ξ_n increases and the corresponding wavelength decreases, hence ξ_n takes on more values in the ultraviolet than in the infrared or visible. The functions Δ_{12} and Δ_{32} are the ratios of the difference and sum of dielectric constants at the frequencies, ξ_n. The Lifshitz constant is a double sum of products of Δ_{qr}'s.

Until recently it was believed that the Lifshitz theory was merely an elegant formalism and the function $\varepsilon(i\xi)$ could not be readily determined. A fundamental achievement of Ninham and Parsegian was to show how to construct the function $\varepsilon(i\xi)$ from available experimental data.

18.3.b(ii) Method of Ninham and Parsegian

The method of Ninham and Parsegian[20,32] is to approximate the absorption spectrum with an infrared peak of zero width and infinite height to characterize the low-frequency absorption spectra and an ultraviolet (or visible) peak again with zero width and infinite height to characterize the high-frequency absorption spectra. The frequency dependence of the dielectric susceptibility can be expressed, therefore, in the terms of Lorentz harmonic oscillators for each absorption peak (infrared, visible or ultraviolet)

$$\varepsilon(i\xi_n) = 1 + \frac{C_{ir}}{1 + \xi_n^2/\omega_{ir}^2} + \frac{C_{uv}}{1 + \xi_n^2/\omega_{uv}^2} \quad (18.31)$$

and

$$D = 1 + C_{ir} + C_{uv} \quad (18.32)$$

where D is the static dielectric constant (zero frequency), and ω_{ir}, ω_{uv} are the frequencies of absorption, and C_{ir} and C_{uv} are dielectric constants to be determined. If more than one absorption in the visible or ultraviolet is significant, additional terms can be added. If absorption in the microwave is significant (as for water), it is approximated by Debye rotational relaxation.

Values of the dielectric constants are determined as follows. The real part of the dielectric response changes only near an absorption peak and is constant between them. Hence for each absorption peak j,

$$C_j = \varepsilon_j^b - \varepsilon_j^a \quad (18.33)$$

where

ε_j^b = the dielectric constant at frequencies just less than that of the absorption peak,

ε_j^a = the dielectric constant at frequencies just greater than that of the absorption peak.

The experimental procedure is to measure the absorption spectrum from the ultraviolet through the infrared (or microwave if necessary) and the dielectric constant (square of the refractive index) at convenient frequencies between the absorption peaks. Using the dielectric constants on each side of the peaks, the quantities C_{ir} and C_{uv} are calculated and using the frequencies of the peaks, the quantities $\varepsilon(i\xi_n)$ are calculated for each value of n by means of Eq. (18.31). From the values of $\varepsilon(i\xi_n)$ the Δ_{qr} are calculated by means of Eq. (18.29) and from those quantities the Lifshitz constant is calculated by means of Eq. (18.28). Sufficient optical constants to make these calculations are listed for some common materials in Table 18.4. Constants for many more materials are gradually becoming available in the published literature. A literature search with the key word "Lifshitz" and the name of the material is often enough to locate the necessary optical constants or the Lifshitz–Hamaker constant itself.[32,33,34,35]

Because of the importance of water in emulsions and suspensions, more care has been taken to characterize its optical properties so as to have a better approximation for $\varepsilon(i\xi_n)$. Parsegian[37] gives the following:

$$\varepsilon(i\xi_n) = 1 + \frac{d}{1 + \xi_n \tau} + \sum_j \frac{f_j}{\omega_j^2 + \xi_n^2 + g_j \xi_n} \quad (18.34)$$

THEORETICAL APPROACHES

TABLE 18.4 Values of the Optical Constants,[32] C_{ir}, ω_{ir}, C_{uv}, and ω_{uv}

Substance	C_{uv}	ω_{uv} ($\times 10^{16}$ rad/s)	C_{ir}	ω_{ir} ($\times 10^{14}$ rad/s)
Water	0.755	1.899	See Table 18.2	
Crystalline quartz	1.359	2.032	1.93	2.093
Fused quartz	1.098	2.024	1.70	1.880
Fused silica	1.098	2.033	1.71	1.880
Calcite	1.516	1.897	5.7	2.691
Calcium fluoride	1.036	2.368	5.32	0.6279
Sapphire	2.071	2.017	8.5	1.880
Polymethylmethacrylate	1.189	1.915	1.2	5.540
Polyvinylchloride	1.333	1.815	0.9	5.540
Polystyrene	1.424	1.432	0.2	5.540
	1.447	1.354		
Polyisoprene	1.255	1.565	0.16	5.540
Polytetrafluoro-ethylene	0.846	1.793	0.25	2.270
Normal alkanes (number of carbons)				
5	0.819	1.877	0.025	5.540
6	0.864	1.873	0.026	5.540
7	0.898	1.870	0.025	5.540
8	0.925	1.863	0.023	5.540
9	0.947	1.864	0.025	5.540
10	0.965	1.873	0.026	5.540
11	0.979	1.853	0.026	5.540
12	0.991	1.877	0.023	5.540
13	1.002	1.852	0.025	5.540
14	1.011	1.846	0.025	5.540
15	1.019	1.845	0.025	5.540
16	1.026	1.848	0.025	5.540

TABLE 18.5 Optical Constants for Water,[a] $d = 74.8$, $1/t = 6.5 \times 10^{-5}$ eV

ω_j (eV)	f_j (eV)	g_j (eV)
2.07×10^{-2}	6.25×10^{-4}	1.5×10^{-2}
6.9×10^{-2}	3.5×10^{-3}	3.8×10^{-2}
9.2×10^{-2}	1.28×10^{-3}	2.8×10^{-2}
2×10^{-1}	5.69×10^{-3}	2.5×10^{-2}
4.2×10^{-1}	1.35×10^{-2}	5.6×10^{-2}
8.25	2.68	0.51
10	5.67	0.88
11.4	12	1.54
13	26.3	2.05
14.9	33.8	2.96
18.5	92.8	6.26

[a] The first five frequencies are in the infrared, the rest are in the ultraviolet.[37]

where the second term is for the microwave contribution, and the third term is the sum of the infrared and the ultraviolet contributions. The constants d, t, f_j, ω_j and g_j are given in Table 18.5.

An atomic-force microscope has been used successfully to measure the force between a single colloidal particle of rutile and a rutile crystal as a function of separation distance, pH, and electrolyte concentration.[38] An experimental nonretarded Lifshitz constant of $6 \pm 2 \times 10^{-20}$ J was determined at the pH of the isoelectric point of the TiO_2, 5.6. At the isoelectric point there is no electrostatic interaction between the two rutile surfaces. This value may be compared to a value of $7 \pm 1 \times 10^{-20}$ J calculated from the spectroscopic data given in Table 18.6 using Eq. (18.34). Further agreement between observation and theory is shown in Figure 18.1.

TABLE 18.6 Spectroscopic Data Used in the Calculation of the Nonretarded Lifshitz Constant for Rutile[38a]

Direction	$\varepsilon(0)$	ω_{ir} (rad/s)	C_{ir}	ω_{uv} (rad/s)	C_{uv}
Perpendicular	86	1×10^{14}	80	7.49×10^{15}	4.77
Parallel	170	1×10^{14}	163	7.24×10^{15}	6.01

[a]The parallel and perpendicular orientation refers to the direction of the applied electric field.

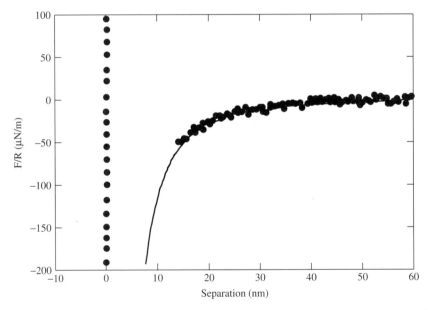

Figure 18.1 Force-separation data taken at the isoelectric point of rutile on the atomic-force microscope. The full curve is the theoretical interaction using the nonretarded Lifshitz constant.[38]

In biological systems the differences in the dielectric permittivities of the various components at ultraviolet frequencies are small, hence the interaction energies are more dependent on the infrared and microwave frequencies. In applications of Hamaker theory, only ultraviolet frequencies are used, hence biological systems should be analyzed by Lifshitz theory.

18.3.b(iii) Two Half-Spaces, Including Retardation Retardation of the dispersion forces is significant when the time required for fluctuating electric fields to propagate from one particle to another is comparable to the period of the fluctuation. When these effects are important, the Lifshitz "constant" is given by an expression that gradually reduces the contribution from the higher frequencies at longer distances.[32] This adds a computational burden but no fundamental difficulty to the calculations. Typically, the effect of retardation is small when the distance of separation is less than about 5–10 nm. For lipid–water systems of biological interest, the interactions are at low frequencies; that is, the microwave or infrared, hence retardation can be ignored.

18.3.b(iv) Lifshitz Theory for Dispersions in Ionic Solutions When the intervening medium between two half-spaces is an ionic solution, the dielectric properties of the solvent are altered. At room temperature $\xi_n = n \times 2.4 \times 10^{14}$ rad/s. Ions in solution cannot respond to any of these frequencies other than the $n = 0$ term, hence the only term in the calculation of the Lifshitz constant, Equation (18.28), that is different for ionic solutions is the first, $n = 0$, term.[39] If the ionic strength is high, the $n = 0$ term is completely screened and makes no contribution to the interaction free energy.

18.3.b(v) Lifshitz Theory for Other geometries The free energy of interaction for any geometry other than flat plates cannot be factored into a material-dependent term and a geometric term, as it can be in Hamaker theory. Exact, explicit, analytic expressions for other than planar geometries for Lifshitz theory are more difficult to derive than for Hamaker theory. Nevertheless, progress has been made in analyzing other significant geometries. Mitchell and Ninham[40] found simple expressions for the limiting cases of two spheres at close separation (no retardation) and two spheres at large distances (no retardation); the interaction of long thin rods, retarded and unretarded, was calculated by Mitchell, Ninham and Richmond[41]; Ninham and Parsegian[20] calculated the interactions for planar multilayer systems (coated particles).

A possible approximation for interactions other than planar geometries is a combination of the Hamaker and Lifshitz theories, by using Lifshitz "constants" in the Hamaker geometries. This approximation is best when the dielectric differences are small, that is, when the interactions are between similar materials, or when the distances of separation are small. Smith, Mitchell and Ninham.[42] give a careful analysis of the regimes in which Lifshitz theory deviates significantly from Hamaker theory and where this approximation is inappropriate.

18.4 OTHER METHODS TO OBTAIN HAMAKER CONSTANTS

18.4.a Direct Measurements of Forces of Attraction

Attempts have been made to measure experimentally the dispersion forces between macroscopic bodies. The most direct approach might seem to be to measure the work necessary to separate two surfaces; or, equivalently, the centrifugal force necessary to separate a particle from a solid surface. This approach is untenable, however, because clean surfaces in close contact form short-range chemical bonds and the energy needed to break these bonds can be significantly greater than dispersion energies. The successful approach, first reported by Derjaguin et al.[43] in 1954, is to measure the attractive force between carefully prepared surfaces as they are slowly brought together. Another technique measures the disjoining (disjunctive) pressure in a liquid film as the film thins. The early measurements were successful down to separation distances within 100 nm; this distance, however, is still sufficiently large that the dispersion forces are retarded. The results agree with the Lifshitz calculations for retarded interactions. Subsequent measurements, 1968–1972, were made with improved equipment (called Jacob's box) to enable approach to 1.4 nm where the dispersion forces are unretarded. These measurements confirm Lifshitz theory for unretarded interactions.[44,45] Derjaguin and coworkers in 1982, measuring the interactions of crossed glass fibers in ionic solutions, confirmed Lifshitz theory down to molecular distances.[46,47] All these measurements substantiate the theory.

18.4.b Indirect Measurements of the Forces of Attraction

Heats of vaporization or sublimation, adhesion, capillary phenomena, and so on, all depend on intermolecular forces. When the interactions are predominantly dispersion interactions, these phenomena can be used to determine Hamaker constants. The surface tension of a liquid can be thought of as half of the energy necessary to separate two liquid surfaces of unit interfacial area from intimate contact to infinity. Equation (18.21) can be rewritten in the form

$$A_{11} = 24\pi r_1^2 \sigma^d \qquad (18.35)$$

If this approximation is extended to the work of adhesion between two different materials at an interface, Eq. (18.22), then

$$A_{12} = 12\pi r_{12}^2 W_{12}^d \qquad (18.36)$$

where r_{12} is the equilibrium intermolecular distance between the two materials.

Any theory that predicts the dispersion force contribution to the surface tension from other physical properties can be used in conjunction with Eq. (18.35) to calculate the Hamaker constant for that material. A particularly interesting example is due to Croucher[48] who took the Davis and Scriven equation for surface free energy, the corresponding states relations of Prigogine, and Eq. (18.35) to derive a relation between the Hamaker constant and the bulk thermodynamic properties of a material:

$$A_{11} = \frac{3kT}{4\left(1 - bV^{-1/3}\right)} \qquad (18.37)$$

where

$$b = (m/n)^{1/(n-m)} \qquad (18.38)$$

m, n are the exponents of the interaction potential [e.g., $(m, n) = (6, 12)$ for the Lennard–Jones potential], and the reduced volume V is given implicitly by the relation

$$(\alpha T)^{-1} = -\frac{m}{3} + \frac{n-m}{3\left(V^{(n-m)/3} - 1\right)} + \frac{b}{3\left(V^{1/3} - b\right)} \qquad (18.39)$$

where α is the coefficient of thermal expansion. For the 6–∞ potential, $b = 1$ and

$$V = \left(\frac{3 + 7\alpha T}{3 + 6\alpha T}\right)^3 \qquad (18.40)$$

Therefore, from the coefficient of thermal expansion and an estimate of the form of the interaction pair potential (often taken as 6–12), the Hamaker constant can be calculated.

Any process that is strongly influenced by the long-range dispersion attractions can be a resource for the calculation of Hamaker constants. The coagulation rate of dispersions is such a process, where the rate of flocculation (which is measurable) depends, in part, on dispersion-force attractions. The Hamaker constant can be estimated from an analysis of the variation in the stability of a dispersion as electrolyte is added.[49] See also Section 20.3.

Another process influenced by long-range dispersion attractions is the spreading of a liquid film on a solid when the initial spreading coefficient is positive. The Hamaker constant can be estimated from ellipsometric measurements of the equilibrium film thickness (see Section 7.7).

REFERENCES

[1] See the cover of *Science* for 25 October 1985 for a photograph of the effect.

[2] Hirschfelder, J.O.; Curtis, C.F.; Bird, R.B. *Molecular theory of gases and liquids*; Wiley: New York; 1954; p 250.

[3] Hirschfelder loc.cit.; p 22.

[4] London, F. The general theory of molecular forces, *Trans. Faraday Soc.* **1937**, *38*, 8–26.

[5] Hirschfelder loc.cit.; p 30.

[6] Landolt–Börnstein: numerical data and functional relationships in science and technology; Springer Verlag: Heidelberg; 1997–2001.

[7] Gregory, J. The calculation of Hamaker constants, *Adv. Colloid Interface Sci.* **1969**, *2*, 396–417.

[8] Maxwell, J.C. *Electricity and magnetism*, in 2 vols.; Clarendon Press: Oxford; 1873; p 176.

[9] Hamaker, H.C. The London–van der Waals attraction between spherical particles, *Physica (Utrecht)* **1937**, *4*, 1058–1072.

[10] Lifshitz, E.M. The theory of molecular attractive forces between solids, *Sov. Phys–JETP* (Engl. Transl.) **1956**, *2*, 73–83.

[11] Parsegian, V.A.; Ninham, B.W. Application of the Lifshitz theory to the calculation of van der Waals forces across thin lipid films, *Nature* **1969**, *224*, 1197–1198.

[12] van Oss, C.J.; Visser, J.; Absolom, D.R.; Omenyi, S.N.; Neumann, A.W. The concept of negative Hamaker coefficients II. Thermodynamics, experimental evidence and applications, *Adv. Colloid Interface Sci.* **1983**, *81*, 133–148.

[13] Visser, J. The concept of negative Hamaker coefficients, *Adv. Colloid Interface Sci.*, **1981**, *15*, 157–169.

[14] Bargeman, D.; van Voorst Vader, F. van der Waals forces between immersed particles, *J. Electroanal. Chem. Interfacial Electrochem.* **1972**, *37*, 45–52.

[15] Rabinovich, Ya.I.; Churaev, N.V. Results of numerical calculations of dispersion forces for solids, liquid interlayers, and films, *Colloid J. USSR* **1990**, *52*, 256–262.

[16] Israelachvili, J.N. *Intermolecular and surface forces*, 2nd ed.; Academic Press: New York; 1991; pp 186–187.

[17] Visser, J. On Hamaker constants: a comparison between Hamaker constants and Lifshitz–van der Waals constants, *Adv. Colloid Interface Sci.* **1972**, *3*, 331–363.

[18] Croucher, M.D.; Hair, M.L. Hamaker constants and the principle of corresponding states, *J. Phys. Chem.* **1977**, *81*, 1631–1636.

[19] Mahanty, J.; Ninham, B.W. *Dispersion forces*; Academic Press: New York; 1976; pp 10–22.

[20] Ninham, B.W.; Parsegian, V.A. van der Waals interactions in multilayer systems, *J. Chem. Phys.* **1970**, *53*, 3398–3402.

[21] Vold, M.J. The van der Waals interaction of anisometric colloidal particles, *Proc. Indian Acad. Sci., Sect. A* **1957**, *46*, 152–166.

[22] Vold, M.J. The effect of adsorption on the van der Waals interaction of spherical colloidal particles, *J. Colloid Sci.* **1961**, *16*, 1–12.

[23] Vold, M.J. van der Waals' attraction between anisometric particles, *J. Colloid Sci.* **1954**, *9*, 451–459.

[24] Sparnaay, M.J. Four notes on van der Waals forces, *J. Colloid Interface Sci.* **1983**, *91*, 307–319.

[25] Fowkes, F.M. Attractive forces at interfaces, *Ind. Eng. Chem.* **1964**, *56*(12), 40–52; also *Chemistry and Physics of Interfaces*; Ross, S., Ed.; American Chemical Society: Washington, DC; 1965; pp 1–12.

[26] Langbein, D. *van der Waals attraction*; Springer: New York; 1974; pp 72–73.

[27] Casimir, H.B.G.; Polder, D. The influence of retardation on the London–van der Waals forces, *Phys. Rev.* **1948** (2), *73*, 360–372.

[28] Overbeek, J.Th.G. The interaction between colloidal particles, in *Colloid Science*; Kruyt, H.R., Ed.; Elsevier: New York; 1952; Vol. 1, pp 245–277.

[29] Fowkes, F.M.; Pugh, R.J. Steric and electrostatic contributions to the colloidal properties of nonaqueous dispersions, *ACS Symp. Ser.* **1984**, *240*, 331–354.

[30] Clayfield, E.J.; Lumb, E.C.; Mackey, P.H. Retarded dispersion forces in colloidal particles: Exact integration of the Casimir and Polder equation, *J. Colloid Interface Sci.* **1971**, *37*, 382–389.

[31] Dzyaloshinskii, I.E.; Lifshitz, E.M.; Pitaevskii, L.P. The general theory of van der Waals forces, *Adv. Phys.* **1961**, *10*, 165–209.

[32] Hough, D.B.; White, L.R. The calculation of Hamaker constants from Lifshitz theory with applications to wetting phenomena, *Adv. Colloid Interface Sci.* **1980**, *14*, 3–41.

[33] Chen, X.J.; Levi, A.C.; Tosatti, E. Hamaker constant calculations and surface melting of metals, *Surf. Sci.* **1991**, *251/252*, 641–644.

[34] Bergström, L.; Meurk, A.; Arwin, H.; Rowcliffe, D.J. Estimation of Hamaker constants of ceramic materials from optical data using Lifshitz theory, *J. Am. Ceram. Soc.* **1996**, *79*, 339–348.

[35] See www.deconvolution.com for computer programs.

[36] Meurk, A.; Luckham, P.F.; Bergström, L. Direct measurements of repulsive and attractive van der Waals forces between inorganic material, *Langmuir* **1997**, *13*, 3896–3899.

[37] Parsegian, V.A. Long-range van der Waals forces, *Phys. Chem.: Enriching Topics in Colloid Surf. Sci.* **1975**, 27–72.

[38] Larson, I.; Drummond, C.J.; Chan, D.Y.C.; Grieser, F. Direct force measurements between TiO_2 surfaces, *J. Am. Chem. Soc.* **1993**, *115*, 11885–11890.

[39] Mahanty loc.cit.; p 202.

[40] Mitchell, D.J.; Ninham, B.W. van der Waals forces between two spheres, *J. Chem. Phys.* **1972**, *56*, 1117–1126.

[41] Mitchell, D.J.; Ninham, B.W.; Richmond, P. van der Waals forces between cylinders. I. Nonretarded forces between thin isotropic rods and finite size corrections, *Biophys. J.* **1973**, *13*, 359–369.

[42] Smith, E.R.; Mitchell, D.J.; Ninham, B.W. Deviations of the van der Waals energy for two interacting spheres from the predictions of Hamaker theory, *J. Colloid Interface Sci.* **1973**, *45*, 55–68.

[43] Derjaguin, B.V.; Titijevskaia, A.S.; Abricossova, I.I.; Malkina, A.D. Investigations of the forces of interaction of surfaces in different media and their application to the problem of colloid stability, *Discuss. Faraday Soc.* **1954**, *18*, 24–41.

[44] Tabor, D.; Winterton, R.H.S. Surface forces: Direct measurement of normal and retarded van der Waals forces, *Nature (London)* **1968**, *219*, 1120–1121.

[45] Israelachvili, J.N.; Tabor, D. Measurement of van der Waals dispersion forces in the range 1.4 to 130 nm, *Nature (London), Physical Sci.* **1972**, *236*, 106.

[46] Rabinovich, Ya.I.; Derjaguin, B.V.; Churaev, N.V. Direct measurements of long-range surface forces in gas and liquid media, *Adv. Colloid Interface Sci.* **1982**, *16*, 63–78.

[47] Derjaguin, B.V.; Churaev, N.V.; Muller, V.M. *Surface forces*; Kisin, V.I., Trans.; Plenum: New York; 1987, pp 128–146.

[48] Croucher, M.D. A simple expression for the Hamaker constant of liquid-like materials, *Colloid Polym. Sci.* **1981**, *259*, 462–466.

[49] Verwey, E.J.W.; Overbeek, J.Th.G. *Theory of the stability of lyophobic colloids, the interaction of sol particles having an electric double layer*; Elsevier: New York; 1948.

19 Forces of Repulsion

After dispersed particles or drops collide by Brownian motion, interparticle forces determine whether they adhere or separate. Dispersion forces of attraction exist between all particles and drops and tend to make particles adhere. The magnitude of those forces is discussed in Chapter 18.

The forces of repulsion can occur in two different ways: by electrostatic charge or by polymeric layers, charged or uncharged.

Many suspended particles have an electrical charge as described in Chapter 17. If the particles have the same sign charge, the usual case, then they feel both the dispersion-force attraction and a force of electrostatic repulsion. The balance between those two forces determines whether the particles adhere or separate. In this chapter we estimate the magnitude of the electrostatic repulsion.

The surfaces of particles in many dispersions are covered with polymer layers, usually as a result of adsorption of polymer from solution. Polymers that are soluble in a solvent are spontaneously separated from each other by interaction with the solvent. That such polymers remain soluble necessarily implies that they do not cohere. This appears as a kind of repulsion between molecules. The particles do not adhere because the polymer molecules, even when adsorbed, do not cohere. This is called "steric" repulsion.

19.1 ELECTROSTATIC REPULSION BETWEEN FLAT PLATES

The electric potential around a suspended particle decreases exponentially to zero from the electric potential at the surface of the particle (see Section 17.1.e). The exponential decay constant is proportional to the square root of the ionic strength: the higher the ionic strength, the more steeply the electric potential drops. The more steeply the electric potential decays, the closer the particles approach, and the stronger is the dispersion-force attraction.

The electric potential between two particles decreases exponentially with distance at first, to reach a plateau (not necessarily zero) between the particles, and then increases exponentially to the potential of the second particle at its surface. (See Figure 19.1.) The exact relation depends on the electric potential of each particle, the ionic strength of the medium, and the distance between the particles.

The mathematics required is to solve the Poisson–Boltzmann equation with the boundary conditions of electric potentials at the two particle surfaces and

ELECTROSTATIC REPULSION BETWEEN FLAT PLATES

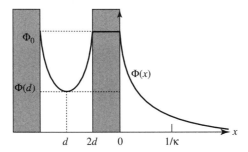

Figure 19.1 Schematic representation of the electric potential between two plates (left) in comparison with that of a single double layer (right).

the assumption of an electric potential plateau between the particles. The Poisson–Boltzmann equation for symmetric electrolytes and planar geometry is

$$\frac{d^2\Phi}{dx^2} = -\frac{\rho}{D\varepsilon_0} = -\frac{2zen_0}{D\varepsilon_0}\sinh\frac{ze\Phi}{kT} \quad (19.1)$$

where Φ is the electric potential, x is the distance from the surface, ρ is the space charge (electric charge per unit volume), D is the dielectric constant of the liquid, ε_0 is the permittivity of free space, z is the valence of the counter ion, e is the charge on the electron, n_0 is the concentration of counter ions at infinite distance, k is the Boltzmann constant, and T is the absolute temperature.

For two parallel plates separated by a distance, $2d$, the following boundary conditions must be met by the solution to the differential equation:

$$\text{at } x = d, \quad \Phi = \Phi_d \quad \text{and} \quad \frac{d\Phi}{dx} = 0$$
$$\text{at } x = 0, 2d, \quad \Phi = \Phi_0 \quad (19.2)$$

The solution for the electric potential at the midpoint between two flat plates of equal charge, Φ_0, in a 1-1 electrolyte of charge, z, and inverse Debye length, κ, is be represented by the integral[1]

$$-\kappa d = \frac{1}{\sqrt{2}} \int_{\Phi_0}^{\Phi_d} \frac{(ze/kT)\,d\Phi}{\left[\cosh(ze\Phi/kT) - \cosh(ze\Phi_d/kT)\right]^{1/2}} \quad (19.3)$$

This integral is an elliptical integral of the first kind for which tables are available; it can therefore be solved numerically for the electric potential at the midpoint, Φ_d, if surface potentials and the inverse Debye length are known.[2] Equation (19.3) is not very illuminating, as it is an implicit function of potential. What is sought is some explicit analytic expression that gives potential as a function of separation so that the force of repulsion can be calculated. To reach this goal the following approximation is usually used.

Langmuir[3] derived an expression for the repulsive force between two charged flat plates by a physicochemical approach. The increase in osmotic pressure as the ionic layers interpenetrate is a secondary effect of the approach of the two plates and so can be equated to the electrostatic repulsion, which is equally a secondary effect. Langmuir proposed that the net repulsion may be calculated either as electrostatic or osmotic in origin and the latter provided a simpler calculation.

The additional osmotic pressure, Π_d, created by excess ions at the electrostatic plateau is kT times the excess concentration of ions. For a 1-1 electrolyte,

$$\Pi_d = kT \left[n_0 \exp\left(\frac{ze\Phi_d}{kT}\right) + n_0 \exp\left(\frac{-ze\Phi_d}{kT}\right) - 2n_0 \right]$$
$$= 2n_0 kT \left(\cosh \frac{ze\Phi_d}{kT} - 1 \right) \tag{19.4}$$

where Φ_d is the electric potential at the plateau. Equation (19.4) is an expression for the *force* per unit area between the plates. An equivalent expression was obtained by Mitchell and Richmond[4] from Lifshitz theory.

The potential energy, found by integration with respect to distance, is the free energy of double layer repulsion, ΔG_R, between the two flat plates,

$$\Delta G_R = \int_d^\infty \Pi(x)\, dx \tag{19.5}$$

The integration requires an expression for Φ_d as a function of plate separation d; but the precise relation is complex. A good approximation is to assume that the potential of the overlapping double layers is the sum of the potentials of the undisturbed double layers [Eq. (17.21)].

$$\Phi_d = \frac{8\gamma kT}{ze} \exp(-\kappa d) \tag{19.6}$$

where γ is given by Eq. (17.19).

Expanding Eq. (19.4) in a Taylor series and keeping only the first term, substituting Eq. (19.6) for the electric potential at the midpoint and integrating gives the following approximation for the repulsive potential for highly charged parallel plates at relatively large separations, $H = 2d$:

$$\Delta G_R = \frac{64 n_0 kT}{\kappa} \gamma^2 \exp(-\kappa H) \tag{19.7}$$

where κ is given by Eq. (17.18).

Equation (19.7) is essential for the understanding of electrostatic repulsion. Among other things it shows that an increase in ionic strength (increasing κ) diminishes the repulsion between particles at any distance. Derivations of ΔG_R with fewer approximations are available but yield more complex expressions and are less often used in practical applications.

Combining Eq. (19.7) with a suitable expression for the attractive energy between dispersed particles gives a relation for the stability of the dispersion in terms of measurable or calculable values, and is the essence of the DLVO theory.

19.2 ELECTROSTATIC REPULSION BETWEEN SPHERES

In many colloidal systems such as emulsions, inorganic sols, soap solutions, and certain polymer solutions, the particles are better represented by a spherical model than by plane parallel plates. In principle, the repulsive energy for spherical particles can be calculated in the same way as for plane parallel plates, starting with the expressions for the isolated spherical double layer. The Poisson–Boltzmann equation, however, has only been solved numerically for a spherical particle, so the repulsive energies for the overlap of spherical double layers must also be found numerically.

Various approximations (to be described next) are available for special cases: (a) where the particles are large compared to the double layer thickness;[5] or (b) where the potential between the particles is approximated as the sum of the potentials of the undisturbed spherical double layers;[6] or (c) where it might be permissible to use the low-potential (the Debye–Hückel approximation) form of the Poisson–Boltzmann equation.[7]

19.2.a The Derjaguin Approximation

Each sphere is considered to have a stepped surface, and the electrostatic repulsion between two charged spheres is approximated as the sum of a series of separate repulsions between parallel rings of graded diameters, facing each other in matched pairs. The interaction between the spheres is thus reduced to the sum of interactions between parallel walls at different distances of separation. By making the rings infinitesimal the summation becomes an integration (see Figure 19.2). For each pair of rings the repulsive interaction is the product of the facing areas, $2\pi y\, dy$ and the repulsion interaction, $\Delta G_R^{\text{plane}}(x)$ per unit area of two flat parallel plates at a distance x. The total repulsion, $\Delta G_R^{\text{sphere}}(H)$, for the two spheres of radius a, a distance H apart, is obtained by integrating from $y = 0$ to $y = \infty$.

$$\Delta G_R^{\text{sphere}}(H) = 2\pi \int_0^\infty \Delta G_R^{\text{plane}}(x) y\, dy \qquad (19.8)$$

From Figure 19.2,

$$\tfrac{1}{2}(x - H) = a - (a^2 - y^2)^{1/2} \qquad (19.9)$$

Differentiating Eq. (19.9) with respect to x and rearranging gives

$$2y\, dy = dH(a^2 - y^2)^{1/2} \qquad (19.10)$$

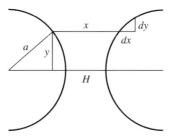

Figure 19.2 The Derjaguin approximation for the integration of two spheres of radius a at a distance H apart.

Since $y^2 \ll a^2$ for the range of significant interaction, that is, for thin double layers between large spheres,

$$2y\,dy \cong a\,dx \tag{19.11}$$

Combining Eqs. (19.8) and (19.11) gives

$$\Delta G_R^{\text{sphere}}(H) = \pi a \int_H^\infty \Delta G_R^{\text{plane}}(x)\,dx \tag{19.12}$$

Equation (19.12) gives the Derjaguin method of calculating the interaction of two equal spheres from the interaction per unit area of two parallel planes, with the added assumptions of thin double layers and large particles.

Substituting Eq. (19.7), the Langmuir approximation for the repulsion between two parallel plates, into Eq. (19.12) and integrating gives an approximation for the repulsion between two large and equal spheres[8]

$$\Delta G_R^{\text{sphere}}(H) = \frac{64 n_0 \pi a k T \gamma^2}{\kappa^2} \exp(-\kappa H)$$

$$= \frac{16 \pi a D \varepsilon_0\, k^2 T^2 \gamma^2}{n_0 z^2 e^2} \exp(-\kappa H) \tag{19.13}$$

This equation applies to emulsion droplets, for instance, but would not be as suitable for particles of colloidal size and high potential. For the latter situation, Overbeek used Derjaguin's method with the exact numerical approximation for the repulsion between parallel planes, solutions to Eq. (19.3), to obtain more exact numerical solutions for the repulsion potential between equal spheres.[2]

19.2.b The Debye–Hückel Approximation

When κa is large, $\kappa a > 10$, the radius of curvature is larger than the thickness of the double layer, and Derjaguin's approximation is reasonable. When κa is small, $\kappa a < 1$, a different approach is required. Here the conditions are favorable for the Debye–Hückel approximation, and the solution of Eq. (19.3) becomes practicable. Verwey and Overbeek[9] found a relation between the charge on spherical particles and the surface potential. From this relation they were able to calculate the energy of interaction for constant surface potential and constant surface charge density. The results were presented in tables. A simple approximate expression is

$$\Delta G_R^{\text{sphere}}(H) = \frac{4\pi a^2 D\varepsilon_0 \Phi_0^2 \exp(-\kappa H)}{2a + H} \quad (19.14)$$

where $2a + H$ is the distance between the centers of the spheres. The conditions for the use of this solution are rarely met in an aqueous medium, but the equation is well suited for suspensions of small particles in nonpolar media where double-layer thicknesses are large.

19.2.c   Constant Surface-Charge Density

When sorption equilibrium between the substrate and its potential-determining ions is rapid compared to the time of particle collisions, the electrical potential, which is determined by the activity of the ions in solution, remains constant as two charged surfaces interact. This would be expected for silver iodide particles. On the other hand, slow equilibrium, such as might occur when soap, pyrophosphate, or proteins are adsorbed, causes the interaction to take place at constant surface charge. For some materials, like metal oxides, neither surface charge density nor surface potential would remain constant during collisions. Fortunately, however, the resulting repulsion at constant surface-charge density does not vary much from the repulsion at constant surface potential, so the difference may be ignored.[2] Neither extreme condition is likely to occur in a real system, but rather some intermediate, with both surface-charge density and surface potential changing slightly.

19.3   REPULSION BY POLYMER LAYERS

The source of steric repulsion between dispersed particles is the increase in free energy resulting from the overlap of adsorbed layers. The thermodynamic stability of the suspension is determined by the work required to concentrate adsorbed polymers as two particles interact. By concentrating molecules osmotic pressure is created, which accounts for the repulsion. If the polymer layers are thick, osmotic effects are significant at large interparticle separation. This is a qualitative argument for the direct dependence of steric repulsion on the thickness of the adsorbed layer. A quantitative argument is given in Chapter 21. For practical stability the required thickness of the adsorbed layer is an order of magnitude less than the radius of the particle. Therefore common surfactants, such as fatty-acid derivatives, which are far too short to be comparable to the size of colloids, are not steric stabilizers except for nanoparticles.

Figure 19.3 shows actual measurements of the repulsion between two polymer-coated mica surfaces as a function of separation. These measurements were obtained on an instrument designed by Israelachvili and known as "Jacob's box." The diagram shows that the force of repulsion builds up steadily as the separation,

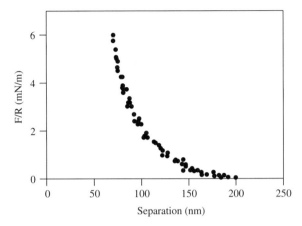

Figure 19.3 Typical data from Jacob's box for a block polymer on crossed cylinders of mica.[10] The adsorbed polymer is PS/PEO with 334/19 in monomer units. F/R is the measured force divided by the radius of the cylinders.

decreases from 200 nm to 80 nm. The polymer layers become incompressible at about 80 nm separation, which corresponds to the distance of closest approach in a collision. The "thickness" of the polymer layer on each surface is 40 nm. Other ratios of polystyrene to poly(ethylene oxide) and their molecular weights were studied to obtain optimum thickness.[10]

Naturally occurring hydrophilic polymers, such as starch, gelatin, milk casein, egg albumen, lecithin, or gum arabic have been exploited for centuries as steric stabilizers. A well-known example of steric stabilization is its use in culinary arts. Another example is the silver halide "emulsion" used in photography, which is stabilized with gelatin. Such substances were called "protective colloids" because they mitigated the deleterious effect of added electrolyte or freezing. The availability of synthetic polymers of high molecular weight has extended the range and type of solid/liquid suspensions, especially nonpolar, that can be sterically stabilized. These suspensions are increasingly important in various technical areas, such as coatings, inks and paints, reinforcement of plastics and rubbers, lubricants, food additives, milk and other dairy products, and biocolloids such as tissue and blood cells and protein solutions.

Only soluble polymers can be used as steric stabilizers. Dissolved polymer molecules move by Brownian diffusion albeit more slowly than solvent molecules. They also collide with each other, but do not stick. Polymer solubility can be described as negative free energy of mixing. The polymer-coated surfaces stay apart for the same thermodynamic reason that polymers stay apart in solution. Of course, whenever the conditions are changed, say of temperature or pressure, or solvent composition, to make the polymer insoluble, the polymer-coated particles are also unstable and flocculate.[11]

The degree to which adsorbed (or attached) polymer on a surface keeps the particles apart depends on two general properties of the polymer, the thickness of the adsorbed layer and its compressibility. The thickness of the polymer layer is primarily determined by the length of the polymer between attachment points. The two extreme cases are homopolymers adsorbed flat on the particle surface, whose layer thickness would only be of molecular dimensions, and the end-functionalized polymer, whose layer thickness would be nearer to its end-to-end length in solution.

The compressibility of a polymer in solution is an area of active research. A number of models have been developed as quantitative theories.[12,13,14,15] For practical use, cruder approximations are adequate. The simplest of these is to assume that the polymer layer forms an impenetrable barrier at some distance from the solid surface. The distance can be estimated from the radius of gyration of the polymer in solution or from intrinsic viscosity measurements. Measuring the intrinsic viscosity has the considerable practical advantage of allowing characterization of polydisperse polymers and batch-to-batch variations. This latter advantage is significant since the specifications on the molecular weight ranges of polymers are often quite wide, and the stability of dispersions can be sensitive to these variations.

REFERENCES

[1] Verwey, E.J.W.; Overbeek, J.Th.G. *Theory of the stability of lyophobic colloids*; Elsevier: Amsterdam; 1948; Chapter IV.

[2] Overbeek, J.Th.G. The interaction between colloidal particles, in *Colloid Science*, Kruyt, H.R., Ed.; Elsevier: New York; 1952; Vol. 1, pp 245–277.

[3] Langmuir, I. The role of attractive and repulsive forces in the formation of tactoids, thixotropic gels, protein crystals and coacervates, *J. Chem. Phys.* **1938**, *6*, 873–896.

[4] Mitchell, D.J.; Richmond, P. A general formalism for the calculation of free energies of inhomogeneous dielectric and electrolytic systems, *J. Colloid Interface Sci.* **1974**, *46*, 118–127.

[5] Derjaguin, B.V.; Landau, L. Theory of the stability of strongly charged lyophobic sols and of the adhesion of strongly charged particles in solutions of electrolytes, *Acta Physicochim. URSS* **1941**, *14*, 733–762.

[6] Reerink, H.; Overbeek, J.Th.G. The rate of coagulation as a measure of the stability of silver iodide sols, *Discuss. Faraday Soc.* **1954**, *18*, 74–84.

[7] Wiese, G.R.; James, R.O.; Yates, D.E.; Healey, T.W. Electrochemistry of the colloid–water interface, in *Physical Chemistry*; Butterworths: London; 1976; Ser. 2, Vol. 6, pp 53–102.

[8] Verwey loc. cit.; Part III.

[9] Verwey loc. cit.; p 152.

[10] Guzonas, D.; Boils, D.; Hair, M.L. Surface force measurements of polystyrene-*block*-poly(ethylene oxide) adsorbed from a nonselective solvent on mica, *Macromolecules* **1991**, *24*, 3383–3387.

[11] Croucher, M.D.; Hair, M.L. Upper and lower critical flocculation temperatures in sterically stabilized nonaqueous dispersions, *Macromolecules* **1978**, *11*, 874–879.

[12] Clayfield E.J.; Lumb, E.C. A theoretical approach for polymer dispersant action. I. Calculation of entropic repulsion exerted by random polymer chains terminally adsorbed on plane surfaces and spherical particles, *J. Colloid Interface Sci.* **1966**, *22*, 269–284.

[13] Napper, D.H. *Polymeric stabilization of colloidal dispersions*; Academic Press: New Yorl; 1983; Chapter 4.

[14] Russel, W.B.; Saville, D.A.; Schowalter, W.R. *Colloidal dispersions*; Cambridge University Press: New York; 1989; Chapter 6.

[15] Evans, D.F.; Wennerström, H. *The colloidal domain: where physics, chemistry, biology, and technology meet*; VCH: New York; 1994; Chapter 7.

20 Dispersion Stability

20.1 LYOPHILIC AND LYOPHOBIC DISPERSIONS

Colloidal dispersions can be divided into two major classes according to their mode of stabilization: (a) lyophilic colloids that acquire stability by solvation of the interface, where the term "solvation" includes all degrees of interaction from mere physical wetting to the formation of adherent thick layers of polymer molecules; (b) lyophobic colloids that acquire stability by an electrostatic repulsion between particles, arising from ions that are either adsorbed by or dissolved from the surface of the solid.

Hauser[1] named the two classes of colloid systems lyocratic (governed by solvation) and electrocratic (governed by electrostatics), respectively. The stability of biocolloids and synthetic macromolecules can be attributed to interaction with the solvent (hydration, solvation). Stable dispersions of lyophilic colloids, including polymer stabilized dispersions, require more than physical wetting: Without a strong anchoring bond the adhesion of the stabilizer is not firm enough to withstand even relatively small shearing forces; consequently an acid–base interaction combined with the steric effect of a large lyophilic group on the stabilizer is required for stability. Hauser's lyocratic dispersions would nowadays be said to be sterically stabilized, but the old name is still suggestive. To form a stable dispersion of an electrocratic colloid, a surface charge and a surrounding electric double layer are required. In water the two classes of dispersion differ in their sensitivity to added electrolyte. A small quantity of soluble salts added to an electrocratic dispersion causes the particles to coagulate. Lyocratic colloids are less sensitive to added electrolyte, although they too can be coagulated by a high concentration of salts, which changes the hydration forces.

In reality, the two mechanisms are not mutually exclusive; they frequently occur together, although often one evidently predominates and the other is relatively minor. The adsorbed ions that are responsible for the electric charge also provide a hydrated layer on the surface of the particle. The electrostatic repulsion may extend over tens of nanometers, depending on the ionic strength of the solution; the repulsion between solvated layers extends far less. Steric stabilization, provided by adsorbed polymers of high molecular weight, can extend much farther than solvated layers but seldom as far as electrical double layers.

20.2 KINETICS OF COAGULATION

No lyophobic sol is absolutely stable against coagulation. Electrostatic repulsion provides only a finite barrier and the dispersed particles or droplets will, with a sufficient number of collisions, ultimately coagulate and separate by sedimentation or creaming. The relative stability of such a dispersion depends on the rate of particle collisions and the probability of a collision leading to sticking. In this chapter we discuss these two factors.

20.2.a Rates of Flocculation

An approximate kinetic theory for the rate of particle collisions in a dispersion was first proposed by Smoluchowski[2] for particles without interparticle attraction (rapid coagulation). Later, Fuchs[3] included the effects of particle–particle attraction and repulsion (slow coagulation). The approach used by both was to describe the particle–particle collision rate by the same kinetic equation as that for bimolecular reactions. For a monodispersed system of particles, the rate of change in the number of particles per unit volume is given by the second-order rate equation

$$-\frac{dn}{dt} = k_2 n^2 \tag{20.1}$$

where n is the number of particles per unit volume, t is time, and k_2 is the rate constant for the loss of singlets; other derivations use k_2 as the rate constant for the appearance of doublets. The rate at which singlets are lost is twice the rate at which doublets are formed. Equation (20.1) assumes that only doublets are formed. The dispersed particles or droplets will, however, with sufficient collisions, ultimately coagulate and separate by sedimentation or creaming. The rate constant, k_2, is the initial rate of flocculation.

Equation (20.1) has the solution

$$\frac{1}{n} = \frac{1}{n_0} + k_2 t \tag{20.2}$$

where n_0 is the initial concentration of particles. The time $t_{1/2}$ for the concentration to reach to one-half of its initial value is given by

$$t_{1/2} = \frac{1}{k_2 n_0} \tag{20.3}$$

Overbeek evaluated k_2 from Fick's first law of diffusion assuming that the particles approach each other by Brownian diffusion plus any forces of interaction such as van der Waals attraction and electrostatic repulsion to obtain a general expression[4]

$$k_2 = \frac{4kT}{3\eta W} \tag{20.4}$$

where k is the Boltzmann constant, T is the absolute temperature and η is the viscosity of the liquid; and where the factor W is called the stability ratio and is given by

$$W = 2a \int_{2a}^{\infty} \exp\left(\frac{U_{121}}{kT}\right) \frac{dr}{r^2} \tag{20.5}$$

where a is the particle radius and U_{121} is the particle–particle interaction energy. When U_{121} is zero everywhere except where particles touch, W is one, and Eq. (20.4) then gives the Smoluchowski rate constant for rapid coagulation. When the particles attract each other, U_{121} is negative, the exponential is everywhere less than one, W is less than one, and the rate of flocculation is fast. When the particles repel each other, U_{121} is positive, the exponential is everywhere greater than one, W is greater than one, and the rate of flocculation is slow.

Substituting Eq. (20.4) into Eq. (20.3) gives the half-life

$$t_{1/2} = \frac{3\eta}{4kTn_0} W \qquad (20.6)$$

or in terms of mass fraction ϕ, particle density ρ_2, and liquid density ρ_1,

$$t_{1/2} = \frac{\pi \eta \rho_2 a^3}{\phi \rho_1 kT} W \qquad (20.7)$$

This half-life is the time required for the number of single particles to decrease by 50%. For aqueous dispersions at 25°C,

$$t_{1/2} = \frac{2 \times 10^{11}}{n_0} W \qquad (20.8)$$

or

$$t_{1/2} = 0.68 \frac{\rho_2 a^3}{\phi} W \qquad (20.9)$$

where the half-life is in seconds, the density of the particles is in gm/cm^3 and the radius, a, is in microns.

Equation (20.6) shows that the half-life of a dispersion depends inversely on the initial concentration of particles. When the concentration is small, dispersions appear to be stable simply because of so few collisions. Some particle-sizing techniques, especially light scattering, require dilution to such low concentrations of particles that the unstable dispersions produced do not flocculate during the time of the measurement.

Overbeek's calculation embodies several simplifying assumptions:

(a) Only binary collisions are taken into account.
(b) The particles are uniform in size.
(c) The relative diffusion during coagulation is the sum of the single-particle diffusion coefficients.
(d) The effect of viscous flow of the medium on close approach of particles is ignored.

These assumptions have been discussed by several writers.[5,6] In general, however, they introduce minor errors compared to the uncertainties that still beset the expressions for attractive and repulsive forces.

20.2.b Rates of Rapid Flocculation

The Smoluchowski model is a limiting case of no interparticle forces; that is, U_{121} is zero and the stability ratio is one. In terms of mass fraction ϕ, particle density ρ_2, and liquid density ρ_1 the half-life is

$$t_{1/2} = \frac{\pi \eta \rho_2 a^3}{\phi \rho_1 kT} \tag{20.10}$$

This half-life is the time required for the number of single particles to decrease by 50%. For aqueous dispersions at 25°C the half-life in seconds is

$$t_{1/2} = 0.68 \frac{\rho_2 a^3}{\phi} \tag{20.11}$$

For example, a 1% dispersion of 1 μm radii particles with a density of 1 gm/cm^3 has a half-life of 68 s. That is, after 68 seconds half of the particles would have flocculated to form pairs if particles stuck on every collision. Dispersions with higher mass fractions or smaller particles last an even shorter time. Obviously stable dispersions depend on strong interparticle repulsions.

20.2.c Interparticle Forces from Rates of Flocculation

An important application of this theory is the ability to calculate interparticle force from the measured half-life of a dispersion. The measured half-life is the length of time for the mass average particle size to increase by 50%. The stability ratio W may be calculated from Eq. (20.7). In principle, the interparticle force, U_{121}, may be calculated from W by means of Eq. (20.5). In practice, a useful approximation is to replace the energy barrier by a rectangular step of height U_{max} and width $1/\kappa$ (see Figure 20.1), so[7]

$$\begin{aligned} U(r) &= -\infty, \quad r < 2a \\ U(r) &= U_{max}, \quad 2a < r < 2a + \frac{1}{\kappa} \\ U(r) &= 0, \quad r > 2a + \frac{1}{\kappa} \end{aligned} \tag{20.12}$$

Substituting this potential into the expression for the stability ratio, Eq. (20.5), gives

$$W \approx \frac{1}{2\kappa a} \exp \frac{U_{max}}{kT} \tag{20.13}$$

If κ is taken to be the inverse Debye length (see Equation 17.6) and the particle size, a, is known, the energy of interaction, U_{max} can be calculated. The calculation of the stability ratio from the more realistic energy curves obtained from DLVO theory (see Section 20.3) can be approximated in other ways: namely,

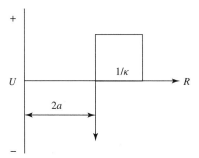

Figure 20.1 A useful approximation for interparticle potential according to Eq. (20.12).

graphically or by a series expansion of the exponential term or by approximating the maximum by a Gaussian function.

Equation (20.13) shows that dependence of the stability of the dispersion increases exponentially with the repulsion between particles. Another relation that may be deduced from Eq. (20.13) is the effect of concentration of electrolyte on the value of W when W is large: since κ is proportional to the square root of the ionic strength then

$$\log W = -\frac{1}{2} \log I + k_2 \qquad (20.14)$$

where I is the ionic strength. This linear equation is verified by experiments as seen in Figure 20.6.

20.3 ELECTROCRATIC REPULSION VERSUS DISPERSION-FORCE ATTRACTION

Numerous products and processes depend on controlling the balance between attractive and repulsive interactions: products such as paints and inks, aerosols, jellies and gels, rubber latexes; processes such as filtration and clarification, emulsification and demulsification, paper making. Examples of reversals of interparticle forces are seen in the mining and processing of kaolin clay. The clay as it is mined cannot be dispersed in water no matter how vigorous the agitation. With the addition of a few tenths of a percent of a pyrophosphate salt, however, the clay is dispersed readily as a free-flowing suspension. The adsorption of the pyrophosphate anions overcomes the attractive forces between the clay platelets and makes them mutually repellent. The addition of an acid to this suspension immediately coagulates it, showing that the acid eliminates the repulsive forces between the platelets, which it does by combining with the adsorbed phosphate anions to form neutral molecules of phosphoric acid.

An outstanding contribution to colloid science was the development of a quantitative theory to explain the mechanisms of stability in many of these systems.

This theory was independently formulated by B. V. Derjaguin and L. Landau in the U.S.S.R. and E. J. W. Verwey and J. Th. G. Overbeek in the Netherlands and is now denoted by the term "DLVO theory." The basic idea of the theory is that the stability of a dispersion is determined by the sum of attractive and repulsive forces between individual particles. The mutual attraction of particles is a consequence of dispersion forces, often called London–van der Waals forces, and the mutual repulsion of particles is a consequence of the interaction of the electrical double layers surrounding each particle. An additional term to describe electrosteric-stabilized systems is added to DLVO theory, when polymers are adsorbed by the charged surfaces.

The significant properties of the electric double layer that are experimentally available are the zeta potential, ζ, and the thickness of the double layer, $1/\kappa$. Neither of these properties can be measured directly: The zeta potential is computed from measurements of electrokinetic motion of particles, and the thickness of the double layer is computed from the ionic composition of the medium. The stability of an electrocratic dispersion depends on both properties. The zeta potential by itself is useful to predict the resistance of a dispersion to flocculation by electrolytes, by determining the "critical zeta potential," that is, the value of ζ below which the suspension is coagulated; but zeta potential alone does not predict the effects of other stresses such as those that affect temporal stability (shelf life), thermal stability, or stability under mechanical shock or shear.

20.3.a DLVO Theory: Two Flat Plates

The determinant of stability is the balance between the electrostatic repulsion as calculated by double-layer theory and the attraction as calculated by dispersion-force theory. This concept is the DLVO theory. The free energy of electrostatic repulsion between two flat plates as a function of the distance between the plates, H, given by the Langmuir approximation, Eq. (19.7), is

$$\Delta G_R = \frac{64 n_0 k T \gamma^2}{\kappa} \exp(-\kappa H) \qquad (20.15)$$

where n_0 is the concentration of counter ions at infinite distance, γ is given by Eq. (17.19) and κ is given by Eq. (17.18). The energy of dispersion-force attraction between two flat plates through material 2, Eq. (18.12), is

$$\Delta G_{121}^d = -\frac{A_{121}}{12 \pi H^2} \qquad (20.16)$$

where A_{121} is the Hamaker constant.

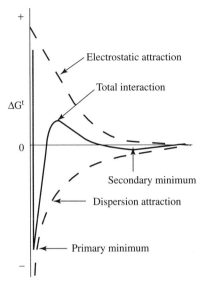

Figure 20.2 DLVO theory for the total interaction between two charged, flat plates per unit area.

Combining these two equations gives the total free energy per unit area of the interaction between two flat plates as a function of distance, H,

$$\Delta G_t = \frac{64 n_0 kT \gamma^2}{\kappa} \exp(-\kappa H) - \frac{A_{121}}{12\pi H^2} \qquad (20.17)$$

Equation (20.17) is illustrated in Figure 20.2, calculated for typical values of the constants. The electrostatic repulsion is described by the first term of Eq (20.17) and upper curve in Figure 20.2. The dispersion attraction is described by the second term of Eq. (20.17) and the lower curve in Figure 20.2. When the maximum value in the total interaction is large compared to kinetic energy, the dispersion is stable.

20.3.b DLVO Theory: Two Equal Spheres

The total energy of interaction between two equal spheres of radius a is also obtained by adding the electrostatic repulsion and the dispersion-force attraction. The energy of repulsion between two spheres with thin double layers at shortest distance H between their surfaces, by the Derjaguin approximation, Eq. (19.13), is

$$\Delta G_R = \frac{64 n_0 \pi a kT \gamma^2}{\kappa^2} \exp(-\kappa H) \qquad (20.18)$$

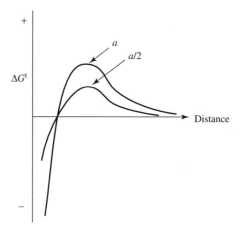

Figure 20.3 The effect of particle size on the interaction between spheres according to DVLO theory. The smaller particle has the lower barrier to flocculation.

and the free energy of attraction for the pair of spheres through material 2, Eq. (18.14), is

$$\Delta G^d_{121} = -\frac{A_{121}a}{12H} \qquad (20.19)$$

Therefore the total energy of interaction is

$$\Delta G_T = \frac{64 n_0 \pi a k T \gamma^2}{\kappa^2} \exp(-\kappa H) - \frac{A_{121}a}{12H} \qquad (20.20)$$

The variation of total energy of interaction with distance has, on the whole, the same characteristics as that for flat plates, with the exception of the effect of particle size, a. The interaction between two flat plates has no size dependence, whereas the interaction between spheres is directly proportional to the particle size. Thus the height of the barrier to coagulation of spherical particles is 10 times greater for particles of 1 μm compared to particles of 0.1 μm. (See Figure 20.3) *In general, electrostatic stabilization is more important the larger the particle size, and in practice it is inoperative for microemulsions or colloidal sols.*[8]

20.3.c Example Application of DLVO Theory

Figure 20.4 shows the variation in total potential energy of interaction (in units of kT) between two 2 μm diameter oil drops in water as a function of the distance of separation. The potential energy was calculated by Eq. (20.20) assuming $A_{121} = 0.5 \times 10^{-20}$ J and $\zeta = -100$ mV at low ionic strength, which is typical for O/W emulsions. (The ionic strength is equal to the concentration of added salt for a 1-1 electrolyte.) As the ionic strength is increased by adding salt, the zeta potential decreases; see Eq. (17.17).

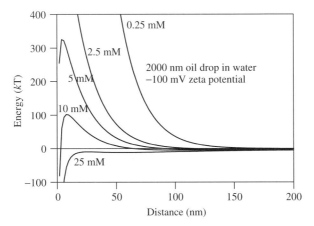

Figure 20.4 Total potential energy A_{121} of interaction between two oil drops in water calculated by Equation (20.20) with $A_{121} = 0.5 \times 10^{-20}$ J. The drops are 2000 nm in diameter with a zeta potential at low ionic strength of -100 mV. The curves correspond to a gradual increase in concentration of 1-1 electrolyte.

At low ionic strength, 0.25 mM, the potential energy is repulsive at all distances, the oil droplets remain apart, and the emulsion is stable. Even at 10 mM, a repulsive barrier equal to nearly 100 times thermal energy, kT, keeps oil drops apart. As the concentration of electrolyte is increased, the energy barrier declines to zero and a concentration can be identified as the value of n_0 at which the potential energy barrier opposing coagulation just disappears; this concentration is called the *critical coagulation concentration* (CCC).

Figure 20.5 shows the variation in total potential energy of interaction (in units of kT) between two 200 nm diameter titania particles in water as a function of the distance of separation, assuming the titania particles are spheres. The potential energy was calculated by Eq. (20.20) assuming $A_{121} = 7 \times 10^{-20}$ J and $\zeta = -100$ mV at low ionic strength. The dispersion force attraction between titania particles in water is much larger than that between oil drops in water. The curves in Figure 20.5 are qualitatively similar to the curves in Figure 20.4. Quantitatively they differ in two important ways. First, the CCC for the titania particles is an order of magnitude less than for the larger oil droplets and second, the secondary minima for these titania particles are deeper and longer range indicating that uncoated titania particles are difficult to stabilize with only electrostatic repulsive forces. The titania used in inks and paints are coated with alumina and silica which greatly reduces the dispersion-force attraction.

20.3.d Critical Coagulation Concentration

An estimate of the CCC can be obtained by using the expressions above for the free energies of repulsion and attraction. When the potential energy barrier just disappears, $d(\Delta G_A + \Delta G_R)/dH = 0$ and $\Delta G_A + \Delta G_R = 0$. Using Eq. (20.17) and

DISPERSION STABILITY

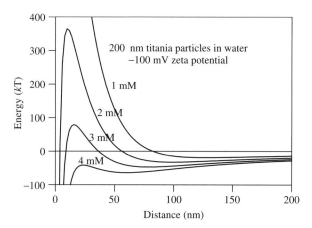

Figure 20.5 Total potential energy of interaction between two titania spheres in water calculated by Eq. (20.20) with $A_{121} = 7 \times 10^{-20}$ J. The spheres are 200 nm in diameter with a zeta potential at low ionic strength of -100 mV. The curves correspond to a gradual increase in concentration of 1-1 electrolyte.

solving these simultaneous equations for H gives $\kappa H = 1$. Substituting $\kappa H = 1$ into Eq. (20.17) and using the expression for the inverse Debye length, κ, Eq. (17.6), gives an expression for the CCC,

$$\text{CCC(molecules/cm}^3) = \frac{(4\pi\varepsilon_0)^3 \left(2^{11} 3^2 D^3 k^5 T^5 \gamma^4\right)}{\pi \exp(4) \, e^6 A_{121}^2 z^6} \quad (20.21)$$

Taking $T = 298$ K and $D = 80$, leads to the critical coagulation concentration as a function of Hamaker constant, A_{121}, valency, and surface potential

$$\text{CCC(moles/liter)} = \frac{8.74 \times 10^{-39} \gamma^4}{z^6 A_{121}^2} \quad (20.22)$$

where the Hamaker constant is in joules. When the surface potential is sufficiently high, γ is near unity and is independent of the valency and the surface potential. The CCC then depends on the inverse sixth power of the valency and is independent of the zeta potential. At lower surface potentials, γ can be approximated by Eq. (17.20) and the CCC then varies with the fourth power of the ζ potential and the inverse square of the valency.

A similar calculation for the stability of equal spheres starting with Eq. (20.20) instead of Eq. (20.17) leads to an equation similar to (20.21) except reduced by a factor of $\exp(2)/2^4 \approx 0.46$ which implies that spheres will flocculate at a little lower concentration of electrolyte than flat plates.[8]

Under freeze-thaw conditions, water-based formulations, such as emulsions or latexes, if stored in a garage or freight car would break and could not be restored to the original. The freezing would separate ice crystals and so increase the ionic strength of the serum which eventually exceeds the CCC. Particles are irreversibly flocculated once they touch. Stabilization can be achieved by adsorbing polymer such as PEO as a protective colloid.

20.3.e Schulze–Hardy Rule

The stability ratio is determined by measuring the half-life of the dispersion by its change in particle size with time. Figure 20.6 shows the effect of added electrolyte on the stability ratio of various electrostatically-stabilized, lyophobic sols. The CCC is the concentration of ions necessary to reduce the stability ratio to one (log $W = 0$). The strong dependence of the CCC on the valence of the ion is noteworthy; the higher the charge per ion, the more effective the flocculation. This strong dependence of the critical coagulation concentration on the valency (z) is known as the Schulze–Hardy rule, found empirically by Schulze[9] and Hardy.[10] They observed that the coagulating power of the potential determining ion increases with its valency roughly in the proportion of 1:100:1000 for uni-, bi-, and trivalent ions. Thus if the univalent ion requires 100 units of concentration for coagulation, the bivalent ion would require 1 unit and the trivalent would require 0.1 unit.

At high surface potentials $\gamma \to 1$, the coagulation values of uni-, bi-, and trivalent ions, according to Eq. (20.22) occur in the ratio $1:(1/2)^6:(1/3)^6 = 100:1.6:0.13$. The DLVO theory provides the theoretical rationale for these observations and their agreement with observation is a major success of the theory.

The effect of electrolyte concentration in reducing the repulsive energy barrier, which is shown clearly in Figure 20.6, has practical applications in understanding the use of alum to coagulate dispersed solids in waste water and in other industrial

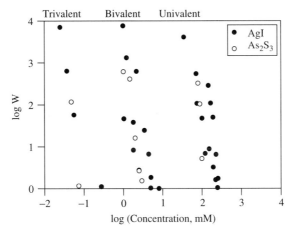

Figure 20.6 Effect of electrolyte concentration on the stability ratio of various lyophobic sols. The three groups of curves refer to univalent, bivalent, and trivalent electrolytes (from the right).[11]

processes such as paper making. The alum provides polyvalent ions in solution and effectively compresses the double layer.

20.3.f Secondary Minimum

Another feature of the potential curves of interaction is shown in the curve for 10 mM ionic strength of Figure 20.5 is the presence of a shallow minimum, known as the secondary minimum, at a relatively large distance of separation (75–100 nm). The presence of a secondary minimum is due to the longer range of the dispersion force attraction compared to that of electrostatic repulsion. The depth of the minimum depends mainly on the attractive force which, for two spheres, is proportional to the size of the particles; therefore the depth of the secondary minimum is also proportional to the size of the particles. Hence effects of the secondary minimum are more likely to occur with larger particles. If the depth of the minimum is several times kT, flocculation results on collision. The nature of flocculation in the secondary minimum differs from the coagulation in the primary minimum in that the equilibrium distance is several times the Debye length, and the particles adhere without uniting. The floc, therefore, is easily redispersed. Practical use is made of this property in paints, where a loosely flocculated structure entails a yield point. This confers two advantages: It prevents running and dripping during application and also prevents a hard sediment from forming in time at the bottom of the can.

20.4 COLLOID STABILITY AND COMPLEX ION CHEMISTRY

The work of Matijević and his school[12] shows that the effect of electrolytes on the stability of dispersions in aqueous media is not fully described unless the possibility of complex ion formation in the solution is included. For example, in using aluminum salts as coagulating agents, the nature of the counterion depends on the pH of the solution: at low pH the counterion is Al^{+3}; as the pH is increased, the counterions are higher order complexes formed by hydrolysis.

Complex ions in the medium have two main effects: They change the ionic strength of the medium, and they can change the charge on the particle by adsorption. The high valency of most complex ions does more to increase the ionic strength than the loss of concentration on complexing does to decrease it. The high molecular weight of complex ions also promotes their adsorption. Coordination sometimes occurs on the surface of the particle (chemisorption) instead of in the medium.

Figure 20.7 shows the CCC of two aluminum salts on a negatively charged silver-iodide sol as a function of pH. The two horizontal sections of these curves represent complex ions that are stable throughout a certain pH range. The previously held idea that Al^{+3} forms lower charged aluminate ions as the pH increases leads to the erroneous conclusion that aluminum salts should be poor flocculants at neutral pHs since the aluminate ions have lesser charge.

The Schulz–Hardy rule predicts that the lower the charge per ion, the less effective the flocculant. The data in Figure 20.7 show, contrary to this expectation, that aluminum salts are excellent flocculants at high pH. The truth is that higher order complexes, such as $Al_8(OH)_{20}^{+4}$ or $Al_7(OH)_{17}^{+4}$, form at higher pH. Because of their higher charge these complexes make effective flocculants (lower CCC).

The presence of complex ions introduces a second phenomenon, namely, that at higher concentrations they may become potential-determining ions. Hence the suspension can become stable again as adsorption of charged counterions reverses the charge on the particle. This concentration is called the critical stabilization concentration (CSC). Because stability depends on both pH and concentration of counterion, a plot of these variables is a kind of phase diagram revealing stability domains. Figure 20.8 represents the behavior of three different sols in the presence of aluminum salts in solution. The shaded regions represent coagulated sols (two phases) and the unshaded regions represent stable sols (one phase). The lower boundaries of the two-phase regions represents the CCC as a function of pH. Similarly, the upper boundaries of the two-phase regions represent the CSC as a function of pH. Note that the CCC varies by orders of magnitude through a relatively small range of pH. Comparing Figure 20.7 and Figure 20.8, note that they agree and show a low CCC at pH of seven. Figure 20.8, however, extends to a pH beyond seven and shows the reversal of charge on the particles.

Matijević warns that many current textbooks and articles quote data in which the chemistry of complex ions is ignored in explaining colloidal stability.

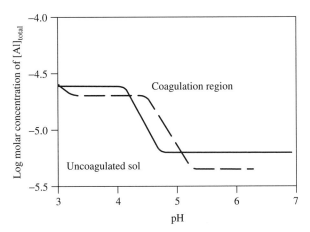

Figure 20.7 Critical coagulation concentrations of aluminum nitrate, acidified with HNO_3, when necessary (———), and of aluminum sulfate, acidified with H_2SO_4 when necessary (- - -), as a function of pH for a negatively charged silver iodide sol.[12]

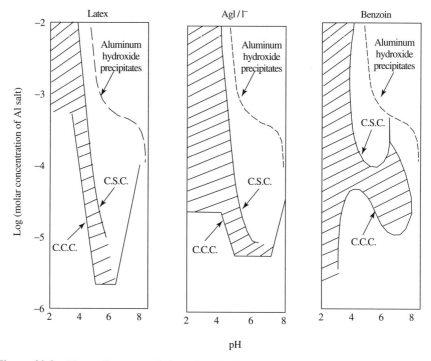

Figure 20.8 Phase diagrams of three lyophobic sols, showing stability domains as a function of Al(NO$_3$)$_3$ or AlCl$_3$ concentration and pH; styrene-butadiene rubber (SBR) latex (left); silver iodide sol (middle); and benzoin sols prepared from powdered Sumatra gum (right).[13]

20.5 ELECTROSTATIC STABILITY IN NONPOLAR MEDIA

Dispersions in nonpolar media are often stabilized by electrostatic charge.[14] DLVO theory, which explains charge stabilization in aqueous media, cannot be applied to nonpolar dispersions because the ionic strength and hence the Debye length, a necessary parameter for DLVO theory, is unknown. The ionic strength is sometimes estimated from electrical conductivity but the values found are so small that the DLVO theory reduces to simple Coulombic repulsion rather than repulsion due to electrical double-layer overlap.[15]

The Coulombic repulsion between two charged, nonconducting spheres is

$$\Delta G_R^{NP} = \frac{4\pi D \varepsilon_0 a^2 \Phi_0^2}{H + 2a} \qquad (20.23)$$

where the superscript NP stands for nonpolar, D is the dielectric constant of the medium, ε_0 the permittivity of free space, a is the particle radius, Φ_0 is the surface potential and $H + 2a$ is the distance between the centers of the particles (see Figure 19.2). The energy of attraction between two spheres is given by the Hamaker expression Equation (18.14)

$$\Delta G_{121}^d = -\frac{A_{121}a}{12H} \tag{20.19}$$

where A_{121} is the Hamaker constant. Therefore the total energy of interaction between two charged spheres at low ionic strength is

$$\Delta G_t^{NP} = -\frac{A_{121}}{12H} + \frac{4\pi D\varepsilon_0 a^2 \Phi_0^2}{H + 2a} \tag{20.24}$$

The stability ratio Equation (20.5) is calculated for the total energy of interaction from the relation

$$W \simeq 2a \int_0^\infty \exp\left(\frac{\Delta G_t^{NP}}{kT}\right) \frac{dH}{(H+2a)^2} \tag{20.25}$$

Equation (20.24) can be substituted into Eq. (20.25) and the integral evaluated to give[15]

$$W \simeq \left(\frac{A_{121}k^2T^2}{3072 D^3 \varepsilon_0^3 a^3 \Phi_0^6}\right)^{1/4} \exp\left(\frac{2\pi D\varepsilon_0 a \Phi_0^2}{kT}\right) \tag{20.26}$$

This is the expression sought, the stability ratio in terms of material-dependent quantities. Equation (20.26) shows the stability to be a strong function of particle size and surface potential but a weak function of the attractive force. Because the Hamaker constants for many nonpolar dispersions are of the order of 10^{-20} J, and the stability ratio depends on the fourth root, not much error is introduced by assuming this value as a general approximation. Furthermore, if a stable dispersion is defined as one whose stability ratio is greater than 10^5, then the necessary values of $Da\Phi_0^2$ for stability can be calculated from Eq. (20.26) to give

$$\Phi_0^2 > \frac{10^3}{Da} \tag{20.27}$$

where a is in μm and Φ_0 is in mV. In hydrocarbons, where the dielectric constant is 2, the required surface potentials to stabilize particles of various radii are shown in Table 20.1.

As in aqueous systems, the smaller particles require larger surface potentials for stability. Surface potentials of 10–200 mV are common in nonpolar media and this calculation shows that those potentials are sufficient to stabilize such dispersions. The number of charges per particle to give surface potentials in this range in media of low dielectric constant is too small to be uniformly distributed across the particle surface. Any nonuniformity in surface charge distribution could result in flocculation; consequently, steric stabilization is probably always

TABLE 20.1 Surface Potentials to Stabilize Particles as a Function of Their Radius

Radius, μm	Surface Potential, mV	Radius, μm	Surface Potential, mV
0.01	224	0.5	32
0.05	100	0.75	26
0.1	71	1.0	22
0.3	41	5.0	10

a necessary factor, especially as the charging mechanism requires polymers or inverse micelles.

20.6 THEORY OF STERIC STABILIZATION

The modern term is "polymer stabilization," sometimes "steric stabilization," and in the older literature, "lyocratic" or "lyophilic" colloids. All these terms point to the interaction of the solvent with the stabilizing material as the key to understanding dispersion behavior.

To stabilize a suspension, the effective thickness of the adsorbed layer, L, should be such that the total free energy of attraction at closest approach, $H = 2L$, is less than kT, the average kinetic energy per particle, so that no sticking occurs on collision (see Figure 20.9). For two spherical particles of the same size at a short distance of separation,

$$\Delta G_{121}^d = -\frac{A_{121}a}{12H} \tag{20.19}$$

where a is the radius of the particle, H is the shortest distance between the two particles, and A_{121} is the Hamaker constant for particles of material 1 immersed in a medium of material 2. The critical distance of separation, H_0, below which

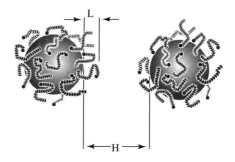

Figure 20.9 Model for steric stabilization.

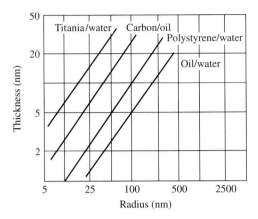

Figure 20.10 Polymer thickness, t_0, of equal spherical particles of radius a to prevent flocculation at 25°C, for titania in water, carbon black in oil, polystryene in water, and oil in water.[17] ($A_{121}/24\ kT = 0.70, 0.28, 0.10, 0.05$ respectively)

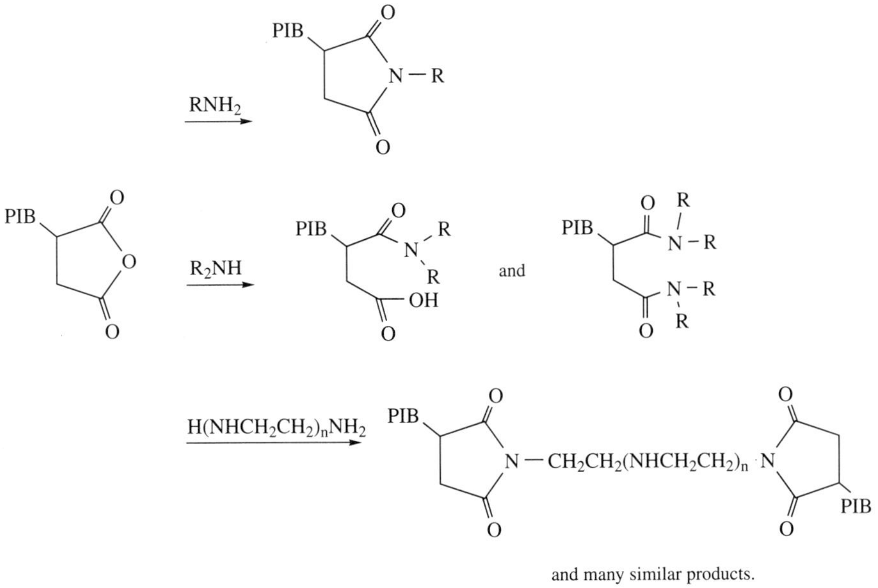

Figure 20.11 Chemical structures in polyisobutylene succinimdes, such as OLOA 371 (formerly known as OLOA 1200.) The commercial product is a reaction of polyethylene amine and polyisobutylene succinic anhydride. The final product contains imide, amine, carbonyls, and carboxylic acid groups incorporated in a polymeric matrix. The oil solutions are mostly micelle-like structures with these active groups forming the micellar core.

the dispersion would rapidly flocculate, is given by setting $\Delta G^d_{121} = -kT$ in Eq. (20.19). The necessary polymer thickness, t_0, on each particle is $H_0/2$ or

$$t_0 = \frac{aA_{121}}{24kT} \tag{20.28}$$

Values of the Hamaker constant, A_{121}, can be obtained as previously described (see Section 18.3). An example of the use of Eq. (20.28) is given by Pugh, Matsunaga and Fowkes[16] for carbon black particles dispersed in dodecane. For these materials, $A_{121} = 2.8 \times 10^{-20}$ J, and the characteristic wavelength (required to calculate the retardation correction f) is 90 nm. The requirement for stability is that a particle of radius a must have a minimum film thickness of adsorbate of t_0 thus, to stabilize particles of carbon black of 25 nm radius in dodecane, a film thickness of 7 nm is required, and particles of 200 nm radius require a film thickness of 56 nm. Clearly, only adsorbed polymers can provide film thicknesses of this magnitude for steric stabilization. Results of similar calculations made for some suspensions are shown in Figure 20.10.

Practical applications of this concept must take into account that the polymer needs to be firmly attached to the substrate or it will be sheared off in collisions. Specific interactions between functional groups on the polymer and on the substrate are required. The strongest of these is acid–base interaction, such as between basic amides and surface acid groups. A widely used oil additive that combines both high molecular weight and basic functionality is polyisobutylene succinimide, marketed as OLOA 371 (formerly OLOA 1200) by Chevron Chemicals Co (See Figure 20.11).

REFERENCES

[1] Hauser, E.A. *Colloidal phenomena: an introduction to the science of colloids*; McGraw-Hill: New York; 1939; p 28.

[2] von Smoluchowski, M. Versuch einer mathematischen theorie der koagulationskinetik kolloider lösungen (Mathematical theory of the kinetics of the coagulation of colloidal solutions), *Z. Physik. Chem., Stoechiom. Verwandschaftsl* **1917**, *92*, 129–168.

[3] Fuchs, N. Zur theorie der koagulation, *Z. Physik. Chem.* A **1934**, *171*, 199–208.

[4] Overbeek, J.Th.G. Kinetics of flocculation, in *Colloid science*; Kruyt, H.R., Ed.; Elsevier: New York; 1952; Vol. 1, pp 278–301.

[5] Honig, E.P.; Roebersen, G.J.; Wiersema, P.H. Effect of hydrodynamic interaction on the coagulation rate of hydrophobic colloids, *J. Colloid Interface Sci.* **1971**, *36*, 97–109.

[6] Vold, R.D.; Void, M.J. *Colloid and interface chemistry*; Addison-Wesley: Reading, MA; 1983; p 261.

[7] Kruyt, H.R., Ed. Colloid science, Vol. 1 *Irreversible systems*; Elsevier: New York; 1952; p 285.

[8] Overbeek, J.Th.G. Stability of hydrophobic colloids and emulsions, in *Colloid science*, Vol. 1 *Irreversible systems*, Kruyt, H.R., Ed.; Elsevier: New York; 1952; p 305.

[9] Schulze, H. Schwefelarsen in wässeriger lösung, *J. Prakt. Chem.* **1882**, (2), *25*, 431–454.

[10] Hardy, W.B. A preliminary investigation of the conditions that determine the stability of irreversible hydrosols, *Proc. Roy. Soc. London* **1900**, *66*, 110–125.

[11] Overbeek loc. cit.; p 320.

[12] Matijević, E. Colloid stability and complex chemistry, *J. Colloid Interface Sci.* **1973**, *43*, 217–245.

[13] Benzoin sol is the exudate from the pierced bark of various trees native to Sumatra, Java, and Thailand. Chemically it is a mixture of aromatic acids, including benzoic acid anhydride and cinnamic acid.

[14] Morrison, I.D. Electrical charges in nonaqueous media, *Colloid Surf. A. Physicochemical Eng. Aspects* **1993**, *71*, 1–37.

[15] Morrison, I.D. Criterion for electrostatic stability of dispersions at low ionic strength, *Langmuir* **1991**, *7*, 1920–1922.

[16] Pugh, R.J.; Matsunaga, T.; Fowkes, F.M. The dispersibility and stability of carbon black in media of low dielectric constant. I. Electrostatic and steric contributions to colloidal stability, *Colloids Surf.* **1983**, *7*, 183–207.

[17] Fowkes, F.M.; Pugh, R.J. Steric and electrostatic contributions to the colloidal properties of nonaqueous dispersions, *ACS Symp. Ser* **1984**, *240*, 331–354.

21 Polymeric Stabilization

Polymers can be used to stabilize suspensions either by providing electrostatic or steric repulsion. Polyelectrolytes are used in aqueous systems to confer electric charge on suspended particles. Polyacrylic acid is widely used as a stabilizer. Polyacrylamide, which has many positive charges, is used as a flocculating agent, which acts by binding together negatively charged particles (Section 24.4).

"Steric stabilization" is the term used to indicate the type of stabilization of a suspension conferred by adsorbed polymers. Most examples are found in nonpolar solvents, where electrostatic stabilization is less common. For adsorbed polymer to provide steric stabilization the molecule must be firmly anchored to the particle surface to withstand shear forces, but only at a few points so that the bulk of the molecule extends into the solvent a significant distance. This requires a careful match of polymer, particle, and solvent interactions.[1]

21.1 ADSORPTION FROM SOLUTION BY SOLIDS

Dispersed particles may be classified with respect to the medium as lyophilic or lyophobic, that is, those that have a specific interaction with the medium and those that lack such an interaction. These terms usually serve to distinguish between particles of colloidal dimensions in solution (lyophilic), such as soluble polymers or micelle-forming solutes, and insoluble particles that do not attain colloidal size spontaneously (lyophobic), such as metallic oxides. The same distinction is sometimes indicated by the terms "reversible" and "irreversible" dispersions. Since the distinction between soluble and insoluble materials is clear, as is that between reversible and irreversible dispersions, neologisms that do no more than make the same distinction are redundant. They need not be wasted, however. We suggest reserving the terms "lyophilic" and "lyophobic" to characterize the affinity of the medium for the *interface*, as measured by its spreading coefficient. We shall, therefore, use *lyophilic* for interfaces, whether associated with soluble particles or not, on which the medium has a positive spreading coefficient and *lyophobic* for those on which the medium does not spread. These terms now refer to the interface rather than to the bulk material, and consequently make it possible to say that the function of a dispersing agent is to convert a lyophobic to a lyophilic interface.

If the medium is water and the interface is lyophilic, it is called hydrophilic; if the medium is hydrocarbon and the interface is lyophilic, it is called lipophilic. On lyophilic interfaces, specific interactions include electrostatic and electron donor–acceptor interactions, either between the interface and the medium or between the interface and a solute component; where water is the medium, the electron donor–acceptor interaction on a hydrophilic interface is known as hydrogen bonding. Examples of adsorbents with hydrophilic interfaces are alumina, barium sulfate, calcium carbonate, glass, ion exchange resins, quartz, silica gel, titanium dioxide and most metallic oxides, and zeolites; these can be further subdivided as electron donors (bases) or electron acceptors (acids). Examples of adsorbents with hydrophobic interfaces are bone char, carbon blacks, charcoals, graphite, organic resins and plastics, paraffin wax, stibnite and most metallic sulfides, selenium, sulfur, and talc. Electron donor–acceptor interactions, or the special case of such interactions known as hydrogen bonding, are stronger than interactions caused by dispersion forces of attraction; but the latter are present even in the absence of the former. Dipole–dipole (Debye) and dipole-induced dipole (Keesom) interactions are known to be less significant than dispersion-force interactions, especially for polymers.

21.1.a Adsorption of Polymers

The adsorption of a soluble polymer from solution by a solid surface is a process in three steps. The first step is the removal of some of the solvent from the solvated polymer in solution. This requires energy ($\Delta G > 0$) because the polymer is spontaneously soluble. Some space on a solid surface needs to be cleared of solvent to attach a polymer. This dewetting also requires energy since solid–solvent interactions are always attractive ($\Delta G > 0$). The polymer also loses some freedom of motion when adsorbed and that also requires energy. When the polymer is adsorbed by the solid, surface energy is released because the process is spontaneous ($\Delta G < 0$). The magnitude of this released energy depends on the degree of polymer–solid interaction. The solvent molecules released from the surface by the dewetting do regain some energy by attractive interactions with the rest of the free solvent. Whether polymer is adsorbed by the solid surface or not depends on the sum of these free energies. While it may be impractical to measure each process quantitatively, a qualitative understanding of the relative strengths of interaction goes a long way toward understanding important parts of polymer stabilization. If one were to establish an experimental program to study all these processes, the experiments would include measurements of adsorption from solution, measurements of heats of emersion, and measurements of solubility. (See Figure 21.1).

Measurements of polymer adsorption from solution are straightforward in principle. A known amount of solid is stirred in a known volume of a known concentration of the polymer. After some time, generally several hours or a day, an aliquot of the dispersion is taken, centrifuged free of particles, and the polymer concentration measured. IR or UV absorption and intrinsic viscosity are common methods. If available, radioisotope labeled polymers can be used. A particularly useful analytic tool is nuclear magnetic resonance (NMR). NMR measurements can be used to determine concentration and composition. Detecting

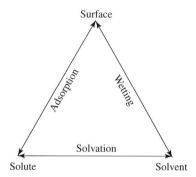

Figure 21.1 Competing processes during polymer adsorption: adsorption from solution, wetting, and solvation.

any unexpected changes in composition can be a clue to particularly active adsorbate components. The measurements of adsorption from solution should be taken over a period of time (often up to a week!) because the rate of adsorption of high molecular weight polymers is slow. Knowing the rate of adsorption for the polymers being used in any practical dispersion gives a guideline to the length of time milling processes need to be run in order to obtain polymer-stabilized dispersions. Since the rate of adsorption is a strong function of molecular weight, the higher the molecular weight, the slower the adsorption. Therefore many successful dispersion formulations use a range of molecular weights. The low molecular weight polymers are adsorbed quickly, but do not give the most stable dispersions. Polymers of higher molecular weight will gradually replace the adsorbed components of lower molecular weight and give better steric stabilization.

The "thickness" of an adsorbed polymer layer is determined by its molecular weight, its solubility, and the concentration of polymer at the surface. Experimental evidence indicates that only the most outward-extending tails in the adsorbed layer, that is, the highest molecular weights, determine the thickness of the adsorbed film, even when the polymer density is quite low.[2] A small fraction of high molecular weight polymer in the adsorbed layer increases the hydrodynamic size and strengthens the steric stability of particles. This is a further argument in favor of using a broad distribution of molecular weights for steric stabilization; only a small fraction of the total polymer, that with the highest molecular weight, provides steric stabilization.

The higher the molecular weight of the polymer, the more mass is adsorbed per unit area. This follows from the consideration that the adsorbed layer is thicker. However, the higher the molecular weight, the less efficient is the packing in the adsorbed film, so the overall effect is that the mass of adsorbed polymer increases approximately with the logarithm of the molecular weight.

For most polymers, the higher molecular weight components replace the lower molecular weight components over time. The exception is the end-functionalized polymer. For these, the lower molecular weights replace the higher

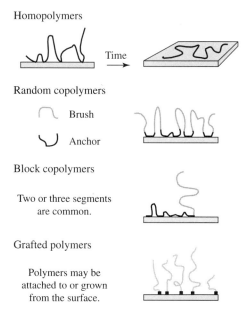

Figure 21.2 Configurations of adsorbed polymers of various kinds.

molecular weight portions. This process is driven by the increase in entropy attained by releasing the high molecular weight components. The number of end-functionalized interactions with the surface remains the same during the exchange.

A further consideration is the configuration of the adsorbed polymer (see Figure 21.2). Homopolymers are first adsorbed forming loops and tails. If any part of the polymer is capable of being adsorbed, then all parts are capable of being adsorbed. Ultimately, when all parts are adsorbed, the polymer is flat on the surface. For this reason the required thickness for steric stabilization is not attained. High molecular weights are required to delay this process and provide sufficient steric thickness over long periods of time. The naturally occurring polymers known as "protective colloids," such as gelatin, are homopolymers of high molecular weight. The use of a high molecular weight homopolymer requires that the dispersion medium be a poor enough solvent to ensure strong adsorption of the stabilizer onto the particle but a good enough solvent to impart solubility. The combination of requirements usually requires a mixture of a good solvent and a poor one, or a nice adjustment of pH, so as to be near the solubility limit of the polymer.

Random copolymers, where the anchoring groups constitute only few percent of their mass, are adsorbed selectively along their length leaving most of their mass to act as a stabilizer. The stabilizing loops remain in the solution and determine the effective thickness of the adsorbate. A guide for applications is to select a monomer that is soluble in the medium, a second monomer to provide an acid–base interaction with the substrate, and a composition of about ten percent of the latter. Commercially available random copolymers consist of more than two monomers in order to give them a wider field of application.

A block copolymer has all its anchoring groups together and all its stabilizing groups together, which makes for a more efficient configuration. Block copolymers are more difficult to synthesize than random copolymers, but the additional cost is often worth the increased efficiency. Various configurations are available from suppliers. This is an active field of research.

The most efficient polymer stabilizers are polymers grafted to the substrate. The chemical bond prevents desorption resulting from collisions, higher temperatures, and changes in solvent. The most common graft is a condensation reaction with groups on the surface, usually hydroxides, such as with silanes, silanols, siloxanes, alcohols, acids, and amines.

21.1.b Acidic or Basic Character of Solid Substrates

(See also Section 8.4). The major interactions of a solid surface with an adsorbate are of two kinds: the universal interactions of the London dispersion forces and the acid-base interactions. London dispersion forces are long range and, even though small per molecule, their sum is a significant portion of the total interaction per unit area across an interface. The London dispersion energies of solid substrates, as determined by their interactions with various liquids, are described in Chapter 8 and, as determined from their electronic properties, are described in Chapter 18. The chief specific interactions are Lewis base–Lewis acid reactions (electron donor–acceptor interactions): this category includes chelation, coordination, hydrogen bonding, and the complexes between nucleophiles and electrophilic sites that are so important in catalysis.[3] The energy per mole of these interactions is large but, because of their short-range character, only those pairs in proximity contribute to the total interaction. Consequently, the dispersion-force interactions and the donor–acceptor interactions are comparable in magnitude. These two types of interactions have proved to be sufficient to account for the solution and interfacial properties of dispersions.[4]

Inverse gas chromatography (IGC) is a technique to study the acid–base character or possibly the combination of both acid and base on solid substrates. The principle of IGC is a reversal of conventional gas chromatography. An empty column is uniformly packed with the solid material of interest, typically a powder, fiber, or film. A pulse of an acidic or basic vapor, carried in a flow of helium, is injected down the column and its retention time is measured by a detector. The stronger the acid–base interaction, the longer the retention time. In principle the retention time could be related to the heat of adsorption but so many parameters are required for this purpose that it is seldom done. The key parameter in the IGC measurements is the specific retention volume, V_N, or the amount of carrier gas to elute a probe from a column containing one gram of solid. The exact quantity in terms of experimental variables is

$$V_N = \frac{273.2}{T} \cdot \frac{(t_p - t_0)}{W} \cdot F \cdot J \cdot C \qquad (21.1)$$

where

$$C = 1 - \frac{P_{\text{water}}}{P_0} \qquad (21.2)$$

and

$$J = 1.5 \frac{(P_i/P_o)^2 - 1}{(P_i/P_o)^3 - 1} \qquad (21.3)$$

where T is the temperature of the flowmeter, F is the carrier gas flow rate, W is the mass of the solid sample, t_p is the retention time of the probe, t_0 is the retention time of a noninteracting molecule such as air, J is the correction for the pressure drop across column, C is the correction for the vapor pressure of water in the soap bubble flowmeter, P_i is the pressure of the carrier gas at the inlet, P_o is the pressure of the gas at the outlet, and P_{water} is the vapor pressure of water at the temperature of the flowmeter.[5]

The difference between retention volumes measured with an acidic or basic probe and measured with a nonpolar probe such as an alkane corresponds to the free energy of adsorption for the acidic or basic probe. For substrates that contain both acidic and basic groups, measurements with basic or acidic probes, respectively, reveal the relative degrees of each type. IGC measurements with acidic or basic probe molecules have been used to investigate batch-to-batch variations in various materials where other techniques have not been able to detect differences. Similarly, differences in surface treatment of adsorbents can be detected.[6,7]

Various empirical approaches to the problem of predicting donor-acceptor interactions are proposed.[8] The Drago correlation predicts the enthalpy of acid–base reactions in the gas phase or in poorly solvating solvents, by a four-parameter equation,

$$-\Delta H_{ab} = E_a E_b + C_a C_b \qquad (21.4)$$

The acid a and base b are each characterized by two independent parameters: an E value that measures its ability to participate in electrostatic bonding, and a C value that measures its ability to participate in covalent bonding. Both E and C are derived empirically to give the best agreement between calculation and experiment for the largest possible number of adducts. A self-consistent set of E and C values is now available for several dozen bases, allowing the prediction of ΔH for over 1500 interactions[9] (see Table 8.4).

Fowkes[4] extended Drago's approach to include the interactions taking place at an interface. For example, he found that the silanol groups on a silica surface are strongly acidic, whereas a glass surface, because of its silicate content, is basic. The acidic silica surface adsorbs more of a basic adsorptive such as pyridine or polyvinyl pyridine than does a basic surface such as glass. The Drago E and C constants for a solid substrate can be evaluated by measuring the enthalpy of interaction of a test acid or base whose constants are already known. Heats of adsorption may be obtained calorimetrically. A useful technique is flow microcalorimetry in which a stream of solution containing the adsorbate at a known concentration is passed over a known weight of adsorbent previously wetted with the flowing stream of the solvent. The concentration of the adsorbate in the effluent stream is monitored as well as the heat evolved until the concentration in the effluent is the same as that in the incoming stream. Simple calculations give the total mass adsorbed and the total heat evolved for adsorption at that concentration. Repeated measurements at varying adsorbate concentrations give the adsorption isotherm and the differential heats of adsorption.

The heat of adsorption may also be obtained from adsorption isotherms measured at two different temperatures. Adsorption from solution is often described by the Langmuir equation (see Chapter 15), which assumes that all the adsorption sites are equivalent. The assumption is valid except for substrates with a wide distribution of site energies, and a wide distribution of specific acidic or basic sites on a substrate is unlikely. The heat of adsorption is then obtained from the Clausius–Clapeyron equation applied to the two isotherms:

$$\Delta H = \frac{RT_1 T_2 \ln(K_1/K_2)}{T_2 - T_1} \quad (21.5)$$

where K_1 and K_2 are the Langmuir constants at absolute temperatures T_1 and T_2, respectively.

By the application of these techniques, Fowkes and his coworkers determined the E and C constants for the SiOH sites of silica, the TiOH sites of rutile, and the FeOH sites of ferric oxide. These three kinds of surface sites are all acidic. Table 21.1 contains the Drago parameters determined for these oxides in terms of moles of active group on the surface assuming 1:1 adduct formation.[10] From these parameters the heats of adsorption of various nitrogen bases, oxygen bases, sulfur bases, aromatic bases, and basic polymers on these substrates may be calculated.

TABLE 21.1 E and C Parameters for Active Groups on Surfaces of Some Solids[10]

Solid	E_a(kcal/mol)$^{1/2}$	C_a (kcal/mol)$^{1/2}$
Silica	4.2 ± 0.1	1.16 ± 0.02
Rutile	5.7 ± 0.2	1.02 ± 0.03
α-Ferric oxide	4.5 ± 1.1	0.8 ± 0.2

21.1.c Traube's Rule Revisited

Traube's rule (Section 13.3.d) for lowering the surface tension of an aqueous solution as a function of the alkyl-chain length of the hydrophobe, is reproduced at the solid/liquid interface by implicating the degree of adsorption of solute, which is readily measured, in place of the interfacial tension, which cannot be measured. For many systems thus studied a similar rule is found: The quantity of solute that is adsorbed by a hydrophobic interface from an aqueous solution increases geometrically as the chain length of the solute increases arithmetically. Nonpolar solutions show an inverse relation of the same sort: A lyophobic interface in a hydrocarbon solvent, such as silica in mineral oil, takes out less solute, by a constant factor, each time the chain length of the solute increases by one methylene group.

The foregoing generalizations of adsorption behavior may be summarized in the following rules to guide prediction of relative adsorption:

(a) The extent of adsorption is usually greater from solvents in which the adsorptive is less soluble (Lundelius's rule.) A corollary to this rule is that for a given adsorbate the better solvent makes the better eluant.

(b) In the absence of specific interactions between adsorbent and adsorbate, the amount adsorbed is never large.

(c) The greater the electron donor–acceptor interaction between adsorbent and adsorbate, the greater the adsorption. This may be the reason why hydrophilic solids, such as silica gel, are more favored for adsorption separation (except special separations such as decolorization with carbon black). A corollary to this rule is that the greater the donor–acceptor interaction between solvent and adsorptive, the less the adsorption.

(d) The extent of adsorption out of aqueous solution changes in a regular manner along a homologous series.

Rule (a) would warn, however, that, since solubility usually decreases (with any solvent) with a sufficiently great increase in molecular weight, this corollary must be applied with discernment since the two operative factors tend to offset each other (see the Ferguson effect, Chapter 13). The simple generalizations expressed by this form of Traube's rule are not applicable where the solute molecules are adsorbed by attraction mechanisms other than dispersion forces. For example, with cationic surface-active solutes, the positive ionic charge of the surface-active ion is attracted by negatively charged substrates, such as quartz or glass, whereby the hydrophobic moiety of the solute makes the interface hydrophobic; at higher concentrations of solute a second layer of surface-active ions is adsorbed with a reverse orientation, so that the interface becomes hydrophilic and positively charged. The second layer of adsorbate would agree with the mechanism of Traube's rule, but the first layer would not.

21.1.d Rates of Adsorption on Solid Surfaces

The rate of adsorption of a surface-active solute at the liquid/vapor surface is diffusion controlled (Section 13.6) and reaches equilibrium in less than a minute. Rates of adsorption at solution/solid interfaces are considerably slower than the rates of arrival by diffusion. The difference between them is shown in Figure 21.3.

The time of arrival at the substrate by diffusion from a 0.5% solution is calculated for comparison with the measured amounts adsorbed from solutions of fatty acids, amines, and alcohols. The amines require an hour for complete adsorption while stearyl alcohol requires a week. Diffusion alone would have completed the monolayer in 0.02s as calculated by an equation given by Ward and Tordai,[12]

$$n = 2000cN_0 \left(\frac{Dt}{\pi}\right)^{1/2} \tag{21.6}$$

where n is the number of molecules/m^2 after time t in seconds, D is the bulk diffusion constant in m^2/s, c is the bulk concentration in mol/L, and N_0 is Avogadro's number. The factors that can slow adsorption so markedly are the presence of cracks, pores, or capillary spaces in the solid substrate, the rate of desorption of solvent, the time required for the adsorbate to rearrange, especially at high

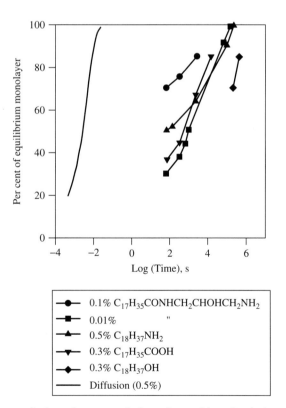

Figure 21.3 Rates of adsorption on steel plates from white-oil solutions compared with rates of diffusion.[11]

coverage, and electrostatic repulsion. Any one of these factors can be expressed as an energy barrier for adsorption; therefore, increasing the temperature usually increases the rate of adsorption.

The rate of adsorption of polymers at a solid substrate is further affected by subsequent rearrangement of the polymer chain from its initial single-point attachment to loop attachment, and then to extended, multiple-site attachment. Also, in a distribution of molecular weights, the low molecular weight fractions that are first adsorbed because of faster diffusion are later displaced by the higher molecular weight fractions. The nature of the solvent affects both the rate of adsorption and the amount adsorbed: the rate is faster and the amount of adsorption greater in a poorer solvent.

The rate of adsorption is the rate-limiting step in many dispersion processes. Newly created surfaces are not stabilized until sufficiently covered by adsorbate. For this reason, machinery that creates new surfaces more rapidly than they can be stabilized uses excess energy, as unstabilized particles or droplets soon flocculate. If the adsorption is slow, prolonged contact of the new surface with the solution is required. A big holding tank and a small-volume grinding unit, such as sand mills or recirculating grinding mills, meet this condition.

21.2 STABILIZATION AND THE PHASE DIAGRAM

Adsorption of a polymer at an interface depends on a balance of attractions between polymer and solvent, polymer and particle, and solvent and particle. Block copolymers can be designed to provide both polymer–particle interaction and polymer–solvent interaction. The polymer–particle interaction is usually an acid–base interaction between the surface of the particle and the groups on the polymer; that part of the copolymer is called the anchoring group. Flocculation of the suspension takes place when conditions are so altered, whether by change of temperature, concentration, or solvent composition, to reduce the solubility of the stabilizing chain in the medium. These conditions are described in the phase diagram of the stabilizing chain; hence the understanding of the flocculation of a sterically stabilized suspension depends on understanding solution thermodynamics.

Steric stabilization differs from electrostatic stabilization in that it is not a question of the balance of forces of attraction and repulsion between particles. The adsorbed polymer layers should be so thick that particles stay far enough apart to make dispersion-force attraction insignificant. The coiled length of the polymer chain provides a long-range repulsion. The stability of the dispersion depends on the thickness of the polymer layer, which in turn is a consequence of its solubility, therefore it is read from the phase diagram. Flocculation is not the result of the overcoming of repulsion by particle–particle attraction but is the consequence of the insolubility and precipitation of the stabilizing chain.

Typical phase diagrams of polymer solutions are shown Figure 21.4, representing polymers of high molecular weight in solvents that show phase separation

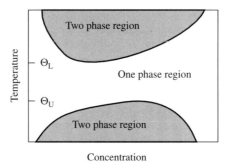

Figure 21.4 Schematic diagrams of a high molecular weight polymer in a solvent that shows phase separation occurring at the UCST and the LCST.[13]

at theta temperatures. The theta conditions of a polymer solution occur when the polymer–polymer interactions are the same as the polymer–solvent interactions. On one side of a theta condition, the polymer is soluble (one phase) and on the other side, it is insoluble (two phases.) Generally, polymers have both an upper critical solution temperature (UCST) and a lower critical solution temperature (LCST).

Polymer-stabilized suspensions are stable at temperatures when the polymer is soluble, that is, described by that part of the one-phase region between the upper and lower critical solution temperatures. Generally, all sterically stabilized dispersions exhibit two flocculation temperatures, known as the upper critical flocculation temperature (UCFT) and the lower critical flocculation temperature (LCFT).[14] These are the same as the critical solution temperatures shown on the two-component phase diagram of polymer and solvent. That confirms the assumption that the particle with adsorbed polymer behaves as the polymer itself. Although the theta conditions vary with concentration, the flocculation temperatures are almost insensitive to concentration.

The same conclusion applies to three-component systems of polymer and two solvents: when the polymer is soluble in a mixed solvent, the polymer-stabilized dispersion is stable.

The practical handling of sterically stabilized dispersions is facilitated by the reversibility of the flocculated condition, which allows the operator to go back and forth between flocculated and deflocculated states. An important variable to effect those conditions is the change of medium. Two or three components may be combined in the solvent, and the addition of one component or another can be used to flocculate or deflocculate the dispersion. The dispersion is flocculated at the solvent composition at which the stabilizing polymer moiety precipitates, and is deflocculated when the stabilizing polymer moiety is soluble as described by the phase diagram of the polymer and solvent.

The limits of stability of a sterically stabilized suspension are obtained by altering the cardinal conditions of temperature, solvent composition, or pressure that control the solubility of the stabilizing polymeric moiety. These limits may be known from the critical points of the polymer–solvent phase diagram, many of which are reported in polymer handbooks or which are easily determined by nephelometry (cloud points). The limits may also be obtained directly by measuring the change in the number of particles as conditions are altered. Specifically, the suspension is titrated with a nonsolvent until coagulation occurs, as shown by a sudden increase in turbidity. Temperature or pressure changes can also be tested in the same manner.

21.3 EFFECT OF FREE POLYMER ON DISPERSION STABILITY

The stabilization of dispersions by polymers depends primarily upon the properties of the adsorbed or grafted polymer on the surface. Soluble, free polymer is not expected to affect the stability of the dispersion because it is not expected to interact with adsorbed polymer. Free polymer and adsorbed polymer are simultaneously soluble only because they do not interact with each other.

The most common effect of free polymer is to increase the viscosity of the dispersion. In fact, a common test for how much stabilizing polymer to add to a dispersion is to measure the viscosity of the dispersion as a function of added polymer. At low concentrations of polymer, the dispersion is not stable and the viscosity of the flocculated dispersion is high. Once sufficient polymer has been added to stabilize the dispersion, the viscosity reaches a minimum. The addition of more polymer results in an increase in polymer concentration in the solvent, which increases the viscosity.

Free polymer in solution can have other effects. Of course, the addition of a second type of polymer can flocculate a dispersion by specific polymer–polymer interactions, the bridging of polymer from particle to particle, or by charge neutralization. These mechanisms are well known. However, if the free polymer is really neutral and noninteracting, flocculation can be caused by a process known as depletion flocculation.

This is a term that refers to a process by which the presence of soluble and unadsorbed polymer is able to flocculate a suspension. An explanation was first suggested by Asakura and Oosawa[15] in 1954 but subsequently received considerable attention at a number of research laboratories.[16,17,18,19] The suggested mechanism is that as suspended particles approach each other the distance between the particles becomes less than the size of a polymer molecule in solution. The suspension therefore contains a region between the particles where the soluble polymer is excluded. A gradient of osmotic pressure forces the particles even closer together. The process is analogous to the dehydration of tissues in a concentrated saline solution.

The depletion force between two surfaces caused by the steric exclusion of free polymer has been measured directly.[16] A stearylated silica particle (7.6 μm diameter) in cyclohexane was brought near a stearylated silica surface in the presence and absence of polydimethylsiloxane. The presence of the silicone polymer increased the attraction between the two surfaces by about 0.2 mN/m. This corresponds to a real, albeit small, change in attraction.

Polyethylene glycol (PEG) is water soluble at moderate temperatures over a wide range of molecular weights. Nevertheless, PEG is widely used for aggregating or fusing cells.[18] By careful analysis of the molecular weight dependence of this aggregation, these authors propose that PEG is excluded from the space between the two surfaces when the surfaces are too close for a molecule of PEG to fit.

Another mechanism for depletion flocculation is the same as that for the phase separation of a mixture of polymer solutions. Mixtures of polymers and proteins have been studied for many years and often show phase separation with increasing concentration. This occurs even without specific polymer–protein interactions, as in mixtures of polysaccharides and proteins.[20] Polymer-coated particles behave like large polymer molecules and are expected to phase separate like the phase separation of polymer mixtures.

Not only is there evidence for depletion flocculation, there is also evidence for depletion stabilization.[19] The improved stability is attributed to the repulsive barrier created by the close approach of particles prior to exclusion of nonadsorbed species from the gap. If this barrier is high enough, the particles do not approach each other closely enough for a depletion region to form and destabilize the dispersion. This alternating inclusion and exclusion of polymer in the interparticle gap may lead to an oscillating force curve, much like the Boscovichean force diagram as a function of distance between atoms.[21]

These effects, whether depletion flocculation or depletion stabilization, are small amounting to a few tenths of a mPa. The effect may be significant in biological systems where polymers of high molecular weight at high concentrations are common.

21.4 STERIC VERSUS ELECTROSTATIC STABILIZATION

The sensitivity to electrolytes that is characteristic of electrostatic stabilization is not displayed in steric stabilization. Further differences are illustrated in Figure 21.5, which shows the total potential energy versus distance of separation for a pair of electrostatically stabilized particles, and Figure 21.6, which shows schematic representations of the total potential energy versus distance of separation for a pair of sterically stabilized particles, under different conditions (*a*) and (*b*). The sudden increase in potential energy near 5 nm separation represents the repulsion as the polymer layers are compressed on collision. The particles in (*a*) are thermodynamically stable because the repulsive barrier is reached before the dispersion forces of attraction are greater than $-kT$; (*b*) represents the same particles in a worse-than-theta solvent or the same particles with a polymeric stabilizer of lower molecular weight. Curve (*b*) shows a potential well of about $-3kT$ which is sufficient to flocculate the particles.

An electrostatically stabilized dispersion coagulates on addition of enough electrolyte and may also flocculate in a looser agglomerate if a secondary minimum is present (see Section 20.3.b). Coagulation in the deep primary minimum

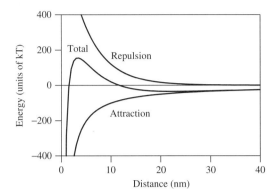

Figure 21.5 Potential-energy diagrams of two 200 nm particles with $A_{121} = 7 \times 10^{-20}$ J and electrostatically stabilized at -100 mV zeta potential at 4 mM ionic strength.

of Figure 21.5 is irreversible. In Figure 21.6 the absence in (*a*) and (*b*) of a deep primary minimum, is a consequence of the limit to which the adsorbed polymer can be compressed on particle–particle collision and means that sterically stabilized dispersions are only capable of being flocculated in a shallow minimum, such as shown in (*b*). This type of flocculation is brought about by a change of conditions, and is readily reversed by reversing the conditions.

An important difference between steric and electrostatic stabilization that should always be kept in mind is their contrasting relation with respect to particle size. We have already shown that steric stabilization is more effective for smaller particles and requires polymers of inordinate molecular weight for larger particles (See Section 20.6) whereas electrostatic stabilization is more effective the larger the particle size (See Section 20.3.b).

21.5 ELECTROSTERIC STABILIZATION

The two types of stabilization can be combined, where it is known as electrosteric stabilization. Steric stabilization alone usually has a shallow minimum at some distance from the particle, but no primary minimum; electrostatic stabilization usually has only a primary minimum at a close distance. When the two types are combined, the particles experience repulsion at all distances, save for the secondary minimum, and the suspension is stable against irreversible flocculation (See Figure 21.7).

The effect of salt on electrosteric stabilized systems is to reduce the electrostatic repulsion and possibly even diminish the solubility of the stabilizing polymer. This usually flocculates the dispersion, but the flocculation can be reversed by reducing the salt concentration.

At concentrations of electrolyte greater than those required to neutralize electrostatic stabilization some systems actually increase in stability on a further increase of salt concentration.[22,23] This increase is extremely odd since all electrostatic stabilization is gone and the solubility of the stabilizing polymer is

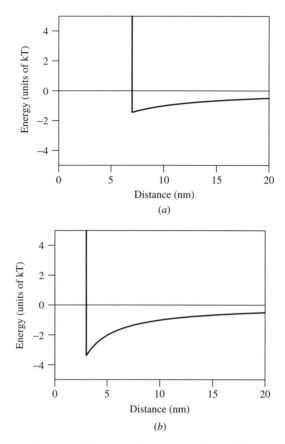

Figure 21.6 Potential-energy diagrams of two sterically stabilized particles: (*a*) thermodynamically stable because of a thick steric barrier; (*b*) thermodynamically unstable because they are in a worse-than-theta solvent or have a lower molecular weight polymer stabilizer. The particles are the same as in Figure 21.5 except with no electrostatic barrier. The polymer thickness in (*a*) is about 3.5 nm corresponding to an interparticle separation of about 7 nm; the polymer thickness in (*b*) is about 1.5 nm, corresponding to an interparticle separation of about 3 nm.

diminished (but not yet poor). Berg describes this unexpected increase in stability as restabilization. Although many examples are reported, the cause remains obscure.[24] One plausible explanation is that restabilization is associated with the presence of micelles or polymer agglomerates, which form barriers between the particles.

21.6 BLOCK COPOLYMERS

The most effective steric stabilizers are block or graft copolymers that contain both anchoring groups and stabilizing chains. The anchoring groups have strong

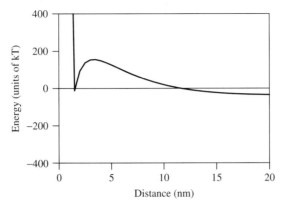

Figure 21.7 Potential-energy diagram for electrosteric stabilization. A repulsive steric layer of about 1 nm thickness is combined with the electrostatic energies of repulsion and dispersion energies of attraction from Figure 21.5.

TABLE 21.2 Typical Anchoring Groups and Stabilizing Chains for Sterically Stabilized Dispersions[25]

Anchoring Groups	Stabilizing Chains
Aqueous Suspensions	
Polystyrene	Polyethylene oxide (PEO)
Polyvinyl acetate	Polyvinyl alcohol
Polymethyl methacrylate	Polyacrylic acid
Polyacrylonitrile	Polymethacrylic acid
Polydimethylsiloxane	Polyacrylamide
Polyvinyl chloride	Polyvinyl pyrrolidone
Polyethylene	Polyethylene imine
Polypropylene	Polyvinyl methyl ether
Polylauryl methacrylate	Poly(4-vinyl pyridine)
Polypropylene oxide (PPO)	
Nonpolar Suspensions	
Polyacrylonitrile	Polystyrene
Polyethylene oxide (PEO)	Polylauryl methacrylate
Polyethylene	Poly(1,2-hydroxystearic acid)
Polypropylene	Polydimethylsiloxane
Polyvinyl chloride	Polyisobutylene
Polymethyl methacrylate	cis-1,4-Polyisoprene
Polyacrylamide	Polyvinyl acetate
	Polymethyl methacrylate
	Polyvinyl methyl ether

418 POLYMERIC STABILIZATION

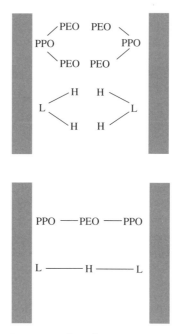

Figure 21.8 Triblock copolymers used as dispersants and as flocculants. L is for lipophile and H is for hydrophile.

affinity for the particle surface and are usually insoluble in the medium. These groups provide strong adsorption (chemisorption) of the polymer molecule on the particle surface, to prevent its being displaced during particle collisions. The stabilizing chains have strong affinity for the medium, to bring the whole polymer molecule into solution, from which adsorption takes place. Some typical polymers for aqueous and for nonpolar suspensions are listed in Table 21.2. Copolymers specifically designed for use as steric stabilizers are available from ICI Americas under the tradename Elvacite® and from Avecia, Ltd. under the tradename Solsperse®.

Block copolymers composed of a single anchor polymer (A) and single stabilizing polymer (B) are called diblocks. Triblocks can be either ABA or BAB. The distinction between those two is that ABA is a flocculant and BAB is a dispersant. This is illustrated for A equal to poly(propylene oxide), PPO, and B equal to poly(ethylene oxide), PEO in Figure 21.8.

REFERENCES

[1] Napper, D.H. *Polymeric stabilization of colloidal dispersions*; Academic Press: New York; 1983.

[2] Stenkamp, V.S.; Berg, J.C. The role of long tails in steric stabilization and hydrodynamic layer thickness, *Langmuir* **1997**, *13*, 3872–3832.

[3] Fowkes, F.M. Characterization of solid surfaces by wet chemical techniques, *ACS Symp. Ser.* **1982**, *199*, 69–88.

[4] Fowkes, F.M. Donor-acceptor interactions at interfaces, *Polym. Sci. Technol.* **1980**, *12A*, 43–52.

[5] Bolvari, A.E.; Ward, T.C.; Koning, P.A.; Sheehy, D.P. Experimental techniques for inverse gas chromatography, *ACS Symp. Ser.* **1989**, *391*, 12–19.

[6] Wesson, S.P.; Allred, R.E. Surface energetics of plasma-treated carbon fiber, *ACS Symp. Ser.* **1989**, *391*, Chapter 15.

[7] Surface Measurement Systems, London, UK; www.smsna.com.

[8] Jenson, W.B. *The Lewis acid–base concepts*; Wiley: New York; 1980.

[9] Drago, R.S. Quantitative evaluation and prediction of donor–acceptor interactions, *Struct. Bonding (Berlin)* **1973**, *15*, 73–139.

[10] Joslin, S.T.; Fowkes, F.M. Surface acidity of ferric oxides studied by flow microcalorimetry, *Ind. Eng. Chem., Prod. Res. Dev.* **1985**, *24*, 369–375.

[11] Fowkes, F.M. Orientation potentials of monolayers adsorbed at the metal–oil interface, *J. Phys. Chem.* **1960**, *64*, 726–728.

[12] Ward, A.F.H.; Tordai, L. Time-dependence of boundary tensions of solutions I. The role of diffusion in time-effects, *J. Chem. Phys.* **1946**, *14*, 453–461.

[13] Croucher, M.D. Effect of free volume on the steric stabilization of nonaqueous latex dispersions, *J. Colloid Interface Sci.* **1981**, *81*, 257–265.

[14] Croucher, M.D.; Hair, M.L. Upper and lower critical flocculation temperatures in sterically stabilized nonaqueous dispersions, *Macromolecules* **1978**, *11*, 874–879.

[15] Asakura, S.; Oosawa, F. On interaction between two bodies immersed in a solution of macromolecules, *J. Chem. Phys.* **1954**, *22*, 1255–1256.

[16] Milling, A.; Biggs, S. Direct measurement of the depletion force using an atomic-force microscope, *J. Colloid Interface Sci.* **1995**, *170*, 604–606.

[17] Smith, N.J.; Williams, P.A. Depletion flocculation of polystyrene latices by water-soluble polymers, *J. Chem. Soc. Faraday Trans.* **1995**, *91*(10), 1483–1489.

[18] Kuhl, T.; Guo, Y.; Alderfer, J.L.; Berman, A.D.; Leckband, D.; Israelachvili, J.; Hui, S.W. Direct measurement of polyethylene glycol induced depletion attraction between lipid bilayers, *Langmuir* **1996**, *12*, 3003–3014.

[19] Ogden, A.L.; Lewis, J.A. Effect of nonadsorbed polymer on the stability of weakly flocculated suspensions, *Langmuir* **1996**, *12*, 3413–3424.

[20] Hoskins, R.; Robb, I.D.; Williams, P.A.; Warren, P. Phase separation in mixtures of polysaccharides and proteins, *J. Chem. Soc., Faraday Trans.* **1996**, *92*, 4515–4520.

[21] Boscovitch, R.J. A *theory of natural philosophy, put forward and explained by Roger Joseph Boscovich*; Open Court: Chicago; 1922.

[22] Einarson, M.B.; Berg, J.C. Effect of salt on polymer solvency: Implications for dispersion stability, *Langmuir* **1992**, *8*, 2611–2615.

[23] Virden, J.W.; Berg, J.C. The steric stabilization of small unilamellar vesicles, *J. Colloid Interface Sci.* **1992**, *153*, 411–419.

[24] Stenkamp, V.S.; McGuiggan, P.; Berg, J.C. Restabilization of electrosterically stabilized colloids in high salt media, *Langmuir* **2000**, *17*, 637–651.

[25] Napper loc. cit.; p 29.

22 Emulsions

22.1 DEFINITIONS AND GLOSSARY OF TERMS

An emulsion is a dispersion of one liquid in another with which it is immiscible. The particle sizes of the dispersed phase lie between a few hundred nanometers and a few tens of micrometers. Stable emulsions require the presence of a third component (the emulsifying agent), but practical emulsions seldom consist of only three components. Polycomponent systems are not readily accessible to mathematical descriptions, so the understanding (technology) of emulsions is largely a matter of empirical rules. Like other specialties, emulsion technology had developed its own language and definitions. The two immiscible liquids that constitute an emulsion are referred to as "oil" and "water," as these are proverbial representatives of two such liquids. Within an emulsion, one liquid phase is in the form of droplets and is therefore distinguished from the other phase. A number of different terms, listed in Table 22.1, are used to express this distinction.

TABLE 22.1 Terminology of Phases

Phase 1	Phase 2
Droplet	Serum
Dispersed	Medium
Discontinuous	Continuous
Internal	External

Emulsions appear as two types: water droplets dispersed in oil, designated W/O, and oil droplets dispersed in water, designated O/W. Read W/O as water in oil, and O/W as oil in water. A common example of an O/W emulsion is milk, and a common example of a W/O emulsion is butter. (See Table 22.2 for other food emulsions.) When one type of emulsion is altered to the other, the process is called inversion. When an emulsion separates into its two constituent phases, it is said to be "broken." Because a density difference may exist between the two phases, the dispersed phase may rise or sink within the medium; these processes are called creaming or sedimenting (or settling), respectively, and they are not the same as breaking of the emulsion.

22.2 DETERMINATION OF EMULSION TYPE

The simplest way to determine the type of an emulsion is to see whether a small volume of the emulsion mixes readily with water; if it does so, the continuous

TABLE 22.2 Typical food emulsions[1]

Food	Emulsion type	Dispersed phase	Continuous phase	Stabilization factors, etc.
Milk, cream	O/W	Butterfat triglycerides partially crystalline and liquid oils Droplet size: 1–10 μm Volume fraction: milk: 3–4% cream: 10–30%	Aqueous solution of milk proteins, salts, minerals, etc.	Lipoprotein membrane, phospholipids, and adsorbed casein
Ice cream	O/W (aerated to foam)	Butterfat (cream) or vegetable, partially crystallized fat Volume fraction of air phase: 50%	Water and ice crystals, milk proteins, carboxydrates (sucrose, corn syrup) Approx. 85% of the water content is frozen at −20°C	The foam structure is stabilized by agglomerated fat globules forming the surface of air cells. Added surfactants act as "destabilizers" controlling fat agglomeration. Semisolid frozen phase
Butter	W/O	Buttermilk: milk proteins, phospholipids, salts. Volume fraction: 16%	Butterfat triglycerides, partially crystallized and liquid oils; genuine milk fat globules are also present	Water droplets distributed in semisolid, plastic continuous fat phase
Imitation cream (to be aerated)	O/W	Vegetable oils and fats Droplet size: 1–5 μm. Volume fraction: 10–30%	Aqueous solution of proteins (casein), sucrose, salts, hydrocolloids	Before aeration: adsorbed protein film After aeration: the foam structure is stabilized by aggregated fat globules, forming a network around air cells; added lipophilic surfactants promote the needed fat globule aggregation

(*Continued*)

TABLE 22.2 (Continued)

Food	Emulsion type	Dispersed phase	Continuous phase	Stabilization factors, etc.
Coffee whiteners	O/W	Vegetable oils and fats Droplet size: 1–5 μm Volume fraction: 10–15%	Aqueous solution of proteins (sodium caseinate), carbohydrates (maltodextrin, corn syrup, etc.), salts, and hydrocolloids	Blends of nonionic and anionic surfactants together with adsorbed proteins
Margarine and related products (low calorie spread)	W/O	Water phase may contain cultured milk, salts, flavors. Droplet size: 1–20 μm Volume fraction: 16–50%	Edible fats and oils, partially hydrogenated, of animal or vegetable origin Colors, flavor, vitamins	The dispersed water droplets are fixed in a semisolid matrix of fat crystals; surfactants added to reduce surface tension/promote emulsification during processing.
Mayonnaise	O/W	Vegetable oil Droplet size: 1–5 μm Volume fractions: minimum 65% (U.S. food standard)	Aqueous solution of egg yolk, salt flavors, seasonings, ingredients, etc. pH: 4.0–4.5	Egg yolk proteins and phosphatides
Salad dressing	O/W	Vegetable oil Droplet size: 1–5 μm. Volume fractions: minimum 30% (U.S. food standard)	Aqueous solutions of egg yolk, sugar, salt, starch, flavors, seasonings, hydrocolloids, and acidifying ingredients pH: 3.5–4.0	Egg yolk proteins and phosphatides combined with hydrocolloids and surfactants, where permitted by local food law

phase of the emulsion is aqueous. This test may be performed under a microscope using a glass rod to mix the water and the emulsion. Conversely, a W/O emulsion mixes readily with oil and not with water. Concentrated emulsions are highly viscous and, even when water is the continuous phase, may not readily mix with water. Such emulsions are likely to defy any test for type that can be suggested.

Another test for emulsion type is the electrical conductivity. O/W emulsions almost always have high conductivity, whereas W/O emulsions have low conductivity. Very concentrated emulsions may invert on addition of more of their internal phase or on other change of conditions. Such inversion is accompanied with a large reduction of viscosity, or "thinning."

22.3 INTERFACIAL TENSION

The methods of measuring surface tension already described are all readily adapted to measure the tension at the interface between two immiscible liquids. Such techniques include the Wilhelmy plate and Du Noüy ring, which measure the force required to make an object traverse the interface, the drop volume method, based on the balance between gravity and surface-tension effects, the spinning-drop method, which measures the distortion of an air bubble or a droplet of an immiscible liquid caused by centrifugal acceleration, and the sessile and pendant drop methods, which measure the equilibrium shapes created by the simultaneous action of gravity and surface tension. All these methods apply to macroscopic samples of several millimeters or larger. But the interfacial tensions so measured do not necessarily reflect the actual tension at the surfaces of the dispersed droplets owing to dissimilar partitioning of adsorbed solute.[2] For the study of emulsions a direct measurement of the interfacial tension *at the droplet interface* is more relevant.

A technique designed for this purpose was recently described by Moran, Yeung and Mislayah[3] An emulsion drop is distorted (elongated) from its spherical shape by being drawn between two suction pipets and the resisting force is measured. For micrometer-sized drops the bond numbers [Eq. (6.48)] are low, so that only capillary forces are significant. By applying the Laplace equation [Eq. (6.35)] to the resulting axisymmetric drop shape, as determined by the fixed parameters of the apparatus, the force measurement provides an absolute value of interfacial tension.

22.4 COALESCENCE OF EMULSION DROPLETS

We have already seen (Section 6.2.a) that the Helmholtz free-energy change for the process of coalescence of two droplets at constant volume, temperature, and composition is negative and that therefore coalescence is spontaneous. The situation is different for emulsion droplets, because the reduction of area consequent on coalescence leaves less room for adsorbed solute, which therefore has to be returned to the solution. But the solute had arrived at the interface spontaneously, and its removal or desorption requires energy. Therefore, for the net change of free energy on coalescence of emulsion droplets to be positive, and the emulsion to be sufficiently stable so as not to coalesce spontaneously, the third component should have a negative free energy of adsorption large enough to overcome any effect of area reduction [see Equation (22.19)]. This means that the necessary third component of a stable emulsion is a surface-active solute. Another consequence of this condition, a concomitant of adsorption, is the reduction of interfacial tension occasioned by the third component. If the net free energy of coalescence of an emulsion is positive, spontaneous emulsification is thermodynamically possible. Nowadays, with many powerful emulsifying agents available, spontaneous emulsification is commonplace and widely accepted; but so improbable did it appear to the savants of seventy years ago that when J. W. McBain announced this discovery (or rather rediscovery, for it had been known since 1878) to the Royal Society, it was dismissed by the chairman, Sir W. B. Hardy, with the comment, "Nonsense, McBain!"

22.5 BANCROFT'S RULE AND ITS EXCEPTIONS

W. D. Bancroft pointed out that the type of emulsion is derived, at least in part, from the nature of the emulsifying agent.[4] He formulated the rule that, in the making of an emulsion, the liquid in which the agent is the more soluble becomes the continuous phase. Thus, for example, a water-soluble substance like sodium oleate emulsifies oil in water, and an oil-soluble substance like calcium oleate emulsifies water in oil. The rule tends to break down when high concentrations of internal phase are emulsified, but it may well be that the larger relative quantity of internal phase leaches emulsifier from the continuous phase. Bancroft provided no explanation of this rule, which yet has been found of wide general validity. The following considerations help to make Bancroft's rule more intelligible.

In an aqueous medium a preponderantly hydrophilic solute will have either an ionic charge or a polyethylene-oxide moiety, which because of its ability to attach water by hydrogen bonding creates a thick gelated layer on the aqueous side of the interface. The electric charge entails an electrostatic field, by which a force of repulsion for like charges extends outward from the interface into the aqueous medium. Therefore, these hydrophilic characteristics confer on the solute molecule either an electrostatic or a steric field of repulsion that extends farther into the aqueous than into the nonpolar medium. The converse holds for molecules of preponderantly lipophilic solute. Here again, both steric and electrostatic fields of mutual repulsion are possible. The mechanism of charge separation in lipophilic solvents is now better understood (see Section 14.3). In nonpolar media, because of the low dielectric constant, the repulsive field of force can extend for a considerable distance from the interface. Steric repulsion in a nonpolar medium is created by the same mechanism as in an aqueous medium, namely, by adsorption of solute molecules in an orientation of minimum potential energy at the interface. If solute molecules are represented schematically in terms of their respective fields of force rather than by a literal interpretation of the actual volumes of their atomic constituents, we should expect to see for the preponderantly hydrophilic molecule a larger territory in the aqueous medium, and for the preponderantly lipophilic molecule, a larger territory in the nonpolar medium. The former molecules are therefore more effective in separating oil drops in an aqueous medium than in separating water drops in an oil medium; the latter molecules are more effective in separating water drops in an oil medium than in separating oil drops in an aqueous medium. These comparisons are shown schematically in Figures 22.1 and 22.2. The destabilization of an emulsion or a dispersion can be achieved by destroying or diminishing the repulsive field of force, or territory, created by the interfacially active solute.

The preceding rationale is postulated on the assumption that electrostatic or steric stabilization lies at the root of Bancroft's rule. This is indirectly confirmed by Ruckenstein, who concludes, by a thermodynamic argument, that the magnitude of the excess surface concentration of the emulsifying agent underlies the behavior summarized by Bancroft's rule.[6]

Exceptions to Bancroft's rule occur under conditions of too weak a surfactant or too low a concentration of a stronger surfactant. In the dynamics of emulsion making, fingers of oil extend into water and fingers of water extend into oil. Under dynamic conditions, diffusion, adsorption, and Marangoni effects occur. These hydrodynamic effects are likely responsible for violation of Bancroft's rule.[7] Exceptions also occur when wedge-shaped molecules affect the curvature of the interface, which applies more particularly to microemulsions where the size of the droplet is comparable to that of the surfactant molecule.

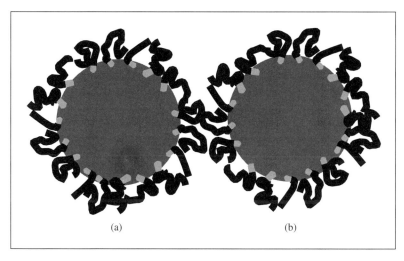

Figure 22.1 A lipophilic solute adsorbed at the surface of a water drop is effective in keeping the water drops apart in a W/O emulsion.[5]

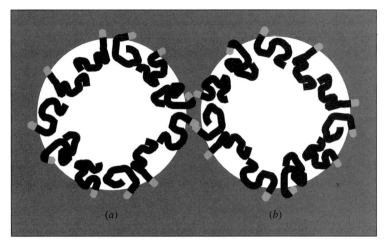

Figure 22.2 A lipophilic solute adsorbed at the surface of an oil drop is not effective in keeping the oil drops apart in an O/W emulsion.[5]

22.6 AMPHIPATHIC PARTICLES AS EMULSION STABILIZERS

Effective resistance to coalescence of emulsion drops is conferred by certain finely divided solids. The basic sulfates of iron, copper, nickel, zinc, and aluminum in moist condition emulsify petroleum oil in water; carbon black, rosin, and lanolin stabilize water in kerosene and in benzene. Any insoluble substance that is wetted more readily by water than by oil, if sufficiently finely divided, will serve to emulsify oil in water. In some cases it is possible to see under a microscope the coating of solid particles that surrounds the oil drop. These emulsions resist spontaneous demulsification since the interface is mechanically stronger than an interface built from soluble components. Excellent emulsions may be formed by adding lime, or limewater, to the normal sulfates of iron or copper; on adding kerosene emulsification is produced by slight agitation.[8] Besides their ease of manufacture and the absence of spontaneous demulsification, these emulsions are not decomposed by adding caustic soda. When the copper salt is used, the emulsion has the insecticidal properties of Bordeaux mixture,* without the disadvantage of settling out as hard-packed sediment.

A general resemblance to Bancroft's rule is demonstrated by the properties described above, as the moist surfaces of the inorganic salts are predominantly hydrophilic and so tend to promote O/W emulsions, while the surfaces of carbon black and rosins are predominantly lipophilic and so tend to promote W/O emulsions. This resemblance can be demonstrated by a simple model of a spherical particle of radius a sited at the interface between liquids 1 and 2. When the particle is at the interface, the area immersed in liquid 1 is A_{1s} the interfacial energy as F_{1s} and the area immersed in liquid 2 is A_{2s} with an interfacial energy

---

* Bordeaux mixture is an aqueous suspension of copper sulfate and lime.

of F_{2s}, and it has displaced an area A_{12} of the interface with an interfacial energy of σ_{12}. The equilibrium position of the particle is determined by the minimum in free energy,

$$F_{1s}dA_{1s} + F_{2s}dA_{2s} + \sigma_{12}dA_{12} = 0 \tag{22.1}$$

The change in area can be calculated by simple geometry and the equation simplified to[9]

$$F_{2s} - F_{1s} = \sigma_{12}\cos\theta \tag{22.2}$$

which is the Young–Dupré equation for contact-angle equilibrium. The particle will seek a position such that θ becomes the equilibrium contact angle. Equation (22.1) tacitly disregards buoyancy forces.

Equation (22.1) can be satisfied by two positions of the sphere, giving two complementary values of θ, of which, however, only one is consistent with the Young–Dupré equation (22.2). The two solutions are given by

$$\cos\theta = 1 - h/r \tag{22.3}$$

Where the true angle of contact is acute, $h/r < 1$, and where the true angle of contact is obtuse, $h/r > 1$. Figure 22.3 shows that the liquid making the smaller angle of contact with the particle contains the bulk of the particle. The emulsion droplets are stabilized by virtue of the steric repulsion of the bulk of the particles in the continuous phase, that is, by the same mechanism described in Chapter 21 for molecular stabilizers of suspensions. The liquid with the smaller angle of contact has the larger work of adhesion to the particle surface. We may, therefore, rephrase Bancroft's rule for applications to emulsions stabilized by solid particles: the liquid with the larger work of adhesion to the particle is the continuous phase.

The angle of contact at the three-phase boundary between two immiscible liquids and a solid substrate can be calculated if the surface tensions of the two liquids in the presence of both of their vapors, the interfacial tension between them, and the contact angles that each of them makes separately with the substrate are all known. If these contact angles are measured with the purpose of using

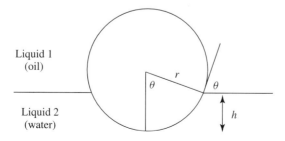

Figure 22.3 Solid sphere at the liquid/liquid interface.[9]

them in this calculation, the substrate should be in equilibrium with the combined vapors of *both* liquids. The works of adhesion of each liquid on the substrate are given by

$$W_1^{adh} = F_{sv} + \sigma_1 - F_{1s} = \sigma_1 (1 + \cos \theta_1) \qquad (22.4)$$

$$W_2^{adh} = F_{sv} + \sigma_2 - F_{2s} = \sigma_2 (1 + \cos \theta_2) \qquad (22.5)$$

The Young–Dupré equation for the three-phase boundary of the two liquids in equilibrium with the substrate is

$$F_{2s} - F_{1s} = \sigma_{12} \cos \theta \qquad (22.6)$$

where θ is contact angle measured through liquid 1. (See Figure 22.3.) Substituting (22.4) and (22.5) into (22.6) gives

$$\sigma_{12} \cos \theta = (\sigma_2 - \sigma_1) - \left(W_2^{adh} - W_1^{adh}\right) \qquad (22.7)$$

from which

$$\sigma_{12} \cos \theta = \sigma_1 \cos \theta_1 - \sigma_2 \cos \theta_2 \qquad (22.8)$$

Equation (22.8) is a form of Neumann's equation (6.28) but applicable to a solid particle at a liquid/liquid interface. As long as θ_1 and θ_2 are finite, θ exists and the particle is sited at the interface.

As an example, consider a sphere of PTFE at the interface between white oil and water. The surface tension of water and its contact angle against PTFE are 72.9 mN/m and 115.7°; the surface tension of white oil and its contact angle against PTFE are 28.9 mN/m and 50.0°; the interfacial tension between white oil and water is 51.3 mN/m. An application of Eq. (22.8) gives the angle of contact of white oil on the PTFE sphere at the three-phase boundary, that is, in the presence of water, as 12°. The sphere would, therefore, be situated at the oil–water interface, immersed in the water to a depth of about 2% of its radius.

For any value of the contact angle, θ, in Figure 22.3, the free energy of desorption required to remove the particle from the interface into the oil phase (liquid 1) is given by[10]

$$\Delta F_1 = 2\pi r^2 (1 - \cos \theta)(F_{2s} - F_{1s}) + \pi r^2 (1 - \cos^2 \theta)\sigma_{12} \qquad (22.9)$$

Substituting (22.6) into (22.9) gives

$$\Delta F_1 = \pi r^2 \sigma_{12} (1 - \cos \theta)^2 \qquad (22.10)$$

The free energy of desorption required to remove the particle from the interface into the water phase is derived (*mutatis mutandis*), by the same argument:

$$\Delta F_2 = \pi r^2 \sigma_{12} (1 + \cos \theta)^2 \qquad (22.11)$$

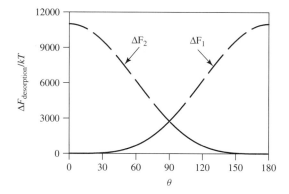

Figure 22.4 Variation of the free energies of desorption, ΔF_1 and ΔF_2 (relative to kT), of a spherical particle radius 10 nm radius at a planar O/W interface of interfacial tension 36 mN/m with a contact angle θ which the particle makes with the interface (measured through the oil phase) at 298 K, calculated using Eqs. (22.10) and (22.11).

Since both free energies are positive, the particle is thermodynamically stable at the interface. When the contact angle is acute the particle is more easily transferred into the oil phase; when the contact angle is obtuse the particle is more easily transferred into the water phase.

Variation of ΔF_1 and ΔF_2, at constant r and σ_{12}, for particles at a toluene/water interface is shown in Figure 22.4, where it can be seen that the particle is most strongly held at the interface for $\theta = 90°$. The free energy falls rapidly on either side of $90°$ such that for θ between 0 and $20°$ or between 160 and $180°$, the energy is only 10 kT or less. Emulsions stabilized by either very hydrophilic or very lipophilic particles for large drops (greater than 0.1 mm) are unstable. The extreme variation of the free energy of desorption with contact angle has a major influence on the ability of particles of different wettability to stabilize emulsions.

22.7 THE HLB SCALE

Bancroft's rule points out the importance of the emulsifier in determining the type of emulsion. An attempt to give the rule quantitative expression was initiated in 1949 by Griffin[11,12] of the Atlas Powder Company (now part of ICI; see www.uniqema.com). William C. Griffin was awarded the 1999 Maison G. deNavarre Award from the Society of Cosmetic Chemists. This company markets a number of nonionic surface-active solutes used as emulsifiers in food industries. Some of these are water soluble and, in accordance with Bancroft's rule, used as stabilizers of O/W emulsions; others are oil soluble and are used as stabilizers of W/O emulsions. Griffin created a continuous series of emulsifying agents, ranging from 100% oleic acid, predominantly lipophilic, to 100% sodium oleate, predominantly hydrophilic, by making mixtures of known composition from these two substances. The relative proportions of the two ingredients determine the balance between the hydrophilic and lipophilic properties, or HLB, of the mixture (see Section 11.2). A value of 1 was assigned arbitrarily to oleic acid and a value of 20 to sodium oleate; intermediate values were based on the relative amounts of each constituent in the composition, as follows:

$$\text{HLB} = 1W_1 + 20W_2 \tag{22.12}$$

where W_1 is the weight fraction of oleic acid and W_2 is the weight fraction of sodium oleate. Each composition was tested as an emulsifying agent by adding 1 g to a mixture of 50 ml of a refined white oil and 50 ml of water and shaking on a standard shaker for a fixed time. The emulsions were poured into viewing tubes, held for a period of time, and the volume and type of stable emulsion remaining were compared. Where two types of emulsion were present in the same tube, only the one of larger volume was noted. A presentation of typical data is shown in Figure 22.5.

The compositions made with oleic acid and sodium oleate serve as reference standards for commercial emulsifiers, which are tested in exactly the same way. The commercial emulsifier is assigned the HLB number corresponding to the reference sample that it most closely matches. Those that form W/O emulsions have a low HLB number, and those that form O/W emulsions have a high HLB number. After much experience, the HLB number was found to be associated with other applications: A particular number was best for a particular application of a surface-active solute. A summary of the HLB range required for different purposes is given in Table 22.3. A number of surfactant manufacturers use the system, particularly for cosmetic and personal-care applications, coatings, and crop protection.

The original method of determining HLB numbers is so profligate of time that shorter routes to the same end have been suggested. One approach is to calculate the HLB number of a molecule by adding contributions from its constituent groups, using the empirical relation obtained by Davies:[13]

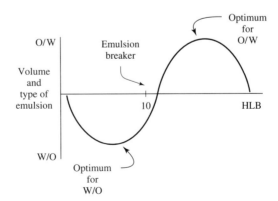

Figure 22.5 Variation of type and amount of residual emulsion with the HLB number of emulsifier at room temperature.

TABLE 22.3 Summary of Applications at Different Ranges of HLB at Room Temperature[13]

HLB Range	Application
3.5–6	W/O emulsifier
7–9	Wetting agent
8–18	O/W emulsifier
13–15	Detergent
15–18	Solubilizer

TABLE 22.4 Davies' HLB Group Numbers[13]

Hydrophilic groups	Group number
$-OSO_3^- Na^+$	38.7
$-COO^- K^+$	21.1
$-COO^- Na^+$	19.1
N (tertiary amine)	9.4
Ester (sorbitan ring)	6.8
Ester (free)	2.4
$-COOH$	2.1
$-OH$ (free)	1.9
$-O-$	1.3
$-OH$ (sorbitan ring)	0.5
$(-CH_2CH_2O-)_n$	$0.33n$
Lipophilic groups	**Group number**
$-CH-$	
$-CH_2-$	0.475
CH_3-	
$=CH-$	
$(-CHCH_3CH_2O-)_n$	$0.15n$

$$HLB = 7 + \sum(\text{hydrophilic group numbers}) - \sum(\text{lipophilic group numbers}) \quad (22.13)$$

The HLB group numbers are given in Table 22.4. A useful discussion of HLB and PIT is given by Shinoda and Kunieda.[14] Extensive bibliographies on HLB are given by Becher and Griffin in *McCutcheon's Detergents and Emulsifiers* (2000), by Becher in *Encyclopedia of Emulsion Technology,* Volume 2 (1985),

Volume 3 (1988) and Volume 4 (1996), and by Flick in *Industrial Surfactants*, 2nd ed. (1993).

The agreement between HLB numbers determined experimentally and those calculated from group numbers is satisfactory. The concept underlying the HLB number, as it is based on Bancroft's rule, is informative about the emulsion type but not necessarily about emulsion stability. Experiments by Boyd et al.[15] and by Berger et al.[16] showed that emulsion stability, not merely emulsion type could depend on HLB, thus extending the original concept beyond Bancroft's rule. Nevertheless, emulsion stability is the result of a complex interaction of properties, which include droplet size, interfacial viscosity, the magnitude of electrostatic and steric repulsion, internal volume, and so on. For this reason the application of the HLB concept cannot be expected to provide the whole answer to practical emulsion problems.

22.8 THE PHASE INVERSION TEMPERATURE (PIT)

The HLB number of an emulsifier corresponds to a composition on a graded scale from those that stabilize W/O emulsions to those that stabilize O/W emulsions. The HLB number of an emulsifier varies with temperature because the relative solubilities of the lipophile and the hydrophile vary with temperature. The variation of solubility with temperature is most profound for nonionic emulsifiers containing a PEO hydrophile because their solubility in water depends on hydrogen bonding. At higher temperatures hydrogen bonding is weakened by thermal forces and the emulsifier is less soluble in water. Common nonionic emulsifiers are water soluble at low temperatures, where they stabilize O/W emulsions and are oil soluble at higher temperatures, where they stabilize W/O emulsions. If anything, ionic surfactants are more water soluble at higher temperatures and therefore more likely to stabilize O/W emulsions, which makes the effective of temperature opposite to that of nonionics.

Figure 22.6 shows a series of 9-mL glass tubes containing equal weights of brine and oil. Each tube is held at a different temperature. The temperature at which mass of the water-rich phase equals the mass of the oil-rich phase is called the *phase inversion temperature* or PIT.[17] The PIT of an emulsifier is the temperature at which its lipophilic nature and its hydrophilic nature just balance. The PIT, therefore, is a measure by which the hydrophile–lipophile balance of emulsifiers, particularly nonionics, can be classified. The advantage of the PIT over the HLB number is that it is easier to determine experimentally. Both the HLB number and the PIT are functions of the composition of the oil and the ionic strength of the water. The study of the effects of these variables on emulsion stability is rendered simpler by use of the PIT.

The HLB number and the PIT are correlated. Figure 22.7 shows the relation of the HLB number to the PIT of a variety of nonionic emulsifiers for cyclohexane and water emulsions. At temperatures above the PIT, the emulsifier stabilizes the W/O emulsion; at temperatures below the PIT, the emulsifier stabilizes the O/W

THE PHASE INVERSION TEMPERATURE (PIT) **433**

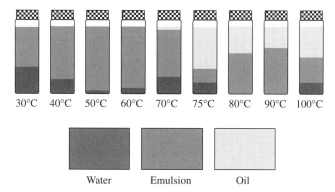

Figure 22.6 Emulsion stability for an nonionic emulsifier according to results obtained by Baglioni et al. At low temperatures water drains out of an O/W emulsion. At the PIT of 75° the emulsion breaks. Above 75° the emulsion inverts and excess oil drains out.[18]

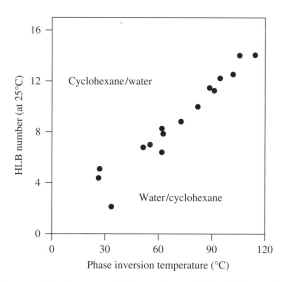

Figure 22.7 The correlation between HLB numbers of 16 nonionic surface-active solutes and their PIT in cyclohexane and water emulsions, at 3% by weight of emulsifier.[19]

emulsion. The relation between the HLB number and PIT was determined for a variety of oils and ionic strengths.[19] The HLB corresponding to a PIT at any temperature is the HLB at the interpolate of the curve shown in Figure 22.7.

As with the HLB numbers, the PIT can be estimated from the knowledge (rarely accurate) of the amphiphile,[20]

$$\text{PIT} = -160 + 15.5(\text{HLB}) + 1.8(\text{ACN}) + b\text{S} \qquad (22.14)$$

where b is a function of the electrolyte. For example, $b = -0.68°C$ L/g for sodium carbonate and $-0.25°C$ L/g for sodium chloride; S is the salinity in grams of electrolyte per liter of aqueous solution; ACN is the alkane carbon number of the oil.[20] Examples of a good agreement between the observed values of PIT and the PIT calculated by Eq. (22.14) are reported.[20]

As temperatures approach the PIT, the interfacial tension decreases continuously. This result conforms to the Ross–Nishioka effect, which calls for greater surface activity as the conditions for a phase separation are approached (see Section 16.1).

22.9 MECHANICAL PROPERTIES OF THE INTERFACE

Emulsifying agents are adsorbed at the oil/water interface, which may become so crowded with solute molecules that it develops rheological properties different from those of either bulk phase. The interface may be non-Newtonian, plastic, for example, while the oil and the water phases remain Newtonian. Great stability is conferred on an emulsion by a plastic interface, which is solid at low shearing stresses and so prevents coalescence of droplets on collision. Many instruments have been designed to measure the rheology of an interface. The simplest of these will hardly do more than register the presence of a plastic interface; a more sophisticated design is capable of providing quantitative information that can be interpreted in terms of absolute film properties.

The rheological behavior of an interface can be measured either by shearing with a ring or disc in the plane of the surface while the area remains constant, or by expanding or contracting the interface. An example of the first method is the use of a torsional pendulum, which is damped by immersing its bob, in the form of a double cone so as to have one cone in the upper liquid, the other cone in the lower liquid, and the widest part of the bob, that is, where the cones are joined, at the interface (see Figure 22.8). The torsion wire is supported at the end of a shaft that can be given a partial turn by a lever. The damping of the oscillations is measured by a light beam reflected from a mirror attached to the wire. The rate at which the amplitude of the oscillation decreases is semilogarithmic when the pendulum is placed in a Newtonian fluid.[22] Normally the two bulk liquid phases in which the cones are immersed are Newtonian fluids, but the interface may not be. The bob is so designed that the damping influence of the interface is maximized. A plastic interface would affect the damping behavior sufficiently to throw it out of its regular semilogarithmic pattern; the amplitudes of the oscillations decrease more rapidly. Plasticity is characterized by a "yield point," that is, the critical shearing stress above which flow begins. At certain portions of its oscillatory cycle, the motion of the pendulum produces a shearing stress that is less than the yield point; the result is as much as if the free oscillation were halted from time to time by an applied force (Figure 22.9).

To obtain interfacial rheometry of more complex systems, especially with polymeric surfactants with nonlinear effects, such as shear-rate dependency of

MECHANICAL PROPERTIES OF THE INTERFACE   435

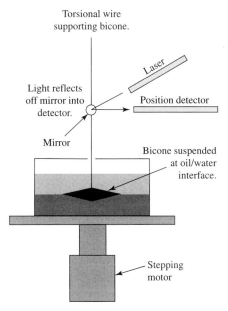

Figure 22.8 Schematic of a interfacial viscosimeter with an oscillating double-cone bob.[21]

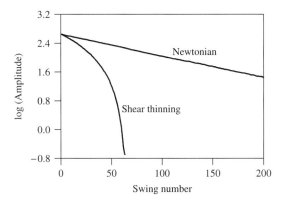

Figure 22.9 The semilogarithmic damping of the torsion pendulum for Newtonian and shear-thinning interfacial flow.

viscosity and elastic effects, more sophisticated instruments are required. The biconical-bob interfacial viscosimeter is adapted to provide a constant shearing stress or a forced sinusoidal oscillation with a small amplitude by driving the bob with a stepping motor controlled by a function generator.[21] The torque communicated to the wire supporting the bob is measured by a strain gauge. An instrument of this type is available from Camtel. Ltd.

Another type of instrument is reported by Biswas and Haydon.[23] It uses an electromagnetic device to drive a ring placed at the interface at constant applied stress or to hold a constant position at varying stress. The current in the electromagnet is a direct measure of the force applied to the ring at any time. The motion imparted is a two-dimensional analogue of Couette flow between concentric cylinders and is analyzed in a similar way. As well as measuring the coefficient of viscosity, the instrument can be used in two types of quasi-static experiments (i.e., those in which the shear rate is so small that inertial effects can be ignored): creep recovery and stress relaxation. Adsorbed films of serum albumin, pepsin, arabinic acid, and poly-α, laevo-lysine of known area per residue were formed at hydrocarbon/water interfaces and their rheology described by means of this instrument.

Instruments of similar design are used to measure the rheology of liquid/vapor surfaces (Chapter 23).

22.10 VARIATION WITH CONCENTRATION OF INTERNAL PHASE

The closest packing of uniform spheres fills about 74% of space; each sphere has 12 nearest neighbors, and about 26% of the space is void. Applied to emulsions, it seemed that some hindrance would be met on trying to emulsify more than 74% by volume of internal phase. This is the stereometric hard-sphere model of emulsions, originated by Walter Ostwald,[24] the younger son of the more famous Wilhelm. So insistent was he and other German chemists on this "critical point" of 73.4% solids and 26.6% serum, postulated at a period of strained relations between Germany and Great Britain, that he stimulated an Englishman, S. U. Pickering, to disprove it. Pickering succeeded in making emulsions containing greater than 99% internal phase. In such systems the continuous phase must be in the form of thin lamellae, and the stability of the emulsion depends on these lamellae being heavily gelated and incapable of flow. This property is secured by the high concentration of soap in the water which brings the system into a mesophase region of the phase diagram. The lamellae are liquid crystals, immobilizing water within their structure and forming a gel. Pickering proved his point, but concentrations near 74% by volume internal phase nonetheless influence the physical properties of an emulsion. Even though emulsions are seldom composed of droplets of uniform size, droplets begin to make close contact at between 60 and 70% volume concentration; and close contact begins to affect some of their sensible properties.

Manegold (1952) has published a useful diagram (Figure 22.10) to represent the dependence of some properties of emulsions on volume concentration. The diagram is a summary of general behavior, which cannot be relied on to predict properties of a specific emulsion. Above a volume concentration of about 70% the properties of an emulsion become either discontinuous (inversion) or remain continuous, with the production of deformed droplets in contact. The deformation of the spheres is created by concentrations of internal phase greater than about

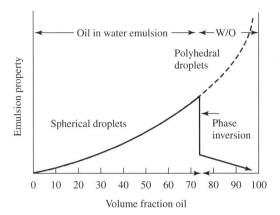

Figure 22.10 The variation of properties of emulsions with changes in composition. If inversion occurs, there is a discontinuity in properties, as they change from one curve to the other. Above 74% there is either a phase inversion or the droplets are deformed to polyhedra.[25,26]

70% by volume. This type of emulsion is called "polyhedral foam type," using an analogy with the structure of drained foam, and, like such a foam, has a large structural viscosity. Ostwald called such emulsions "liquid–liquid foams." Bütschli[27] pointed out their analogy to protoplasm, and Sebba[28] has suggested that the secret of life lies in the behavior of immiscible liquids under the influence of their interfacial forces.

Ruckenstein[6] confirms by thermodynamic reasoning that the inversion behavior is typical of relatively weak surface-active emulsifying agents, while agents that continue to stabilize emulsions to high internal-phase concentrations are stronger. He found that exceptions to Bancroft's rule occur in the former group whereas those in the latter group follow Bancroft's rule.

22.10.a Rheology of Emulsions

A readily observed property of emulsions is viscosity. At concentrations well below 74% the spheres are not in contact, and the flow of the emulsion is not unduly impeded by interference of the droplets with one another. In more concentrated emulsions, interference occurs and the resistance to flow becomes more marked. Finally, the droplets may be packed so close to each other that flow is seriously impeded, which is made manifest by a high viscosity, requiring a large shearing stress to overcome the structure built up by many spheres in contact. Should the emulsion invert on addition of more of the internal phase, as frequently happens, the inversion is marked by a sudden reduction of viscosity. The transition is a remarkable transformation from a consistency like ointment to that of a thin cream.

Measurements of the viscosities of emulsions illustrate the changes that occur with concentration of internal phase. Figure 22.11 shows the data of Richardson[29]

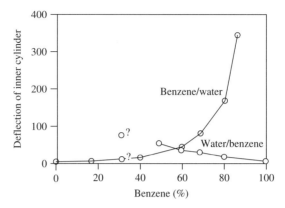

Figure 22.11 The viscosity of two types of emulsion as a function of composition.[25,29]

for both O/W and W/O types. The significant point is the enormous increase of viscosity of the O/W emulsion at concentration above about 70% by volume. The other type of emulsion shows inversion at higher concentrations, with a discontinuity in the observed property.

In emulsions with less than 70% internal phase, the size of the droplets has little effect on the viscosity. As the concentration approaches that of polyhedral foam type where the viscosity would be expected to increase sharply, a wider distribution of droplet sizes makes it possible to include larger volume concentrations before structural viscosity becomes marked. An ordered arrangement of closed-packed spheres forms readily and spontaneously when the spheres are uniform in size; a relatively small number of outsized spheres prevents such an arrangement from forming, thus reducing the average number of nearest neighbors, which increases the free motion of the spheres past each other. Homogenizing such an emulsion narrows the size distribution and results in increased viscosity.

Concentrated emulsions and foams have many practical applications that take advantage of their peculiar rheological properties, such as a high viscosity compared to that of their pure constituents, a yield stress, and shear-thinning behavior. Concentrated emulsions also provide a means to transport highly viscous fluids as emulsified droplets. Asphalt emulsions, used as protective coatings in road paving and as a means of mixing asphalt with fibers and other forms of particulate matter, are more convenient to handle than bulk asphalt. Concentrated emulsions are used in the cosmetic industry as skin conditioners, hand creams, hair-grooming gels, suntan lotions, adhesives, and so on and in the food industry as gravies and salad dressings.

Princen[30] modeled monodisperse, concentrated emulsions, and foams as infinitely long, uniform cylindrical drops (or bubbles.) He took into account the volume fraction of the dispersed phase, the droplet radius, the interfacial tension, the thickness of the interstitial films, and the contact angle associated with the films. He found the yield stress and the shear modulus to be proportional to the interfacial tension and inversely proportional to the radius of the droplets.

The yield stress increases sharply with increasing volume fraction; the shear modulus increases as the square root of the volume fraction. The effect of a finite contact angle is to decrease the shear modulus and, in most cases, to increase the yield stress. Finally, the effect of a finite film thickness is to increase both the yield stress and the shear modulus. These results are in general agreement with rheological measurements of concentrated emulsions and foams, and are valuable in suggesting scaling laws.

In addition to the foregoing, the following trends in the rheology of emulsions have been identified:[30]

1. The yield stress and apparent viscosity increase with the volume fraction of the dispersed phase and with decreasing drop (or bubble) size.
2. The yield stress of typical emulsions with volume fractions of dispersed phase above 0.90 is of the order of a few hundreds to a few thousand milli-Newtons per square meter. It is much smaller for foams, presumably because of the larger size of the dispersed unit.
3. The observed rheology varies with the nature of the wall surface in contact with the sample depending on whether the continuous phase spreads on the wall or not.
4. Both emulsions and foams can be destroyed at sufficiently high rates of shear.

22.10.b Electrical Conductivity of Emulsions

The electrical conductivity of an emulsion can be used to follow changes in emulsion composition and to determine emulsion type. If interfacial conductivity is significant, the electrical conductivity of the emulsion will vary with droplet size and distribution, and may then be used to follow any change of these properties with time. Many emulsions, however, show electrical conductivity that is independent of particle size, and thus may be treated by simple electrical theory. Maxwell[31] developed the following expression (ignoring interfacial effects) for the specific electrical conductivity L of a two-phase system, of which the specific electrical conductivity of the outer phase is L_1 and of the inner phase is L_2:

$$\frac{L_1 - L}{L + 2L_1} = \frac{V(L_1 - L_2)}{L_2 + 2L_1} \tag{22.15}$$

or in its equivalent form,

$$L = \frac{L_1[2L_1 + L_2 - 2(L_1 - L_2)V]}{2L_1 + L_2 + (L_1 - L_2)V} \tag{22.16}$$

where V is the volume fraction. Since this relation specifies the inner and the outer phase, it yields two equations for any pair of liquids, one for the O/W type and another for the W/O type of emulsion that they can form between them.

For example, if the liquids have values of L that are 0.20 and 0.02 Ω^{-1} m^{-1}, respectively, then the conductivities of the W/O and O/W types of emulsion are

$$L_{W/O} = 0.02 \left(\frac{4+6V}{4-3V}\right) \Omega^{-1} m^{-1} \qquad (22.17)$$

$$L_{O/W} = 0.20 \left(\frac{7-6V}{7+3V}\right) \Omega^{-1} m^{-1} \qquad (22.18)$$

The values chosen here as illustrations are not far from those of two actual liquids investigated by Eucken and Becker:[32] an aqueous solution of potassium iodide and water-saturated phenol, at 19.6°. In Figure 22.12 the two theoretical equations, indicated by dashed lines, are compared with experimentally observed values. The observations move from the theoretical values for an O/W emulsion to those for a W/O emulsion, indicating inversion. The inversion is not abrupt but exists through a zone of concentration. Theory and experiment agree well with each other. The formation of multiple emulsions, that is, O/W/O or W/O/W, would invalidate the application of this theory.

The specific conductivity of a material characterizes its steady-state electrical properties. If an AC field is applied to the emulsion the electrical properties are found to vary with the frequency of the electric field. The time dependencies of various conduction processes cause large changes in the conductivity with frequency. Most of these time dependencies result from the motion of ions in the electrical double layer, including the Stern layer. The application of AC

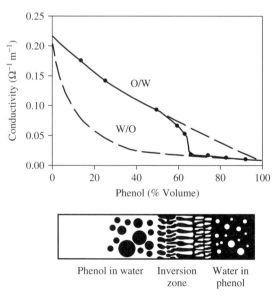

Figure 22.12 The specific conductivity of aqueous potassium iodide and phenol emulsions as a function of composition.[33]

conductivity with varying frequency is called dielectric spectroscopy[34] and has been usefully applied to the measurement of the flocculation and coalescence in emulsions, particularly crude oil emulsions.[35]

22.11 MEASUREMENT OF EMULSION STABILITY

An indication of the thermodynamic work done in creating an emulsion is provided by the area of interface produced. As the emulsion ages, the area of interface decreases. The most fundamental way to express emulsion stability is by means of the variation of interfacial area with time. The interfacial area may be expressed as square meters per cubic meter of emulsified liquid. For an average size droplet of diameter 2 μm, the interfacial area would be 3×10^6 m^2/m^3, and would decrease to 2×10^6 m^2/m^3 when the average droplet size had increased to 3 μm. Rather than measure area directly, it may be calculated from the particle size.

A direct way to get the average particle size is by means of the microscope. Size-distribution curves may be obtained by measuring about 400 droplets, but better accuracy is obtained if the number measured is even larger. Levius and Drommond[36] found a Bausch and Lomb camera lucida more convenient to use than photomicrography. They traced the outlines of from 400 to 800 drops per determination and they report that "after some practice this operation could be performed in less than 45 minutes." Digital imaging techniques are now available. Results reported by this optical method vary from 28,000 to 12,000 cm^2/cm^3. Emulsions were prepared by premixing in a Waring blender for 10 min and then passing them through a hand-operated lever-type homogenizer. Some degree of uniformity is produced by this method, but the emulsions are relatively coarse. Emulsions prepared by the method of phase inversion, for example, produce much smaller particles, but they are then not visible under the microscope and some other method such as quasi-elastic light scattering (see Section 4.5.f) must be used.[37]

The advantages of using the microscope are that it is direct, inexpensive, and yields a size-frequency distribution as well as a value for the interfacial area. Some methods give only an average particle size, and an interfacial area, if wanted, is based on that value. Where it is applicable, the use of the microscope is preferable to any other method.

The electrical conductivity of an emulsion depends on the concentration of the dispersed oil phase. An unstable emulsion will have a variation in dispersed oil concentration from bottom to top. Therefore the stability of an emulsion can be checked by comparing the electrical conductivity at the top and the bottom of a container. A great advantage of this method is the ease of measurement. The detection of a difference in conductivity does not require any sample preparation, especially dilution. Equipment for such a measurement is available from Krüss (www.kruss.de).

Similarly, the speed of sound in an emulsion depends on the concentration of the dispersed oil phase. The speed of sound is measured by transmitting a short pulse of sound and measuring the length of time required for the pulse to reach a detector opposite to the source.[38] This is called a time-of-flight measurement. Two advantages of determining emulsion stability by time-of-flight is that the sample can be measured without dilution and the container and the sample can be optically opaque. The usual procedure is to make a calibration curve of the speed of sound and the volume fraction of the oil. The speed of sound depends on the composition but not the droplet size. A useful technique is to mount the transmitter and the receiver on a moving stage and scan an emulsion from top to bottom.[39] The variation in the speed of sound with height is a measure of the variation in composition with height. The degree to which the composition varies with height and with time measures the stability of the emulsion. The composition will vary as the emulsion creams even if the droplets are stable. Coagulation of emulsion droplets increases the rate of creaming. Figure 22.13 shows the variation in composition with height, horizontal axis, and with time, the multiple curves, for an O/W emulsion of an alkane, n-tetradecane.

Figure 22.13 describes the creaming of an emulsion in terms of the volumes of water, emulsion, and oil phases with time as the emulsion creams. The middle phase is the volume of the original emulsion which declines steadily with time until its upper and lower boundaries meet (i.e., its volume equals zero) while the volumes of the oil and water phases increase.[40] The volume of the water layer increases steadily from the bottom of the tube and the volume of the cream layer increases steadily from the top of the tube as indicated by the changing length of the discontinuity in the curve.

Particle size counters, such as the electrozone sensors and the photozone sensors, work quite well for O/W droplets larger than a micron in diameter.

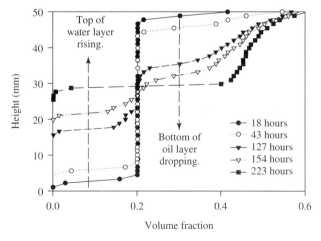

Figure 22.13 Creaming of n-tetradecane-in-water emulsion (15% oil, 0.75% protein, pH 7) containing 0.05% by weight xanthan. Height is plotted against volume fraction of oil on storage at 20°C for various times.[41]

We must turn to other methods for smaller emulsion drops. Van der Waarden[42,43] reported the preparation of emulsions of medicinal oil in water, stabilized with from 5 to 35% of alkylbenzene sulfonates in the oil, prepared by the method of phase inversion. As the content of emulsifying agent was increased, average particle sizes decreased to the range of 20–90 nm. The interface contained all the emulsifying agent and simply increased in area as more agent was present in the system. These emulsions were centrifuged for half an hour at an acceleration of 25,000g without any visible creaming of emulsified oil. The particle sizes were below the resolving power of a microscope, but could be measured by light-scattering methods. The methods of light-scattering, however, require extreme dilution of the emulsion. The dilution may have a destabilizing effect, unless the emulsion serum is used as the diluent.

The centrifuge is used for accelerated stability testing of an emulsion. Another method is to store the emulsion at a higher temperature as long as the phase-inversion temperature is not crossed.[44]

22.12 MAKING EMULSIONS

The use of high-shear mixers, colloid mills, homogenizers, and sonic and ultrasonic dispersers to make emulsions is described in Chapter 5. Additional methods are described below.

22.12.a The Method of Phase Inversion

The method of phase inversion takes advantage of interfacial tension as a mode of creating small particles out of extended liquid films. When the internal phase is at high concentration, the external phase is attenuated into a continuous thin sheet. If the emulsifier is weak, the internal phase coalesces and nips off the interstitial continuous phase into small drops. The emulsion is now inverted. The droplets of the final emulsion are much smaller than those of the parent emulsion. If this technique is to be used, the emulsifier must be able to stabilize, at least temporarily, the type of emulsion opposite to the one that is finally desired. The amphipathic nature of the emulsifier makes this possible. As was pointed out above, inversion is less likely to occur if the emulsifier used is too powerfully surface active. The advantage of the technique is that an emulsion of fine droplets can be made with the least amount of mechanical action and its attendant heat.

To make an O/W emulsion, the emulsifier is dissolved in the oil, and water is added slowly as the emulsion circulates through a colloid mill. As more water is added, the W/O emulsion acquires ointment-like consistency at concentrations of water about 60–70%. The thick emulsion is transferred to a mixing tank and the final amount of water is added with gentle stirring. The viscosity drops suddenly, marking the transition to an O/W emulsion.

The inversion of the W/O emulsion is found occasionally to take place at concentrations much less than 60–70% internal phase, even as low as 25%. When this occurs, the inverted emulsion has a high concentration and consequently a high viscosity. The usual loss of viscosity on inversion is here a gain of viscosity, but the inverted emulsion has water as the external phase and so can imbibe additional water more readily than before its inversion. The driving force for inversion in such cases is not the close contact of water drops but the massive migration of a predominantly hydrophilic emulsifier out of the oil and into the water.[45]

22.12.b Phase-Inversion Temperature (PIT) Method

Small oil droplets can be formed by emulsifying just below the PIT of the emulsifier. As the temperature is raised toward the PIT, the interfacial tension becomes continuously lower: an example of the Ross–Nishioka effect. When oil is emulsified in water 2–4 degrees below the PIT, the low interfacial tension enables small droplets to form. Once the fine emulsion is made, it is cooled quickly to stabilize it at room temperature.[46] The interfacial tension is now greater but the emulsion has been made.

22.12.c Condensation Methods

Colloidal suspensions of insoluble solids, such as arsenic sulfide, barium sulfate, and silver iodide, are made by precipitation reactions in which particle size is limited by promoting a large number of nucleating centers. Fine suspensions are better made by precipitation from solution than by comminution of macroscopic particles. An analogous method to make a fine emulsion is to solubilize an internal phase in micelles. The internal phase may be introduced as a vapor, which nucleates heterogeneously on dust or in micelles, or as a liquid. To obtain small droplets, a large concentration of micelles is required. Since this method depends on a degree of solubility of the internal phase in the medium, in order that molecules may reach the micelles, the same solubility promotes Ostwald ripening, leading to instability. We have already seen how this instability is overcome (see Section 13.3.f) by adding an insoluble coemulsifier to set up a counteracting osmotic pressure within the droplets.

22.12.d Intermittent Milling

An emulsion can be made by shaking two phases together in a test tube. Briggs[47,48] found that emulsification of some systems is much more efficient if the shaking is interrupted by rest periods. For instance, 60% by volume of benzene in 1% aqueous sodium oleate is completely emulsified with only five shakes by hand in about two minutes, if after each shake an interval of 20–25 seconds is allowed. If the shaking is not interrupted, about 3000 shakes in a machine, lasting about seven minutes, is required. Figure 22.14 shows that the time required to make an emulsion with continuous mechanical shaking is much greater than the time required to make the same emulsion with double shakes every 30

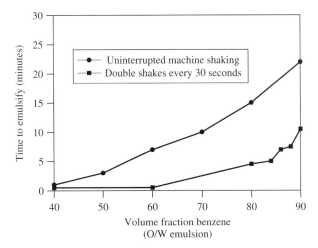

Figure 22.14 Time required to make a stable emulsion as a function of volume fraction benzene in water with 1% sodium oleate at room temperature. Emulsions by intermittent double shakes by hand every 30 seconds, lower curve. Emulsions by continuous mechanical shaking, upper curve.[48]

seconds. The rest intervals provide the time required for the stabilizer to diffuse to newly created interface; without rest intervals, shaking promotes coalescence by collisions of unstabilized droplets.[49] In industrial processing, intermittent milling is frequently used. Continuously recirculating systems, which have large holding volumes and small milling volumes, produce intermittent milling automatically, with efficiency and economy.

22.12.e Electric Emulsification

Mechanical methods of emulsification extend the internal phase as threads or films, which are then pinched off as droplets by surface tension (Rayleigh instability). Charging the interface electrically produces electrohydrodynamic instability, which promotes the formation of more threads and films as electric charges repel each other. The technique is to eject the internal phase through a fine capillary into the medium. The capillary is held at a high potential with respect to ground. A spray of fine droplets emerges from the tip and is dispersed vigorously throughout the medium (see Section 5.4).

22.12.f Special Methods

A stable emulsion can be produced by means of a chemical reaction to produce a stabilizer at the interface. If a fatty acid is dissolved in the oil phase and a base is dissolved in the aqueous phase, these reactants combine to form a soap at the interface where the two phases meet. This method ensures that the stabilizer is concentrated at the interface. A variant of this technique is to perform polymerization at the interface by having a water-soluble monomer in the aqueous phase along with the initiator and an oil-soluble monomer in the oil phase. Besides interfacial polymerization there is interfacial coascervation, which is the formation of thick polymer film by precipitating the polymer from aqueous solution usually by varying the pH or temperature or both.[50]

Another method is to dissolve the stabilizer in the *internal* phase before bringing the two phases together. Bancroft's rule states that the stabilizer is more soluble in the external phase, and therefore it migrates through the interface during the mixing process. This operation also ensures that the stabilizer is concentrated at the interface, especially during the processing.

22.13 BREAKING EMULSIONS

Many situations arise where the separation of the two phases of an emulsion is desired. Coarse or macroemulsions (droplet diameters greater than 1000 nm) separate on standing, especially if the density difference between the two liquid phases is large. If the droplets come together without coalescence, the emulsion becomes more concentrated and the process is called *creaming,* by analogy with the separation of cream from milk. Complete phase separation requires that the droplets coalesce, which may indeed occur spontaneously in the cream if the droplets are not well stabilized. Further steps have to be taken if coalescence is to be promoted. These steps may be mechanical, thermal, or chemical. The subject is reviewed by Menon and Wasan[51] and by Lissant.[52]

22.13.a Mechanical Demulsification

The rate of demulsification can be promoted by agitation, as in a blender. Many emulsions are sensitive to high shear, which throws the droplets into one another with consequent coalescence. Centrifugation is another mechanical method to accelerate creaming or breaking.

Passing the emulsion through a filter bed whose surfaces are wetted by the internal phase often leads to a separation. This process is more effective with W/O emulsions and hydrophilic filter beds than with O/W and lipophilic beds. The same effect can sometimes be obtained more readily by mixing particles of a hydrophilic solid with a W/O emulsion or particles of low surface energy with an O/W emulsion. The latter procedure, again, is less often successful. Ultrafilters employ membranes with pores less than 5 nm in diameter. An O/W emulsion is passed along a tubular membrane under pressure; water diffuses through the membrane and flows out.[53]

22.13.b Thermal Demulsification

Most emulsions are less stable at higher temperatures, as the adsorption of the stabilizer decreases with temperature. In some cases the emulsifying agent is thermally decomposed. Emulsions stabilized with agents containing polyethylene oxide or polypropylene oxide are sensitive to an increase in temperature, as these polymeric moieties become insoluble at higher temperatures. The PIT is the temperature at which the emulsion inverts, which is also the temperature at which it is least stable. Demulsification can then occur by heating or cooling the emulsion to the PIT.

O/W emulsions stabilized electrostatically, but without protective colloid, are subject to breaking on freeze–thaw cycling. The separation of ice allows the electrolyte in the aqueous medium to become concentrated, thus reducing the electrostatic repulsion between droplets. The freeze–thaw process usually needs to be done slowly and repeatedly for best results.

22.13.c Chemical Demulsification

An emulsion will often break if the emulsifying agent is chemically altered. Emulsions stabilized with alkali-metal soaps are broken on adding acid or metal ions, such as iron or calcium, which convert soaps to water-insoluble fatty acids or metallic salts. Emulsions stabilized with anionic agents can be broken by adding a cationic detergent, such as a quaternary ammonium salt, which converts the agent to a water-insoluble complex. A more subtle chemical effect is to alter the HLB of the emulsifying agent by adding a surface-active solute with a very different HLB number. This procedure depends on the two solutes being able to comicellize, which allows them to blend so intimately that the mixture behaves as a unit of intermediate HLB. Thus an O/W emulsion stabilized with a agent of high HLB may be vulnerable to the addition of an agent of low HLB.

Another way to attack a stable emulsion is to replace the emulsifying agent with a surface-active solute of greater adsorptive potential but less stabilizing effect. For example, petroleum emulsions owe their great stability to the mechanical strength of the interface, provided by asphaltenes of high molecular weight. These can be displaced by the more surface-active petroleum sulfonaphthenic acids, which weaken the interface sufficiently to break the emulsion.[54] Canevari and Fiocco applied nonionic surface-active solutes such as the polyethoxyalkenes to break Athabasca-tar-sand froths.[55]

The emulsifying agent can be made to desorb from the droplets by the addition of a water-miscible organic solvent, such as methanol, ethanol, or acetone, to an O/W emulsion, or the addition of an oil-soluble Lewis base or Lewis acid, whichever is appropriate to interact with the emulsifying agent and solubilize it. In an analogous way, emulsions stabilized with adsorbed solid particles can be broken by the addition of a solvent that wets the particles and removes them from the interface.

Among chemical methods to destabilize emulsions is the addition of electrolyte to increase the ionic strength of the medium, thus reducing the repulsion between droplets of an electrostatically stabilized emulsion. The most effective electrolyte for this purpose is a salt with multivalent ions.

A useful method to characterize and predict the performance of emulsion breakers would be the discovery of a relation between the chemical property

448   EMULSIONS

of the demulsifier and that of the oil. Such a relation for the demulsification of crude W/O petroleum emulsions is postulated by Berger et al.[16] Parameters of the demulsifier that were studied were the chemical type, molecular weight, degree of branching, partition coefficient between the water and oil phases, reduction of interfacial tension, and changes of interfacial viscosity. The importance of each of these properties to performance was discussed.

22.14 MICROEMULSIONS AND MINIEMULSIONS

Large amounts of two immiscible liquids can be brought into a single phase that is macroscopically homogeneous but microscopically heterogeneous, by addition of an appropriate surfactant or surfactant mixture. This optically clear mixture is called a microemulsion. In general a microemulsion consists of four components: water, oil, surface–active solute, and coemulsifier such as a fatty alcohol, although other suitable nonionic surface-active solutes may be effective. The droplets are too small to be effectively stabilized by electrostatic repulsion, but they are small enough that an adsorbed layer 2–3 nm thick, provided by conventional nonpolymeric emulsifiers, confers steric stabilization. The volume fraction of the dispersed phase varies over a fairly wide range (20–80%). The droplets have diameters from 10 to 150 nm and more often 10 to 60 nm. In this range the microemulsion is monodisperse. Bicontinuous structures with rapidly fluctuating curvatures are also known. The debate on the thermodynamic stability of microemulsions continues, but the definition of a microemulsion as a combination of water, oil, and surfactants that is macroscopically a single phase and thermodynamically a stable isotropic solution is widely accepted. The widespread use of microemulsions is based primarily on their high capacity for lyophobic materials. They find use in analytic determinations, biotechnology, pharmacy, foods, cosmetics, agrochemicals, dyeing, nanotechnology, enhanced oil recovery, extraction of contaminants, detergency, and metalworking.[56]

Microemulsions result from a large free energy of adsorption of the surface-active components combined with a low interfacial tension. The Helmholtz free-energy change for any process that alters the interfacial area at constant volume and temperature is given by

$$\Delta F = \sigma_{12}\Delta A - W^{des}\Delta n \qquad (22.19)$$

where σ_{12} is the interfacial free energy, ΔA is the change of interfacial area, W^{des} is the work of desorption in mJ/mole, and Δn is the number of moles desorbed. The work of desorption is the result of various components, such as changes of entropy, changes of surface–charge density, and molecular interactions between constituents of the interfacial film. When the work of desorption is sufficiently large and σ_{12} is sufficiently small, $dF/dA < 0$, and the interfacial area increases spontaneously. A spontaneous increase in total interfacial area leads to a spontaneous decrease in average droplet size. For this reason microemulsions are thermodynamically stable. Their small particle size accounts for their being translucent.

A large work of desorption and a small interfacial tension is achieved practically by combining an ionic surface-active solute with an insoluble fatty derivative. This insoluble component contains electron donor or acceptor groups, such as hydroxyl or chloride, that allow it to partition itself between the core of the droplet and the interface. The component is now appropriately named a coemulsifier although its insolubility is still a sine qua non; nevertheless the old adage that bodies do not react unless they are in solution still holds.* The coemulsifier comicellizes with the soluble emulsifier and the micelles provide the molecules of coemulsifier that are adsorbed at the interface. The essential feature of this microsystem is the spontaneous adsorption of coemulsifier at the interface. Such materials are strongly adsorbed because of an acid–base interaction with water at the interface, and because they reduce the electrostatic repulsion between the ionic heads of the primary emulsifier.

Adsorption of the coemulsifier may move the composition of the interface into a liquid-crystal regime that confers additional stability to the microemulsion. For styrene emulsions the stability conferred by a fatty-alcohol coemulsifier decreases in the following order with respect to the chain length of the alcohol: C16 > C18 > C14 > C12 > C10 (similar to the order of the Ferguson effect and perhaps for the same reason; See Section 13.2). This order reflects the way in which the liquid-crystal regime in a polycomponent system changes with the chain length of the coemulsifier.

Swollen micelles, microemulsions, miniemulsions, and (macro)emulsions form a continuum of properties determined by the size of the droplets, which in turn is determined by the nature of the adsorbed solute. Microemulsions combine a property of micelles, inasmuch as they form spontaneously, and a property of emulsions, inasmuch as the interior of the droplet is bulk phase. Symposia on micellization, solubilization, and microemulsions are held frequently.[57]

Miniemulsions droplets occur in the range 100–1000 nm. They may be produced by extreme comminution. Because of their large interfacial area, high concentrations of emulsifier are required. If the internal phase has low solubility in the external phase, miniemulsions may be stabilized by the usual electrostatic or steric mechanisms. If the internal phase is slightly soluble in the medium, however, miniemulsions, even though stabilized with respect to coalescence, are subject to degradation by diffusion. When, for example, a component such as styrene, which is slightly water soluble, is homogenized in water containing a large concentration of emulsifier, microdroplets are formed at first; the styrene then diffuses through the water from smaller to larger droplets because of the differences in Laplace pressures (Ostwald ripening). Higuchi and Misra[58] discovered that a small amount of a water–insoluble component, such as hexadecane, incorporated into the styrene droplet arrests the Ostwald ripening. The mechanism is as follows: styrene dissolves in the droplets containing the water-insoluble

---

* Corpora non agunt nisi soluta.

component, which itself cannot diffuse appreciably; meanwhile Laplace pressure (also called capillary pressure because it arises from differences in interfacial curvature) causes the styrene to diffuse from smaller to larger droplets. The concomitant reduction of styrene concentration in the smaller droplets reduces their osmotic pressure with respect to the larger ones. The difference in osmotic pressure thus created, counteracts the capillary pressure. Miniemulsions of styrene, and in fact of all monomers that are slightly soluble in water (as most of them are, so the application of this principle to emulsion polymerization is evident), may be stabilized in this way. Without the added insoluble component (costabilizer) the smaller drops could not persist. This stabilizing mechanism has been termed osmotic stabilization.

These effects may be treated by simple thermodynamics.[59] Consider the case of a miniemulsion of a completely water-insoluble compound 2 to which is added a slightly water-soluble compound 1. Starting with an miniemulsion of, say, hexadecane (2) in water, how much styrene (1) would be absorbed by the preformed droplets of hexadecane? For an ideal solution in the droplet

$$\frac{RT}{V_m} \ln \Phi_i + \frac{2\sigma}{r_i} = 0 \qquad (22.20)$$

where V_m is the partial molar volume of styrene, Φ_i is the mole fraction of styrene in the ith droplet of equilibrium radius r_i, and σ is the interfacial tension. Equation (22.20) is the thermodynamic condition for the stability of droplets of various sizes as a function of the amount of hexadecane in each droplet. Introducing corrections for nonideality gives the Morton equation,[60]

$$\frac{RT}{V_m} \ln \Phi_1 + \frac{RT}{V_m}\left(1 - \frac{1}{j_2}\right)\Phi_2 + \frac{RT}{V_m}\Phi_2^2 X_1 + \frac{2\sigma}{r} = 0 \qquad (22.21)$$

where j_2 is the ratio of the molar volumes of 2 to 1, X_1 is the Flory interaction parameter related to the enthalpy of mixing, and Φ_1 and Φ_2 are the volume fractions.

Equation (22.21) demonstrates the possible effect on the equilibrium droplet size of varying the molar volume of the insoluble component. For simplicity, take a uniform dispersion with $T = 300\ K$, $X_1 = 0.5$, $V_{m\cdot} = 0.1\ L$, $\sigma = 5$ mN/m, and $r = 100$ nm. The ratio of the volume of styrene-absorbed V_1 to the volume of hexadecane V_2 as calculated by Eq. (22.21) for various values of j_2 is given in Table 22.5.

Table 22.5 shows that the swelling capacity r/r_0 is strongly affected by the value of j_2. The smaller the molecular weight of the insoluble component, the greater the swelling. An explanation of the swelling effect is that the large energy of mixing, mainly entropy of mixing, due to the interaction of component 1 with the insoluble component 2, balances the increase in interfacial energy of the enlarged droplet. Ugelstad et al.[61] demonstrated experimentally the strong effect of hexadecane on the stability and particle size of miniemulsions of styrene. Styrene was added with mild stirring to miniemulsions of small amounts of

TABLE 22.5 Effect of the Ratio of Molar Volumes of Emulsified Components on the Swelling Capacity of Droplets.[59]

j_2	V_1/V_2	r/r_0
1	4000	15.9
2	1350	11.1
5	355	7.1
10	125	5.0
∞	4.5	1.7

hexadecane in water, stabilized with excess of an anionic or cationic emulsifier and with initial particles in the submicron size. The styrene diffused rapidly through the water and was absorbed in the hexadecane droplets to form a stable miniemulsion with droplets in the range of 200–1000 nm in radius. The amount of styrene absorbed per unit volume of hexadecane was 100–200 times.

Two comprehensive reviews of Ostwald ripening in emulsions are recommended.[62,63] A quantitative model for stability, valid even for emulsions with polydispersity in both the droplet size and the number of trapped molecules they contain has been derived.[64] A further paper by the same authors discusses the osmotic stabilization of concentrated emulsions and foams containing trapped insoluble molecules in the dispersed phase.[65] The use of nitrogen as a trapped species to stabilize foams has been investigated with theory, experiments, and computer simulations. The osmotic stabilization of concentrated emulsions can be treated as similar to foams in which the trapped, insoluble gases are treated as ideal, but the work is also applicable to dense emulsions whose drops contain an ideal mixture of soluble and trapped liquids. At high-volume fractions of internal phase the pressure differences driving coarsening are small because many of the interfaces are planar. Consequently, osmotic pressures that would be insufficient to stabilize spherical drops would stabilize foams as age or emulsions at high internal concentrations.

22.15 EMULSION POLYMERIZATION

Emulsion polymerization originated in the wartime need for a substitute for rubber, which comes in the form of a dispersion of natural rubber in an aqueous serum and bears a strong resemblance in appearance to milk, whence arises the name "latex" (Latin for "milk"). A fairly evident approach to make a synthetic latex is to emulsify a monomer or a mixture of monomers in water using conventional emulsifying agents. For gaseous monomers, such as butadiene, the emulsification is done under sufficient pressure to liquefy them. Polymerization is initiated by a water-soluble initiator, such as sodium or potassium persulfate. The quality of the product depends critically on the concentration of emulsifier (soap.) A typical formulation contains about 5–8% soap, almost all of which, during the course of the polymerization, leaves the aqueous phase for the polymer/water interface.

At the start of the reaction, the components are present in the following forms:

(a) Soap micelles swollen with solubilized monomer to an average diameter of about 5–10 nm;
(b) Emulsion droplets of monomer stabilized by soap, about 1–3 µm;
(c) The aqueous medium containing the initiator, dissolved ions to control final stability, and a small amount of dissolved monomer.

Free radicals are generated in the aqueous phase and migrate to the interfaces. Polymerization starts in the aqueous phase to form macroradicals that are absorbed by micelles and continue to polymerize on or in them. About 10^{24} micelles/m^3 and about 10^{17} emulsion droplets/m^3 are present so that almost all the free radicals are captured by micelles rather than by emulsion drops. As polymerization continues, the micelles get larger by diffusion of monomer from the more concentrated emulsion to the erstwhile micelle. The reaction is stopped by "stripping" the latex of excess monomer under vacuum. In the final product the latex is often found to be "soap starved," that is, the final large area per molecule of the soap at the polymer water shows that interface is far from saturated. More emulsifier is then added as a post-stabilizer. In general, the amount of surface-active solute determines the number of micelles and hence the number of particles; the amount of monomer determines the size of the particles, and the amount of initiator controls the molecular weight of the polymer.

REFERENCES

[1] Krog, N.J.; Riisom, T.H.; Larsson, K. Applications in the food industry, in *Encyclopedia of emulsion technology*, Vol. 2. Applications; Becher, P., Ed.; Marcel Dekker: New York; 1985; pp 321–365.

[2] Yeung, A.; Dabros, T.; Masliyah, J. Does equilibrium interfacial tension depend on the method of measurement? *J. Colloid Sci.* **1998**, *208*, 241–247.

[3] Moran, K.; Yeung, A.; Masliyah, J. Measuring interfacial tensions of micrometer-sized droplets: A novel micromechanical technique, *Langmuir* **1999**, *15*, 8497–8504.

[4] Bancroft, W.D. The theory of emulsification, *J. Phys. Chem.* **1913**, *17*, 501–520; **1915**, *19*, 275–309.

[5] Ross, S. Adhesion versus cohesion in liquid–liquid and solid–liquid dispersions, *J. Colloid Interface Sci.* **1973**, *42*, 52–61.

[6] Ruckenstein, E. Thermodynamic insights on macroemulsion stability, *Adv. Colloid Interface Sci.* **1999**, *79*, 59–76.

[7] Ruckenstein, E. Microemulsions, macroemulsions, and the Bancroft rule, *Langmuir* **1996**, *12*, 6351–6353.

[8] Pickering, S.U. Emulsions, *J. Chem. Soc.* **1907**, *91*, 2001–2021.

[9] Becher, P., Ed. Encyclopedia of emulsion technology, Vol. 1 *Basic Theory*; Marcel Dekker: New York; 1983; pp 274–275.

[10] Binks, B.P.; Lumsdon, S.O. Influence of particle wettability on the type and stability of surfactant-free emulsions, *Langmuir* **2000**, *16*, 8622–8631.

[11] Griffin, W.C. Classification of surface-active agents by 'HLB', *J. Soc. Cosmet. Chem.* **1949**, *1*, 311–326.

[12] Griffin, W.C. Calculation of HLB values of nonionic surfactants, *J. Soc. Cosmet. Chem.* **1954**, *5*, 249–256.

[13] Davies, J.T. A quantitative kinetic theory of emulsion type, I. Physical chemistry of the emulsifying agent, *Proc. Int. Congr. Surf. Act., 2nd 1957* **1957**, *1*, 426–438.

[14] Shinoda, K.; Kunieda, H. Phase properties of emulsions: PIT and HLB, in Becher loc. cit.; 1983; pp 337–367.

[15] Boyd, J.; Parkinson, C.; Sherman, P. Factors affecting emulsion stability and the HLB concept, *J. Colloid Interface Sci.* **1972**, *41*, 359–370.

[16] Berger, P.D.; Hsu, C.A.; Arendell, J.P.; Designing and selecting demulsifers. SPE (Society Petroleum Engineers) 16285, 1987; Elsevier Engineering Information: Hoboken, N.J.

[17] Broze, G. Solubilization and detergency in *Solubilization in surfactant aggregates*; Christian, S.D.; Scamehorn, J.F., Eds.; Marcel Dekker: New York; 1995; pp 493–516.

[18] Baglioni, P.; Berti, D.; Bonini, M. Preparation and stability of emulsions. Chemica & Industria; September 2000, 1–8. (www.bias-net.com/chimica/pdf/set_baglioni.pdf)

[19] Shinoda, K.; Friberg, S. *Emulsions and solubilization*; Wiley: New York; 1986; p 63.

[20] Buzier, M.; Ravey, J.-C. Solubilization properties of nonionic surfactants. I. Evolution of the ternary phase diagrams with temperature, salinity, HLB, and CAN, *J. Colloid Interface Sci.* **1983**, *91*(1), 20–33.

[21] Nagarajan, R.; Chung, S.I.; Wasan, D.T. Biconal bob oscillatory interfacial rheometer, *J. Colloid Interface Sci.* **1998**, *204*, 53–60.

[22] Grist, D.M.; Neustadter, E.L.; Whittingham, K.P. The interfacial shear viscosity of crude oil/water systems, *J. Can. Pet. Technol.* **1981**, *20*, 74–78.

[23] Biswas, B.; Haydon, D.A. The rheology of some interfacial adsorbed films of macromolecules. I. Elastic and creep phenomena, *Proc. Roy. Soc. London Ser. A* **1963**, *271*, 296–316.

[24] Ostwald, Wa. Beiträge zur kenntnis der emulsionen, *Kolloid Z.* **1910**, *6*, 103–109; **1910**, *7*, 64.

[25] Manegold, E. Emulsionen; Straßenbau, Chemie und Technik: Heidelberg; 1952; p 23.

[26] Ross, S. Toward emulsion control, *J. Soc. Cosmet. Chem.* **1955**, *6*, 184–192.

[27] Bütschli, O. *Untersuchungen uber mikroskopishe Schäume und das Protoplasma*; Engelmann: Leipzig; 1892; *Investigations on protoplasm and microscopic foams*; Minchin, E.A., *Trans.*; Black: London; 1894.

[28] Sebba, F. Macrocluster gas-liquid and biliquid foams and their biological significance, *ACS Symp. Ser.* **1975**, *9*, 18–39.

[29] Richardson, E.G. Emulsions, in *Flow properties of disperse systems*; Hermans, J.J., Ed.; Interscience: New York; 1953; pp 39–60.

[30] Princen, H.M. Rheology of foams and highly concentrated emulsions. I. Elastic properties and yield stress of a cylindrical model system, *J. Colloid Interface Sci.* **1983**, *91*, 160–175.

[31] Maxwell, J.C. *A treatise on electricity and magnetism*; 3rd ed., Clarendon Press: London; 1891; Vol. 1; Dover: New York; 1954; p 440.

[32] Manegold loc. cit.; pp 27–31.

[33] Manegold loc. cit; p 30.

[34] Clausse, M. Dielectric properties of emulsions and related systems, in *Encyclopedia of emulsion technology*; Becher, P., Ed.; Marcel Dekker: New York; 1983; Vol. 1, pp 481–715.

[35] Sjöblom, J.; Føderdal, H.; Skodvin, T. Flocculation and coalescence in emulsions as studied by dielectric spectroscopy, in *Emulsions and emulsion stability*; Sjöblom, J., Ed.; Marcel Dekker: New York; 1996; pp 237–285.

[36] Levius, H.P.; Drommond, F.G. Elevated temperature as an artificial breakdown stress in the evaluation of emulsion stability, *J. Pharm. Pharmacol* **1953**, *5*, 743–756.

[37] Herb, C.A.; Berger, E.J.; Chang, K.; Morrison, I.D.; Grabowski, E.F. The use of quasi-elastic light scattering in the study of particle size distributions in submicrometer emulsion systems, in *Magnetic resonance and scattering in surfactant systems*; Magid, L., Ed.; Plenum: New York; 1987.

[38] Povey, M.J.W. Determination of emulsion stability by ultrasound profiling, in *Ultrasonic techniques for fluids characterization*; Academic Press: New York; 1997; pp 76–90.

[39] Suitable equipment is available from Applied Sonics, Inc., 10092 S. Stratford Rd., Littleton, CO, 80126.

[40] Povey loc. cit.; p 77.

[41] Cao, Y.; Dickinson, E.; Wedlock, D.J. Influence of polysaccharides on the creaming of casein-stabilized emulsions, *Food Hydrocolloids* **1988**, *5*, 443–454. Also in Povey loc. cit.; p 78.

[42] van der Waarden, M. The process of spontaneous emulsification, *J. Colloid Sci.* **1952**, *7*, 140–150.

[43] van der Waarden, M. Viscosity and electroviscous effect of emulsions, *J. Colloid Sci.* **1954**, *9*, 215–222.

[44] Bjerregaard, S.; Vermehren, C.; Söderberg, I.; Frokjare, S. Accelerated stability testing of a water-in-oil emulsion, *J. Dispersion Sci. Tech.* **2001**, *22*, 23–31.

[45] Herb, C.A. Private communication, 1987.

[46] Shinoda loc. cit.; p 129.

[47] Briggs, T.R.; Schmidt, H.F. Experiments on emulsions II. Emulsions of water and benzene, *J. Phys. Chem.* **1915**, *19*, 478–499.

[48] Briggs, T.R. Experiments on emulsions III. Emulsions by shaking, *J. Phys. Chem.* **1920**, *24*, 120–126.

[49] (i) Clayton, W. *Theory of emulsions and their technical treatment*, 5th ed. (by Sumner, C.G.); Blackiston: New York; 1954; pp 497–505; (ii) Gopal, E.S.R. Principles of emulsion formation, in *Emulsion science*; Sherman, P., Ed.; Academic Press: New York; 1968; p 7.

[50] Bungenberg de Jong, H.G. Crystallisation–coacervation–flocculation; in *Colloid science*; Kruyt, H.R., Ed.; Elsevier: New York; 1949; Vol 2, pp 232–258; and Complex colloid systems; pp 335–432.

[51] Menon, V.B.; Wasan, D.T. Demulsification, in *Encyclopedia of emulsion technology*; Becher, P., Ed.; Marcel Dekker: New York; 1985; Vol. 2; pp 1–75.

[52] Lissant, K.J. Making and breaking emulsions, in *Emulsions and emulsion technology*, Part 1; Lissant, K.J., Ed.; Marcel Dekker: New York; 1974; pp 71–124.

[53] Bailey, P.A. The treatment of waste emulsified oils by ultrafiltration, *Filtr. Sep.* **1977**, *14*, 53–55.

[54] Asianova, M.A. Demulsification and desalting of Emba crude oils (in Russian), *Vost. Neft* (Eastern Petroleum) **1940**, No. *5–6*, 59–64.

[55] Canevari, G.P.; Fiocco, R.J. Treatment of Athabasca tar sands froth, U.S. Patent 3, 331, 765, **1967**.

[56] Solans, C.; Pons, R.; Kumieda, H. Overview of basic aspects of microemulsions in *Industrial applications of microemulsions*; Solans, C.; Kunieda, H., Eds.; Marcel Dekker: New York; 1997; pp 1–19.

[57] Mittal, K.L., Ed. *Micellization, solubilization and microemulsions*; Plenum: New York; 1977; Kumar, P.; Mittal, K.L., Eds. *Handbook of microemulsion science and technology:* Marcel Dekker: New York; 1999.

[58] Higuchi, W.I.; Misra, J. Physical degradation of emulsions via the molecular diffusion route and possible prevention thereof, *J. Pharm. Sci.* **1962**, *51*, 459–466.

[59] Ugelstad, J. Swelling capacity of aqueous dispersions of oligomer and polymer substances and mixtures thereof, *Makromol. Chem.* **1978**, *179*, 815–817.

[60] Flory, P.J. *Principles of polymer chemistry*; Cornell University Press: Ithaca; 1953; p 512, Eq. 30.

[61] Ugelstad, J.; El-Aasser, M.S.; Vanderhoff, J.W. Emulsion polymerization: Initiation of polymerization in monomer droplets, *J. Polym. Sci., Polym. Lett. Ed.* **1973**, *11*, 503–513.

[62] Taylor, P. Ripening in emulsions, *Adv. Colloid Interface Sci.* **1998**, *75*, 107–163.

[63] Kabalnov, A. Ostwald ripening and related phenomena, *J. Dispersion Sci. Technol.* **2001**, *22*, 1–12.

[64] Webster, A.J.; Cates, M.E. Stabilization of emulsions by trapped species, *Langmuir* **1998**, *14*, 2068–2079.

[65] Webster, A.J.; Cates, M.E. Osmotic stabilization of concentrated emulsions and foams, *Langmuir* **2001**, *17*, 595–608.

23 Foams

23.1 PROPERTIES OF FOAMS

Foams are coarse dispersions of gas in relatively small amounts of liquid. Foam cells vary in size from about 50 μm to several millimeters. Foam densities range from nearly zero to about 700 g/L, beyond which gas emulsions rather than foams are found. The liquid films between bubbles, called lamellae, range in thickness from 10 to 1000 nm. Foams are related to *concentrated* emulsions, sometimes called biliquid foams, in which the dispersed phase is another liquid rather than a gas. In solid foams the continuous phase is a solid; examples are foamed plastics, breads, and cakes. Certain characteristics of foam are desired for particular applications; for example, shampoos, shaving creams, and bubble-bath compositions form slow-draining and long-lived foams. For extinguishing gasoline fires, foams should resist destruction during contact with the fuel and on exposure to high temperatures.[1] The low density of foam ensures that it floats on burning oil or gasoline. In other processes, for example, machine laundering, distillation, fractionation, and solvent stripping, too much foam has to be avoided. The control, inhibition, or destruction of foam is often important in industrial processes. Foam properties depend primarily on the chemical composition and properties of the adsorbed films and are affected by numerous factors such as the extent of adsorption from solution at the liquid/gas surface, the surface rheology, diffusion of gas out of and into foam cells, size distribution of the cells, surface tension of the liquid, and external pressure and temperature.

The rheological behavior of foams is usually examined under the following postulates: (1) foams are highly viscous, (2) foams exhibit shear thinning, (3) foams exhibit yield points, and (4) foams appear to "slip" at a solid boundary. Various two- and three-dimensional structural models of foam have been used by Wasan as foam models for rheological behavior.[2]

Much of the scientific work on foams deals with the behavior of liquid films. In examining the interference colors in soap films, in 1672, Hooke observed depressions in the films, and Newton described black films of different shades. In 1869 Plateau's studies of flow behavior in films led to the concept of a surface shear viscosity.

23.2 GEOMETRY OF BUBBLES AND FOAMS

The two laws of bubble geometry (Plateau's laws) that hold for all assemblies of bubbles and for the morphology of foams are based on the minimizing of surface area of liquid films, which is the direct result of the tension of liquid surfaces. These laws are (a) along an edge, three and only three liquid lamellae meet; the three lamellae are equally inclined to one another all along the edge; hence, their mutual or dihedral angles of inclination equal 120°; (b) at a point, four and only four of those edges meet; the four edges are equally inclined to one another in space; hence the angle at which they meet is the tetrahedral angle (109° 28′ 16″.) Plateau's laws describe all foam structures where nonspherical bubbles are in contact. If the structure is disturbed by the rupture of a liquid lamella, the bubbles rearrange in such a way as to reconform to Plateau's laws. Clusters of a few bubbles, known as composite bubbles, demonstrate the laws clearly. They also show striking structural similarities to assemblies of biological cells produced by mitosis.[3] See Figure 23.1.

When large numbers of small bubbles are produced together, a foam is formed. Foams that retain a relatively large amount of liquid between the bubbles are said to be "wet"; the bubbles are then spherical. As the liquid drains away, the bubbles form plane or nearly plane surfaces of contact and, except for rounding at the angles, become irregular polyhedra called foam cells. Because of the variety of cell sizes, these polyhedra are not uniform, but nevertheless the rules of bubble geometry still apply throughout. If the foam is made by blowing bubbles one at a time with gas at uniform pressure so that every bubble is the same size, the spheres thus produced are mobile and will tend to arrange themselves in a close-packed array, in which each bubble has 12 nearest neighbors. As the liquid drains away, the bubbles become almost regular polyhedra. The condition of 12 nearest neighbors means that the polyhedral form they take is that of a dodecahedron. A foam morphology composed of uniform dodecahedra cannot

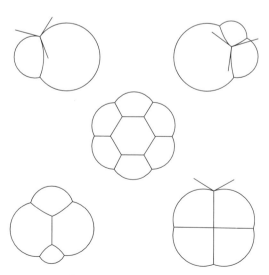

Figure 23.1 The structures of some composite bubbles obeying Plateau's laws. The tangents show that three film meet at an angle of 120°.

meet the laws of bubble geometry.[4] Consequently, a foam constructed in this way readjusts its structure to entrain some additional bubbles that destroy its regularity. A structure of uniform polyhedra all of the same shape and size can be built, however, but not if the foam is produced initially to entail 12 nearest neighbors. We can imagine such an idealized foam. Kelvin[5] pointed out that the appropriate polyhedron for this function, one that divides space into equal cells with minimum surface and without voids, is the "minimal tetrakaidecahedron," which has curved edges meeting at the vertices at the tetrahedral angle, six plane faces that are curvilinear squares, and eight hexagonal faces that are nonplanar but have zero net curvature, as shown in Figure 23.2.

A foam made of such elements would have the lowest potential energy with respect to the form of the cells; but observations of the structure of foams reveal that the majority of the lamellae are pentagonal, whereas the Kelvin cell has only quadrilateral and hexagonal faces. The probable explanation is that in the usual course of foam production, going from spherical to polyhedral bubbles, the structure is influenced toward 12 nearest neighbors rather than the 14 nearest neighbors of a foam composed of Kelvin cells.

Two enjoyable popular books on bubble geometry were written about 100 years apart.[6,7]

Foam is a system that stores potential energy in the compressed gas inside the bubbles and in the extended surface of the liquid that encloses the gas. Foams can change with time, losing their potential energy both by expansion of the contained gas and by contraction of the liquid films. This loss is brought about

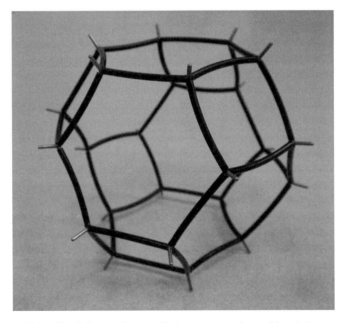

Figure 23.2 Kelvin's unit foam cell that can tesselate with minimum area.[5]

spontaneously by the rupture of the lamellae between foam cells, and also by the transfer of gas from smaller to larger cells by diffusion. Some liquid foams are resistant to lamellae rupture but none is immune to slow attenuation by gas diffusion.

23.3 EQUATION OF STATE OF FOAM

The state variables of a gas are four: moles, temperature, volume, and pressure. When the gas is incorporated in a foam, two further state variables are operative: surface area and surface tension. The equation of state of a gas relates the four state variables so only three are independent; were an equation of state of a foam to exist, it would relate the six state variables so that only five are independent. The question has more than theoretical interest because one of the state variables, the area of the extended liquid surface, is not readily measured and is an important structural property of foam. Also, the variation with time of the area of liquid surface in a foam is a fundamental index of the stability of the foam. A universal equation of state would prove, therefore, of some practical interest to facilitate the determination of the surface area of liquid in a foam and its rate of decay.

The simplest derivation of an equation of state of foam is the application to a single spherical bubble or radius r, in which p denotes internal pressure, P denotes external pressure, σ denotes surface tension, n denotes number of moles of gas, and T denotes temperature in K. Laplace's equation gives

$$p - P = 4\sigma/r \qquad (23.1)$$

For a sphere of volume V and area A,

$$V/A = r/6 \qquad (23.2)$$

For an ideal gas,
$$pV = nRT \qquad (23.3)$$

Combining the above three equations gives

$$PV + \tfrac{2}{3}\sigma A = nRT \qquad (23.4)$$

Equation (23.4) is an equation of state, but proved only for a special case. Derjaguin[8] and Ross[9,10] found the same equation holds for several other special cases. While it is likely that this equation applies to all foam in thermal equilibrium, a general proof is still lacking.[11]

Equation (23.4) can be applied in several different ways. For any spontaneous isothermal process of foam degradation, as long as no gas is lost to the atmosphere and the external pressure remains constant,

$$3P\Delta V + 2\sigma\Delta A = 0 \qquad (23.5)$$

460 FOAMS

For foam degradation at constant volume, the equation gives

$$3V\Delta P + 2\sigma \Delta A = 0 \tag{23.6}$$

The changes of pressure in a closed vessel containing foam can be monitored readily, and by means of Eq. (23.6) the rate of change of area in the foam can be found:

$$\frac{dA}{dt} = -\frac{3V}{2\sigma}\frac{dP}{dt} \tag{23.7}$$

The measurement of the pressure change and the use of Eq. (23.7) to give the change of surface area is more convenient and precise than direct observation of the area.[12]

23.4 THEORIES OF FOAM STABILITY

Pure liquids do not form stable foams: They allow entrained air to escape with no delay other than what is required for the Stokesian rate of rise, which is controlled by the diameter of the bubble of dispersed air and the viscosity of the bulk liquid. Certain solutes are able to stabilize thin lamellae of liquid: If these solutes are present, the escape of entrained bubbles is more or less retarded, and a head of foam is produced. Theories of foam postulate plausible mechanisms to account for this behavior, with the ultimate objective of understanding the phenomenon so thoroughly that predictions can be made about the expected influence of a given solute prior to actual observation. One may say at the outset that this final goal is not yet attained.

23.4.a Thermodynamic Stability

Foams are coarse dispersions of gas in relatively small amounts of liquid. Pure liquids cannot sustain a stable foam. A simple thermodynamic argument makes this statement understandable. For a two-component system (pure liquid plus a completely insoluble gas) that has sufficient surface area to make the surface energy a significant contribution to the total energy, the Helmholtz function is given by

$$dF = -SdT - pdV + \sigma dA + \mu_1 dn_1 + \mu_2 dn_2 \tag{23.8}$$

where σ is the surface tension of the liquid and A is the surface area. Integrating Eq. (23.8) at constant T, V, and n_i gives

$$\Delta F = \sigma \Delta A \tag{23.9}$$

where ΔF is the change of the Helmholtz free energy for bubble coalescence.

If we apply Eq. (23.9) to the coalescence of foam cells, we see that a decrease of the Helmholtz function results from a decrease of area; hence, a foam composed of an insoluble gas in a pure liquid is thermodynamically unstable. Experience shows that such foams have only a momentary existence. The foams we are accustomed to see, therefore, which can remain stable for minutes, hours, or even indefinitely if carefully protected, cannot be created from pure liquids. So universal is this principle that one can estimate the purity of distilled water by watching bubbles that form in it: If the bubbles are stable when they reach the surface, the water is contaminated. To ensure the thermodynamic stability of foam, the component that acts as the foam stabilizer must have some specific properties: Eq. (23.9) requires additional terms besides the $\sigma \Delta A$ term; also, these terms must be negative so that they ultimately change the sign of the whole expression for ΔF. These requirements can be met by a solute that is spontaneously surface active. The surface activity of a solute is defined as its ability to lower the surface or interfacial tension of the solvent. According to the Gibbs adsorption theorem, this ability requires that the solute be positively adsorbed at the surface or interface. Positive adsorption implies that work must be done to transfer solute molecules from the surface to the bulk phase; this work provides the additional free-energy terms that are required. When bubbles or droplets coalesce, surface area is reduced, and at the same time the solute molecules segregated at the surface are transferred back into the bulk liquid. If the loss in free energy resulting from the $\sigma \Delta A$ term were more than made up by the gain in free energy arising from the transfer of solute from surface to bulk phase, spontaneous coalescence would not occur, and the foam would be thermodynamically stable. It remains mechanically fragile, nevertheless, unless the walls between the cells are solid, such as those that exist in polyurethane and similar polymer foams.

Nakagaki[13] has advanced a thermodynamic theory of foam formation and stability in terms of foam volume (foaminess) and foam life. For dilute solutions of low viscosity, foam stability and foaminess are directly proportional to the work required to transfer excess solute from the film surface into the bulk liquid. This argument, although it shows one of the conditions necessary for the production of stable foam, is still incomplete; it lacks a term to express the effects of gravity, and it does not take into account the diffusion of gas from smaller to larger foam cells, due to differences of pressure that necessarily exist because of curvature of the film surfaces. A complete thermodynamic description of foam has, in fact, never been written. Each factor that contributes to foam stability may, however, be identified and discussed in a qualitative way.[14]

The presence of a surface-active solute is, as we have seen, required for thermodynamic stability of a liquid film and hence of a foam. But it does not by itself ensure the persistence of the film against the pervasive action of gravity, as well as other stresses that might tend to destroy the film. Rayleigh[15] pointed out a second consequence of having a surface-active solute present in the solution, namely, local gradients of surface tension induced by gravity:

> Imagine a vertical soap film. Could the film continue to exist if the [surface] tension were equal at all its parts? It is evident that the film could not exist for more than a moment; for the interior part, like the others, is acted on by gravity, and if no other forces are acting, it will fall 16 feet in a second. If the [surface] tension

above be the same as below, nothing can prevent the fall. But observation proves that the central parts do not fall, and thus that the [surface] tension is not uniform, but greater in the upper parts than in the lower. A film composed of pure liquid can have but a very brief life. But if it is contaminated, there is then a possibility of a different [surface] tension at the top and at the bottom, because the [surface] tension depends on the degree of contamination.

23.4.b Gibbs-Marangoni Effect

A thermodynamic "explanation" does not provide any physical mechanism. Having stated that adsorption of the solute is a requirement for stability, it leaves unsaid a description of *how* the adsorbed layer acts to stabilize foam. The earliest attempt to offer such an explanation, based on ideas developed successively by Marangoni,[16,17,18] Gibbs,[19] and Rayleigh,[15] has best withstood criticism through the years. This theory refers the stability of foam to an elasticity or restoration of liquid lamellae which depends on the existence of an adsorbed layer of solute at the liquid surface and the effect of this adsorbed layer in lowering the surface tension of the solution below that of the solvent. The two effects, surface segregation or adsorption and the lowering of the surface tension, are concomitant: An observed reduction of surface tension due to the addition of a solute is evidence, admittedly indirect but no less certain than were it given by direct observation, that the solute is segregated at the surface. The degree of segregation in a two-component system is measured as excess moles of solute per unit area of surface, designated Γ_2^G and is equal to the variation of the surface tension with the chemical potential of solute, that is,

$$\Gamma_2^G = -\frac{d\sigma}{d\mu_2} \qquad (23.10)$$

where $d\sigma$ is the change of the surface tension caused by changes in the chemical potential of solute, $d\mu_2$. For dilute solutions $d\mu_2 = RTd \ln c_2$ where c_2 is the concentration of the solute. Equation (23.10) is derived in Section 15.1.a.

When local areas of a foam lamella are expanded, as would happen, for example, when a bubble of air pushes through a liquid surface, new areas of surface are created where the instantaneous surface tension is large, because the adsorbed layer has not had sufficient time to form. The greater surface tension in these new areas of surface exerts a pull on the adjoining areas of lower tension, causing the surface to flow toward the region of greater tension. The viscous drag of the moving surface carries an appreciable volume of underlying liquid along with it, thus offsetting the effects of both gravitational and capillary drainage (explained below) and restoring the thickness of the lamella. The same mechanism explains how liquid lamellae withstand mechanical shocks, such as the passage through the foam of lead granules, mercury drops, iron filings, or steel spheres, all of which have been used by one or another investigator to test the resiliency of lamellae.[20] The lamella survives because the local increase of surface tension, where the impinging solid deforms the surface, causes flow toward the weakened region. Aged lamellae are less fluid than younger and thicker ones and so are more readily broken, as are lamellae made from solutions in which the surface tension gradient is small, for example, oil solutions or very dilute aqueous solutions of surface-active solutes.

What force checks the downward flow of the central liquid in Lord Rayleigh's example? The required force originates in the higher surface tension in the upper part of the film. Where a higher tension appears, movement of the surface at that point immediately ensues: Surrounding areas of lower tension surface are pulled in toward the high-tension spot until the difference has been effaced. The movement of the surface layer from areas of low to areas of high surface tension is accompanied by the motion of relatively thick layers of underlying fluid, amounting to several microns in depth, so that the downward flow of the central liquid is offset to some extent by a counterflow of liquid at the surface. Flow in response to a surface tension gradient is called the Marangoni effect (see also Chapter 6); it was also recognized by Gibbs. A simple experiment readily demonstrates this effect. On a metal sheet, such as a silver tray, pour out a thin layer of water. When a piece of ice is held below the tray, local cooling of the water above causes its surface tension to increase. Immediately, the motion of the surrounding areas of lower tension becomes evident by the heaping up of the water just above the spot where the ice is held. When the ice is moved slowly across the bottom of the tray, the little mound of water follows it faithfully.

23.4.c Mutual Repulsion of Overlapping Double Layers

The adsorption of ionic surface-active solutes into the surface layer is evident in aqueous solutions and readily leads to the formation of charged surfaces of the lamellae in foams.[21,22,23] The counterions in the liquid interlayer of the lamellae are the compensating charges. When the thickness of the lamellae is on the order of magnitude of 20 times the Debye thickness of the electrical double layer, the counterions adjacent to the two opposite surfaces repel each other more or less according to an exponential decline of electric potential with distance. This repulsion prevents further thinning of the lamellae, and so preserves them from imminent rupture.

The mechanism of charge separation that operates in water does not function in nonionizing solvents. Until relatively recently, it was believed, therefore, that electrostatic repulsion of overlapping electrical double layers could not be a factor in stabilizing liquid lamellae in oil or other nonpolar foams. But we now recognize that mechanisms of charge separation other than electrolytic dissociation are possible, and indeed must operate; zeta potentials of 25 to 125 mV have been observed for various kinds of particles dispersed in nonpolar media of low conductivity. Nevertheless, no evidence has yet been reported to suggest that foams are stabilized by electrostatic repulsion in nonpolar solutions or in cellular foams. Such foams are more probably stabilized by steric repulsion between polymers adsorbed at the air/liquid interface.

23.4.d Drainage of Foams

Stresses that create areas of higher tension on a liquid film are always at hand. The first of these is gravity-induced drainage. The drainage of liquid through the foam structure takes place between two coherent layers provided by the surface-active solute at the liquid/gas surface. The layers themselves are not completely static. They, too, respond to the gravitational tug and then reverse their direction of flow by the Marangoni effect. The coherent surface layers act as a membrane or skin that can stretch and relax in response to the lateral forces acting on it. In doing so, the surface tension changes so as to give a force opposing the motion. This result is, in effect, an elasticity of the surface, a variation of the surface stress with surface strain. It is named "dilatational elasticity." Single pure liquids, which in the absence of a surface-tension gradient are not able to develop a Marangoni effect, therefore show no dilatational elasticity.

As already mentioned, the rate of drainage of liquid through such films is at first induced by gravity. If V is the volume of liquid remaining in the foam after time t, the rate of gravitational drainage[24] is given by

$$V = V_0 \exp(-kt) \qquad (23.11)$$

where the amount of liquid drained from the foam approaches V_0 (the total amount of liquid in the foam at $t = 0$) as t becomes large. This equation has only a single arbitrary constant and describes drainage from different types of foam reasonably well. The agreement is better, however, with a slower draining foam, for example, protein hydrolyzate, than with faster draining foams. This result is to be expected because of the way in which the equation was derived, namely, that the liquid flows down between two solid walls (i.e., infinite surface viscosity) which approach each other as the liquid between them flows out. Subsequent flow depends on dissipation of the nodes. A theory based on this model is in close agreement with data for a variety of drainage experiments.[25] More sophisticated expressions for the rate of drainage result from data obtained by forced drainage experiments in which the foam structure is stabilized by dripping the liquid medium onto the top of the foam after which it thickens the lamellae by drainage and so stabilizes the foam structure. A constant input of liquid at the top of the foam maintains a constant flow throughout the foam.

At first, the drainage of the central liquid, taking place between two membranes at the surface, is entirely induced by gravity; but once spherical bubbles are in contact, flat walls develop between them, and polyhedral cells appear in the foam. When this occurs, the rate of drainage ceases to be determined by hydrodynamic flow under gravity because capillary action becomes significant. The foam films are flat in some places and have convex curvature at the places where the liquid accumulates in the interstices between the foam cells. The line along which three films meet is called the Plateau border after J.-A.-F. Plateau (also called the channel), and the point at which four lines meet is called the tetrahedral angle or the Gibbs angle after J. W. Gibbs (also called the vertex or the node). (See

Figure 23.2.) The Plateau border is not a mathematical line but a thin vein of liquid whose cross section is approximately a curvilinear equilateral triangle; the tetrahedral angle is not a mathematical point but a small volume of liquid whose shape is approximately a tetrahedron with concave faces. Most of the liquid in a foam resides in the channels and nodes. The convex curvature inside the tetrahedral angles creates a capillary force that draws liquid out of the connecting Plateau borders; and the convex curvature of the Plateau borders draws liquid out of the adjoining lamellae. For convenience, let us call this the "Laplace effect"; it is also known, less happily because of ambiguity, but with logical consistency, as capillary flow. As the liquid flows, some of the surface layer is dragged along, and areas of higher tension are created. Then the Marangoni effect comes into play; the central liquid that is lost by the lamella is restored by counterflow along the surface, one effect thus creating the conditions for its reversal by the other. Ultimately, after a period that may be more or less prolonged, depending on the presence of dust, the ease of evaporation, and any departure from thermal equilibrium, the liquid film becomes so thin (20–30 nm) that the central liquid is itself affected by surface forces and so loses its fluidity because of the formation of gelatinous surface layers. X-ray diffraction studies of drained foam films show that a hydrous gel structure extends from the surface to a depth of about 90 nm. Water makes up the principal portion of this surface film (at least 97% by weight). We do not know sufficiently well, however, the principles of the formation of such gels to be able to predict the combination of solutes required to produce them, although we do have some empirical information. It is known, for example, that the addition of relatively small amounts of nonionic surface-active solutes to solutions of certain anionic detergents enhances foam stability by the formation of gelatinous surface layers. The effectiveness of these additives depends on the detergent, the more effective combinations being those in which the additive preponderates in the adsorbed surface layer. The most stable foams are found[26] with anionic nonionic pairs having 60–90% of the nonionic in the adsorbed layer, even though the actual relative amounts of the two in the whole composition is anionic : nonionic = 100:1.

The required gradient in surface tension for the Marangoni effect to operate now cannot be so readily achieved, and the resilience of the liquid film is gradually replaced by increasing brittleness. At this stage in the life of the film, it is readily ruptured by relatively slight mechanical shocks, which it had previously been able to withstand. The foam structure, therefore, begins to collapse at the top of the foam because the older lamellae are located there. A gravitational field promotes instability of foam structure. One of the interesting features of weightlessness in space is the modification of this and other capillary phenomena with which we are familiar on Earth.

To investigate the flow of foams, Gopal, Durian, and Weitz[27,28,29] used multiply scattered light to probe foams flowing under applied shear. Foams behave as elastic solids for small applied shear stress and yet flow like viscous liquids at large applied shear stress. A pronounced change in the dynamics occurs once the distortion exceeds a critical value, requiring bubbles to move beyond their local neighbors. Just above the yield point, the flow is mediated by nonlinear events in which several neighboring bubbles hop from one tightly packed configuration to another, creating a stick-slip motion which decreases sharply when an avalanche-like rearrangement of the neighboring bubbles occurs.[30]

23.4.e Elasticity of Lamellae

The elasticity arising from the variation of the surface tension during deformation of a liquid lamella may be manifested both in equilibrium (when a surface layer under forces leading to deformation is in equilibrium with its bulk phase) and in nonequilibrium conditions. The first case refers to the Gibbs elasticity and the second to the Marangoni elasticity.[31] The Marangoni elasticity is a dynamic, nonequilibrium property, larger in value than the Gibbs elasticity that could be obtained in the same system. The elasticity is defined as the ratio of the increase in the tension resulting from an infinitesimal increase in the logarithm of the area. For a lamella with adsorbed solute on both sides, the elasticity E is given by[32]

$$E = \frac{2d\sigma}{d \ln A} = -\frac{2d\pi}{d \ln A} \tag{23.12}$$

where σ is the surface tension and A is the area of the liquid surface. The factor 2 is required because the stretching of the lamella increases the area on both of its sides. The factor would not be required for the definition of elasticity of the surface of a solution or a liquid substrate holding an insoluble monolayer. The Gibbs elasticity of a monolayer is measured from the equilibrium π versus A curve as previously described (Section 13.5). The Marangoni elasticity of a monolayer is determined by the dynamic (i.e., nonequilibrium) surface tension as the surface is abruptly extended, or pulsated.

The effects described by the Rayleigh–Gibbs theory depend therefore on a combination of two physical properties of the solution. The solute should be capable of lowering the surface tension of the medium, but this alone is not enough; a rate process is also required, by which a freshly created liquid surface retains its initial, high, nonequilibrium surface tension long enough for surface flow to occur. Many instances are known in which the mere reduction of surface tension by the solute does not lead to the stabilization of foam, presumably because it is not accompanied by the relatively slow attainment of equilibrium after a fresh surface is made, which is the second requirement for the ability to stabilize bubbles.

In general, the surface tension of a freshly formed solution of a surface-active solute changes with time until it reaches a final equilibrium value. Equilibrium may be reached in a fraction of a second or it may require several days. Adsorption may be considered as a two-step process involving: (a) the diffusion of the solute molecules from the bulk phase to the subsurface (i.e., the layer immediately below the surface) and (b) the adsorption of the solute molecules from the subsurface to the surface.

In concentrated emulsions and polyhedral foams the stability and the rheology depend on dynamic processes. Wasan and coworkers[33] developed a mathematical model of adsorption of surface-active solutes at a gas/water interface that takes into account both the diffusion from the bulk phase and the energy barrier to adsorption, and later developed a versatile interfacial film tensiometer to measure dynamic surface tension and the capillary pressure of the Plateau borders.[34] Their automated apparatus makes it possible to change the interfacial area in a variety of modes (expansion or contraction at various rates.) Noskov has reviewed the adsorption of surfactants with short hydrocarbon chains, mainly alcohols, at a gas/water interface.[35]

Some systems are diffusion controlled; that is, the activation energy barrier to adsorption is negligible in such cases. A diffusion-controlled rate of adsorption makes itself evident, therefore, by a rapid approach to surface tension equilibrium (on the order of seconds or less). Surfaces that age more slowly imply an activation energy of adsorption (see Section 13.6). The different types of instruments developed to measure dynamic surface tension vary with respect to the age of the surface that they are designed to handle: The vibrating jet deals with surface ages of milliseconds, and conventional techniques to measure surface tension, which react to changes more slowly, can detect surface aging of several minutes, hours, or days.

These considerations apply equally to aqueous and nonpolar solutions. A major difference between the two lies, however, in the magnitude of the effects produced by surface-active solutes. The surface tension of water is reduced from 73 to 25 mN/m quite readily by amphipathic organic solutes, but the surface tension of most organic solvents is already in the low range of 25 to 30 mN/m, so that only a small reduction remains to be achieved by an organic solute. Thus, although Marangoni effects may arise in nonpolar solutions, they are usually much less pronounced than in aqueous solutions of soaps or detergents. Oil lamellae, therefore, have a relatively low resistance to mechanical shock; consequently, oil foams are transient or evanescent, resembling the foam produced from very dilute aqueous solutions of detergents or more concentrated solutions of weakly surface-active solutes. Special solutes have been developed to stabilize nonpolar foams for application in the field of cellular plastics. These solutes incorporate polyalkylsiloxane or perfluoroalkyl moieties in solute molecules, which are able to reduce the surface tension of organic liquid monomers by 12 to 15 mN/m and boost the Marangoni effects. The result is obvious in an increased stability of the foam.

23.4.f Enhanced Viscosity or Rigidity

The persistence of the liquid lamella depends not only on surface tension gradients but also on the presence of a coherent surface layer on each side of the lamella, between which layers the central portion of the sheet of liquid flows downward. Given the presence of a surface-active solute, surface layers certainly would be there. But these layers are not necessarily coherent, and if they should lack this feature, the whole liquid film, surface and central parts alike, drops simultaneously. This requirement for the stability of a liquid film is often overlooked, but it explains, for instance, why the mere ability of a solute to lower surface tension is not enough by itself to ensure that it will also stabilize a liquid film. The surface layer may lack the necessary cohesion. Cohesion and elasticity of a surface film are directly related. A dilute solution of a surface-active solute has a low surface elasticity, which is evident from a π-A curve at low values of π (Figure 13.18). Such solutions have poor cohesion and do not stabilize bubbles well.

A single surface-active species in solution does not usually confer any increase of the viscosity, much less rigidity, to the surface layer of the solution. Although foam is capable of being produced by such a solute, the foam is of brief duration. That kind of foam is called "evanescent foam," but it can nevertheless be a cause of concern; if produced rapidly, it can reach a large expansion ratio and so flood any container. While the first stabilizing factor in foam is the elasticity of the film provided by the Marangoni effect, in special cases additional surface layer phenomena are significant, namely, gelatinous surface layers, low gas permeability, and stability of black films. These phenomena are known to occur in water with certain mixtures of solutes or with certain polymers, both natural and synthetic. Well-known examples in aqueous systems are solutions of water soluble proteins, such as casein or albumen, as in the stable foams produced with whipping cream, egg white, beer, or rubber latex. The highly viscous surface layer is sometimes made by having present one or more additional components in the solution. An example is the surface plasticity of a mixture of tannin and heptanoic acid in aqueous solution, compared to the lack of any such effect of the two constituents separately. Surface plasticity adds enormously to the stability of foam and may be exemplified in meringue, whipped cream, fire-fighting foams, and shaving foams. Solutions of saponins or of proteins provide surface layers that are plastic, that is, that remain motionless under a shearing stress until the stress exceeds a certain yield value, which may be greater than the small gravitational or capillary stresses usually acting on the surface layer.

In nonpolar liquids, particularly in bunker oils and crude oils, surface layers of high viscosity have been observed; porphyrins of high molecular weight have been indicated as a possible cause. In a hydrocarbon lubricant, the additive calcium sulfonate, for example, creates a plastic skin (or two-dimensional Bingham body) at the air surface; it also acts as a foam stabilizer.[36] Thus viscous or rigid layers in nonpolar liquids enhance the stability of foam, just as they do in aqueous solutions.[37,38,39] Cellular plastics are made from a monomer foam, which on polymerization displays high viscosity, finally becoming solid; but the growth of the viscosity is not confined to the surface, nor does it even show preferential development at the surface.

Different kinds of surface viscosity can be distinguished:

   (a) Innate surface viscosity   This viscosity is the resistance to flow that is innately associated with the presence of a liquid surface, whether or not there are additional sources of resistance such as those described below.[40]

(b) Surface shear viscosity This viscosity is associated with the presence of a pellicle or skin, such as an insoluble monolayer, but not restricted to that example, at the liquid surface. A layer of denatured protein that stabilizes the foam of meringue or of whipped cream or of beer is a common example.

(c) Dilatational (or compressional) viscosity The surface elasticity that arises from local differences of surface tension is simultaneously associated with a resistance to surface flow by initiating a counterflow. The local difference of surface tension is caused by dilatation (or compression) of the surface of the solution, so the resistance to flow that results from Marangoni counterflow is known as dilatational (or compressional) surface viscosity. This type of surface viscosity is a dynamic or nonequilibrium property.

Djabbarah and Wasan[41] showed that the presence of a small amount of lauryl alcohol [equal to 1/500% of the sodium lauryl sulfate (SLS)] causes significant decrease in the surface excess concentration of SLS. For example, at an SLS bulk concentration of 2.0×10^{-3} M and without LOH, $\Gamma_{SLS} = 2.47 \times 10^{-6}$ mol/m^2. At the same SLS bulk concentration and in the presence of 6.2×10^{-3} M of LOH, $\Gamma_{LOH} = 2.90 \times 10^{-6}$ mol/m^2, while Γ_{SLS} drops to 2.0×10^{-6} mol/m^2. This result shows that LOH is preferentially adsorbed at the surface, replacing SLS molecules that would otherwise be there. The replacement of SLS by LOH is not a simple 1:1 replacement, as is indicated by the drastic change in molecular packing at the surface from about 0.67 nm^2/molecule without LOH to about 0.34 nm^2/molecule with LOH. Molecules containing long unbranched hydrocarbon chains and a small terminal polar group are those most likely to form a closely packed state. Molecular packing by itself does not account for high surface viscosity.[41] There is, however, a strong dependence of surface viscosity on the relative amount of the more surface-active substance at the surface, which is compatible with the suggestion that the altered composition of the surface may segregate a gelatinous three-component liquid-crystal phase while the bulk solution is still isotropic and of low viscosity. Friberg and his coworkers have pointed out that foam stability ensues when the total composition of a multicomponent system is such that two phases, an isotropic solution and a liquid crystal, are in equilibrium.[42,43] It seems likely that the liquid-crystal phase should occur at the surface where so much of the surface-active components are concentrated (see Section 23.4.i).

An important feature of the gelatinous surface film is its transition to a freely flowing film over a narrow range of temperature.[44,45] The transition is sharp and reversible and is compatible with the variation with temperature of a multicomponent isothermal phase diagram, where the exercise of another degree of freedom without change of overall composition can move the equilibrium with a sharp transition from a condition of two phases to that of a single phase. The sharp transition of the film from a gelatinous or plastic structure to a structureless Newtonian fluid as the temperature is raised is accompanied by an equally sharp transition in the foaming properties. A practical example occurs annually with certain fuel oils that foam excessively in wintertime when pumped into delivery trucks or household oil tanks but that show no sign of this unwanted property during the summer months. It was found that the surface of the oil is gelatinous below $-4°C$ and is fluid immediately above that temperature; consequently, trouble because of foaming is encountered only in cold weather. Another example is provided by Bolles,[46] who describes a foaming problem in a hydrocarbon separation unit. The system that proved to be so troublesome showed a sharp loss of foam stability with increasing temperature. This behavior is typical of foams stabilized with gelatinous films that have a melting point characteristic of a separate phase at the surface.

23.4.g Foam Boosters in the Presence of Oil Drops

Sometimes the foaming medium contains dispersed oil drops. For example, lanolin or silicone oil are introduced into shampoos to improve their conditioning properties. These cause deterioration of the foamability and foam stability, which is undesirable in such products. Betaines (a class of zwitterionic surfactants, e.g. $R(CH_3)_2N^+CH_2COO^-$ where R is a long chain alkyl) suppress this defoaming tendency, even though the surface and interfacial tensions of the oil and the medium are not affected. These values are such as to promote entry of oil drops to foam films, with consequent antifoaming action, and indeed do so in the absence of the foam booster.[47] Evidently the foam booster creates a barrier to the entry of the oil drop to the surface of the foam films. A detailed molecular mechanism responsible for the effect was not suggested by these authors, but solubilization of the oil by the micelles containing betaine is a likely mechanism.[48]

23.4.h Black Films

When a glass or wire frame is dipped into a solution of a surface-active solute and then drawn upward, a thin liquid lamella is formed, which shows interference colors in reflected light. Upon draining, the lamella becomes thinner until it looks gray or black, indicating that its thickness is much less than the wavelength of light. Upon further draining, the lamella either breaks or reaches an equilibrium thickness consisting of two adsorbed layers at the air/liquid surface with a water core between them. The behavior of the lamella depends on the amount and the surface properties of the adsorbate. At more than a minimum concentration of solute (varying for different solutes), the black film is obtained and is stable with respect to rupture.[49,50] These films range in thickness from about 100 nm to the thickness of a bimolecular leaflet, which in the case of sodium oleate is 4.2–4.5 nm. The bimolecular leaflet can be achieved, however, only if enough electrolyte is present in the solution to contract the electrical double layers and thus reduce to very small distances the electrostatic overlap between the two sides of the film. In certain cases, notably that of sodium oleate, the bimolecular leaflet is particularly strong and persistent. Dewar succeeded in preserving such a film for over a year in a closed container.[51] The bimolecular leaflet can be described as a two-dimensional crystal; it is no longer subject to liquid flow nor can it exhibit any phenomena that depend on surface tension effects. Foams in which the stability of black films is prolonged are characterized by an extremely fragile structure at the top of the foam, the structure persisting long after all the unbound water has drained out of the films. Gelatinous surface layers, on the other hand, usually remain thick and immobilize large amounts of water.

The stability of long-lived foams has been attributed to the stability of black films, and this view is put forth as an alternative theory of stability.[52] A black film is composed of two non-Newtonian surface layers in close approach. Even a black film of limiting thinness, that is, a bimolecular leaflet, is the result of contact of two concentrated and coherent monolayers when sufficient electrolyte is present to collapse the electrical double layer and prevent the immobilizing of water in the overlap of extended electrical double layers. We need not, however, consider black films as the only kind of film produced by the close approach of two surface layers.

23.4.i Influence of Liquid-Crystal Phases

A lipophilic solute capable of forming micelles, that is, one with an amphipathic structure, when dissolved in a hydrocarbon forms inverse micelles that are capable of solubilizing water. At higher concentrations, the solute and water plus a small amount of hydrocarbon form a lamellar liquid-crystalline phase (Figure 23.3). The difference between the micellar solution and the liquid-crystalline phase is that the former is an isotropic liquid containing micelles of colloidal size; the latter is a lamellar anisotropic association of the three components in a single phase. Both structures may solubilize considerable amounts of water.

Nonpolar foams, as well as aqueous foams, can be stabilized by the presence of a lamellar liquid crystal in conjunction with an isotropic hydrocarbon solution. If the system contains amphipathic molecules, the liquid-crystal phase may first

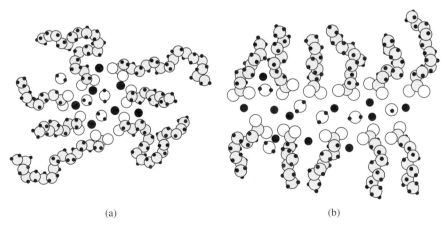

(a) (b)

Figure 23.3 The inverse spherical micelle (*a*) and the lamellar liquid crystal (*b*) in nonpolar media.[53]

form at the surface out of an isotropic solution, by segregation of components. The lamellar liquid crystal is surface active with respect to the hydrocarbon solution and provides a layered or structured film at the surface. This structured film has high viscosity and reduces drainage from the lamellae, thus conferring stability on the foam. The liquid crystal as a separate phase does not foam; foam stability appears only in the two-phase region.[53]

A typical phase diagram is shown in Figure 23.4. Stable foams in such a system are found in the area T, a two-phase area containing isotropic solution and liquid crystal. No other portion of the phase diagram denotes systems with any foam stability. Foam films are stabilized by the presence of the liquid-crystal phase at the surface, which confers surface plasticity. An interesting feature of this nonpolar foam is that the addition of water, by moving the system into another region of the phase diagram, acts as a foam breaker.

In view of this mechanism, the presence of surface-plastic films, which promote foam stability, is probably another example of a liquid-crystal phase formed by a soluble surface-active solute, an insoluble cosolute, and water. The addition of lauryl alcohol to sodium lauryl sulfate is well known to produce a

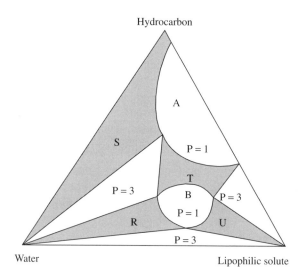

Region	Number of phases	Composition
A	1	Isotropic solution with inverse micelles
B	1	Lamellar liquid crystals
R	2	Water phase plus liquid crystal phase
S	2	Water phase plus solution containing inverse micelles
T	2	Solution containing inverse micelles plus lamellar liquid crystal phase
U	2	Excess emulsifier plus lamellar liquid crystals

Figure 23.4 Typical phase diagram for a lipophilic solute, hydrocarbon, and water in which stable foams are found in the two-phase region marked T. R, S, T, and U are two-phase regions.

surface-plastic film and to stabilize foam (see Section 2.3.4.f above). The surface plasticity disappears sharply at a temperature of about 45°C. This sharp change suggests a phase boundary. Other examples of a similar effect of long-chain, often water-insoluble, polar compounds with straight-chain hydrocarbon groups of approximately the same length as the lipophile of the surface-active solute are lauryl alcohol for use with sodium dodecyl sulfate, lauric acid for use with potassium laurate, N,N−bis(hydroxyethyl)laurylamide for use with dodecylbenzene sulfonate, and N,N−dimethyldodecylamine oxide for use with dodecylbenzene sulfonate and other anionics.[54]

23.4.j Effect of Dispersed Particles on Foam Stability

Ottewill et al.[55] found experimentally that the presence of colloidally stable, suspended, solid particles increases the tendency to form stable foams over and above that of the matrix in the absence of such particles. The increase in foam stability is linked to the increased bulk viscosity of the dispersion with solids content, which is described by a relation of the form

$$\eta_d = \eta_0 \left(1 + k_1 \phi + k_2 \phi^2 \cdots \right) \qquad (23.13)$$

where η_0 is the viscosity of the liquid matrix, η_d is the viscosity of the dispersion, and ϕ is the volume fraction of dispersed solid. The coefficient k_2 was larger than predicted by purely hydrodynamic factors, being enhanced by the electrostatic repulsions between the solid particles, which effectively enlarges each particle and so creates a larger volume fraction of solids than is calculated from the density of the solid. In addition, the presence of a minimum in the pair interaction energy curve introduces some association between the particles with increase in volume fraction, which leads to a viscosity enhancement at the low rates of shear experienced in a slowly draining lamella. The effect of bulk dispersion viscosity on the ripples formed in the lamella surface by thermal fluctuations is not known with certainty, but it seems likely that this would have a damping effect on the magnitude of the ripples and so lead to further enhancement of foam stability.

Even greater stabilizing action conferred by solid particles is achieved if the particles are not wholly hydrophilic. They are then kept at the air/water surface by surface forces, in the same way as lipophilic molecules are adsorbed there (see Chapter 22). If some surface-active solute is also present, the foam lamellae carry these particles upward with the foam, a phenomenon on which is based the process of ore flotation. The particles at the surface of the lamellae add considerably to foam stability by absorbing mechanical shocks that would otherwise destroy the lamellae.

However, a higher degree of water repellency (lipophilia) of the surface of solid particles begins to introduce a new effect, namely, their dewetting effect on the liquid medium, which, if the number of particles is small* converts foam-stabilizing action into destabilizing action (see Section 23.5.f). The juxtaposition

---

*The concentration, size, and degree of water repellency are collective factors in determining how solid particles affect foam properties: All three factors contribute to observed behavior.

of profoaming and defoaming actions resulting from a relatively small change of the hydrophile–lipophile balance of the solid surface strongly resembles a parallel action on the molecular level of surface-active solutes (see Chapter 13) and suggests a similarity of mechanism.

23.4.k Foaminess and Phase Diagrams

A long-accepted theory of foaming in fractionation or distillation towers traces the cause to Marangoni flow, induced by a difference in composition, and hence of surface tension, between thin films of liquid at the bubble caps and in the bulk liquid phase from which they originate.[56] Liquid films on walls of the container, or foam lamellae, because of their extended surfaces, evaporate faster than does a bulk liquid phase; if the loss of the more volatile constituent causes the surface tension of the residual solution to increase, then the liquid of greater surface tension draws liquid away from that of smaller surface tension, and so the stability of the liquid lamella is maintained. The mental imagery is identical with the model used to describe brandy tears (see Chapter 6).

This rule is far too inclusive: In normal liquids, volatility and low surface tension stem from a common cause, namely, relatively small forces of intermolecular attraction and therefore occur together. Thus, admittedly with certain exceptions (see Table 6.1), the usual behavior of solutions leads to the surface tension of the residual solution being greater after the loss of its more volatile components. According to this rule, therefore, foaming within a fractionation tower would be the common condition. In practice, the problem of foaming is less prevalent than this rule predicts.

A more selective prediction can be obtained by studying the phase diagram of the system. Surface activity, and hence a propensity toward foaminess, is specifically inherent in solutions only at certain temperatures and compositions that are related to solubility curves and other features of the phase diagram, as well as to the relative surface tensions of the components. Although Marangoni effects may arise as a result of the volatility of a component of small surface tension, these are less significant in stabilizing foam than are Marangoni effects derived from adsorption of solute at the liquid/gas surface, that is, surface activity.

Adsorption from the bulk solution to its surface is well known to occur with amphipathic solutes such as soaps and detergents. Less well known is the adsorption of more ordinary solutes, not specifically amphipathic, that takes place under conditions of incipient phase separation. An early intimation of such an effect was Lundelius's rule, which may now take its place in history as a glimpse of an even larger generalization. Lundelius's rule is to the effect that a given solute is more adsorbed from a solvent in which it is less soluble. The next development to be reported along the same lines was the Ferguson effect, which is the variation of the effectiveness of members of an homologous series of soaps or detergents in aqueous solution with the length of the hydrocarbon chain. Every increase in the length of the chain both augments the surface activity and reduces the solubility of the solute. These effects are described in detail in Chapter 6. They are precursors of the discovery, by Ross, Nishioka and coworkers, of the incidence of capillarity (i.e., various manifestations of surface activity) in systems described by certain defined locations within phase diagrams (the Ross–Nishioka effect.) A characteristic of this effect, particularly in systems of two components or more whose phase diagrams show a region of partially miscible liquids, is the gradual growth of surface activity of solutions as their compositions approach the limit of their solubility, followed by their separation into two phases. On the nucleation of the conjugate solution, however, the increasing foaminess of the solutions, indirectly a reflection of their increasing surface activity, is replaced by a dramatic break to zero foam stability. In fact, for reasons we discuss below, the insoluble conjugate liquid of lower surface tension, once nucleated, acts as an inhibitor of foam.

We have, therefore, in historical sequence, the Lundelius rule, the Ferguson effect, and the Ross–Nishioka effect, describing the gradual disclosure of the related phenomena of capillarity and solubility to include an increasingly wider range of systems and materials.

Experimental observations of the foaminess of binary and ternary solutions in systems in which a miscibility gap exists show that the foaming of a solution reaches its maximum under conditions of temperature and concentration where a transition into two separate liquid phases is imminent. Figure 23.5 shows one such diagram, for the system ethylene glycol in methyl acetate. Of the two, methyl acetate has the lower surface tension. As the compositions move from 100% ethylene glycol to the onset of phase separation, the foam stability steadily increases. When the phase separation is achieved the methyl acetate is dispersed in the ethylene glycol, represented on the diagram by the symbol MA/EG. In this narrow range of compositions the system is essentially defoamed by the insoluble

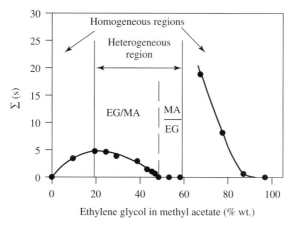

Figure 23.5 Foam stability, $\Sigma(s)$, versus weight percent ethylene glycol in methyl acetate at room temperature. The circles represent foam stability at a certain concentration. The vertical lines mark phase boundaries.

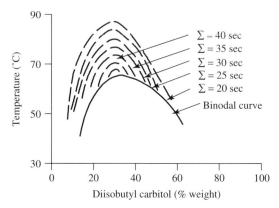

Figure 23.6 Phase diagram and interpolated isaphroic lines of the two-component system, 2,6-dimethyl-4-heptanol (diisobutyl carbinol) and ethylene glycol, showing maximum foaminess at an epicenter.[57]

droplets of methyl acetate. When the phase inversion occurs and EG is dispersed in MA the antifoaming action ceases and is replaced by a slight growth in foam stability, which reaches its maximum at the opposite phase boundary and then declines to the zero foaminess of pure liquid methyl acetate.

A phase diagram of the 2,6-dimethyl-4-heptanol and ethylene glycol is shown in Figure 23.6.[57] Superimposed on the diagram by means of dotted lines are interpolated contours of equal foam stability (isaphroic lines, from *aphros,* Greek for "foam"). The isaphroic lines center about a point close to the critical point as a maximum and decrease in value the farther they are from it.

We now review applications of this principle that throw light on observations made in the past as well as on some current problems.

The stability of bubbles historically provided a measure of the alcohol concentration in distilled alcoholic beverages. Davidson[58] traced the origin of the U.S. proof scale to the relatively sudden onset and decline of bubble stability in whiskey with alcohol concentration, which occurs at about 50% alcohol by volume, and which provided a readily observable test of strength that was used by distillers for hundreds of years and is still used by illegal distillers or "moonshiners" in the United States. Observations of the bubbles makes it possible, with a little practice and without any instrument, to tell if a whiskey is above or below "proof." When Congress established a legal definition of proof, it essentially accepted this empirical usage by fixing it at a concentration near the concentration of maximum bubble stability, which happened to be close to 50% by volume alcohol in whiskey. The official U.S. proof scale is twice the volume percentage of alcohol (100 proof is 50% alcohol by volume.)

Pure ethanol and water solutions at 50% alcohol by volume show no bubble stability, but whiskey is not as simple as that. When overproof whiskey is diluted with water, some secondary organic constituents, such as tannin, fusel oil, acids and aldehydes approach their limit of solubility, and surface activity

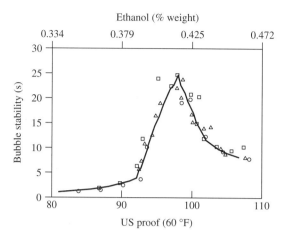

Figure 23.7 The Ross–Nishioka effect in fortified bourbon whiskey on dilution with water. Reading from right to left, as the solutions approach a phase boundary, bubble stability increases until phase separation creates a foam inhibitor.[58]

is made manifest by bubble stability. Figure 23.7 shows the variation of bubble stability with water content of a bottled bourbon to which ethanol had been added to bring it above 140 proof. The bubble stability was measured as it was progressively diluted with water. As some of the secondary organic constituents began to approach the limit of their solubility, bubble stability increased. When further addition of water brought them out of solution, bubble stability declined precipitously. Maximum bubble stability occurred at 100 proof, which of course is not a coincidence, because it forms the original basis for defining 100 proof. It may be mentioned, however, that Scotch whisky contains fewer tannins, acids, and fusel oils than bourbon and consequently this effect is less pronounced with Scotch whisky.

Any series of discrete compounds within which series there is a gradation of a chemical substituent, such as the homologous series that demonstrates the Ferguson effect, or the degree of variation of the proportion of one substituent in a related series of discrete amphipathic solutes (e.g., the degree of hydrophile in the ratio of hydrophile to lipophile balance, which is the basis of the HLB concept for selecting emulsifiers), or of a physical property, such as is described in the next example, that is, a gradation of molecular weights of a polymer, as long as insolubility in a given solvent occurs or may reasonably be inferred to take place at some point within the series, will display the same onset of capillarity as occurs in systems described by phase diagrams with their continuously varying compositions.

A recent example of this principle in a practical application occurred when the U.S. Air Force changed its lubricant from hydrocarbon oil to synthetic esters. Polydimethylsiloxane (PDMS) had been discovered as an antifoam in lubricating oil during World War II and has continued to be used in polyester (PE) aviation lubricants, but the foaming returned. Center[59] discerned that the average molecular weight of the polymer, or that which measures its average molecular weight, the viscosity of the polymer mixture, is the significant variable, and by so doing he provided a rationale of what had been a number of puzzling, and at times apparently contradictory, results.

The foam behavior of PDMS dispersed in a polymer-ester (PE) turbine-engine lubricant, as a function of increasing viscosity of the PDMS, at a constant concentration of 10 μg/g, follows a familiar pattern in almost every detail (Figure 23.8), the sole exception being the more gradual onset of the antifoam action, which may be attributed to the presence of a range of molecular weights in the polymer, rather than the more abrupt onset seen with simpler and better defined systems.

Very-low-viscosity PDMS (less than about 10^{-5} m^2s^{-1} or 10 cSt) is very soluble in the ester and hence is too little adsorbed at the liquid–air surface to reduce the surface tension; it therefore causes no foaminess. PDMS mixtures of intermediate viscosities (15 to 5000×10^{-6} m^2s^{-1}) act as profoamers, increasingly so with increasing viscosity. The behavior is similar to that of a series of solutions as their concentration approaches solubility saturation at the phase boundary. Finally, 10 μg/g PDMS dispersions with viscosities greater than 5000×10^{-6} m^2s^{-1} are very effective foam inhibitors: these correspond in behavior to the insoluble droplets of the conjugate phase of lower surface tension in the matrix of its higher surface tension conjugate.

Thanks to Center's insight, the PDMS-PE system can be classified as another of the now numerous examples of capillarity associated with the imminence of a phase change.

Two-component systems show maximum foam stability at a temperature and composition near that of the critical point; and three-component systems show maximum foam stability at compositions near that of the plait or consolute point, but only as long as the systems are maintained as homogeneous one-phase solutions. The slightest degree of separation of liquid phases produces a conjugate

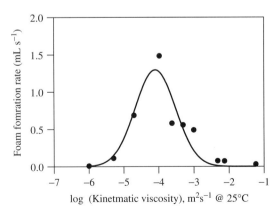

Figure 23.8 The foam behavior of PDMS dispersed in a synthetic-ester turbine-engine lubricant.[59]

solution that can defoam its foamy conjugate. Both these effects, the foaming enhancement in the one-phase solution and the foam inhibition in the two-phase solutions, can be ascribed to the surface activity of the component of lower surface tension in the system, which reveals itself in the one-phase solution by adsorption at the surface and, in the two-phase solutions, by the positive spreading of one conjugate on the other. The occurrence of surface activity in the vicinity of critical or consolute points is well established experimentally by these and many other similar types of observation.

The resemblance of the foaminess (isaphroic) contours shown on Figure 23.6 and the cosorption shown on Figure 16.1 is striking, even though the diagrams pertain to different systems. Later work[60] compared the cosorption contours[61] of Figure 16.1 with the isaphroic contours for the same system, and the resemblance confirmed the relation of surface activity and its manifestation in foaminess, to the character of the phase diagram.

Gas–liquid contacting is the basis of many production processes, in which an adequate interfacial area is required for good mass transport, but excessive foaming leads to instability, especially in equipment such as sieve trays. Industrial foaming problems were the stimulus to investigate connections between capillarity and phase diagrams. Foaming is indeed so prevalent in this type of industrial equipment that design engineers routinely provide extra capacity in gas–liquid contact towers. Problems of this kind may be anticipated and avoided by studying the phase diagram of the system.

Foaming in distillation and fractionation towers, for example, by which the liquid is carried into spaces intended for the vapor, is called "foam flooding," and is often encountered. The processes of degasification after gas absorption or during "stripping" of a monomer in emulsion polymerization are prone to the same effect. The cause of the foam stability may occasionally be traced to the presence of minute concentrations of unintended contamination by substances of strong surface activity; but usually the profoaming solute is a legitimate component of the system, present in substantial concentration. Such a component is not a conventional surface-active solute but becomes surface active under certain conditions of temperature and concentration in a medium in which these conditions are conducive to decreasing its solubility. These conditions may occur in the process of fractionation, stripping, and so on; even so, the surface activity thus elicited is minimal, capable of stabilizing foam only for short times. Although such evanescent foams last no more than several seconds, they are the source of severe foam-flooding problems in distillation or fractionation towers, where rapid evolution of vapor may build up a large volume of dynamic foam. For example, in the Girdler sulfide process for producing heavy water, liquid water and gaseous hydrogen sulfide are brought into contact at high temperatures and at pressures of several atmospheres over sieve trays.[62,63,64] In certain parts of the exchange tower the conditions are such that the overall composition approaches the phase boundary at which liquid hydrogen sulfide separates from its conjugate aqueous solution. In accordance with observations of the systems with more accessible conditions described in this section, the hydrogen sulfide in aqueous solution would be expected to be surface active, and indeed foam does appear. The foaming was sufficiently severe to require drastic reduction of throughput in the exchange tower and was sufficiently abstruse to delay remedial action for over a year. Again, in accord with generalizations drawn from cognate systems, the conjugate liquid-hydrogen-sulfide phase would be expected to inhibit foam, as in fact it was found to do.

23.4.1 Evanescent Foams and Stable Foams

Summarizing the discussion thus far, we have pointed out three distinct mechanisms of foam stabilization: the Marangoni effect, the formation of gelatinous surface layers, and the immobilizing of interstitial liquid by electrostatic effects. These mechanisms accord with the observed fact that two distinct ranges of foam stability differing considerably in magnitude exist for two different types of foam. The relatively unstable foams (e.g., champagne foam) result from the Marangoni effect alone; they drain rapidly to a critical thickness on the order of 20 to 30 nm, at which point the Marangoni effect can no longer operate and the lamellae are brittle and soon rupture. The very stable foams (e.g., beer foam) begin their lives in the same way, being preserved from instantaneous destruction only by the existence of the Marangoni effect. This effect is soon replaced, however, either by the slow growth of gelatinous surface layers or by the overlap of diffuse double layers, which takes over the task of preserving the lamellae by immobilizing the fluid inside so that gravitational and capillary stresses are both insufficient to thin them immediately.

23.5 MECHANISMS OF ANTIFOAMING ACTION

23.5.a Impairment of Foam Stability

We distinguish between impairment of foam stability and the inhibition of foam formation. Foam stability may be impaired by the presence of certain substances of low molecular weight that by their presence in the surface layer reduce its coherence. Such a substance does not need to be surface active to affect the surface layer if enough of it is mixed with the solution to act as a cosolvent. This effect has an obvious explanation because the surface-active solute that is responsible for foaminess may be rendered more soluble, and hence less surface active, in a mixture of solvents than in one solvent alone.

The term "cosolvent" implies a concentration of 10% or more of a second solvent. Occasionally, the addition of such a large proportion of, for example, methanol, ethanol, or acetone to an aqueous solution may be practicable and would thereby solve a troublesome foaming problem. Customarily, a much smaller quantity of agent is desirable. Excluding this case then, foam stability may still be impaired by solubilizing small amounts of nonionic organic liquids in the aqueous solution, using the foam-producing solute as a solubilizing agent. This operation has previously been mentioned as a means of increasing the stability of foams by the formation of plasticized (gelatinous) surface layers, but the same operation may produce quite different results, depending on the nature of the solubilizate. If the molecular structure of the nonionic agent is such that it does not form a coherent surface layer, its presence will impair rather than enhance the stability of the foam. Thus, for example, while a small quantity of solubilized dodecanol will increase the foam stability of an anionic agent in water, a like quantity of solubilized octanol will have a contrary effect. The octanol, which is less surface active than dodecanol, is not present in the surface in amount sufficient to create a liquid-crystal phase and also, by virtue of its more rapid diffusion to the surface, reduces the dynamic surface tension more rapidly and thus allows less time for the Marangoni effect to function.[37] Nevertheless, some coherence of the surface layer and some Marangoni effect still remain, so although the foam is rendered less stable, its formation is not entirely inhibited. This result is "foam impairment."

23.5.b Surface and Interfacial Tension Relations

Certain chemical agents are well known to do more than merely impair the stability of foam; they totally inhibit the forming of foam. These foam inhibitors are always insoluble materials, usually liquids, although hydrophobic solid particles, such as Teflon®, waxes, and silane-treated inorganic oxides also are used.

Not every insoluble substance can act as a foam inhibitor; there are additional requirements. The nature of these additional requirements is readily comprehended if the mechanism of the foam-inhibiting action is first understood. Such an understanding of the underlying mechanism was inferred gradually from observations and experiments with foam-producing systems to which various agents were added so their effects could be studied.[38] The solution of practical problems came before theoretical explanations, the latter being elicited only rather slowly and not without several hypotheses later found to be erroneous. At first, so spectacular was the phenomenon and so unfamiliar were the principles of surface activity, that one often heard it described as magic. As in so many other cases, the magic was banished by science.

Consider the successive operations that are required. First, the droplet or particle of insoluble agent has to be admitted to the surface; that is, when the agent approaches the vicinity of the surface as a result of its random movement in the foamy liquid, the medium should withdraw so that the agent has an exposed surface toward the vapor phase. A factor that would be helpful to promote this action would be a slight difference in density between the agent and the medium so that a buoyant force directed toward the surface would be added to the random movements of the droplet that are the result of thermal currents, stirring, or Brownian motion. This by itself would not be enough; droplets of the agent would indeed float to the surface, but they might still remain covered with a thin liquid film of the medium. The medium would not withdraw and expose the surface of the droplet unless, by so doing, the net surface-free energy were reduced, that is, only if the surface tension of the agent σ_a were less than the sum of the surface tension of the medium σ_m and the interfacial tension σ_{int}. This condition requires that the agent have a relatively low surface tension, too low to be wetted by the foamy medium. Robinson and Woods[39] were the first to point out this condition, which they defined by means of a function named the *entering coefficient E*,

$$E = \sigma_m + \sigma_{\text{int}} - \sigma_a \qquad (23.14)$$

The agent enters the surface only if the value of E is positive. But the term "entering coefficient" seems to be an unnecessary neologism when we already possess a well-known coefficient to serve as a criterion for wetting, namely, the spreading coefficient S.[65] The condition is stated more expressively if we say that the droplet of the agent enters the surface only if it is dewetted by the medium, that is, if the medium is incapable of spreading on the agent. Let S_1 be the spreading coefficient of the medium on the agent; then,

$$S_1 = \sigma_a - \sigma_m - \sigma_{\text{int}} \qquad (23.15)$$

The agent would be dewetted by the medium if the value of S_1 were negative. The two statements of the necessary condition are, of course, identical, since $S_1 = -E$.

The droplet of the agent enters the surface by virtue of withdrawal of the medium, a process aptly described as "dewetting." A foam lamella has two surfaces; consequently, dewetting can occur also on the second surface once the lamella has sufficiently thinned, either by capillary suction or hydrodynamic drainage. The second withdrawal, at the opposite surface, causes the rupture of the lamella, whether due to the mechanical agitation caused by the spontaneous motion or to the poor adhesion of the medium to the droplet, which makes it ill-adapted to hold the lamella together by bridging the gap.

The surface having been entered, the next operation that may be required of a foam-inhibiting agent is for the droplet of agent to spread spontaneously across the surface of the medium. Let S_2 be the spreading coefficient of the agent on the medium; then,

$$S_2 = \sigma_m - \sigma_a - \sigma_{\text{int}} \qquad (23.16)$$

The agent spreads spontaneously if the value of S_2 is positive. The second condition is perfectly compatible with the first condition; S_1 being negative does not prevent S_2 from being positive. The only prohibition is that both S_1 and S_2 cannot be positive; therefore, if S_2 is positive, S_1 *must* be negative. Thus, if the interfacial tension between the medium and the foam-inhibiting agent is sufficiently low, the droplet will spread after it enters the surface of the lamella. The critical analysis of the literature on this subject by Garrett[66] has shown that a positive value of E is a necessary but insufficient condition for an effective antifoam. If spreading can occur, it inhibits bridging. Examples of antifoaming by spreading are the action of a spray, or even the presence of a vapor, of a volatile liquid such as diethyl ether or acetone when introduced as a foam breaker above the surface of a foam already formed.

Ross predicted in 1950 that "the film is thinnest precisely at the place where it is composed entirely of the antifoaming agent and it is therefore at this spot that rupture of the film can be expected to take place."[67] The mechanism of foam-inhibiting action as observed by a high-speed video camera was studied by Denkov et al.[68,69,70,71] Foam films of an aqueous solution of sodium dioctyl sulfosuccinate were destabilized with an antifoam composed of silicone oil and hydrophobic silica. The results showed that in this system the films are destroyed by the formation of unstable oil bridges, which afterwards stretch. The film thins because of capillary pressure and eventually ruptures. (See Figure 23.9.) Denkov et al. observed rupture of the biconcave oil bridge to occur at its center, as predicted.

The dewetting mechanism is used in the application of dispersed droplets of polydimethylsiloxane as a foam inhibitor in lubricating oils. The surface tensions of liquid hydrocarbons are between 25 and 30 mN/m; a liquid able to enter on such a low-energy substrate is required to have a surface tension lower than these by several milli-Newtons per meter. Such liquids are usually volatile, which makes them unsuitable in many applications. Only special polymers such as polydimethylsiloxane or perfluorinated hydrocarbons combine the usually disparate properties of low surface tension and low volatility.

23.5.c Foam Inhibition: Aqueous Systems

The above principles are applied in compounding a foam-inhibiting agent for a particular use. An aqueous solution or dispersion that is the source of troublesome foam usually has dissolved organic materials that act as the profoamer. Their presence is evident from the surface tension, which in such a case is less than that of pure water at the same temperature. Occasionally, when imperfectly wetted solid particles are present, the stabilizing of the foam may be caused by the solid particles holding entrapped air between their hydrophobic surfaces, but even then

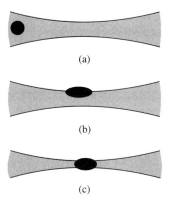

Figure 23.9 Schematic representation of an antifoam drop in a foam lamella. (*a*) Entering the surface. (*b*) Bridging the lamella. (*c*) Leading to rupture of the lamella.

some dissolved profoamer is usually also present to initiate film formation. If surface tension is less than that of water by no more than 20 mN/m, the selection of a foam-inhibiting agent of abnormally low surface tension, for example, a polydimethylsiloxane, is hardly justified. A less unusual and, incidentally, less expensive agent serves just as well. One could begin by trying glyceryl esters in the form of lard, pork fat, or butterfat. These common substances have many of the desired qualities: insolubility in water, lower density than water, relatively low surface tension, and relatively low interfacial tension against aqueous solutions. For many practical purposes, they are effective foam inhibitors where the systems being treated have surface tensions greater than that of the fat.[72] If the surface tension of the aqueous system is too low, however, the fat is emulsified and is then not effective as a foam inhibitor, at least by the mechanism described above. The system may stop foaming if enough fat is emulsified to deplete the solution of its profoamer; but this result has the disadvantage, for some purposes, of incorporating a quantity of fat in the system.

If readily available natural fats prove to be ineffective agents, a number of synthetic esters of polyhydric alcohols, which are marketed under various trade names, may be tried. These substances extend the range of choice of materials and offer variations in the balance of the lipophilic and hydrophilic portions of the molecule. The lipophilic portion confers low surface tension; the hydrophilic portion confers low interfacial tension, both within the limits of a suitable balance. Completely lipophilic substances, for example, paraffin oil or kerosene, besides being very insoluble in water, are not even able to spread as a monolayer on a water surface. Their interfacial tension against water is large, which means that they cannot be wetted by water. They will, therefore, enter a water surface (S_1 negative) but are too inert to act as antifoams. If the substance has a small degree of hydrophilia added to it, for example, fatty alcohols or fatty acids, the interfacial tension is the first property to respond, and it does so by becoming smaller. We now find that although S_1 is still negative, S_2 may become positive; in terms of behavior, the substance will enter the aqueous surface and also spread on it. These materials are of the class in which excellent foam inhibitors are to be found. Let the substance be made a little more hydrophilic, and the interfacial tension is still further lowered to a very small value so that S_1 may become positive. These materials are not dewetted by water and consequently are unable even to enter the surface. Still more hydrophilic substances, for example, sucrose or glycerol, are soluble without limit in water (See Table 11.2 for a description of a graded series of lipophilic to hydrophilic compounds).

Evidently, an optimum balance is desirable between the lipophilic and hydrophilic portions; but the optimum varies with the nature of the medium and with the nature of the profoamer dissolved therein. The more lipophilic the profoamer, the more lipophilic is the optimum foam-inhibiting agent. It seems, therefore, that a certain constant difference in the HLB scale must be maintained between a profoamer and its optimum foam inhibitor. The point has not been investigated experimentally, although it would be well worth the effort. If it were verified, we would have a basis for selecting the optimum foam inhibitor once the HLB of the profoamer were ascertained. At present, no single test has been widely accepted whereby the HLB of a surface-active solute may be measured, but such a test is the objective of a number of workers.[73] One of the fruits of success in this endeavor would be to diminish the empiricism that at present characterizes the search for an optimum foam inhibitor.

23.5.d Compounded Foam Inhibitors

A practical discovery of emulsion technology is that a mixture of emulsifying agents is frequently more effective to stabilize an emulsion than a single ingredient. As a result of this knowledge, the practice was derived of compounding agents to a desired HLB by mixing ingredients lying on each side of the average. Foam-inhibiting agents can be compounded in the same way, as long as all ingredients are miscible. A paraffin oil, for example, would not inhibit the foaming of an aqueous system; the surface tension, which is about 30 mN/m, may not be low enough for the medium to dewet the droplet so that it could enter the surface. To make it more effective, the interfacial tension must be reduced. One way to do this is to add a surface-active solute to the aqueous medium, but this offers too narrow a margin to work with because a very small amount may be enough to reduce the surface tension of the medium as well as the interfacial tension, thereby emulsifying the paraffin oil and keeping it from entering the surface. A better way is to compound an oil-soluble surface-active agent with the paraffin oil. Such agents usually do not lower the surface tension of organic liquids but exhibit their surface activity by reducing the interfacial tension at the oil/water interface.

Substances suitable for this purpose include alkyl benzenes, natural fats, fatty acids, and metallic soaps. Even the moderate hydrophilia conferred by aromatic substituents in an alkane hydrocarbon is sufficient to reduce the oil water interfacial tension; more hydrophilic substituents, such as a carboxylate group, make highly effective agents for this purpose. An oil-soluble polymeric silicone can also be used as the second ingredient, thereby reducing the cost of using an uncompounded silicone as the sole agent.

Another example of the same principle is the use of homopolymers of propylene oxide or butylene oxide and copolymers of ethylene, propylene, and butylene oxides, when compounded with metal salts of higher fatty acids, for the inhibition of foam in proteinaceous glues.[74]

The same principles can be applied *mutatis mutandis* to the problem of foam in nonpolar media. For example, the foaming of lubricating oil in aircraft at high altitudes was an urgent problem during World War II. A foam-inhibiting agent for lubricating oil was made of glycerol, selected because of its low volatility and insolubility in the oil, with 2% of a surface-active agent (e.g., Aerosol OT) dissolved in it.[75] The function of the Aerosol OT is to reduce the surface tension of the glycerol below that of the lubricating oil and also to reduce the interfacial tension between the glycerol and the oil; when these tensions are sufficiently low, the composition functions as a foam-inhibiting agent. Other glycerol-soluble surface-active agents can be substituted for the Aerosol OT, such as a block copolymer of polyoxyethylene and polydimethylsiloxane.

23.5.e Silicone Foam Inhibitors

Polydimethylsiloxanes are close to being universal agents for inhibiting foam because they combine two properties seldom found together, namely, involatility and low surface tension; in addition, they are chemically inert, thermally stable, and are insoluble in water and in lubricating oil. They are effective in concentrations in the range of 10 µg/g or less, whereas other foam-inhibiting agents are often used in the range of 100 to 1000 µg/g of bulk fluid. On the other hand, the silicone polymers cost 10 to 20 times that of the common organic substances.[76]

Reduction of the particle size of the composition to approximately the thickness of a foam lamella is a requirement for its effectiveness as a foam inhibitor. Unfortunately, the potentially more effective agents are precisely those most likely to coalesce on standing. Silicone antifoams are sometimes provided by manufacturers in the form of emulsions in an aqueous phase, which will then mix readily with the aqueous foamy liquid. This form meets the need for small droplets by predispersing the fluid before it is added to the foam-producing medium. Ultimately, the low concentrations of antifoam used are effective in preventing coalescence because of the small probability of two particles colliding. Silicones are used as foam inhibitors[77] in a wide range of industrial foaming problems, including those in distillation towers. The standard polydimethylsiloxane molecule may be modified to decrease its solubility for certain applications and thereby improve its effectiveness as a foam inhibitor; for example, polytrifluoropropylsiloxane defoamers are used with some organic systems in which the polydimethylsiloxanes are too soluble to be effective.[78]

The science and technology of silicone foam inhibitors is discussed by Kulkarni et al.[79]

23.5.f Hydrophobic Particles

Silicone oil, polydimethylsiloxane, is an effective foam inhibitor of oil foams, and is used to prevent the foaming of hydrocarbon lubricants in aircraft engines.[80] But it is far from being equally effective to control foaming in aqueous solutions. It was found, however, that after adding finely divided silica in proportions of 3 to 6% the mixture acts as a highly effective antifoam, even when only a few parts per million is put into the foamy liquid. Scores of patents have been issued describing progressively better methods to incorporate silica into the formulation, with more effective foam-inhibiting action marking each advance in the art of preparation. The accumulated evidence reported in those patents leaves no doubt of the reality of the "activation" by the presence of the silica; it contributes nothing to our understanding of why it should occur. The improvement of the antifoam by the addition of silica is, however, now so well established that few, if any, commercial formulations designed to suppress or destroy foam of aqueous systems by means of silicone oil are marketed without the added silica.

Silica normally has a hydrophilic surface, which is to say it is perfectly wetted by water. But the surface of silica can be drastically altered to become hydrophobic by chemical treatment. Since finely divided particles necessarily have a large surface area, and as we are discussing silicas with 200 to 400 m^2 of surface area per gram, this alteration of the surface character implies marked alterations of the gross behavior of the material. A particle of hydrophobic silica is not wetted by water; it floats on the surface of the water, just as an oiled needle would, in spite of its density being greater than that of water. Water may be said to roll off its back, just as it rolls off a duck's back, and fundamentally for the same reason. When mixed with a hydrophobic liquid, such as silicone oil, however, hydrophobic silica is readily wetted by the oil and disperses easily. In such media it is now the turn of hydrophilic silica to show reluctance to be wetted; it does not allow the oil to penetrate freely along the surface and especially into the narrow spaces between particles. If many such hydrophilic particles are stirred into silicone oil, they create an interconnected or network structure throughout the oil and so give it the viscoelastic properties of a jelly.

Ross and Nishioka[81] prepared suspensions of silica in polydimethylsiloxane. These suspensions were of two types, designated alpha and beta, depending on how much heat or agitation is put into the system during its preparation. The suspension is very viscous and elastic and resembles a jelly; this is the α suspension. When ball-milled or heated to 150°C for a few hours, or to a higher temperature for a shorter time, the jelly-like structure breaks down irreversibly and the suspension becomes more fluid. The suspension is now sensibly different both to the eye and to the touch; this is the β suspension.

The α and β suspensions differ in the following respects:

(a) The obvious and immediately apparent difference is the jelly-like consistency of the α suspension compared with the fluid, though still viscous, β suspension. If, for example, the concentration of silica is made more than about 5%, the α suspension is too stiff to be stirred conveniently by hand or by ordinary laboratory mixing equipment, but on conversion to the β suspension it flows like a heavy lubricating oil.

(b) On diluting the α suspension with n-hexane, the suspended silica becomes unstable and soon separates and sinks to the bottom. This behavior reveals that the suspension lacks true stability and is maintained only by virtue of its jelly-like consistency. The β suspension on the other hand is evidently well stabilized because on dilution with hexane to a low viscosity the aggregates of silica remain suspended indefinitely. This difference in stability argues in favor of steric stabilization of the aggregates in the β suspension and the lack of such stabilization of those in the α suspension; that is, the presence of a thick adsorbed layer of polymer around aggregates in the β suspension.

(c) Tested as foam-inhibiting agents, the α suspension hardly differs from polydimethylsiloxane taken by itself without added silica, whereas the β suspension destabilizes foam, an effect well recognized in the commercial compositions of polydimethylsiloxane as being conferred by silica.

(d) No surface-active solute is present in this system to stabilize these suspensions, which must therefore be stabilized by adsorption of the medium itself. The usual techniques to measure adsorption in a three-component system, consisting of a solid, a solvent and a solute, cannot be applied. The properties of polydimethylsiloxane, which is involatile and spreads as a monolayer on water, suggest the use of the Langmuir film balance as an instrument of investigation. The quantitative determination of bound polymer on the silica substrate can be calculated directly from the differences between pressure area isotherms of monomolecular layers of polydimethylsiloxane with and without dispersed silica. Bound polymer on the α suspension is found to be about 0. l0 g/g of silica; on the β suspension it is about 1.0 g/g of silica.[82]

(e) When spread as a compressed monolayer on a water surface, the α suspension stabilizes single bubbles blown under it exactly as does the monomolecular layer spread from polydimethylsiloxane taken by itself without added silica. A monomolecular layer spread from the β suspension does not stabilize a bubble at any degree of compression.

These differences demonstrate that the silica in the α suspension is only partly converted to hydrophobic silica, and in that partially converted form it adds nothing at all to the foam-inhibiting properties of the polydimethylsiloxane. The following commercial preparations of the polydimethylsiloxane containing silica, all marketed as foam inhibitors, were tested by the preceding five criteria. Each one was found to correspond to the β suspension by the previous tests: (a) Dow Corning Antifoam M, (b) Dow Corning Antifoam MSA, (c) Rhodorsil Antimousse 454.

The results of the measurements of bubble stability under a monolayer are particularly revealing with respect to the mechanism of rupture of foam lamellae. Trapeznikoff and Chasovnikova[83] showed that polydimethylsiloxane when spread as a monolayer on water actually stabilizes a water lamella. A spread layer of polydimethylsiloxane on water stabilizes bubbles blown under it in a manner similar to the bilayers of Trapeznikoff and Chasovnikova. Significantly, a film of the α suspension on water also stabilizes bubbles, whereas a film of the β suspension does not. The similarity in behavior of bubbles created under a spread film of polydimethylsiloxane taken by itself without added silica and a spread film of α suspension, and the dramatic difference displayed when the β suspension is substituted for either of them, demonstrates the strong destabilizing effect produced by the hydrophobic silica particles of the β suspension in the polydimethylsiloxane film.

A comprehensive review by Garrett of the mechanisms of foam inhibition by silicone foam inhibitors containing hydrophobic particles nevertheless concludes that "no published hypothesis concerning the role of the particles in synergistic oil-particle antifoams is entirely satisfactory."[84]

The mechanism by which dispersed solid hydrophobic particles are so effective in destroying bubbles depends on the degree of hydrophobia of the particle. An oily surface makes a film of water recede from it, and in the same way the surface of a hydrophobic solid particle makes water recede, or dewet, where the two make contact. The withdrawal of the water of the bubble lamella from the surface of the particle may by itself be enough of a mechanical shock to rupture the lamella and release the enclosed gas, or the rupture may be due to the poor adhesion between the water and the hydrophobic particle, causing the water to fall away from the particle as it withdraws.[82] Roberts et al.[85] published high-speed photographs that seem to demonstrate the latter mechanism for rupture of soap lamellae by insoluble droplets. Garrett,[86] Dippenaar,[87] Kurzendoerfer,[88] and Aronson[89] have also demonstrated that various chemically inert dispersed solids act as antifoams and have proposed that the particles rupture the foam film by a bridging-dewetting mechanism. "Bridging" implies that the particle touches both surfaces of the lamella, but whether just one or both sides of the lamella have to be dewetted in order to cause rupture remains uncertain. The finding by Ross and Nishioka that the presence of hydrophobic silica is crucial to foam-breaking was confirmed by a completely different technique of investigation, namely, fluorescence labeling of the silica and examination of defoamer-induced coalescence of bubbles by scanning laser confocal microscopy.[90]

23.6 METHODS OF MEASURING FOAM PROPERTIES

23.6.a Stability

The measurement of foam stability is made either on static or dynamic foams. A static foam is one in which the rate of foam formation is zero: The foam once formed is allowed to collapse without regeneration by further agitation or input of gas. A dynamic foam is one that has reached a state of dynamic equilibrium between the rates of formation and decay. The typical measurement of a dynamic foam is the volume of foam at the steady state; the typical measurement of a static foam is its rate of foam collapse. Dynamic-foam measurement is applicable to evanescent or transient foams; static-foam measurement is applicable to foams of high stability, such as are generated from solutions of detergents or proteins.

Stability of dynamic foams is measured by passing gas through a porous ball or plate into a solution, which may be contained in either a cylindrical or a conical vessel. Gas is passed through a suitable volume of the solution at measured rates, V/t(m^3/s), and the steady-state volume of foam, v, is measured. The depth of the liquid layer above the porous plate should be adequate to ensure a result independent of that depth. The ratio of the flow rate of the gas, whether incoming or outgoing, to the steady-state volume of the foam has the units of time and represents the average lifetime of gas in the foam.[91] It is designated Σ and is used as a unit of foam stability. Frequently, however, the ratio is not constant through an extensive range of flow rates, or the foam is so stable that it floods the container even at low flow rates. The substitution of a conical container instead of a cylindrical one improves those situations and extends the use of the method.[92] A suitable apparatus is shown in Figure 23.10. A conical container avoids flooding. The area of the top level of the foam increases as it rises and so

490 FOAMS

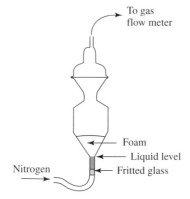

Figure 23.10 An apparatus to measure the stability of dynamic foam.

promotes the escape of gas to reach an equilibrium before the volume of foam becomes excessive.[93]

The stability of static foams is fundamentally the rate at which the total area of the liquid lamellae disappears. These measurements can be obtained photographically[94] or by digital image analysis. An indirect technique is available that uses the equation of state of a foam as its theoretical basis:

$$3V\Delta P + 2\sigma\Delta A = 0 \qquad (23.6)$$

where ΔP is the change of external pressure and ΔA is the change of the area of the liquid lamellae during the decay of a static foam. Values of absolute area of the liquid lamellae in the foam are obtainable as follows. The area of liquid lamellae in foam is

$$A(t) = A_0 + \Delta A \qquad (23.17)$$

where $A(t)$ is the area at time t and A_0 is the initial area. When the foam has completely collapsed,

$$A_\infty = A_0 - \frac{3V\Delta P_\infty}{2\sigma} \qquad (23.18)$$

where ΔP_∞ is the final pressure change in the head space when the foam is gone; therefore

$$A_0 = \frac{3V\Delta P_\infty}{2\sigma} \qquad (23.19)$$

and

$$A(t) = \frac{3V}{2\sigma}[\Delta P_\infty - \Delta P(t)] \qquad (23.20)$$

Interfacial area of a foam can be measured, therefore, simply by monitoring the change in pressure external to the foam in a container of constant volume and

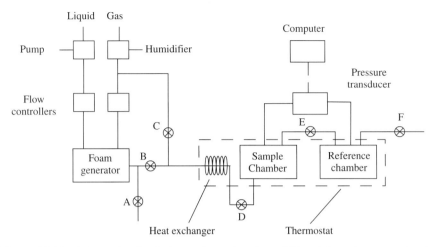

Figure 23.11 Schematic of a device to determine the stability of foam by measuring the decay of area with time obtainable from H and N Instruments.[95]

constant temperature, if the total volume of the system and the surface tension of the foamed liquid are known. The value ΔP_∞ can be obtained by letting the foam decay for a sufficiently long time or by injecting a small quantity of antifoam through a septum in the container.

Values for the area of a decaying foam by use of Eq. (23.20) were found to agree with photographic estimates of the area.[12]

Nishioka[95] developed an apparatus to measure foam stability, using a mechanical foam generator, an Oakes Foamer with a two-inch mixing head, and a pressure transducer (Dynisco Model PTI4-03) to measure the growth of the pressure, which was about 0.2 kPa over a period of 10 h. Figure 23.11 is a schematic of the equipment, which consists of controllers to regulate the flow of the liquid and gas into the generator, the foam generator, the foam-measuring system, and the data-collecting system. The results obtained by Nishioka by this equipment are reproducible with an error of 3–4%. The foams produced by the generator initially have about 20 m² of area, which decays to about 2 m² within an hour.

A condition of the application of this method, pointed out by Nishioka et al.[96] is that Eq. (23.20) is not valid for spherical foam cells (except under conditions of microgravity), because hydrostatic effects create additional pressures. The method of measuring pressure in the head space is therefore limited to the study of dry, thermostatic foams.

23.6.b Surface Rheology

The presence of a highly viscous layer at the surface of an aqueous soap solution can be observed by sprinkling flowers of sulfur on the surface and attempting to move them by blowing gently. On a perfectly fluid surface, such as that of

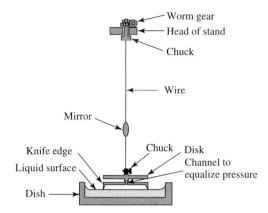

Figure 23.12 Surface viscosimeter with an oscillating circular knife edge suspended by means of a torsional pendulum.

pure water, the particles may be moved readily; but on certain soap solutions they move only a short distance and spring back when the blowing stops. The phenomenon does not occur with solutions of a single surface-active solute but arises when a sparingly soluble cosolute is also present. Current belief is that the surface is a separate liquid-crystal phase produced within the adsorbed layer of solute. While the *surface* may show extreme non-Newtonian flow behavior, the *bulk solution* retains its low Newtonian-type viscosity.

Various instruments to measure the rheology of the surface have been described of which the simplest is the torsional pendulum[97] (see Figure 23.12). A circular knife edge, supported by a torsion wire, is placed so as just to touch the surface of the solution. The torsion wire is supported at the end of a shaft that can be given a rotational twist so as to cause the knife edge to oscillate in the plane of the surface. The damping of the oscillation is followed by observing a light beam reflected from a mirror attached to the wire. The amplitude of the oscillation decreases semilogarithmically when the knife edge is immersed in a perfectly fluid surface. If the surface is plastic, the amplitude decreases more rapidly. At certain portions of its oscillation the shearing stress produced by the pendulum is less than the yield point; the result is much as if the free oscillation were arrested from time to time by an external force. While this instrument can detect and even give relative values of the surface plasticity, the measurements do not lend themselves to deduce rheological coefficients.

The instrument designed by Burton and Mannheimer[98,99,100] allows absolute values of surface shear viscosity and yield strength to be obtained. The solution is contained in a stainless-steel dish into which is placed an annular canal formed by two concentric cylinders. A small gap (approximately 130 μm) is left between the bottom of the canal walls and the bottom of the dish. The walls of the canal are held stationary as the dish is rotated on a turntable at a fixed angular velocity (see Figure 23.13). A few nonwetted particles are dropped onto the surface within

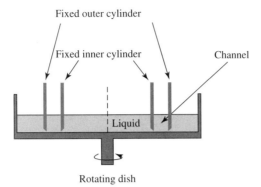

Figure 23.13 The deep-channel surface viscosimeter.[101,102]

the canal. The velocity communicated to the surface is measured by the motion of the particles. The yield strength of a film is determined by measuring the maximum angular velocity at which the surface flow is zero.

Brooks et al. describe the design and use of a surface rheometer to study rheological transitions of an insoluble monolayer at the air/water interface.[103] A magnetized needle is floated at the surface and is moved by means of a variable magnetic field. The position of the needle is followed by means of a linear photodiode array. The advantages of such a device over rotating discs, knife-edge devices, or channel-flow devices include that (1) the frequency of the strains can be changed easily, (2) the strain rate is determined from the position of the needle and not by floating particles, and (3) the surface pressure of the monolayer can be changed easily without having to change the apparatus.

REFERENCES

[1] Briggs, T. Foams for firefighting, in *Foams: theory, measurements, and applications*; Prud'homme, R.K.; Khan, S.A., Eds.; Marcel Dekker: New York; 1996; pp 465 –509.

[2] Edwards, D.A.; Wasan, D.T. Foam rheology: The theory and role of interfacial rheological properties, in *Foams: theory, measurements, and applications*. Prud'homme, R.K.; Khan, S.A., Eds.; Marcel Dekker: New York; 1996; pp 189 –215.

[3] Thompson, D.W. *On growth and form*, 2nd ed.; Cambridge University Press: London; 1942.

[4] Ross, S.; Prest, H.F. On the morphology of bubble clusters and polyhedral foams, *Colloids Surf.* **1986**, *21*, 179–192.

[5] Thomson, W. (Lord Kelvin) On the division of space with minimum partitional area, *Phil. Mag.* **1887**, (5), *24*, 503–514.

[6] Boys, C.V. *Soap bubbles, their colours and forces which mould them (1890)*; Dover: New York; 1959.

[7] Isenberg, C. *The science of soap films and bubbles* (1978); Dover: New York; 1992.

[8] Derjaguin, B. Die elastischen eigenschaften der schäume, *Kolloid Z.* **1933**, *64*, 1–6.

[9] Ross, S. Bubbles and foam: A new general law, *Ind. Eng. Chem.* **1969**, *61*(10), 48–58; *Chemistry and physics of interfaces II*; Ross, S., Ed.; American Chemical Society: Washington, DC; **1971**; pp 15–25.

[10] Ross, S. Cohesion of bubbles in foam, *Am. J. Phys.* **1978**, *46*, 513–516.

[11] Morrison, I.D.; Ross, S. The equation of state of a foam, *J. Colloid Interface Sci.* **1983**, *95*, 97–101.

[12] Nishioka, G.; Ross, S. A new method and apparatus for measuring foam stability, *J. Colloid Interface Sci.* **1981**, *81*, 1–7.

[13] Nakagaki, M. A new theory of foam formation and its experimental verification, *J. Phys. Chem.* **1957**, *61*, 1266–1270.

[14] Weaire, D.; Hutzler, S. *The physics of foam*; Oxford University Press: New York; 1999.

[15] Rayleigh, (John William Strutt) Foam, in *Scientific papers by Lord Rayleigh*; Cambridge University Press: Cambridge; 1902; Vol. 3, pp 351–362; also in 6 vols. bound as 3; Dover: New York; 1964; Vol. 2, pp 351–362.

[16] Marangoni, C. Difesa della teória dell'elasticità superficiale dei liquidi: Plasticità superficiale, *Nuovo Cimento* **1878**, (3), *3*, 50–68.

[17] Marangoni *ibid.*; 97–123.

[18] Marangoni *ibid.*; 193–211.

[19] Gibbs, J.W. Liquid films, in *Scientific papers of J. Willard Gibbs: Thermodynamics*; Longmans, Green: London; 1906; Vol. 1, pp. 300–314; Dover: New York; 1961.

[20] Thomas, T.B.; Davies, J.T. On the sudden stretching of liquid lamellae, *J. Colloid Interface Sci.* **1974**, *48*, 427–436.

[21] Derjaguin, B.V.; Titijevskaya, A.S. Static and kinetic stability of free films and froths, *Proc. Int. Congr. Surf. Act., 2nd, 1957* **1957**, *1*, 211–219.

[22] Mysels, K.J. Soap films and some problems in surface and colloid chemistry, *J. Phys. Chem.*, **1964**, *68*, 3441–3448.

[23] Overbeek, J.Th.G. Black soap films, *J. Phys. Chem.* **1960**, *64*, 1178–1183.

[24] Ross, S. Foam and emulsion stabilities, *J. Phys. Chem.* **1943**, *47*, 266–277.

[25] Koehler, S.A; Hilgenfeldt, S.; Stone, H.A. A generalized view of foam drainage: experiment and theory, *Langmuir* **2000**, *16*, 6327–6341.

[26] Sawyer, W.M.; Fowkes, F.M. Interaction of anionic detergents and certain polar aliphatic compounds in foams and micelles, *J. Phys. Chem.* **1958**, *62*, 159–166.

[27] Gopal, A.D.; Durian, D.J. Nonlinear bubble dynamics in a slowly driven foam, *Phys. Rev. Lett.* **1995**, *75*, 2610–2613.

[28] Durian, D.J. Foam mechanics at the bubble scale, *Phys. Rev. Lett.* **1995**, *75*, 4780–4782.

[29] Durian, D.J.; Weitz, D.A.; Pine, D.J. Multiple light-scattering probes of foam structure and dynamics, *Science* **1991**, *252*, 686–688.

[30.] Weitz, D.A. Foams flow by stick and slip, *Nature* **1996**, *381*, 475–476.

[31] Rusanov, A.I.; Krotov, V.V. Gibbs elasticity of liquid films, threads, and foams, *Prog. Surf. Membr. Sci.* **1979**, *13*, 415–524.

[32] Van den Tempel, M.; Lucassen, J; Lucassen-Reynders, E.H. Application of surface thermodynamics to Gibbs elasticity, *J. Phys. Chem.* **1965**, *69*, 1798–1804.

[33] Borwankar, R.P.; Wasan, D.T. The kinetics of adsorption of surface active agents at gas liquid surfaces, *Chem. Eng. Sci.* **1983**, *38*, 1637–1649.

[34] Kim, Y.-H.; Koczo, K.; Wasan, D.T. Dynamic film and interfacial tensions in emulsion and foam systems, *J. Colloid Interface Sci.* **1997**, *187*, 29 –44.

[35] Noskov, B.A. Fast adsorption at the liquid-gas interface, *Adv. Colloid Interface Sci.* **1996**, *69*, 63 –129.

[36] Mannheimer, R.J.; Schechter, R.S. Shear-dependent surface rheological measurements of foam stabilizers in nonaqueous liquids, *J. Colloid Interface Sci.* **1970**, *32*, 212–224.

[37] Ross, S.; Haak, R.M. Inhibition of foaming. IX. Changes in the rate of attaining surface-tension equilibrium in solutions of surface-active agents on addition of foam inhibitors and foam stabilizers, *J. Phys. Chem.* **1958**, *62*, 1260–1264.

[38] Ross, S.; McBain, J.W. Inhibition of foaming in solvents containing known foamers, *Ind. Eng. Chem.* **1944**, *36*, 570–573.

[39] Robinson, J.V.; Woods, W.W. A method of selecting foam inhibitors, *J. Soc. Chem. Ind., London* **1948**, *67*, 361–365.

[40] Maru, H.C.; Mohan, V.; Wasan, D.T. Dilational viscoelastic properties of fluid interfaces. I. Analysis, *Chem. Eng. Sci.* **1979**, *34*, 1283–1293.

[41] Djabbarah, N.F.; Wasan, D.T. Relationship between surface viscosity and surface composition of adsorbed surfactant films, *Ind. Eng. Chem., Fundam.* **1982**, *21*, 27–31.

[42] Friberg, S.; Ahmad, S.I. Liquid crystals and the foaming capacity of an amine dissolved in water and *p*-xylene, *J. Colloid Interface Sci.* **1971**, *35*, 175.

[43] Friberg, S.; Saito, H. Foam stability and association of surfactants, in *Foams, Proc. Symp.*, 1975; Akers, R.J., Ed.; Academic: New York; 1976; pp 33–38.

[44] Epstein, M.B.; Wilson, A.; Jakob, C.W.; Conroy, L.E.; Ross, J. Film drainage transition temperatures and phase relations in the system sodium lauryl sulfate, lauryl alcohol, and water, *J. Phys. Chem.* **1954**, *58*, 860–864.

[45] Epstein, M.B.; Ross, J.; Jakob, C.W. The observation of foam drainage transitions, *J. Colloid Sci.* **1954**, *9*, 50–59.

[46] Bolles, W.L. The solution of a foam problem, *Chem. Eng. Prog.* **1967**, *63*(9), 48–52.

[47] Basheva, E.S.; Ganchev, D.; Denkov, N.D.; Kasuga, K.; Satoh, S.; Tsujii, K. Role of Betaine as foam booster in the presence of silicone oil drops, *Langmuir* **2000**, *16*, 1000–1013.

[48] DelaMaza, A.; Parra, J.L. Solubilization of phospholipid bilayers by C-14 alkyl betaine/anionic mixed surfactant systems, *Colloid Poly. Sci.* **1995**, *273*, 331–338.

[49] Exerowa, D.; Lalchev, Z.; Marlnov, B.; Ognyanov, K. Method for assessment of fetal lung maturity, *Langmuir* **1986**, *2*, 664–668.

[50] Mysels, K.J.; Shinoda, K.; Frankel, S. *Soap films, Studies of their thinning and a bibliography*; Pergamon Press: New York; 1959.

[51] Dewar, J. Soap bubbles of long duration, *Proc. R. Inst. G.B.* **1917**, *22*, 179–212.

[52] Sheludko, A. *Colloid chemistry*; Elsevier: New York; 1966.

[53] Friberg, S.E.; Cox, J.M. Stable foams from nonaqueous liquids, *Chem. Ind. (London)* **1981**, *17* Jan., 50–52.

[54] Rosen, M.J. *Surfactants and interfacial phenomena*, 2nd ed.; Wiley: New York; 1989; pp 297–299.

[55] Ottewill, R.H.; Segal, D.L.; Watkins, R.C. Studies on the properties of foams formed from nonaqueous dispersions, *Chem. Ind. (London)* **1981**, *17* Jan., 57–60.

[56] Zuiderweg F.J.; Harmens, A. The influence of surface phenomena on the performance of distillation columns, *Chem. Eng. Sci.* **1958**, *9*, 89–103.

[57] Ross, S.; Nishioka, G. Foaminess of binary and ternary solutions, *J. Phys. Chem.* **1975**, *79*, 1561–1565.

[58] Davidson, J. Foam stability as an historic measure of the alcohol concentration in distilled alcoholic beverages, *J. Colloid Interface Sci.* **1981**, *81*, 540–542.

[59] Centers, P. Behavior of silicone antifoam additives in synthetic ester lubricants, *Tribol. Trans.* **1993**, *36*, 381–386; Effect of polydimethylsiloxane concentration on ester foaming tendency, *Tribol. Trans.* **1994**, *37*, 311–314; Modeling and prediction of foaming tendencies of polydimethylsiloxane polyol ester mixtures, *Tribol. Trans.* **1995**, *38*, 183–187; Turbine engine lubricant foaming due to silicone basestock used in nonspecification spline lubricant, *Lub. Eng.* **1995**, *51*, 368–371.

[60] Ross, S.; Townsend, D.F. Foam behavior in partially miscible binary systems, *Chem. Eng. Commun.* **1981**, *11*, 347–353.

[61] Nishioka, G.M.; Lacy, L.L.; Facemire, B.R. The Gibbs surface excess in binary miscibility-gap systems, *J. Colloid Interface Sci.* **1981**, *80*, 197–207.

[62] Haywood, I.R.; Lumb, P.B. The heavy water industry, *Chem. Can.* **1975**, *27*(3), 19–21.

[63] Sagert, N.H.; Quinn, M.J. Influence of high-pressure gases on the stability of thin aqueous films, *J. Colloid Interface Sci.* **1977**, *61*, 279–286.

[64] Bancroft, A.R. *Heavy Water GS Process: R&D Achievements*; Atomic Energy of Canada Ltd., Report 6215, October, 1978.

[65] Harkins, W.D. A general thermodynamic theory of the spreading of liquids to form duplex films and of liquids or solids to form monolayers, *J. Chem. Phys.* **1941**, *9*, 552–568.

[66] Garrett, P.R. The mode of action of antifoams, in *Defoamimg: theory and industrial applications*; Garrett, P.R. Ed.; Surfactant Science Series, Vol. 45; Marcel Dekker: New York; 1993; pp 1–117.

[67] Ross, S. A mechanism for the rupture of liquid films by antifoaming agents, *J. Phys. Colloid Chem.* **1950**, *54*, 429–436.

[68] Denkov, N.K.; Cooper, P.; Martin, J.-Y. Mechanisms of action of mixed solid–liquid antifoams. 1. Dynamics of foam-film rupture, *Langmuir* **1999**, *15*, 8514–8529.

[69] Denkov, N.K. Mechanisms of action of mixed solid-liquid antifoams. 2. Stability of oil bridges in foam film, *Langmuir* **1999**, *15*, 8530–8542.

[70] Denkov, N.K.; Marinova, K.G.; Christova, C.; Hadjiiski, A.; Cooper, P. Mechanisms of action of mixed solid–liquid antifoams. 3. Exhaustion and reactivation, *Langmuir* **2000**, *16*, 2515–2528.

[71] Marinova, K.G.; Denkov, N.D. Foam destruction by mixed solid-liquid antifoams in solutions of alkylglucoside: electronic interactions and dynamic effects, *Langmuir* **2001**, *17*, 2426–2436.

[72] Ross, S. The inhibition of foaming, *Rensselaer Polytech. Inst., Eng. Sci. Ser.* **1950**, *63*, 1–40.

[73] Becher, P. HLB: Update III, in *Encyclopedia of emulsion technology*, Vol. 4; Becher, P., Ed.; Marcel Dekker: New York; 1996; pp 337–356.

[74] Stephan, J.T. Combination polyglycol and fatty acid defoamer composition, U.S. Patent 2, 914, 412, 1959.

75 Robinson, J.V. The rise of air bubbles in lubricating oils, *J. Phys. Colloid Chem.* **1947**, *51*, 431–437.

76 Bergeron, V.; Cooper, P.; Fischer, C.; Giermanska-Khan, F.; Langevin, D.; Pouchelon, A. Polydimethylsiloxane (PDMS) based antifoams, *Colloids Surf. A: Physico. Eng. Asp.* **1997**, *122*, 103–120.

77 Rauner, L.A. Antifoaming agents, *Encyclopedia of polymer science technology* Interscience, New York, N.Y. **1964–1972**, *2*, 164–171.

78 Whipple, C.L.; Oppliger, P.E.; Schiefer, H.M. *Abstracts of papers*; 138th National Meeting American Chemical Society: New York; Sept. 11–16, 1960.

79 Kulkarni, R.D.; Goddard, E.D.; Chandar, P. Science and technology of silicone antifoams, in *Foams: theory, measurements, and applications*; Prud'homme, R.K.; Khan, S.A., Eds.; Marcel Dekker: New York; 1996, 555–585.

80 McBain, J.W.; Ross, S.; Brady, A.P.; Robinson, J.V.; Abrams, I.M.; Thorburn, R.C.; Lindquist, C.G. Foaming of aircraft-engine oils as a problem in colloid chemistry I, *Nat. Advis. Comm. Aeronaut., Rep. ARR 4105*, **1944**.

81 Ross, S.; Nishioka, G. Experimental researches on silicone antifoams, in *Emulsions, Latices, and Dispersions*; Becher, P.; Yudenfreund, M.N., Eds.; Marcel Dekker: New York; 1978; pp 237–256.

82 Ross, S.; Nishioka, G. Monolayer studies of silica/poly(dimethylsiloxane) dispersions, *J. Colloid Interface Sci.* **1978**, *65*, 216–224.

83 Trapeznikov, A.A.; Chasovnikova, L.V. Stabilization of bilateral films by monolayers and thin films of poly(dimethylsiloxanes), *Colloid J. USSR (English trans.)* **1973**, *35*, 926–928.

84 Garrett, P.R. The mode of action of antifoams, in *Defoaming, theory and industrial applications*; Garrett, P.R., Ed.; Marcel Dekker: New York; 1993; pp 1–117, esp. p. 98.

85 Roberts, K.; Axberg, C.; Österlund, R. Emulsion foam killers in foams containing fatty and rosin acids, in *Foams, Proc. Symp.*, 1975; Akers, R.J., Ed.; Academic: New York; 1976; pp 39–49.

86 Garrett, P.R. The effect of poly(tetrafluoroethylene) particles on the foamability of aqueous surfactant solutions, *J. Colloid Interface Sci.* **1979**, *69*, 107–121.

87 Dippenaar, A. The destabilization of froth by solids. I. The mechanism of film rupture; II. The rate-determining step, *Int. J. Miner. Process.* **1982**, *9*(1), 1–14; 15–22.

88 Kurzendoerfer, C.P.: Mechanisms of foam inhibition by trialkylmelanines, Tr.-Mezhdunas, *Kongr. Poverkhn.-Abt. Veshchestvain, 7th2, 1976* **1978**, *2*(I), 537–538.

89 Aronson, M.P. Influence of hydrophobic particles on the foaming of aqueous surfactant solutions, *Langmuir* **1986**, *2*, 653–659.

90 Wang, G.; Pelton, R.; Hrymak, A.; Shawafaty, N.; Heng, Y.M. On the role of hydrophobic particles and surfactants in defoaming, *Langmuir* **1999**, *15*, 2202–2208.

91 Bikerman, J.J. The unit of foaminess, *Trans. Faraday Soc.* **1938**, *34*, 634–638.

92 Watkins, R.C. An improved foam test for lubricating oils, *J. Inst. Pet., London* **1973**, *59*, 106–113.

93 Ross, S.; Suzin, Y. Measurement of dynamic foam stability, *Langmuir* **1985**, *1*, 145–149.

94 Savitskaya, E.M. Analysis of the dispersity of foams, *Kolloidn. Zh.* **1951**, *13*, 309–313.

95 Nishioka, G. Stability of mechanically generated foam, *Langmuir* **1986**, *2*, 649–653.

[96] Nishioka, G.; Ross, S.; Kornbrekke, R.E. Fundamental methods for measuring foam stability, in *Foams: theory, measurements, and applications*; Prud'homme, R.K.; Khan, S.A., Eds; Marcel Dekker: New York; 1996; Chapter 6, pp 275–285.

[97] Grist, D.M.; Neustadter, E.L.; Whittingham, K.P. The interfacial shear viscosity of crude oil/water systems, *J. Can. Pet. Tech.* **1981**, *20*, 74–78.

[98] Burton, R.A.; Mannheimer, R.J. Analysis and apparatus for surface rheological measurements, *Adv. Chem. Ser.* **1967**, *63*, 315–328.

[99] Mannheimer, R.J. Surface rheological properties of foam stabilizers in nonaqueous liquids, *AIChE J.* **1969**, *15*, 88–93.

[100] Mannheimer, R.J.; Schechter, R.S. An improved apparatus and analysis for surface rheological measurements, *J. Colloid Interface Sci.* **1970**, *32*, 195–211.

[101] Edwards, D.A., Brenner, H.; Wasan, D.T. *Interfacial transport processes and rheology*; Butterworth-Heineman: Stoneham, MA; 1991; Chapter 7, pp 213–226.

[102] Kim, Y.H.; Koczo, K.; Wasan, D.T. Dynamic Film and Interfacial Tensions in Emulsion and Foam Systems, *J. Colloid Interface Sci*, **1997**, *187*, 29–44.

[103] Brooks, C.F.; Fuller, G.G.; Frank, C.W.; Robertson, C.R. An interfacial stress rheometer to study rheological transitions in monolayers at the air–water interface, *Langmuir* **1999**, *15*, 2450–2459.

24 Technology of Suspensions

Suspensions are dispersions of fine particles in liquids. Practical applications are to inorganic and organic pigments, printing inks, plastics and rubber, paper coatings, strength enhancement of materials, ceramics, magnetic recording materials, pharmaceuticals, fertilizers, pesticides, fungicides, and processed foods.[1] The stability has already been discussed in terms of electrostatic and steric stabilization. Various techniques to determine particle size of particles in suspension are treated in Chapter 4 and theories to describe the stability of suspensions are treated in Chapter 20. The techniques described in the present section refer to physical properties of whole systems and not merely to the solid/liquid interface.

24.1 DISPERSING AGENTS

Successful formulations for a given powder depend on the selection of suitable media and suitable suspending agents. In nonpolar media where adsorption depends on acid–base interactions, the selection of a suitable solvent depends on a balance of its interaction with the agent and the powder. This balance is illustrated in Figures 14.2 and 14.3.

Suspensions of solid particles in water require preliminary sample preparation before being used either for particle-size analysis or for testing suspending agents. Typically a powder is composed of clumps of primary particles, called agglomerates when they are loosely adherent and aggregates when they are more tightly bound. Preparation of samples for various purposes is described in ISO 14887.[2] This publication also contains a procedure for selecting suspending agents in water or in organic liquids. Powders are categorized under about a dozen descriptive terms such as metals, metal oxides, hydrogen-bonding organics, such as cellulose, and ionic salts; liquid media are categorized as aqueous, or as various types of nonpolar such as highly polar, polar, weakly polar, or nonpolar organics. Having characterized the powder and the liquid, the tables list the various combinations of solid and liquid categories along with generic suspending agents, such as phospholipids, PEO-PPO copolymers, and mercaptans. For example, for a metal oxide such as titania in water with an ionic strength less than 0.1 N (Normal) the suggested suspending agent is simply to adjust the pH either two units below the isoelectric point or two units above. For ionic strengths greater than 0.1 N, polyions such as meta or pyrophosphate are suggested and for organic media an organic acid or an organic amine is suggested.

This publication may be supplemented by Nelson's tables[3] where specific dispersants are suggested under headings indicating the nature of the particles and the required properties of the suspension such as stability in acids or bases or high ionic strengths. These tables introduce the practitioner to the use of less familiar agents including alkanolamides, acetylenic glycols, phospholipids, and taurates. If, for example, agents are required to disperse metal particles in water, the suggestions are fatty acids, PEO-mercaptans, alkylamines, or taurates. A selection between these candidates is based on whether requirements include nonfoaming, stability to acidic or basic conditions, or stability to high ionic strengths. All this information is included in this reference.

Another useful source of information is R. F. Conley's book[4] which discusses and gives separate treatment to each of the following dispersants: inorganic acid salts (phosphates, silicates, aluminates, and borates), organic polyacid salts (polyacrylates and polymaleates), polar nonionics (e.g., ethylenediamine and morpholine), and polypolar nonionics (polyethylene oxide and polypropylene oxide), along with methods of creating dispersions and instrumentation for measuring various properties.

24.2 EXPERIMENTAL TECHNIQUES FOR SUSPENSIONS

The number of experimental methods for suspensions is far less than that for emulsions because the solid/liquid interface is less amenable to research than the liquid/liquid interface. Consequently the techniques that are available are all the more important.

24.2.a Adsorption from Solution

Adsorption from solution is undertaken to determine the degree of adsorption of a given solute, in an attempt to unravel the mechanism of stabilization. The method is to bring a finely divided solid into contact with a solution of the adsorbate at constant temperature and to measure the decrease of concentration of solute when equilibrium is reached and after removing the solid. Various methods are available to measure concentration change: these include the optical density of a dye solution, the surface tension of the solution, UV absorption, radioactive tracers, and NMR. The most efficient technique is to use multiple samples at different initial concentrations. Known weights of solid and solution are brought together, allowed to equilibrate at constant temperature, and the resulting change of composition of the solution is measured. The solution may be separated from the solid by filtration or by centrifugation. Methods of analysis depend upon the nature of the solute: optical absorption and radiotracers are favorite methods. The experiments must be so designed, by the choice of the ratio of the mass of the adsorbent to the volume of the solution, that the initial and final concentrations are within the sensitive range of the analytic method.

Let the mass of solid be m and the amount of solution of mole fraction X_2^0 be n_0 (mass or volume). At equilibrium let the composition of the solution be X_2. The surface excess amount of component 2 is

$$n_2 = n_0 \left(X_2^0 - X_2 \right) = n_0 \Delta X_2 \tag{24.1}$$

The adsorption by unit mass of solid, or the specific surface excess, is

$$n = \frac{n_2}{m} = \frac{n_0 \Delta X_2}{m} \tag{24.2}$$

The specific surface excess is the preferred form to report experimental data.[5] If the specific surface area of the solid, Σ, is known, the excess surface concentration, Γ_2, is

$$\Gamma_2 = \frac{n_0 \Delta X_2}{m \Sigma} \left(\frac{\text{moles of component 2}}{\text{m}^2} \right) \tag{24.3}$$

It also follows for a two-component system that since $X_1 + X_2 = 1$, $\Delta X_2 = -\Delta X_1$; hence,

$$\Gamma_1 = -\Gamma_2 \tag{24.4}$$

Equation (24.4) points out that solute and solvent compete for room on the substrate.

Adsorption from solution at constant temperature is expressed as a plot of the amount of solute adsorbed against its equilibrium concentration in solution. Figure 24.1 shows a series of Langmuir adsorption isotherms that describe the adsorption of the selenite ion (SeO_3^{-2}) by goethite (a hydrous mineral oxide of iron) as a function of pH. These isotherms show that the amount of selenite ion taken up by goethite at constant pH approaches a maximum value at higher concentrations of solute.

These plots are often described by the Langmuir equation,

$$\frac{n}{n_m} = \frac{c_2/K}{1 + c_2/K} \tag{24.5}$$

where n is the amount of solute taken up per gram of the adsorbent when present at concentration c_2 in the solution; n_m is the number of moles of solute in the saturated monolayer per gram of solid and K is related to the solute-solid adsorption potential. [See Chapter 4 for the use of Eq. (24.5) to determine the specific surface area of the adsorbent.]

To test whether experimental data are adequately described by the Langmuir equation the quantity $1/n$ is plotted versus $1/c_2$, which has an intercept of $1/n_m$ and a slope of K/n_m. A straight line through the points indicates a fit by the Langmuir equation. Figure 24.2 shows the Langmuir plots of the data in Figure 24.1.

The Langmuir equation is based on the model of a uniform surface or at least assumes that the heat of adsorption does not vary with surface concentration.

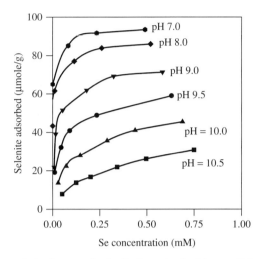

Figure 24.1 Langmuir isotherms of selenite ion adsorbed by goethite at 20°C at various values of pH.[6]

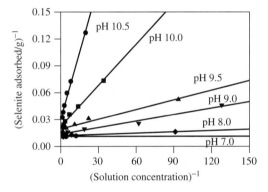

Figure 24.2 Langmuir plots of the data from Figure 24.1.

The special model by which the Langmuir is derived has not proved to be valid even when the resulting equation describes the experimental data. This equation is, therefore, better regarded as empirical, unless additional information besides adsorption data is available about the system.

24.2.b Preferential Adsorption

At higher concentrations of solute the Langmuir equation is inapplicable; frequently the adsorption of solute appears to go through a maximum, then decreases and may become negative, as illustrated by the adsorption isotherms shown in Figure 24.3.

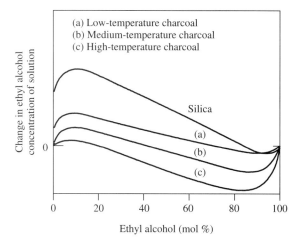

Figure 24.3 Preferential adsorption of ethyl alcohol from benzene as a function of charcoal treatment.[7]

For each adsorbent, ethyl alcohol is adsorbed from a benzene solution at low concentration; similarly, at low concentrations of benzene in ethyl alcohol, a preferential adsorption of benzene takes place, which also passes through a maximum. The peculiarity of an apparently negative adsorption stems from the customary method of calculating the amount adsorbed, which is taken as proportional to the change in the concentration of solute in the solution as measured before and after adsorption has occurred. If the solution is found to be *more* concentrated after reaching equilibrium with the adsorbent, then the solvent rather than the solute has been preferentially adsorbed. When the simultaneous adsorption of the two components of the solution is taken into account, the apparent anomaly of negative adsorption disappears.

Kipling and Tester have shown how the adsorption isotherm for the complete range of concentration, as shown in Figure 24.3, can be related to the vapor-adsorption isotherms.[8] They used the Langmuir equation to describe simultaneous adsorption from solution of the two components; they also assumed that the whole surface of the adsorbent is covered by the two adsorbed components, that is, $\theta_1 + \theta_2 = 1$. Under these conditions, the fractions of the surface covered by each component are

$$\theta_1 = \frac{K_2 C_1}{K_2 C_1 + K_1 C_2} \tag{24.6}$$

and

$$\theta_2 = \frac{K_1 C_2}{K_2 C_1 + K_1 C_2} \tag{24.7}$$

The constants K_1 and K_2 in Eq. (24.6) and (24.7) can be derived from the vapor-adsorption isotherms of the two components, which were measured by placing the adsorbent in the vapor phase of the solution; by analysis of the mixed adsorbate both (x_1/m) and (x_2/m) were determined separately, even though mixed vapors were used. The vapor-adsorption measurements provide an independent means of obtaining K_1 and K_2 for use in the mathematical description by means of Eqs. (24.6) and (24.7) of adsorption from solution. The derivations of these equations require that the two components form ideal solutions. For nonideal solutions, more complex relations are obtained.[9]

The heterogeneity of the adsorbent remains the unknown factor in all investigations of adsorption from solution. The presence of this factor nullifies all efforts to treat the Langmuir equation as anything more than an empirical description. Different combinations of the solute–adsorption isotherm and the distribution of adsorptive potential energies can lead to the Langmuir equation, the Freundlich equation and to several other shapes of the adsorption isotherm. Giles and coworkers[10] attempted to classify solution adsorption isotherms for use in diagnosing adsorption mechanisms and measuring specific surface areas of solids. The shape of the adsorption isotherm, however, is determined by an unknown combination of both lateral interactions and surface heterogeneity, and whatever the one lacks in describing the data can be supplied by the other. The system of Giles et al. is, in effect, to equate the heterogeneity to zero and to interpret all differences as due to variations in the adsorption mechanism. The system will, therefore, be successful with near-homotattic surfaces but could be completely wrong for surfaces with a wide distribution of adsorption potentials. The latter type of surface, unfortunately, is by far the more common. The system of Giles et al. can, however, be applied with more probability of success to the class of adsorption isotherms pertaining to the adsorption of dyes and other large molecules.

24.2.c Adsorption of Dyes

The effect of surface heterogeneity can be suppressed by using adsorbates of large molecular size. This circumstance is, probably, the basis of the validity of dye adsorption as a technique to determine specific surface areas of solid adsorbents. The adsorption isotherm is always determined in the dilute range of concentrations where competitive adsorption of the solvent is not significant; such isotherms frequently show a saturation plateau at high equilibrium concentrations of the free dye. Sheppard and his coworkers,[11,12] who investigated the adsorption of cyanine dyes by silver halides in connection with the study of optically sensitized photographic emulsions, concluded that the saturation plateaus observed in the adsorption isotherms of a number of cyanines adsorbed by silver bromide microcrystals correspond to the formation of a close-packed monolayer of essentially planar cations, oriented with the planes of the molecules steeply inclined to the substrate, that is, a configuration in which the edge of the molecule is presented to the substrate (edge-on adsorption.)

When the dye molecule is nonplanar or has a relatively high solubility in water, two distinct factors that reduce adsorption from aqueous solution, the type of isotherm shown in Figure 24.4, curve 2, frequently results.[13] This isotherm shows poor adsorption of the tetramethyl dye (curve 2 in the diagram) compared with its planar counterpart (curve 1), and curves 3 and 4 show poor adsorption of 2,2′-cyanines whose molecules have been forced from planarity by bulky substituents in the methine bridge. The discontinuity exhibited in the adsorption isotherm of the tetramethyl dye is accompanied by a change in the absorption spectrum of the adsorbed dye. In the low-concentration "foot" of the isotherm, the spectrum is that of the isolated molecule as modified by its adsorption, and probably corresponds to unassociated molecules in flat orientation with respect to the crystal surface; in the other region, the absorption maximum undergoes a bathochromic shift to a wavelength (J band) similar to that of the oriented aggregates of planar dyes in solution (micelles) For the nonplanar dye, if the adsorption plateau is identified with the completion of a monolayer, the area per molecule is found to be consistent only with edge-on adsorption, although the average intermolecular distance, 0.598 nm, is greater than for the corresponding planar molecules, as might be expected from the twisted configuration of the molecule.

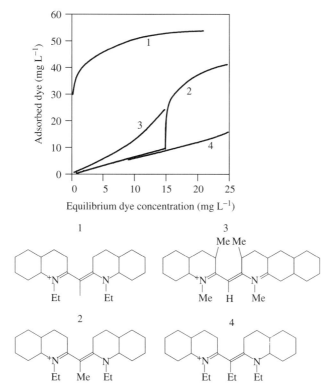

Figure 24.4 Adsorption isotherms of cyanine dyes on silver bromide, showing the effect of nonplanarity of the adsorbate molecule.[13]

For dyes such as 1,1'-diethyl-2,2'-cyanine, whose adsorption isotherm is shown in Figure 24.4, curve 1, the critical concentration at which lateral interactions become strong enough to induce cooperative edge-on orientation occurs in such a dilute solution that the isotherm appears to be continuous on the scale shown. Nevertheless, a small foot, corresponding to noncooperative adsorption, can sometimes be found in the adsorption isotherm of well-adsorbed dyes; the effect can be magnified by making the adsorption conditions less favorable, for example, by introducing a competing adsorbate or an unfavorable concentration of silver ions.

With systems such as these, the large size of the adsorbate molecule so masks the smaller scale heterogeneity of the substrate that conclusions about adsorption mechanisms drawn from the shapes of the isotherms have more authority.

24.2.d Surface-Tension Titrations

Latexes made by emulsion polymerization contain 8–10% of surface-active solute, which in this context is simply referred to as "soap." The finished latex has a surface tension not much reduced from that of pure water, despite the large amount of soap in the system. Most of the soap is evidently retained at the polymer–liquid interface, leaving very little unadsorbed soap in the solution. The latex is titrated with a standard solution, approximately 0.01 M, of the same soap that was used as stabilizer.[14] As this solution is added, the soap equilibrates between the solid/liquid interface and the serum, its presence in the serum can be detected by the change of surface tension at the liquid/air surface. As the titration proceeds, the surface tension decreases, until the concentration of free soap in the solution is equal to the CMC, which is, effectively, the largest possible concentration of soap *ions* in solution, and which is, therefore, in equilibrium with the largest amount of adsorbed soap that the particular system of soap + water can retain on that particular interface. (Figure 24.5). The endpoint of the titration occurs when minimum surface tension, corresponding to the CMC, is reached. The amount of soap added at the endpoint equals the quantity adsorbed at the solid/liquid interface during the titration plus the quantity required to reach the CMC in the solution. The total amount of soap adsorbed on the latex particles is the sum of the amount originally present and the amount added to the particles in the titration. The specific interfacial area is then calculated from the total amount of adsorbed soap per gram of polymer and the limiting area of each adsorbed soap molecule. The latter quantity can be obtained from Table 24.1 for a variety of soaps.

24.2.e Flocculated and Deflocculated Suspensions

The surfaces of silica and other inorganic oxides are hydrophilic; the aqueous medium releases ions from the crystal lattice and so confers ionic charge on the surface and establishes a diffuse double layer extending into the solution. Thus, although the system has only two components, electrostatic repulsion between the particles can stabilize the suspensions. The "rule" that two pure components

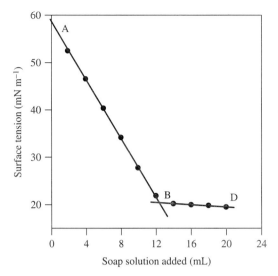

Figure 24.5 Surface tension titration of a latex with soap at 10°C. The point of intersection of the lines, B, gives the quantity of added soap at which micellization occurs in the aqueous phase and is thus the end point of the titration.[14]

TABLE 24.1 Effective Molecular Areas of Potassium and Sodium Soaps[14]

Soap	Molecular area (nm^2)
Laurate	0.414
Myristate	0.341
Oleate	0.282
OSR[a]	0.265
Palmitate	0.251
Stearate	0.233

[a] A mixture of oleate, palmitate, and stearate soaps.

cannot form a stable dispersion does not apply to situations where the medium reacts with the dispersed phase (i.e., a lyophilic colloid). This consideration also explains why water dissolves sucrose: Each sucrose molecule becomes hydrated, which detaches it from the crystal lattice. A solution is the ultimate deflocculated dispersion. Two-component systems of an insoluble solid and a nonpolar liquid are much less likely to be lyophilic, and so, in spite of positive spreading coefficients, tend to flocculate. Spreading of a liquid at a solid/air interface does not necessarily imply that the same liquid will be able to separate particles immersed in the liquid; the essential difference lies in the Hamaker constants, A_{123} (air/liquid/solid) for the former and A_{323} (solid/liquid/solid) for the latter. Most organic liquids have low surface tensions and will spread on inorganic solids such as ferric oxide, gold, or selenium, yet, if they lack a specific interaction such as electron donor–acceptor, deflocculated suspensions are not formed without the aid of a third component to keep particles sufficiently apart so that their kinetic energy overcomes the energy of attraction.

Some nonpolar systems are made lyophilic as silica dispersed in polydimethylsiloxane, which is made stable by heating the two chemicals together because of a reaction that then occurs at the silica surface.

Table 24.2 contains general properties of deflocculated and flocculated suspensions, showing the major differences between them. Differences between flocculated and deflocculated dispersions also show in many practical effects such as the reinforcement of rubbers and plastics and the UV absorption by paints, where a well-dispersed pigment should be used for high durability.

24.2.f Sediment Volumes

A qualitative method to distinguish degrees of flocculation is to measure the specific volume of sediment (volume ÷ mass). Deflocculated particles are free to move around each other and can settle to a densely packed sediment with minimum void volume. Flocs, on the other hand, which are irregular structures in the suspension, are less able to pack closely in the sediment; between them greater voids are left, which enlarge the volume of the sediment. The effect is augmented the finer the subdivision of the particles. The variation of the sedimentation volumes of Acheson graphites in different liquids is shown in Figure 24.6 as a function of their specific surface area.[15] The regular variation shown is a characteristic of this solid and is not necessarily to be expected in experiments with other solids.

Lyophilic particles are wetted by the medium and are sufficiently deflocculated to settle to a dense sediment. Complete deflocculation requires electrostatic or steric stabilization, and this is shown in Figure 24.6 where the effect of suspending agents is to reduce sedimentation volumes. The volume of sediment reflects which media are effective for wetting the particles. An acid–base interaction

TABLE 24.2 A Comparison of Deflocculated and Flocculated Suspensions

Deflocculated suspensions	Flocculated suspensions
Lyophilic interfaces	Lyophobic interfaces
$F_{sv} > \sigma_1 + F_{sl}$	$F_{sv} < \sigma_1 + F_{sl}$
Slow rate of Stokes settling	Rapid rate of Stokes settling
Small volume of sediment	Large volume of sediment
Dilatant rheology	Shear-thinning rheology
Low or zero-yield value	High-yield value
High heat of immersion	Low heat of immersion
High work of adhesion of medium on solid	Low work of adhesion of medium on solid
Superior opacity of coatings	Low opacity of coatings

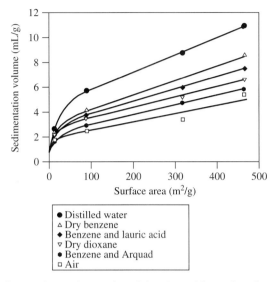

Figure 24.6 Sedimentation volumes in mL/g of graphites of various specific surface areas suspended in different liquids.[15]

TABLE 24.3 Sedimentation Volumes of Zinc Oxide Pigments[15]

Sample label	Surface area (m²/g)	Sedimentation volume (mL/g)	
		In water	In dry benzene
Kadox Black Label-15 F-1601	9.5	3.45	2.40
XX Red-72 K-1602	8.8	2.70	2.20
XX Red-78 G-1603	3.9	3.00	2.63
Reheated superfine KH-1604	0.66	1.50	0.65

between particle and medium will inevitably lead to wetting. For a series of zinc oxide pigments, Table 24.3 shows the same effect of finer subdivision leading to larger sedimentation volumes.[15]

Figure 24.7 reports the results obtained from observations of Attapulgite clay in dioxane–water mixtures showing flocculation in water and deflocculation in dioxane. The mass of Attapulgite is the same in each vessel. The effect of even 4% by volume of dioxane is already apparent; further additions make little difference. Reductions in the sedimentation volume on addition of dioxane to water

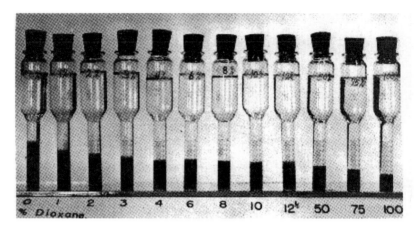

Figure 24.7 Sedimentation tubes containing Attapulgite clay in dioxane and water solutions, after standing for 20 days.[15]

are caused by a nonpolar solid substrate. The surface is thereby shown to be nonpolar.

The technique is as follows: In preparing samples, the solid is dried carefully. To a weighed amount of solid is added a small quantity of the liquid in which the dispersion is to be made. The paste is thoroughly mixed with a spatula and more liquid is gradually added with constant stirring until 25 mL has been added. The dispersion is transferred to a stoppered, graduated tube and shaken, and the volume of the sediment is recorded as a function of time. Final readings should not be made until all the supernatants are clear, which may take several days.

For optical microscopy where the size or shape of an ultimate particle is desired, various liquids are tried until deflocculation is achieved. The powder to be examined is shaken with various liquids. The liquid for which the sediment volume is least is chosen as the medium for sample preparation.

The sediment volumes of a pigment in a graded series of acidic and basic solvents, or even water at low and high pH, can be used to determine whether a powdered material is flocculated or deflocculated in a solvent. For example, rutile powder was so determined with the results shown in Figure 24.8, which is a correlation with the Gutmann acceptor and donor numbers. In this dispersibility map for rutile, well-dispersed powders are shown to lie predominantly with the high donor numbers, showing the surface of rutile to be acidic.[16] The method was used to characterize various surface treatments. The scatter of the data is caused by interactions other than acid–base, such as dipole–dipole interactions.

24.2.g Ultrasonic Techniques

24.2.g(i) *Ultrasonic Spectroscopy of Suspensions* Ultrasonic spectroscopy of dispersions has two fundamental advantages that make it a superior tool to study emulsions and suspensions: It can be used to measure properties at high volume

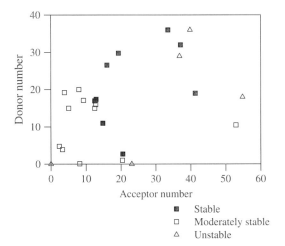

Figure 24.8 Dispersibility of rutile as determined by sediment volume in various acidic and basic solvents.[16]

loadings and it can be used with opaque materials. It can even be used to probe dispersions on the other side of metal walls. An enormous range of materials have been characterized by ultrasound spectroscopy: alcohol and sugar solutions, edible oils and fats, cell suspensions, crystallizing solids, margarine, chocolate, emulsion inversions, and particle size and charges.[17]

The two properties of sound most commonly measured are the velocity of sound in the dispersion and the frequency dependence of the absorption of sound. The velocity of sound in mixtures and suspensions is determined by the average density and compressibility of the whole sample. The average density, ρ, is the volume average of the components and the average compressibility, κ, is the volume average of the compressibilities.

$$\rho = \sum_j \phi_j \rho_j \quad \text{and} \quad \kappa = \sum_j \phi_j \kappa_j \qquad (24.8)$$

where the ϕ_j are volume fractions. The speed of sound, v, is

$$v = \frac{1}{\sqrt{\kappa \rho}} \qquad (24.9)$$

When the densities or compressibilities are not known, a calibration curve is determined by measuring the speed of sound in known mixtures. The measured speed of sound on a sample of unknown composition is determined by reference to the calibration curve. This determination is accurate for nearly the full range of concentrations and can be measured inside pipes and tanks. For example, the butterfat content of milk is readily measured by the speed of sound; the variation in percent solids from the top to the bottom of a tank can also be measured by the speed of sound.

Particle-sizing information is obtained by measuring the acoustic absorption coefficient as a function of frequency (see Chapter 4). The advantages are that the measurements are possible from about 0.1% to about 60% by volume, the equipment is robust enough to use in a factory, it is sensitive to a wide particle size range, 0.01 to 1000 µm, and the measurement is fast enough to obtain kinetic information, especially changes in particle size due to flocculation. Further, commercial equipment is available from a number of sources. (See Table 4.10.)

24.2.g(ii) Electroacoustic Techniques   When a high frequency electric field is applied to a dispersion, dispersed particles move in phase with the field. The vibrating particles generate a pressure wave at the same frequency as the electric field. The average charge or zeta potential of the particles can be calculated from the magnitude of the pressure generated. The higher the volume fraction of particles, the stronger the signal. Therefore, electroacoustics is an ideal technique to determine the zeta potential of suspended particles without the trouble of dilution. Electroacoustics is discussed in Section 17.4.d.

At frequencies less than about a megahertz, the inertia of submicron particles is small enough that the particles move in phase with the applied electric field (see Chapter 3 for a discussion of inertial effects in suspensions). As the frequency of the applied electric field is increased, the larger particles do not stay in phase with the applied electric field. Particles that are somewhat out of phase with the electric field move less and therefore generate less sound. The attenuation in sound amplitude with increasing frequency gives information about the particle size and size distribution. Since this measurement is made without diluting the dispersion, useful information about the stability of high-volume-loaded dispersions can be obtained.

Commercial equipment capable of making these measurements is available from Colloidal Dynamics, Dispersion Technology, Malvern Instruments, and Matec Instruments (see Table 4.10 for contact information.)

24.3 RHEOLOGY OF SUSPENSIONS

Whether a suspension is deflocculated or flocculated can be determined from its rheogram. Deflocculated suspensions are generally Newtonian until they reach high shear rates where particle clustering under high shear conditions leads to dilatancy. Flocculated dispersions have a structure owing to particle–particle adhesion. These structures collapse as the shear is increased. Therefore flocculated dispersions are shear thinning. Of course many dispersions exhibit a wide variety of flow behavior reflecting different particle microstructures. A careful analysis of the rheograms as a function of shear stress, time, and particle concentration gives abundant information about the microstructure and interparticle forces of the suspended particles.[18] For example, some aqueous clay dispersions show a yield stress (microstructures that break with a minimum stress), followed by shear thinning (slowly decaying microstructures) and, finally, an increase in apparent viscosity (creation of microstructure).[19]

24.3.a Dilute Suspensions

At low percent solids, a suspension, whether flocculated or not, continues to behave as a Newtonian fluid with a slightly higher viscosity coefficient than that of the medium. The simplest case of the flow of a dispersion of noninteracting spheres was analyzed by Einstein to give the viscosity, η, as a linear function of volume fraction, Φ,

$$\eta = \eta_0 (1 + 2.5\Phi) \tag{24.10}$$

where η_0 is the viscosity of the medium. The Einstein equation is a limiting law, valid only as Φ tends to zero.

24.3.b Flocculated Suspensions

Flocculated suspensions exhibit yield values at low shearing stresses, because interparticle attractive forces prevent flow until sufficient energy is imparted to the system to overcome their resistance. When the structure is broken, the rate of flow is sometimes linear with the applied stress. A rheogram of this type is known as a Bingham body. Oil-based paints, which are suspensions of pigments in oil, are typical Bingham bodies. These bodies are characterized by a yield point and a plastic viscosity measured by the slope of the linear portion of the rheogram. Other shear-thinning materials show no yield point (or at least an indeterminately small one) without a linear region in the rheogram. These are known as shear-thinning or pseudoplastic materials. This behavior corresponds to a progressive deformation and disruption of flocs in the shear field.[20]

Rheological measurements can also be used to study the rates of flocculation. A complicating factor is that the shear stress imparted by the measurement can break flocs apart as they form. Another is that the volume fraction of particles varies from top to bottom of the cell as flocs sediment. These difficulties apart, the rate of flocculation can be followed by the rate in increase of the apparent viscosity.[21]

In general the rheology of flocculating suspensions is difficult to measure because the flocs tend to settle during the measurement and produce an inhomogeneous composition in the measurement cell. One method to overcome this difficulty is to make the bob for a Couette rheometer short enough to remain in the homogeneous regime, with openings in the top of an open-ended bob to let sediment pass.[22]

Some flocculated suspensions of high volume fraction exhibit such a high yield point that the material holds its shape outside a container for a time before flowing to level. This is seen in fresh concrete. A simple method to measure

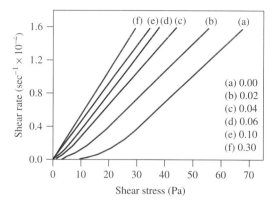

Figure 24.9 Rheograms of 20% by weight deionized kaolin slurries at several levels of tetrasodium pyrophosphate addition: The figures on the curves indicate percent TSPP per weight of clay. An extrapolation of the linear region of each rheogram to zero rate of shear determines an apparent yield point.[24]

the magnitude of the yield point is the slump test.[23] The slump measurement consists of filling a conical frustrum with the material to be tested, removing the frustrum, for example, a pail, allowing the material to collapse under its own weight. The height of the final deformed, or slumped, material is measured. The difference between the initial and final heights is termed the slump height.

24.3.c Deflocculated Suspensions

An example of the transition of a flocculated to a deflocculated system is the change of a plastic kaolin suspension to a fluid slurry, brought about by adding the dispersing agent, tetrasodium pyrophosphate (TSPP). The rheograms that accompany this change are shown in Figure 24.9.

At higher shear rates, the flow becomes more and more impeded, as the inertia of the particles hinders their response to shearing stresses. The concentration of suspended solids at which dilatant flow is observed varies with the particle size and shape and the quantity and nature of the deflocculating agent added. Suspensions of red iron oxide in an aqueous solution of sodium lignin sulfonate at 10% concentration begin to show dilatancy at about 10% solids. Over a small concentration range, 11.3–12.7% solids, the calculated apparent viscosity increases from 0.23 to 7.6 Pa-s for the freshly prepared suspension measured at a low rate of shear (Figure 24.10).

The reason why dilatancy appears at such low concentration of solids is that asymmetric particles during flow sweep out larger volumes than their actual volume, which means that the apparent volume concentration is much larger than the actual volume concentration. A second regime occurs in the rheogram at high shear rates when asymmetric particles entangle and at still higher shear

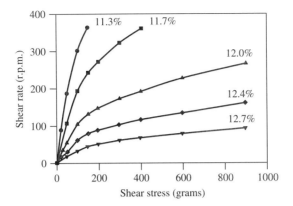

Figure 24.10 Dilatant flow of a series of suspensions of red iron oxide in an aqueous solution of sodium lignin sulfonate at 10% concentration, at 30°C. The volume concentrations of solids are noted on the curves.[25]

rates a third regime might occur if the particles align with the flow and cause the viscosity to drop.

24.3.d Concentrated, Deflocculated Suspensions

Spherical or near spherical particles do not show dilatant behavior until they begin to approach the close packing limit, 60–70% by volume.[26] Deflocculated spheres after settling still have layers of fluid around them and so are still capable of flow. They form structures of layered particles that can slide over each other but require an extra force to do so. That extra force is the cause of dilatancy. Various kinds of interparticle forces, dispersion force attractions, double layer repulsions, and steric interactions all have different distance dependencies and strengths. For instance, a change in solvent properties changes the thickness and interaction between adsorbed polymer layers and hence changes the onset and magnitude of dilatancy.

Flow instability associated with shear thickening is sometimes observed during the coating of aqueous polymer dispersions. This type of dilatancy is characterized by an abrupt, sometimes discontinuous, rise in apparent viscosity above a critical shear rate. As the dispersion does not flocculate during this transition, the sudden increase in viscosity is postulated to be caused by a shear-induced ordering of the particles.[27] The transition does not seem to involve any particle flocculation since the dilatancy is instantaneously reversible (Figure 24.11).

The lack of thixotropy, that is, instantaneous healing, of the dispersion in Figure 24.11 is unusual and arises because this dispersion is sterically stabilized and the particles are nearly monodisperse spheres. The rheopectic structure formed under stress disintegrates quickly when the stress is removed. Another noteworthy behavior of this system is the sudden onset of dilatancy at a concentration about 56.7% solids (see Figure 24.12). The propensity for ordering of the particles is related to their near monodispersity.

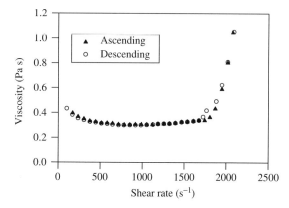

Figure 24.11 Shear-rate sweep of a 59.5% (volume fraction) polyacrylic ester dispersion HYCAR, with the ascending sweep (triangles) followed immediately by a descending sweep (circles).[27]

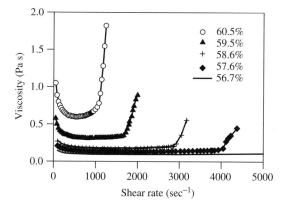

Figure 24.12 Shear-rate sweep of HYCAR at different solid concentrations (volume fractions).[27]

24.3.e Pigments in Polymer Solutions

Paints, inks, and all kinds of coatings are pigmented polymer dispersions. They are generally formulated at low viscosity, applied, and dried. The pigment is often the component that provides the desired optical, electrical, or magnetic property for the coating. The usual role of the polymer is to bind the pigment particles to each other and to the substrate for mechanical stability. Additionally, the polymer may be the dispersing agent.

Formulations for dispersions are usually specified in terms of weight fractions for each component, as these are the most convenient units for measuring. The volume fractions, however, are more convenient for understanding the mechanical properties such as the rheology. Another property that is best understood in terms of volume fractions is the critical pigment volume concentration, CPVC, of a two-component, pigment in polymer film. This concentration is the volume of pigment corresponding to sufficient polymer to displace all the air between the pigment particles. Below the CPVC, that is, excess polymer, well-dispersed pigment particles have lost contact with each other. Above the CPVC, that is, insufficient polymer, some particles are still separated by air gaps. Patton gives examples of fifteen mechanical properties of paint films that have abrupt changes in value at the CPVC, such as density, tensile strength, porosity, reflectance, and hiding power.[28]

Rheology is the tool of choice in understanding and improving these compositions. The complex interactions between all the components have strong effects on the rheology, especially the elasticity and the shear-rate dependence of the viscosity. Rheological studies are undertaken for two ends: first, to be able to control the coating and drying, since the dependence of flow on shear rate and time of applied stress is an essential input for the selection of a coasting method,[29] second, to improve formulations by understanding the interactions between the components that determine the colloidal stability of the dispersion.[30,31,32]

When particle–particle interactions dominate the flow of a suspension, flocculated suspensions will be shear thinning and deflocculated suspensions will be dilatant. But since structures form and break on some time scale, the investigation should be extended to see what those time scales are. One method is to measure the rheogram at various sweep rates, where the sweep rate is the time to go from rest to maximum shear rate. Another method is to follow the change in viscosity at a constant rate of shear rate. A third method is to measure the rate of recovery after the destruction (or creation) of structure with shear, which discloses the thixotropy or rheopexy of the system. All these effects are greatest when the number of interparticle interactions is greatest, and that is at high solid loadings.

When the dissolved polymer dominates the flow of the dispersion, then the dispersion has some of the classical properties of polymer solutions such as significant viscoelasticity. It is easy to see the difficulty of coating a viscoelastic dispersion; a wet coated film will recoil after being stretched to a thin film. The effects are greatest when the polymer concentration is high.

A systematic study of the rheological properties of solutions of polyesters containing quantities of inorganic and organic pigments has been carried out by Lara and Schreiber.[33] Two of the conclusions of that paper are:

1. Dispersions of inorganic and organic pigments in solutions of polyester resins display non-Newtonian flow, with viscosity-shear-rate relations well fitted by a modified form Casson equation,[34,35]

$$\eta^{1/2} = \eta_\infty^{1/2} + \left(\frac{\tau_0}{\dot\gamma}\right)^{1/2} \qquad (24.11)$$

where η is the apparent viscosity (given by $\tau/\dot\gamma$) at shear-rate $\dot\gamma$, τ_0 is the yield point, and the viscosity at infinite shear-rate, η_∞, designates a

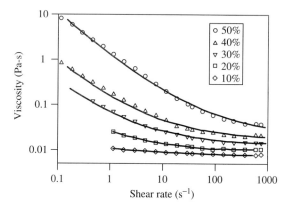

Figure 24.13 The flow behavior of dispersed rutile at various weight percent as a rheogram.[33]

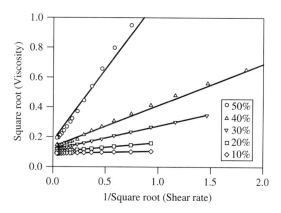

Figure 24.14 The flow of dispersed rutile at various weight percent plotted as the square root of the apparent viscosity, η, versus the inverse square root of $\dot{\gamma}$. Solid lines are calculated from Eq. (24.11).

condition where all structural effects due to pigment flocculation have been eliminated.

Figure 24.13 shows flow data for rutile dispersed in a linear polyester, modeled on resins used as protective coatings in the automotive industry. Non-Newtonian flow is observed at all pigment concentrations. The same data plotted as the square root of apparent viscosity versus the inverse of the square root of the shear rate are shown in Figure 24.14. The solid lines are calculated from Equation (24.11).

2. The effect of the volume fraction of pigment on the relative viscosity of dispersions may be represented by a modification of the Maron–Pierce expression,[36] taking into account the adsorption of polymers on pigment

surfaces,

$$\left[\frac{\eta}{\eta_0}\right]_{\dot\gamma\to\infty} = \eta_r(\phi) = \left(1 - \frac{f\phi}{\phi_m}\right)^{-2} \quad (24.12)$$

where the relative viscosity, η_r, is the ratio of the apparent viscosity to the viscosity of the unpigmented polymer solution, both at infinite shear rate, ϕ is the volume fraction, ϕ_m is the maximum packing factor and f is the correction factor that corrects ϕ as computed from the weight and density of the dispersed solids to account for polymer adsorption. For spherical particles ϕ_m ranges from 0.52 for simple cubic packing to 0.74 for tetrahedral packing. Lara and Schreiber[33] calculate the thickness of the adsorbed polymer layers on the pigment particles and ascribe the adsorption to an acid–base interaction between them.

Equation (24.11) is one of many empirical rheology equations useful in describing the flow of coatings. Another empirical relation is the Herschel–Bulkley equation, which fits the flow behavior of a wide variety of materials over a useful range of deformation conditions. It describes a power-law for shear thinning above a yield stress:

$$\tau - \tau_0 = K\dot\gamma^n \quad (24.13)$$

where τ is the shear stress, τ_0 is the yield stress, $\dot\gamma$ the shear rate, K and n empirical constants.

A theoretical analysis of the rheology of materials over a wide range of shear rates results in a generalized equilibrium flow curve.[35] The analysis assumes that any time-dependent or relaxation effects have been removed experimentally by taking each datum after reaching steady state. Figure 24.15 shows the generalized flow curve, consisting of:

1. A low-shear-rate Newtonian regime, region I;
2. A shear-thinning regime, the power law regime, region II;
3. A high-shear-rate Newtonian regime, region III;
4. A possible shear-thickening regime, region IV.

Each regime has an applicable empirical model. At low shear rates, region I, any structure in the dispersion gives rise to a high viscosity, but the stress is low enough that the structure remains intact. At higher shear rates any structure is steadily destroyed with increasing stress, region II. At high enough shear rates, any structure is totally eliminated and the apparent viscosity is constant with a low value, region III. At still higher shear rates the deflocculated particles do not respond quickly enough to the rate of shear and the system is dilatant, region IV.

An advantage to considering this model for the flow behavior of dispersions in general is that it anticipates radical changes in flow when the shear rate crosses from one regime to another. Eley gives a thorough description of the wide range

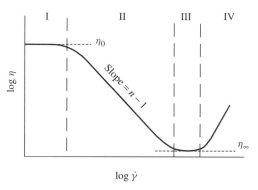

Figure 24.15 Generalized non-Newtonian equilibrium flow curve. η_0 is the zero-shear viscosity; η_∞ is the high-shear limiting viscosity, region I is the first Newtonian plateau; region II is the power-law regime; region III is the second Newtonian plateau; and region IV is the dilatant regime.

of non-Newtonian behavior in coating processes, including time-dependent or viscoelastic behavior.[35]

Paper coatings are also dispersions of pigments in polymer solutions. These complex dispersions contain both clay and calcium carbonate pigments, a variety of polymeric materials, and a range of electrolyte compositions. Rheological studies show clearly which polymers are dispersing agents and the flocculation effects of high ionic strengths.[37]

24.3.f Rheology for Thin-Film Coatings

Paints and protective films are common thin-film coatings. Other thin-film coatings are adhesives, sealants, printing processes, paper coatings, photographic film, magnetic coatings, electronic circuit boards, compact discs, and so on. and almost all include dispersed phases.[38] Table 24.4 gives some approximate apparent viscosities required for various coating processes. The viscosities of the complex fluids used as thin-film coatings are shear-rate (or stress) dependent and often time dependent. The recommended viscosities in Table 24.4 are only guidelines.

To design coating procedures that take shear-rate-dependent viscosities into account is a challenging problem in fluid dynamics. For example, in roll coating the shear rates are high in the nip and low on the coated film as it leaves the nip. The final film thickness and coating defects, such as ripples, are determined by the flow in the nip. The leveling of defects in the coated layer is required to heal at the low shear rates of the free film on the substrate.[40]

24.3.g Particles as Thickeners

Various kinds of clays, particularly montmorillonites, hectorites,[41] and the synthetic clay, Laponite,[42] are able to form three-dimensional structures in aqueous

TABLE 24.4 Some Coating Methods and the Required Viscosity[39]

Process	Viscosity (Pa-s)
Single layer	
Rod (wire wound)	0.02–1
Reverse roll	0.1–50
Forward roll	0.02–1
Air knife	0.005–0.5
Knife over roll	0.1–50
Blade	0.5–40
Gravure	0.001–5
Slot	0.005–20
Extrusion	50–5000
Multilayer	
Slide	0.005–0.5
Curtain, precision	0.005–0.5

dispersion. Under the proper conditions the faces are negatively charged and the edges are positively charged. The clay particles flocculate edge-to-face to form an extended structure like a house of playing cards.[43] The gel (or "hydrogel") formed is homogeneous looking and displays some rigidity and elasticity. In the gel the particles are flocculated into one single floc which fills the entire liquid. The concentration of particles sufficient to gel a liquid can be as low as two percent.[44] The structure formed by Laponite is brittle and disintegrates suddenly under shearing stress as indicated by the yield points in the rheogram (see Figure 24.16.) Most gels are thixotropic. Once the structure is broken it reforms slowly.

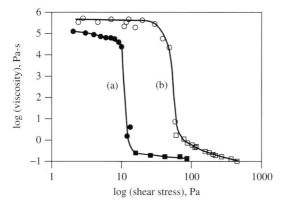

Figure 24.16 Apparent viscosity versus shear stress as obtained by creep tests and capillary viscosimetry. (*a*) 1%. (*b*) 2% w/w dispersions of Laponite RD in 9 mM NaCl solution.[45]

24.3.h Powder Flow

Granular materials are large agglomerates of discrete macroscopic particles. A sand pile at rest with a slope lower than the angle of repose behaves like a solid. The pile remains at rest even though gravitational forces create minute stresses. If the pile is tilted several degrees above the angle of repose, grains start to flow. This flow is not that of an ordinary liquid because it exists only in a boundary layer at the surface.[46]

If the grains are dry, the rate of flow of a granular solid is determined by particle shapes and interparticle forces. The relevant energy scale for the flow of gases and liquids is kT, where k is the Boltzmann constant and T is the absolute temperature. In contrast, the relevant scale for the flow of granular materials is the potential energy of a grain in a gravitational field, mgd, where m is the mass of a particle, g is the acceleration due to gravity, and d is the particle diameter. For typical sand, this energy is at least 10^{12} times as great as kT.[46]

A practical experimental technique to characterize powder flow is proposed by Kaye et al.[47] The powder is placed into a slowly rotating, vertically oriented, transparent disc. As the disc rotates the powder builds at one edge up to an unstable state and then avalanches to form a steady state with constantly changing surface. The equipment is shown in Figure 24.17. The powder flow is characterized by the profile of the powder bed just before avalanche and measures the cohesiveness of powders. The data are used to show the influence of particle size distribution and roughness of the surface of the particle.[48] A coarse powder is free flowing and a fine powder is more cohesive.

Silica flow agents are frequently added to dry powders to improve their flow. As the concentration of flow agent is increased, the improved flow is reflected in the decreasing mean time to avalanche.[49]

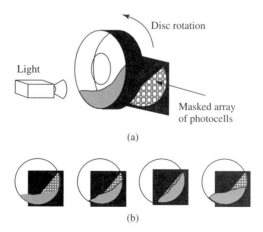

Figure 24.17 The rotating disc avalanching equipment. Avalanching is followed by monitoring the amount of light reaching the photocells placed behind the disc. (*a*) Sketch of the equipment. (*b*) Sketch of the progress of single avalanche in the disc.[47,48]

24.4 FLOCCULATION OF SUSPENSIONS

Suspensions can be destabilized by eliminating electrostatic or steric repulsion or by adding polymer flocculants. This can be done by different methods such as double-layer compression, specific-ion adsorption, enmeshment of flocs, heteroflocculation, bridging by polymers, or changes of temperature, pressure or solvent composition.[50,51] Descriptions of some of these methods follow.

When particles are small, their collisions and consequent flocculation come about by a combination of Brownian motion and interparticle attraction. Stirring has little effect because small particles follow the flow of the liquid. When the floc sizes are larger than a few microns, their motions are inertial; that is, they tend to move in straight lines even when the fluid changes direction. Flocculation that is not affected by stirring is called perikinetic flocculation. Flocculation that is affected by stirring is called orthokinetic flocculation.

Once the particles are large enough for orthokinetic flocculation, stirring increases the growth of large flocs by the accretion of smaller flocs. A transition from laminar to turbulent flow can cause a change in the nature of the flocs formed.[52]

The type of shear produced by various mixer designs influences floc size. The removal of solid particles from a suspension is sometimes enhanced by increasing the density of the initial flocs by slow stirring. The initial floc structure formed from the random collision of individual particles is loosely packed. As the flocs grow larger, they become susceptible to partial breakage by shear. The partial breakage and reforming of large flocs produces a denser floc. These settle more quickly to a dense sediment.[53]

24.4.a Double Layer Compression

Aqueous suspensions stabilized by electrical charges are destabilized by increasing the ionic strength in the medium, which reduces the electrostatic repulsion. The effectiveness of adding an electrolyte for this purpose increases with the charge on the counterion. As described by the Schulze–Hardy rule, the concentration of electrolyte required for coagulation decreases with the sixth power of the charge on the ion. The Schulze–Hardy rule applies to dispersions with one kind of counterion. In most dispersions, however, a variety of solute species is present, particularly so with polyvalent ions, as these tend to form complexes with ions of opposite charge or even with nonionic solutes[54] (Chapter 19). Freezing an electrocratic dispersion concentrates the electrolyte in the remaining liquid and flocculates the dispersion by double-layer compression. This method of flocculation is useful to obtain an equilibrium serum.

Changing the charge on the particle by adsorption of potential-determining ions has a much larger effect on stability than any change of ionic strength obtained by adding indifferent electrolyte. Changes of pH are particularly effective as H^+ and OH^- are often potential-determining ions. An example is a suspension of phosphate-stabilized kaolin. The magnitude of ionic charge on the phosphate ion is directly determined by the pH. Consequently the suspension is flocculated by acid and deflocculated by base. Similarly, suspensions stabilized with polyacrylic acid are flocculated by acid and deflocculated by base. The stability of a silver iodide sol is determined by the relative concentrations of Ag^+ and I^- in solution (see Chapter 17). The surface charge need not be reduced to zero for flocculation to occur; it need only be lowered enough to allow close approach of particles. Adding excess potential-determining ions of opposite sign may repeptize the sol by reversing its charge.

In the treatment of waste water, added aluminum or iron salts hydrolyse to give a fluffy, amorphous hydroxide that enmeshes the flocculated particles of the suspension as it settles. The aluminum salt commonly used is $Al_2(SO_4)_3 \cdot xH_2O$ where x is about 14. It is called papermakers' alum or filter alum.

24.4.b Flocculation by Neutral Polymers

Flocculation of dispersions by neutral polymers is caused by the adsorption of the polymer on two particle surfaces, a phenomenon called bridging. Linear polymers of high molecular weight can be adsorbed by more than one particle at a time, forming a polymer bridge. If the polymer is a polyelectrolyte, some reduction of electrostatic repulsion also occurs and greatly enhances the rate of flocculation. The kinetics of this process is complex as the rates of several processes determine structure of the floc:[55]

1. Mixing of the polymer throughout the dispersion;
2. Adsorption of polymer chains on particles;
3. Rearrangement of the adsorbed polymers to an equilibrium conformation. The bridging must be completed rapidly because in time it spreads on the particle surface and does not extend outwards far enough to form a bridge. The flocculation must be quick and the particles settled or separated before the polymer flocculant loses its effectiveness.[56]
4. Collisions between particles having adsorbed polymer to form flocs, either by bridging or charge effects;
5. Breakup of flocs by stirring.

The efficiency of flocculation by polymers is very sensitive to molecular weight and to particle concentration. A small amount of high molecular polymer creates polymer bridges between particles readily and at long interparticle distances. Polymers of lower molecular weight are less effective because bridging requires particles to approach each other more closely. The optimum polymer dosage need not increase proportionately with percent solids. Bridging of adsorbed polymer between particles is more effective the higher the particle concentration because of the closer proximity of particles.[57]

Flocculation of dispersions by polymers is carried out at low polymer concentrations. The reasons are two. First, at low polymer concentrations, the particles are only partially covered with adsorbed polymer and the chances of polymer bridging between the particles are increased. Second, if the polymer is charged, an excess of polymer adsorption charges the particles and confers electrostatic stabilization.

24.4.c Flocculation by Polyelectrolytes

The most common polyelectrolytes are copolymers with polyacrylamide: Acrylic acid is one possible monomer to make an anionic polyelectrolyte, and a tertiary ammonium ester of acrylic acid is one possible monomer to make a cationic polyelectrolyte. When a polyelectrolyte is added to a suspension of particles with opposite charge, it is strongly adsorbed, reversing the charge on *patches* of the surface. The polymer behaves in this way because its charge density is greater than that of the surface of the particle; thus the whole charge on the particle can be neutralized without coating the entire surface. The resulting patchwork of uneven distribution of charge allows flocculation by direct electrostatic attraction between the polyelectrolyte-coated patch on one particle and the oppositely charged surface of another. Also effective is charge neutralization as demonstrated by the fact that at the optimum concentration of flocculant, the electrophoretic mobility of the particles is close to zero.[55] Optimum flocculation occurs when sufficient charge has been neutralized, regardless of the molecular weight of the polymer.

The various mechanisms of flocculation by polyelectrolytes have been summarized by Gregory:[55]

1. Bridging flocculation — where one polymer chain is adsorbed by two or more particles;
2. Mosaic flocculation — where charged polymers are adsorbed as patches on one particle and attract the oppositely charged portions of other particles;
3. Charge neutralization — where the adsorption of charged polymers neutralizes the charge of a particle and eliminates electrostatic stabilization forces.

24.4.d Flocculation versus Stabilization

Polymer flocculants often show a concentration above which the dispersion becomes restabilized.[57] The flocculation of particles depends upon the adsorption of the polymer just as does the stabilization of particles. The difference is sometimes only one of degree. The adsorption of a small amount of oppositely charged polymer can neutralize the particle charge and lead to flocculation. The adsorption of even more of the polymer, however, can charge the particle with the opposite sign and lead to restabilization.

A similar effect can be found with uncharged polymers. A small amount of adsorption of polymer on the surfaces of particles may permit the bridging of particles leading to flocculation. Higher amounts of adsorption leads to stabilization. The total amount of polymer added for flocculation should not exceed half coverage of the particle surfaces.

In testing polymers either as flocculants or dispersants a range of concentrations is important because the desired feature is not disclosed by testing at a single concentration.

The possibility of both particle flocculation and particle stabilization by polymer adsorption points to the necessity of being cautious about the addition of polymer to a suspension. On the first addition, especially from a concentrated polymer solution, the concentration is high locally, which tends to stabilize suspensions. If the intention is to flocculate, the dilute polymer solution should be added slowly with constant stirring.

24.4.e Heteroflocculation

Heteroflocculation refers to the flocculation of one colloid by another. A classic example is given by two electrocratic colloids of opposite charge, such as arsenic trisulfide (negative) and ferric hydroxide (positive). More recently, heteroflocculation has been studied with monodisperse latexes prepared without added stabilizers, at a pH between their isoelectric points.[58,59]

24.5 NUCLEATION AND CRYSTAL GROWTH

Precipitation occurs only from supersaturated solutions. Homogeneous solutions are supersaturated at solute concentrations greater than the equilibrium concentration of the two-phase system consisting of excess solute and saturated solution. Typically, supersaturated solutions are produced by sudden changes in temperature or pressure, or by mixing of two soluble materials that react to form an insoluble one. The supersaturated solution is unstable because it is at a higher free energy than that of the two-phase system. Nevertheless, supersaturated solutions may be stable for a significant time. This metastability is maintained by a barrier that is the energy required to form the first crystal nucleus. Small crystal nuclei have a high specific free energy because small size implies large curvature, and a particle of large curvature is at a high chemical potential [see Gibbs–Kelvin Eq. (6.45)]. Once a few nuclei form, however, precipitation proceeds rapidly since energy is released both by the precipitation from supersaturated solution and by the reduction of the curvature of the surface as the crystal grows. If a substrate appropriate for crystallization is introduced into a supersaturated solution, as for example, a roughened surface or a few "seed" crystals, the energy barrier to nucleus formation is removed, and precipitation follows at once. Precipitation on nascent nuclei is called homogeneous precipitation and precipitation on extraneous nuclei is called heterogeneous precipitation.

Theories of homogeneous nucleation are as yet incomplete, but the general idea is that as the degree of supersaturation increases, the time to onset of homogeneous nucleation decreases and the number of nuclei formed increases. That is, for a solution just barely beyond saturation, a long time passes before any nucleation occurs and only a few nuclei form; for a highly supersaturated solution, nucleation occurs quickly and a large number of nuclei form. After the first burst of nucleation, subsequent precipitation occurs on the already formed nuclei. Hence, from solutions just barely beyond saturation, a few large crystals form; while from highly supersaturated solutions, many small crystals form.

La Mer and coworkers[60,61] demonstrated these concepts. Figure 24.18 shows the buildup of reaction product beyond saturation, followed by the formation of nuclei in a burst, and then precipitation on the already formed nuclei. The ascending curve in region I describes a reaction that continuously generates molecules in solution. The concentration of molecules increases steadily, passes the point of saturation A, and reaches a point B at which the rate of self-nucleation becomes appreciable; the rate of nucleation rises rapidly to the point C, but the partial relief (C to D) of supersaturation is so rapid and effective that the time of nucleation (region II) is brief and no new nuclei are formed after the initial burst BCD. The nuclei produced grow uniformly by a diffusion process (region III). A uniform suspension can be obtained by this process if the homogeneous nucleation is confined to a single burst, which can be done if the rate of generation of molecules is slow. The rate of generation is controlled by the conditions of mixing.

Theories of nucleation and particle growth are most useful when they can predict the number distribution of particles from addition rates and concentrations, reaction rates, and diffusion constants. The most successful models are even useful in continuous stirred reactors.[62] An example of a model of crystal formation that relates the number of stable crystals formed to the precipitation conditions and to the growth mechanism of the crystals is provided by Leubner.[63] The prediction of the rate of nucleation is more difficult than the prediction of the rate of growth. The rate of crystal growth depends on two separate rates: the rate of diffusion of material to the surface and the rate of incorporation of material into the crystal. The way in which the particle-size distribution changes with time is a clue to the mechanism of crystal growth.

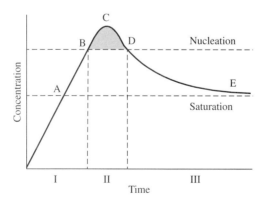

Figure 24.18 The formation of a monodisperse system by controlled nucleation and growth.[60]

1. If the incorporation of material is proportional to the area of the particle (as in diffusion-controlled reactions), the absolute width of the particle size distribution decreases as the average particle size grows, and the relative width of the size distribution decreases even more rapidly.
2. If the rate of incorporation of material into the particle is the same for all sizes (as in the Smith–Ewart model for emulsion polymerization), the size distribution narrows even more markedly.
3. If the rate of incorporation of material into the particle is proportional to the particle volume, the relative width of the particle size distribution remains constant.

Common to these three mechanisms is the decrease in polydispersity during growth; that is, the size distribution becomes narrower. Ultimately this tendency is reversed by the process of Ostwald ripening, unless the solubility is too low for that process to be significant. For this reason, uniform dispersions made by precipitation are of highly insoluble materials, such as sulfur, selenium, and ferric oxide.

The surface energies of various possible crystal faces influence the habit of the crystal, that is, its external form. Surface planes of higher energy grow most rapidly, but they are superseded in time by the slower growing planes of lower surface energy. For example, the {111} faces of sodium chloride first appear as an octahedral crystal; on each of its six vertices the slower growing {100} faces gradually develop and finally produce the stable cube crystal. The high energy of the {111} faces may be reduced by adsorption: thus, the presence of urea stabilizes the {111} faces of sodium chloride and so allows the octahedral habit to be retained. Similarly, the growth of crystals may be inhibited by surface-active solute adsorbed on all faces. Technologically important nucleation and growth processes often use "ripeners" or "growth restrainers" to produce commercial products. An industrial application is the prevention of scaling in pipes by process water that has become supersaturated by evaporation, by adding agents to retard further growth of crystals. The growth of calcium-carbonate scale, for example, is inhibited by adding polyphosphates, phosphonates, or polyacrylic acid.

REFERENCES

[1] McKay, R.B., Ed. *Technological applications of dispersions*; Marcel Dekker: New York; 1994.

[2] Available on the internet for a fee at www.iso.ch.

[3] Nelson, R.D., Jr. *Dispersing powders in liquids*; Elsevier: New York; 1988.

[4] Conley, R.F. *Practical dispersion: a guide to understanding and formulating slurries*; VCH Publishers: New York; 1996.

[5] Everett, D.H. Manual of symbols and terminology for physicochemical quantities and units. Appendix II: Definitions and symbols in colloid and surface chemistry. Part I, *Pure Appl. Chem.* **1972**, *31*, 577–638.

[6] Hingston, F.J.; Posner, A.M.; Quirk, J.P. Adsorption of selenite by geothite, *Adv. Chem. Ser.* **1968**, *79*, 82–90.

[7] Quoted by Ross, S. in *Adsorption, Kirk–Othmer encyclopedia chemical technology*, 2nd ed.; Wiley Interscience: New York; 1963–1971; Vol. 1, pp 421–459.

[8] Kipling, J.J.; Tester, D.A. Adsorption from binary mixtures; Determination of individual adsorption isotherms, *J. Chem. Soc.* **1952**, 4123–4133.

[9] Ościk, J. *Adsorption*; Ellis Horwood: Chichester; 1982.

[10] Giles, C.H.; MacEwan, T.H.; Nakhwa, S.N.; Smith, D. Studies in adsorption. Part XI. (A system of classification of solution adsorption isotherms, and its use in diagnosis of adsorption mechanisms and in measurement of specific surface areas of solids), *J. Chem. Soc.* **1960**, 3973–3993.

[11] Sheppard, S.E.; Lambert, R.H.; Walker, R.D. Optical sensitizing of silver halides by dyes I. Adsorption of sensitizing dyes, *J. Chem. Phys.* **1939**, *7*, 265–273.

[12] Sheppard, S.E. The effects of environment and aggregation on the absorption spectra of dyes, *Rev. Mod. Phys.* **1942**, *14*, 303–340.

[13] West, W.; Carroll, B.H.; Whitcomb, D.L. The adsorption of dyes to microcrystals of silver halide, *Ann. N.Y Acad. Sci.* **1954**, *58*, 893–909.

[14] Maron, S.H.; Elder, M.E.; Ulevitch, I.N. Determination of surface area and particle size of synthetic latex by adsorption, *J. Colloid Sci.* **1954**, *9*, 89–103.

[15] Ross, S.; Schaeffer, H.F. Nonsoap thickeners of lubrication oils. I. Sedimentation volumes, *J. Phys. Chem.* **1954**, *58*, 865–868.

[16] Lee, Y.J.; Feke, D.L.; Manas-Zloczower, I. Dispersibility maps for treated titanium dioxide powders, *Colloids Surf.* **1992**, *64*, 235–244.

[17] Povey, M.J. *Ultrasonic techniques for fluids characterization*; Academic Press: New York; 1997.

[18] Motyka, A.L. An introduction to rheology with an emphasis on application to dispersions, *J. Chem. Ed.* **1996**, *73*, 374–380.

[19] Pignon, F.; Magnin, A.; Piau, J.-M. Thixotropic colloidal suspensions and flow curves with minimum: Identification of flow regimes and rheometric consequences, *J. Rheol.* **1996**, *40*, 573–587.

[20] Chanamai, R.; Herrmann, N.; McClements, D.J. Probing floc structure by ultrasonic spectroscopy, viscometry, and creaming measurements, *Langmuir* **2000**, *16*, 5884–5891.

[21] Wolthers, W.; van den Ende, D.; Duits, M.H.G.; Mellema, J. The viscosity and sedimentation of aggregating colloidal dispersions in a Couette flow, *J. Rheol.* **1996**, *40*, 55–67.

[22] Klein, B.; Laskowski, J.S.; Partridge, S.J. A new viscometer for rheological measurements on settling suspensions, *J. Rheol.* **1995**, *39*, 827–840.

[23] Pashias, N.; Boger, D.V. A fifty cent rheometer for yield stress measurement, *J. Rheol.* **1996**, *40*, 1179–1189.

[24] Olivier, J.P.; Sennett, P. Electrokinetic effects in kaolin-water systems. I. The measurement of electrophoretic mobility, *Clays Clay Miner.* **1967**, *15*, 345–356.

[25] Fischer, E.K. *Colloidal dispersions*; Wiley: New York; 1950; p 200.

[26] Frith, W.J.; d'Haene, P.; Buscall, R.; Mewis, J. Shear thickening in model suspensions of sterically stabilized particles, *J. Rheol.* **1996**, *40*, 531–548.

[27] Xu, J.; Jamieson, A.M.; Wang, S.Q.; Qutubuddin, S. Shear thickening and time-dependent rheological behavior in aqueous polyacrylate ester dispersions, *J. Colloid Interface Sci.* **1996**, *182*, 172–178.

[28] Patton, T.C. *Paint flow and pigment dispersion, a rheological approach to coating and ink technology*, 2nd ed.; Wiley: New York; 1979; p 172.

[29] Cohen, E.D.; Gutoff, E.B. *Modern coating and drying technology*; VCH Publishers: New York; 1992.

[30] Rohn, C.L. *Analytic polymer rheology*; Hanser Publishers: New York; 1995; Chapter 10.

[31] Russel, W.B.; Saville, D.A.; Schowalter, W.R. *Colloidal dispersions*; Cambridge University Press: New York; 1989; Chapter 14.

[32] Larson, R.G. *The structure and rheology of complex fluids*; Oxford University Press: New York; 1999; Part III.

[33] Lara, J.; Schreiber, H.P. Specific interactions and the rheology of pigmented polymer solutions, *J. Polym. Sci.: Part B: Polym. Phys.* **1996**, *34*, 1733–1740.

[34] Casson, N. A flow equation for pigment-oil suspensions of the printing ink type, in *Rheology of disperse systems*; Mill, C.C., Ed.; Pergamon Press: London; 1959; pp 84–104.

[35] Eley, R.R. Rheology in Coatings: Principles and methods, in *Encyclopedia of analytic chemistry*; Meyers, R.A., Ed.; Wiley: Chichester; 2000; pp 1839–1869.

[36] Maron, S.H.; Pierce, P.E. Application of Ree-Eyring generalized flow theory to suspensions of spherical particles, *J. Colloid Sci.* **1956**, *11*, 80–95.

[37] El-Saied, H.; Basta, A.H.; Elsayad, S.Y.; Morsy, F. Some aspects of the rheological properties of paper coating suspension and its application: 2. Influence of pigment composition, binder level, co-binder and simple electrolytes on flow properties, *Polym.* **1995**, *36*, 4267–4274.

[38] Cohen, E.; Gutoff, E.B. *Modern coating and drying technology*; VCH Publishers: New York; 1992.

[39] Cohen loc. cit.; p 18.

[40] Cohu, O.; Magnin, A. Rheometry of paints with regard to roll coating processes, *J. Rheol.* **1995**, *39*, 767–785.

[41] van Olphen, H. *An introduction to clay colloid chemistry*; Wiley: New York; 1963; p 103 ff.

[42] Available from Laporte plc. (www.laporteplc.com).

[43] van Olphen loc. cit.; pp 28–29.

[44] www.paintcoatings.net/lapo.htm

[45] Willenbacher, N. Unusual thixotropic properties of aqueous dispersions of Laponite RD, *J. Colloid Interface Sci.* **1996**, *182*, 501–510.

[46] Jaeger, H.M.; Nagel, S.R.; Behringer, R.P. Granular solids, liquids, and gases, *Rev. Mod. Phys.* **1996**, *68*, 1259–1273.

[47] Kaye, B.H.; Gratton-Liimatainen, J.; Faddis, N. Studying the avalanching behavior of a powder in a rotating disc, *Part. Part. Syst. Charact.* **1995**, *12*, 232–236.

[48] Kaye, B.H. Characterising the flowability of a powder using the concepts of fractal geometry and chaos theory, *Part. Part. Syst. Charact.* **1997**, *14*, 53–66.

[49] www.aerosizer.com

[50] Halverson, F.; Panzer, H.P. Flocculating agents, *Kirk–Othmer encyclopedia of chemical technology*, 3rd ed.; Wiley Interscience: New York; **1978–1984**; Vol. 10, pp 489–523.

[51] Heitner, H.I. Flocculating agents, *Kirk–Othmer encyclopedia of chemical technology*, 4th ed.; Wiley: New York; **1994**; Vol. 11, pp 61–80.

[52] Serra, T.; Colomer, J.; Casamitjana, X. Aggregation and breakup of particles in a shear flow, *J. Colloid Interface Sci.* **1997**, *187*, 466–473.

[53] Spicer, P.T.; Keller, W.; Pratsinis, S.E. The effect of impeller type on floc size and structure during shear-induced flocculation, *J. Colloid Interface Sci.* **1996**, *184*, 112–122.

[54] Matijevic, E. Colloid stability and complex chemistry, *J. Colloid Interface Sci.* **1973**, *43*, 217–245.

[55] Gregory, J. Polymer adsorption and flocculation in sheared suspensions, *Colloids Surf.* **1988**, *31*, 231–253.

[56] Chaplain, V.; Janex, M.L.; Lafuma, F.; Graillat, C.; Audebert, R. Coupling between polymer adsorption and colloidal particle aggregation, *Colloid Polym. Sci.* **1995**, *273*, 984–993.

[57] Hoogeveen, N.G.; Stuart, M.A.C.; Fleer, G.J. Can charged block copolymers act as stabilisers and flocculants of oxides, *Colloids. Surf.* **1996**, *117*, 77–88.

[58] James, R.O.; Homola, A.; Healy, T.W. Heterocoagulation of amphoteric latex colloids, *J. Chem. Soc., Faraday Trans. I* **1977**, *73*, 1436–1445.

[59] Homola, A.; James, R.O. Preparation and characterization of amphoteric polystyrene latices, *J. Colloid Interface Sci.* **1977**, *59*, 123–134.

[60] La Mer, V.K.; Dinegar, R.H. Theory, production and mechanism of formation of monodisperse hydrosols, *J. Am. Chem. Soc.* **1950**, *72*, 4847–4854.

[61] La Mer, V.K. Nucleation in phase transitions, *Ind. Eng. Chem.* **1952**, *44*, 1270–1277.

[62] Leubner, I.H. Particle nucleation and growth models, *Cur. Opin. Colloid Interface Sci.* **2000**, *5*, 151–159.

[63] Leubner, I.H. Crystal formation (nucleation) under kinetically and diffusion controlled growth conditions, *J. Phys. Chem.* **1987**, *91*, 6069–6073.

25 Special Systems

25.1 MODEL COLLOIDAL SYSTEMS

The properties of a suspension are the sum of all the individual interactions between its particles. When these are of uniform size and chemistry, the sum is over identical interactions and the macroscopic properties are then linearly related to the properties of individual particles. Such systems come close to the simplest theoretical models and are therefore used to verify theories. Theories of light scattering, adsorption on surfaces, forces across phase boundaries, and stability of dispersions have all been tested by comparison with the behavior of model systems. Two types of systems approaching ideal have been synthesized: the first, suspensions of uniform particles of known size and surface composition; the second, uniform nonporous surfaces. Suspensions of uniform particles have been used to test models of Brownian motion, theories of rapid coagulation, theories of light scattering, and theories of rheology. The uniform surfaces of soap films and adsorbed films on flat mica surfaces have been used to measure forces across phase boundaries. Carefully prepared uniform solid substrates have been used to test models of adsorption of gases.

Uniform dispersions and uniform surfaces also find practical applications: for example, as supports for agglutination tests for pregnancy and rheumatoid factors, as calibration standards for electron microscopes and particle counters, as tests for filter efficiencies, as "seeds" for laser Doppler velocimetry, as packing for chromatographic columns, as catalysts, in ceramics, as pigments with enhanced optical properties, and as media for recording devices.

25.1.a Particles of Uniform Size

Dispersions of uniform size may form spontaneously, as in some protein solutions, some micellar solutions, and some microemulsions. Such dispersions of proteins were used by Svedberg in his studies with the ultracentrifuge; for which he was awarded the Nobel Prize in Chemistry in 1926. Dispersions of uniform size can be obtained by repeated fractionation. Such a dispersion of gamboge particles was isolated by Perrin for use in the experiments that verified Einstein's theory of Brownian motion. Perrin received the Nobel Prize in Physics in 1926. Dispersions of uniform size can be prepared by chemical precipitation. A nearly uniform dispersion of gold particles was prepared in this way by Zsigmondy to test his invention of the ultramicroscope. He received the Nobel Prize in Chemistry in 1925. Opals are formed naturally by the sedimentation of nearly uniform particles of silica about 150–350 nm in diameter. "Opalescence" is the diffraction of light by regions of uniformly spaced layers of the particles.[1]

Can uniform systems be prepared from any desired substance? Probably yes, if sufficiently low solubility can be obtained. The synthetic method is to control particle growth by a two-step process, of which the first is a nucleation or seeding phase; the second is a growth phase, during which the average particle size increases but the size distribution narrows. Some celebrated examples follow.

(i) *Gold Sols.* In 1857 Faraday reported the results of his study on "diffused" gold.[2] By "diffused" he meant a suspension of particles rather than a molecular solution. Suspended particles too small to be seen by any microscope were detected by light scattering. He adumbrated a relation between the size of the suspended gold particles and the colors of the transmitted and scattered light. His gold sols were produced by electrical evaporation of gold wire and by precipitation of gold from a solution of gold chloride. Zsigmondy developed the "seed" method for consistently producing uniform and stable gold sols. By demonstrating, with extremely clean systems, that the number of particles produced is proportional to the amount of seed solution used, he proved that precipitation takes place on already formed nuclei.[3] By analyzing Zsigmondy's data, Mie in 1908 interpreted quantitatively the vivid colors of colloidal gold sols,[4] thus completing the line of research started 50 years earlier by Faraday. More recently, on the basis of his own data on light scattered by monodisperse gold sols, Doremus[5] verified Mie's classical theory for gold particles as small as 8.5 nm. The kinetics of flocculation of gold sols were studied in detail by measuring fractal structures as flocs developed.[6,7] Most of the work on gold sols has been with nanoparticle dispersions. Goia and Matijević describe a procedure whereby monodisperse gold dispersions with diameters from 80 nm to 5 μm can be produced.[8]

(ii) *Sulfur Sols.* La Mer and coworkers prepared monodisperse sulfur sols from acidified thiosulfate solutions.[9,10] They allowed a large number of nuclei to form in a first precipitation and then had those nuclei grow by the slow addition of more reactant, keeping the supersaturation below the level at which new nuclei would form. The preparation and properties of these sols were thoroughly investigated by this school. Their sulfur sols show higher order Tyndall spectra when illuminated with white light, a dramatic effect obtained only with highly uniform systems. By this and other work they demonstrated the wide potential of the method of controlled nucleation and growth to prepare various monodisperse systems. La Mer sulfur sols were prepared by Kerker et al. for use as model systems to study Mie scattering, which gives a more precise determination of average particle size than the method of higher order Tyndall spectra. The results of the two determinations were found to agree quite closely; analysis of Mie scattering also gives the width of the distribution.[11]

(iii) *Selenium Sols.* Watillon et al.[12,13,14] produced monodisperse selenium sols of 4–50 nm by precipitation of selenium from solution onto gold nuclei.

The use of gold nuclei makes for better control of ultimate particle size by eliminating homogeneous precipitation. To test the Mie theory of scattering by absorbing spheres, particle size was first determined by ultramicroscope counting and electron microscopy; the Mie theory was then confirmed by direct comparisons with those observations of model systems.

(iv) *Metal Oxides.* Synthetic methods to obtain a variety of metal-oxide colloids, uniform in size and shape, were developed by Matijević and coworkers.[15,16] The general techniques are precipitation from homogeneous solutions, phase transformations, and reactions within uniform aerosol droplets. Their most used technique is the slow hydrolysis of metal salts at elevated temperatures in aqueous solutions, usually in the presence of complex-forming sulfate or phosphate ions. Hydrolysis is controlled by the temperature or by release of anions or cations, to ensure that only one burst of nuclei is produced, followed by slow growth on the nuclei. By this method, uniform oxides were produced of aluminum, copper, iron (spherical, cubic, and other shapes), cobalt, and nickel, as well as zinc sulfide, lead sulfide, cadmium selenide, and cadmium carbonate. By the method of phase transformation, the monodisperse precipitate is changed to another monodisperse form by crystallization, recrystallization, or dissolution followed by reprecipitation. By this method, uniform suspensions were produced of cobalt ferrite, nickel ferrite, or cobalt–nickel ferrites by reaction of cobalt and nickel oxides with ferric hydroxide. These products are significant in the corrosion of steel. Uniform particles are also produced by the reaction of uniform aerosol droplets with a reactant in the vapor phase. The advantage of this method is that surface-active solutes or other reagents are not required. Titanium oxide, alumina, titanium silicate, and polymer colloids as large as 30 μm have been produced.

(v) *Silver Halide Sols.* Silver iodide sols figured prominently in the development of the theory of lyophobic colloids. They were extensively investigated in the schools of Kruyt and Overbeek[17] and of Težak.[18] Monodisperse silver iodide and silver bromide sols were obtained by carefully diluting a solution of a complex silver halide ion to form supersaturated solutions of the silver halide.[19] All the sols prepared in this way exhibited higher order Tyndall spectra. Particle size was about 400 nm. Silver halide sols are well suited as model systems to test double-layer-stability theory because their potential-determining ions are either silver or halide, the concentrations of which can be adjusted, the particles have surfaces of known crystal structure, and their sizes and shapes are readily measured.

(vi) *Monodisperse Latex Particles.* Aqueous suspensions of polymers are called "latexes" (or "latices") because they evolved from the work on synthetic rubber production, and the suspensions have a milky look. Three different synthetic routes were developed to produce uniform particles. Particles with average diameters less than about 5 μm are produced by emulsion polymerization with a coefficient of variance of about 1%. Particles with diameters between about 2 and 20 μm are produced by the swollen-emulsion-polymerization technique of Ugelstad with a coefficient of variance less than 3%. Particles with larger diameters are produced by dispersion polymerization, but are not nearly so uniform, having coefficients of variance in diameters of 16–30%. The differences in the uniformity of particles produced by the different techniques are direct consequences of the different mechanisms of particle formation.

Emulsion polymerization is carried out in a solution containing emulsified monomer droplets, sufficient surface-active solute to form a large number of micelles, and a soluble initiator. The initiation of polymerization is generally believed to occur in the homogeneous solution, which contains a low concentration of monomer. The growing oligomers gradually become water insoluble and are taken up by the micelles, inside which subsequent growth occurs. Because polymer growth occurs slowly after the initiation of polymerization, the narrow distribution of micellar sizes narrows even further as the particles grow. The optimum conditions for the formation of monodisperse suspensions are determined empirically.[20,21,22,23]

Larger particles are formed by the addition of more monomer to the suspension. The added monomer swells the already formed particles and is incorporated into the existing polymer chains. The ultimate size that particles can reach and still remain uniform on addition of more monomer is limited. Any slight variation in particle size is increased by Ostwald ripening, the larger particles growing at the expense of the smaller ones.

The upper size limit for particles grown by emulsion polymerization was extended by the work of Ugelstad and coworkers.[24] After a routine initial emulsion polymerization step, a swelling agent, such as dodecyl chloride, is added, which causes the particles to swell by imbibing a large amount of monomer, while still maintaining uniform size. Large particles grown by this technique are nearly as uniform as those grown in space under conditions of microgravity at much greater expense.

The largest latex particles are grown by suspension polymerization. In this process an emulsion of the monomer is made and an oil-soluble initiator is added to polymerize the emulsion droplets. The final average particle size and size distribution of the polymerized particles is nearly identical with the average particle size and size distribution of the initial emulsion droplets.

Croucher et al. synthesized a number of thermodynamically stable nonpolar suspensions.[25,26,27] In every case the particles are copolymers: One moiety forms the insoluble core of the particle and the other moiety forms the soluble outer layer that stabilizes the suspension. When the particle is in a worse-than-theta solvent, which may be brought about by changing solvent composition or by changing temperature or pressure, the outer polymer collapses against the core and loses its ability to stabilize. The flocculation, however, is into a shallow minimum. When the solvent is then restored to better-than-theta conditions, the outer polymer reextends, enabling the particles to be easily redispersed. These model systems are ideal to study mechanisms of polymer stabilization as well as to provide useful practical materials. These uniform latex suspensions are a subset of suspensions called polymer colloids, which are used as adhesives, coatings, and sealants, as well as in paints, papermaking, and medical science. The science of polymer colloids is the overlap of polymer chemistry and colloid science. "We may think of a Venn diagram of two circles, one labeled polymers and the other colloids."[28] The subject is reviewed in monographs.[29,30,31]

(vii) *Monodisperse Emulsions.* These can be produced by various techniques, such as fractionated crystallization,[32] driven Rayleigh-jet breakup,[33] membrane extrusion,[34] and by shearing a polydisperse emulsion in a viscoelastic medium.[35,36] This last technique is by far the easiest. The mechanism of the emulsion fragmentation is not yet understood, but a practicable process, both convenient and rapid, has been achieved. A series of papers by Bibette and coworkers report the process and describe a suitable mechanical mixer. Important preconditions are that the initial emulsion have a viscoelastic medium and the applied shear be carefully controlled. Taking these features into account, a number of monodisperse emulsions were produced, independently of the chemical nature of the system, in which drop sizes were tuned from 0.3 to 10 μm. The authors speculate that the monodispersity results from the suppression of the capillary instability until the droplet is sufficiently elongated so that its Laplace pressure overcomes the medium's elasticity, thereby allowing the internal phase to flow and rupturing to occur.

A rather similar technique is described by Perrin[37] who stabilized the ordered structure by chemically cross-linking the polymeric emulsifier macromolecules that are dissolved in the medium. Again, the process begins by shearing disordered premixed emulsions and takes advantage of the tendency towards self-assembly when droplets of uniform size are in close proximity. Perrin does not require viscoelastic media and the production of monodispersity in its absence tends to cast doubt on the generality of the speculation by Mason and Bibette about its cause.

25.1.b Surfaces of Uniform Composition

25.1.b(i) Soap Films Soap films have the advantage, common to any liquid surface, of being energetically uniform as long as they are not subject to changes of temperature, area, or composition. They have proved useful for direct measurements of disjunctive (disjoining) pressures since they can be thinned to tens of nanometers without rupture. The thickness is made to vary by applied pressure, which can be measured directly.[38,39] The soap film is supported, as shown in Figure 25.1 (inset), by a porous porcelain ring connected to a reservoir of soap solution. When pressure is applied, the soap film thins out and the excess solution flows into the reservoir. When equilibrium is reached, the thickness of the film is measured by means of an interferometer. Thus, by controlling the air pressure within the cell, the resistance to compression between the two charged surfaces of the soap film is observed as a function of distance. The analysis of the data gives a direct measurement of the Debye length, $1/\kappa$, and a test of theories of electrostatic stabilization. For instance, a plot of the logarithm of the pressure against thickness has a slope of $-\kappa$.

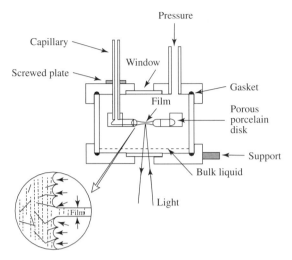

Figure 25.1 Schematic cross-section of the apparatus used to compress soap films. The inset shows how air is retained but liquid allowed to pass through the porous solid. A layer of liquid on the bottom of the cell prevents evaporation from the soap film and allows light to pass.[39]

25.1.b(ii) Homotattic Solid Surfaces Uniform solid substrates are the key to understanding the behavior of adsorbed vapors. To be useful, substrates must be fine powders because the amount of adsorption varies directly with the specific area; and on large flat surfaces sufficient vapor is not adsorbed to be accurately measured. Observed monolayer adsorption at constant temperature depends on both the equation of state of the adsorbed (two-dimensional) film on each patch and on the distribution of patches that compose the solid substrate. On a heterogeneous substrate, more than one combination of an equation of state and of a degree of heterogeneity may be found to fit the data. If the true equation of state on each energetically uniform (homotattic) patch were known, then the true heterogeneity could be deduced. The usefulness of finely divided solids with nearly uniform surfaces is, therefore, to establish experimentally the actual equation of state of an adsorbed film, unobscured by substrate heterogeneity, so that thenceforth any arbitrary substrate can be analyzed with only heterogeneity as the unknown factor.

The behavior of an adsorbed monolayer bears a remarkable analogy to that of matter in the bulk. A vapor condenses to a liquid with the appearance of a discontinuity in the $p - V$ isotherm, and above a critical temperature such condensation is no longer possible. Analogous behavior is observed with adsorbed monolayers on solid substrates, including first-order phase transitions and two-dimensional critical temperatures. However, these effects are screened from observation unless the substrate is homotattic. An excellent test for uniformity of a substrate, therefore, is whether the adsorbed film shows a first-order phase transition, as revealed by a discontinuity in the adsorption isotherm. Even small degrees of heterogeneity of the substrate obscure this evidence. Ross and coworkers,[40,41,42] starting in 1947, published a series of adsorption isotherms of ethane adsorbed by cube crystals of carefully prepared sodium chloride at various temperatures above and below what is clearly a two-dimensional critical temperature. Subsequently Ross and his school[43] made similar observations with krypton on graphitized carbon black.

Analysis of the adsorption isotherms of these homotattic adsorbents identified a useful equation of state as the two-dimensional analogue of the van der Waals equation of a gas. More precise data led to the use of a two-dimensional virial equation of state,[44] with refinements of the numerical analysis to determine the distribution of surface energies of adsorbents (CAEDMON).[45]

A significant outcome of this work on model solid adsorbents is that it fails to confirm the widely used Langmuir adsorption isotherm as a valid description of vapor adsorption on homotattic substrates, although much theorizing is based on that assumption.

25.1.c Colloidal Crystals

The most studied feature of colloid suspensions is their stability. The transition from a dispersed to a flocculated state can be thought of as a phase change, the flocculated state being the more concentrated state. The flocculated state is usually amorphous, but colloids can self-assemble into highly ordered states. These states occur in monodisperse systems of low ionic strength. If particles are all the same size and carry the same charge they repel each other to the same degree and find equilibrium positions in an array that is equivalent to hexagonal close packing. An opalescent display results from the Bragg diffraction of light. Addition of relatively small amounts of salt collapses the double layer and completely destroys the opalescence by introducing some randomness in the spacing. One striking example is the structure formed when needle-like particles, such as beta ferric oxyhydroxide rods, are concentrated into a moderately dense suspension. The layers of rods diffract light and are therefore iridescent.[46] A similar example is the formation of hexagonally packed micelles forming liquid crystals, which also diffract light (see Chapter 13). Monodisperse, charged spheres develop similar crystalline order at low ionic strengths, owing to long-range electrostatic repulsions. These dense phases are highly iridescent and beautiful to examine. Careful studies of these dense phases has led to increased understanding of the interactions between dispersed particles.[47,48]

Closely packed colloidal crystals display a number of potentially useful characteristics such as light diffraction and photonic band gaps, high packing density and surface-to-volume ratio, maximal structural stability, and high catalytic activity. Structures of close-packed particles of various shapes have been constructed by placing drops of aqueous suspensions on the surface of perfluoromethyldecalin and allowing the dispersion to dry.[49] Interesting porous materials can be fabricated by dispersing uniform particles in a material that will solidify without disturbing the structure of a ordered colloidal phase. The colloidal material is removed, say by calcination, to leave a highly structured, porous material.[50,51] These materials hold promise for use as photonic crystals, advanced catalysts, and as templates for structured porous polymers and metals.

25.2 CLOUDS

Clouds are naturally occurring dispersions of minute droplets of water in air (aerosol). By Stokes law a droplet of water 25 μm in diameter will fall through air at a rate of 0.02 m/s. If the droplet is one-tenth that size, its rate of fall would be 100 times smaller, or about 200 μm/s. The settling of a cloud formed of such particles is unnoticeable; indeed the settling of very small particles suspended in air, such as the fine spray of waves or waterfalls and all kinds of dust and smoke, is very slow. Stokes law, however, is not the last word on the motion of airborne particles. Solar radiation is absorbed by the particles, which radiate heat to the surrounding air, thus reducing its density and causing it to rise, carrying the particles with it. If the particles are water, the heat of the sun causes evaporation. Water vapor is less dense than air, and moist air is lighter than dry air at the same temperature and pressure. The water droplets make the air damp and, if the density is less than that of the surrounding dry air, the cloud will rise. Moist air over a body of water absorbs heat and rises, then condenses at the lower temperature of the higher altitude, falls, and rises again in the heat of the sun.

25.3 SMOKES

The air over an industrial city contains 100–500 kg of solid matter per cubic kilometer. Dust particles, even more than water droplets, absorb sunshine and generate thermal gradients in the surrounding air; consequently, in spite of their higher density, solid particles remain suspended in air and may be carried hundreds of kilometers by air currents. Finely divided particulate matter is carried into the upper atmosphere and can be transported around the world. Smoke is also responsible for "pea-soup" fogs by nucleating moisture that would otherwise remain as a vapor. These fogs carry a high concentration of dirt and sometimes sulfuric acid.

25.4 LUBRICATING GREASES

The function of a lubricant is to prevent direct contact between surfaces in relative motion. Grease is designed to flow into bearings by the application of pressure and stay in place without dripping when the machinery is stationary. Therefore, grease must have a yield point. Another desideratum of a grease is to retain plastic viscosity at temperatures up to 200°C.

Lubricating greases are stable solid or semifluid colloidal dispersions of a base oil, thickeners, and other additives.[52] Simple greases are made from lubricating oil by dissolving lithium, calcium, or sodium soaps in hot oil; on cooling, the soaps crystallize as a mass of ramified fibers from 200 nm to 1 mm in length and from 10 to 1000 nm in width, enmeshing the oil. Nonsoap inorganic thickeners, including colloidal silica, Attapulgite clay, and Montmorillonite clay treated with long-chain amines, make greases that can be used at higher temperatures. Thickeners are almost always the minor component; their function being to alter the rheological behavior of the lubricant.

Some greases useful for high-pressure applications contain dispersions of metaborate, rare-earth fluorides, and rare-earth complexes in mineral oil and microemulsions of metaborate or $La(OH)_3$ in mineral oil.[52] The mechanism appears to be the formation of organic complexes of inorganic cations in the boundary layers.

The existence of a yield point depends on the flocculated structure of the thickener, called the matrix, and on the particle size and aspect ratio of its primary units. The lubricant is held in the solid matrix by capillarity and by adsorption, for which a large interfacial area is advantageous. The matrix is held together by short-range forces of attraction between its primary units. Water has a large effect on the rheology of greases, suggesting that the fiber–fiber interactions are modulated by acid–base exchanges.

25.5 GLUES

TABLE 25.1 Types of Glue

Type	Examples
Solvent glues	Adhesive base and solvent (usually toluene)
	Liquid solders
	Contact cements
Water-based glues	White glue and water
	Casein powder and water
Two-part glues	Epoxies
Animal hide glues	Mixed with water and heated to form a thermoplastic
Thermosetting adhesives	Cyanoacrylate polymers

25.6 INKS FOR XEROGRAPHY

In 1938 Chester Carlson demonstrated xerography, a new copying process where dry pigment particles are attracted to an electrostatic image formed by exposing a charged photoconductive film to light reflected from the original. Nowadays the light used is a scanning laser beam and the equipment is called the laser printer. Seven steps are common to all copiers and laser printers: the uniform electrical charging of a photoconductive roll in the dark, the exposure of the charged photoconductor to a light image that creates an electrostatic image, the electrophoretic deposition of charged particles on the electrostatic image, the transfer of the powder pattern to paper, the fusing of the powder to paper, the cleaning of surface of the photoconductor, and then the complete exposure of the photoconductor to erase any residual electric charge, after which it is ready to be used again (see Figure 25.2).

The various steps in xerography depend on the electrical and mechanical properties of particles and dispersions.[53] Modern xerographic copiers and printers scan documents and store the image electronically, at least temporarily. The scanning laser is directed by the information from this stored image or from information created by a computer directly. Color images are created by making this xerographic cycle repeat for each primary color and black. The physics, especially electrostatics fields and their effects on charging, image formation, and development, are well understood.[54]

The image is developed by electrodeposition of a charged aerosol (dry ink or toner). Xerography or "dry writing" is so named because dry, pigmented powder is carried to the charged portions of the photoconductor surface and so develops the electrostatic image. The dry powder image on the photoconductor surface is then transferred to paper, which is heated to make a permanent print.

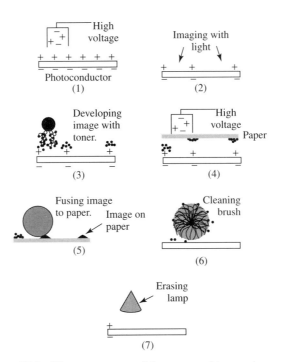

Figure 25.2 The seven steps of the xerographic copying cycle.

This step in the xerographic cycle depends on particulate properties. The dry toner particles are pigmented (usually black) polymer beads. The function of the polymer is to attach the pigment to the paper. The finer and the more uniform the particles, the higher resolution the image, but particles much less than about 10 μm pose a health hazard. The electric charge on the toner particles is controlled by adding an oil-soluble, surface-active agent when the pigment is mixed with the polymer resin. A common charging agent is a long-chain alkyl quaternary ammonium halide. The toner particles are then shaken with much larger metallic beads which generates the charge separation by leaving the cation on the metal and the anion on the toner particle.

A strong electric field is required for electrodeposition. This is created as follows. The metallic beads are held on a development roll by a magnetic field. When the electrostatic image on the surface of the photoconductor rotates under this biased roll, a strong electric field is created. The high electric fields in the nip separate the toner particles from the metallic beads and carry the toner to the photoconductor surface. The magnitude of the electric field is determined by the gap, typically about 10^{-3} m, and the voltage drop, typically a few hundred volts, producing an electric field of a few hundred thousand V/m.

The number of colloid and interface properties significant in this development process are many: the dispersion of the pigment into the polymer resin, the selection of a charging agent that is sufficiently surface active to remain near the toner surface and have the proper acid–base characteristics, the dynamics of charge exchange, the toner adhesion to metallic beads, to the photoconductor surface and to the paper, and the thermal and mechanical properties of the toner as it is melted and fused to the paper surface.

A less used, but interesting xerographic process, is one in which the toner particles are dispersed in a nonpolar liquid, usually an aliphatic hydrocarbon. The advantage of having the toner particles in a liquid dispersion rather than as a powder cloud is that small particles can be used safely. The use of smaller toner particles enables higher resolution imaging. The disadvantage is that the hydrocarbon needs to be removed from the paper after the image has been transferred. This inevitably produces annoying odors, not to mention hazards. The use of a liquid toner in xerography is called liquid-immersion development or LID.

The electrical charging of the toner in the hydrocarbon is accomplished by what are now standard procedures. (See Chapter 17.) While in dry xerography the toner exchanges ions with the metal bead, in liquid xerography the toner exchanges ions with the micelles. The charge exchange is probably with the pigment on the surface of the toner particles, but this is sometimes not enough to provide adequate charge. A higher charge can be attained by previous incorporation in the toner of charge adjuvants, namely, Lewis acids or Lewis bases. A Lewis acid in the toner makes negatively charged particles. The mechanism is described in Figure 14.4.

25.7 ELECTROPHORETIC DISPLAYS

Electrophoretic displays, sometimes referred to as EPIDs, are related to the liquid-immersion development of inks for xerography as described above. A simple EPID consists of toner particles dispersed in a nonpolar solvent, usually an aromatic or aliphatic hydrocarbon with an oil-soluble dye, encased between two parallel plate electrodes. The electrode on the viewing side is transparent, usually glass or plastic coated with indium-tin oxide and is grounded. When the rear electrode is charged with the same sign as the particles, they move by electrophoresis to plate the viewing electrode. The viewer sees the color of the particles, usually white titania. When the rear electrode is charged to the opposite sign of the particles, they migrate to the rear electrode. The viewer then sees the dye color as incident light travels through the front electrode, through the color dye solution, scatters off the particles plated on the rear electrode, and exits back through the front electrode. When the rear electrode is segmented, an arbitrary image can be displayed by using a computer to direct the electric field to all the segments, moving particles to the front where the image is bright and moving particles to the rear where the image is the color of the dye (see Figure 25.3).

An advanced EPID would have a matrix-addressable rear electrode to be able to display high resolution images. Some EPIDs have been built with a photoconductive layer on the rear electrode so an image of toner can be displayed when the rear is exposed to a light image. This is particularly interesting for

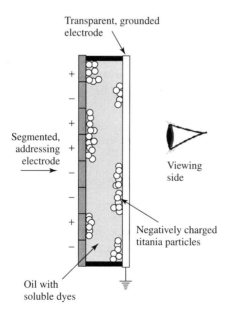

Figure 25.3 An electrophoretic display with a segmented, addressing electrode. When particles are on the viewing electrode, the display is bright; when particles are on the rear electrode, the display is dark.

x-ray diagnostics. For a chest x-ray the EPID is placed in front of the patient and the x-ray source behind him. The emergent x-rays expose the photoconductor and the image is displayed. Once activated, the pattern remains after the x-ray source is turned off and then can be examined immediately without processing. Photoconductors are more sensitive than x-ray film, so a lower dose of x-rays is achieved by this equipment.

EPID technology was developed about the same time as liquid-crystal displays, LCDs. The EPIDs have higher contrast and better viewing angles and so are much better looking displays. EPIDs, however, suffer from two instabilities. The first is gravitational sedimentation. The pigment of choice is titania because it is so bright, but it is also so much more dense than hydrocarbons that the titania eventually settles. Further, the electric fields between parallel plate electrodes are not linear when the electrodes are too highly segmented and the electrophoretic motion of the toner particles is therefore uneven. Nevertheless, much of the early work on the electrical properties of nonpolar dispersions was done to understand the origin of electric charges in EPIDs. Claims that electric charges are insignificant in insulating medium are easily dismissed when whole technologies are based on the significance of those charges.

A similar technology is based on the dispersion of bichromal balls in small, spherical cavities, one bichromal ball to a cavity.[55] A typical bichromal ball is a sphere that is half black and half white, say a toner particle where the black pigment is only on one side of the toner particle. When one half of a toner is of a different composition from the other, it is not surprising that the electric charge on one side would be different from the other, thus giving the bichromal ball a permanent electric dipole. If the ball and its surrounding oil, usually a hydrocarbon or silicone oil, is constrained in a cavity only slightly larger than itself, it will rotate in an applied electric field. A sheet is manufactured with nearly a close-packed arrangement of bichromal balls in their individual oil-filled cavities by dispersing them in a plastic at high volume loading and then swelling the sheet with oil. The cavities expand a little as the sheet is saturated with oil, leaving the bichromal balls free to rotate. When this sheet is held between two parallel-plate electrodes, the one on the viewing side being transparent, the display looks and acts like an EPID. The balls will turn so that their dipoles line up with the electric field lines of an applied field. The viewer sees one side. When the electric field is reversed, the balls turn round, and the viewer sees the other side and, hence, the other color. Just as in EPIDs, if a pattern of electric fields is applied across a sheet of bichromal balls, they all turn in the appropriate orientation to display an image. The great advantage of this technique over the original EPIDs is that the bichromal balls cannot move out of their individual cavities and hence the image is stable against sedimenting.

An improved technology is to combine the idea of the EPID display of electrophoretic particles moving between parallel plate electrodes with the idea of the rotating-ball display where the balls are prevented from settling by being held in small capsules. These displays are referred to as encapsulated, electrophoretic displays.[56] The procedure is to take the titania dispersions in a dyed hydrocarbon, as used in the EPIDs, and encapsulate them in capsules of a few hundred microns diameter or less. One electrode is coated with the capsules and then laminated to a second electrode to form a structure similar to an EPID except that the liquid phase is now subdivided into small, independent regions. The appearance of the display is nearly the same as the EPIDs as long as the capsule walls are thin compared to the capsule diameter. One great advantage gained by this encapsulation is that the suspended pigment particles can only settle a hundred microns or so to the capsule wall. From there they are easily redispersed when the electric field is turned on. The display is now stable, overcoming the grave shortcoming of the original EPIDs. A second advantage is that the capsule layer acts like a spacer between the two electrodes so the electrodes can be made of flexible materials and the whole display can be flexible, lightweight, and physically robust. A third advantage is that the display is constructed by coating and lamination rather than by high vacuum techniques in clean rooms.

25.8 NANOPARTICLES AND NANOTECHNOLOGY

Chemistry has traditionally been the manipulation of materials on the atomic and molecular level. Materials science or engineering has traditionally been the manipulation of materials on the macroscopic scale. Nanotechnology is the manipulation on the nanoscale.[57] Nanoparticles* are generally taken to be solids whose maximum diameter is less than about 100 nm. Every physical property, such as optical absorption or electrical conductivity, has a critical size and if the particle is smaller than that critical size, the property becomes a function of size. By controlling the particle size, the surface chemistry, and the assembly, it is possible to engineer properties and functions in unprecedented ways.

Nanotechnology is not an entirely new field; colloidal sols and supported platinum catalysts are nanoparticles.[58] Nevertheless, a recent explosion of interest in the nanoscale has produced materials that are used as thermal and optical barriers, imaging enhancement, ink-jet materials, coated abrasive slurries, information-recording layers, chemical and electrical energy storage devices, sensors, porous membranes and sieves, drug delivery, tailored catalysts, sorption–desorption materials, magnetic devices, single molecule DNA sizing and sequencing, biomedical sensors, lasers, as fillers in metal, in ceramic, and in polymer matrices, grit on cutting tools, nanocomposite cements, magnetic refrigerants, and so on.

Conductors have free electrons. The energy levels, or resonance states, of these free electrons depend on the size of the particle. Therefore, each size of particle has a characteristic absorption at the frequency of the resonance state, called a surface plasmon resonance (collective oscillations of electron density).[59] Each characteristic absorption shows a characteristic color. For example gold nanoparticles are red; when flocculated they are purple or black. This property is used in

---

* "Nano" derives from the Greek word for dwarf. When used as a prefix it means "one billionth." A nanometer is one billionth of a meter, about the length of a few atoms lined up.[58]

pregnancy tests by attaching suitable gold particles to suitable antibodies. When the antibodies bind to the corresponding proteins, the gold particles flocculate and change color. Nanoparticles of the noble metals, gold, silver, and platinum, are chemically stable compared to nanoparticles of more active metals and so are more often used.

Semiconducting materials have a significant energy gap between the highest occupied molecular orbitals and the lowest unoccupied molecular orbitals. The width of this energy gap depends on the size of the particle; the smaller the particle, the wider the gap. The width and position of the energy gap in semiconductors is important in determining their role in electronic circuits.[60]

25.8.a Synthesis

Nanoparticles are synthesized from ionic or molecular precursors and rarely by attrition or by etching larger particles.[61] Nanoparticles of conductors like gold and silver are produced by precipitation from solution, the condensation of aerosols, and electrolysis. The advantages of chemical processes are that both the size and the size distribution can be controlled. Other processes have been suggested, but have not been as effective.[62]

Monodisperse colloidal gold can be synthesized with diameters between 30 and 100 nm by hydroxylamine seeding of a colloidal gold sol.[63] This technique improves the monodispersity beyond the one-step reduction of Au^{3+} with citrate. Nanoparticles of gold from 1.4 to 5 nm can be obtained by ligand exchange.[64] Gold nanoparticles can be reacted with aliphatic thiols to produce a preparation that can be dried and handled as a simple chemical compound. The dry product is dark brown and waxy; it is "soluble" in toluene, pentane, and chloroform.[65] The octyl thiol is sufficiently large to confer steric stabilization on nanoparticles. Nanoparticle gold of 2.5 to 7 nm is precipitated in toluene by the reduction of $AuCl_4^-$ in a two-phase water–toluene system in the presence of a linear alkane amine.[66] Nanoparticle platinum has also been produced in a one-phase system with tetrahydrofuran as the solvent and lithium triethylborohydride ("superhydride") as the reducing agent.[67] The nanoparticle of platinum has a face-centered cubic structure with diameter from about 2 to 5 nm. The preparation of nanoparticle platinum is motivated by its catalytic properties and for biomedical application.

Spherical nanoparticles of semiconducting materials are called quantum dots since their electronic states depend on one dimension, the diameter of the sphere.[68,69] Quantum dots are of interest in electronics and optics, particularly in nonlinear optical effects from highly polarizable excited states and novel photochemical behavior.[70] Some quantum dots are overcoated with higher bandgap materials to prevent quenching of excited electronic carriers by the ambient.[71] Particle sizes from 2.3 to 5.5 nm fluoresce at narrow wavelength peaks from blue to red with quantum yields of 30–50%.

Two variations have been studied, nanoparticles of different shapes and composite nanoparticles. Silver nanoparticles in the form of triangular prisms are created by vapor deposition in the interstices between a close-packed monolayer of monodisperse polystyrene spheres on a mica surface.[72] These silver nanoparticles are useful in surface-enhanced Raman scattering because their surface plasmon resonances can be tuned by varying their size. Composite nanoparticles consisting of a dielectric core with a metallic shell of nanometer thickness, have optical resonances that can be designed in a controlled manner. For example, nanoparticle gold is precipitated in the presence of larger nanoparticle silica. The gold particles coalesce to form a composite nanoparticle with a core-shell structure.[73] These composite particles are also useful in surface-enhanced Raman scattering and as biological tags.

25.8.b Assembly

A useful concept is to think of nanoparticles as large molecules. These "molecular" units can be assembled into condensed structures with unique bulk electrical, mechanical, magnetic properties.[74] The electronic properties depend strongly on particle size and interparticle spacing which can be controlled with angstrom accuracy by the choice of dithiol spacers.[75] Monolayers of organically functionalized silver nanocrystals on water have been monitored as a continuous function of interparticle separation distance. As the monolayer is compressed from an average separation between the surfaces of the metal cores of 1.2 (± 0.2) nm to ~ 0.5 nm a sharp, reversible insulator-to-metal transition is observed.[76]

A clever method to assemble one-dimensional arrays of nanoparticles is to align the nanoparticles in the pores of Al_2O_3 or polycarbonate filtration membranes.[77] These arrays can be thought of as nanowires. Of course the most familiar nanowires are the single-atom-walled carbon nanotubes during arc discharge of a carbon rod.[78] Two-dimensional arrays of gold nanoparticles have been formed by chemisorption to the functionalized surfaces of larger silica nanoparticles.[79] Gold nanoparticles have also been built into patterns by preferential adsorption on individual phases of block copolymer coatings.[80]

Thicker layers of nanoparticles have been prepared by a one-step exchange, cross-linking-precipitation route that is simpler than laying layers down one at a time.[81] Another method to assemble nanoparticles into larger structures is to use DNA as a molecular template.[82]

25.8.c Self-Assembly

Interparticle forces control the behavior of nanoparticles at close approach and therefore are used in the self-assembly of structures. "Self-assembly" is a frequent term in the nanotechnology literature. For instance, properly surface-treated nanoparticles form "monomolecular" films at the air–water interface and can then be transferred to solid surfaces in thin, ordered states by the Langmuir–Blodgett technique. Cadmium selenide (CdSe) nanocrystallites self-assemble into three-dimensional arrays whose interparticle spacings can be controlled with angstrom precision by varying the molecular weight of the steric-stabilizing molecule.[83]

Surface-active solutes form many different types of phases at high concentrations, especially in three or more component systems. These structures "self-assemble." (See Section 13.23.) Since hydrophilic and hydrophobic mesoregions alternate, the reactions that produce nanoparticles can be carried out in the appropriate phase. This creates a self-assembled array of nanoparticles. The surface-active solute may sometimes be removed to leave an ordered array of nanoparticles.

An expansion of the idea of nanoparticle self-assembly is to consider building nanomachines or even nanobiochemical machines. After all, the cell is a self-assembly of even smaller nanostructures.

25.8.d Applications

Dilute gold-nanoparticle dispersions are ruby red. When the interparticle spacing decreases to less than approximately the particle diameter, the color becomes blue.[84] This phenomenon has been used as a detector for specific polynucleotides.[85,86,87] The gold nanoparticles, 13 nm in diameter, are chemically attached to an oligonucleotide by means of a mercaptoalkylnucleotide. The oligonucleotide is chosen to bind selectively to a polynucleotide critical to the diagnosis of a genetic or pathogenic disease. In the presence of the target polynucleotide, the dispersion changes from red to pinkish-purple. Combinations of conductive polymers and nanoparticle gold are similarly useful for biomolecular recognition.[88] Gold nanoparticles have also been attached to evaporated gold layers to provide enhanced sensitivity in surface-plasmon-resonance biosensing.[89,90]

Semiconducting quantum dots have been used to fabricate transistors.[91] Nanocrystals have been reported to melt at 1000 °C lower than the melting temperature of their bulk phase, the reason being that a larger fraction of the total number of atoms is on the surface.[92] Surface bonds more readily dissociate than bonds in the bulk phase. As a result, 2 nm cadmium selenide particles can be processed into a functioning semiconductor at 350 °C.

The feasibility of a quantum dot laser has been demonstrated by the optical gain of strongly confined CdSe nanocrystal quantum dots assembled as a close-packed film. Stimulated emission is spectrally tunable with dot size and has a clear threshold behavior.[93]

REFERENCES

[1] Sanders, J.V. Diffraction of light by opals, *Acta Cryst.* **1968**, *A24*, 427–434.

[2] Faraday, M. Experimental relations of gold (and other metals) to light, *Phil. Trans. Roy. Soc.* London **1857**, *147*, 145–181; also *The foundations of colloid chemistry*: Hatschek, E., Ed.; Benn: London; 1925, pp 65–92.

[3] Zsigmondy, R.; Thiessen, P.A. *Das kolloide gold*; Akademische Verlagsgesellschaft: Leipzig; 1925.

[4] Mie, G. Contributions to the optics of diffusing media, particularly of colloidal metals, *Ann. Physik* **1908**, *25*, 377–445.

[5] Doremus, R.H. Optical properties of small gold particles, *J. Chem. Phys.* **1964**, *40*, 2389–2396.

[6] Weitz, D.A.; Oliveria, M. Fractal structures formed by kinetic aggregation of aqueous gold colloids, *Phys. Rev. Lett.* **1984**, *52*, 1433–1436.

[7] Weitz, D.A.; Huang, J.S.; Lin, M.Y.; Sung, J. Dynamics of diffusion-limited kinetic aggregation, *Phys. Rev. Lett.* **1984**, *53*, 1657–1660.

[8] Goia, D.V.; Matijević, E. Tailoring the particle size of monodisperse colloidal gold, *Colloids Surf. A: Physicochem. Eng. Aspects* **1999**, *146*, 139–152.

[9] La Mer, V.K.; Dinegar, R.H. Theory, production and mechanism of formation of monodisperse hydrosols, *J. Am Chem. Soc.* **1950**, *72*, 4847–4854.

[10] La Mer, V.K. Nucleation in phase transitions, *Ind. Eng. Chem.* **1952**, *44*, 1270–1277.

[11] Kerker, M.; Daby, E.; Cohen, G.L.; Kratohvil, J.P.; Matijević, E. Particle size distribution in La Mer sulfur sols, *J. Phys. Chem.* **1963**, *67*, 2105–2111.

[12] Watillon, A.; van Grunderbeeck, F.; Hautecler, M. Preparation et purification d'hydrosols de selenium stables et homeodisperses, *Bull. Soc. Chem. Belg.* **1958**, *67*, 5–21.

[13] Dauchot, J.; Watillon, A. Optical properties of selenium sols I. Computation of extinction curves from Mie equations, *J. Colloid Interface Sci.* **1967**, *23*, 62–72.

[14] Watillon, A.; Dauchot, J. Optical properties of selenium sols II. Preparation and particle size distribution, *J. Colloid Interface Sci.* **1968**, *27*, 507–515.

[15] Matijević, E. Monodispersed metal (hydrous) oxides — a fascinating field of colloid science, *Acc. Chem. Res.* **1981**, *14*, 22–29.

[16] Matijević, E. Production of monodisperse colloidal particles, *Ann. Rev. Mater. Sci.* **1985**, *15*, 483–516.

[17] Kruyt, H.R., Ed. *Colloid science*, Vol. 1; Elsevier: New York; 1952.

[18] Težak, B.; Matijević, E.; Schulz, K.F.; Kratohvil, J.; Mirnik, M.; Vouk, V.B. Coagulation as a controlling process of the transition from homogeneous to heterogeneous electrolytic systems, *Faraday Discuss. Chem. Soc.* **1954**, *18*, 63–73.

[19] Ottewill, R.H.; Woodbridge, R.F. The preparation of monodisperse silver bromide and silver iodide sols, *J. Colloid Sci.* **1961**, *16*, 581–594.

[20] Woods, M.E.; Dodge, J.S.; Krieger, I. M.; Pierce, P.E. Monodisperse latices. I. Emulsion polymerization with mixtures of anionic and nonionic surfactants, *J. Paint Technol.* **1968**, *40*, 541–548.

[21] Dodge, J.S.; Woods, M.E.; Krieger, I.M. Monodisperse latices. II. Seed polymerization techniques using mixtures of anionic and nonionic surfactants, *J. Paint Technol.* **1970**, *42*, 71–75.

[22] Papir, Y.S.; Woods, M.E.; Krieger, I.M. Monodisperse latices. III. Cross-linked polystyrene latices, *J. Paint Technol.* **1970**, *42*, 571–578.

[23] Tamai, H.; Hamada, A.; Suzawa, T. Deposition of cationic polystyrene latex on fibers, *J. Colloid Interface Sci.* **1982**, *88*, 378–384.

[24] Ugelstad, J.; El-Aasser, M.S.; Vanderhoff, J.W. Emulsion polymerization: Initiation of polymerization in monomer droplets, *J. Polym. Sci., Polym. Lett. Ed.* **1973**, *11*, 503–513.

[25] Croucher, M.D.; Hair, M.L. Upper and lower critical flocculation temperatures in sterically stabilized nonaqueous dispersions, *Macromolecules* **1978**, *11*, 874–879.

[26] Croucher, M.D.; Hair, M.L. Selective flocculation in heterosterically stabilized nonaqueous dispersions, *Colloids Surf.* **1980**, *1*, 349–360.

[27] Croucher, M.D.; Lok, K.P. Stability of sterically stabilized nonaqueous dispersions at elevated temperatures and pressures, *ACS Symp. Ser.* **1984**, *240*, 317–330.

[28] Fitch, R.M. *Polymer colloids: a comprehensive introduction*; Academic Press: New York; 1997; p 1.

[29] Fitch, R.M., Ed. *Polymer colloids*; Plenum Press: New York; 1971.

[30] Buscall, R., Corner, T.; Stageman, J.F., Eds. *Polymer colloids*; Elsevier: New York; 1985.

[31] Fitch, *loc. cit. passim*

[32] Bibette, J. Depletion interactions and fractionated crystallization for polydisperse emulsion purification, *J. Colloid Interface Sci.* **1991**, *147*, 474–478.

[33] Timm, E.E.; Leng, D.E. Process for preparing uniformly sized polymer particles by suspension polymerization by vibratorily excited monomers in a gaseous or liquid stream, U.S. Patent Number 4,623,706, **1986**.

[34] Yamano, Y.; Kagawa, Y.; Kim, K.H.; Ghotani, S. Stability and uniformity of oil droplets in preparation of O/W agar gel emulsion, *Food Sci. Technol. Int., Tsukuba, Jpn.* **1996**, *2*(1), 16–18. *C.A.* **1996**, *125*, 140842a.

[35] Mason, T.G.; Bibette, J. Emulsification in viscoelastic media, *Phys. Rev. Lett.* **1996**, *77*, 3481–3484.

[36] Mabille, C.; Schmitt, V.; Gorria, Ph.; Calderon, F. Leal; Faye, V.; Deminière, B.; Bibette, J. Rheological and shearing conditions for the preparation of monodisperse emulsions, *Langmuir* **2000**, *16*, 422–499.

[37] Perrin, P. A one-step method to create and immobilize the ordered structures of fluid dispersions, *Langmuir* **2000**, *16*, 4774–4778.

[38] Derjaguin, B.V.; Martynov, G.A.; Gutop, Yu. V. Thermodynamics and stability of free films, *Colloid J. USSR* (Engl. Transl.) **1965**, *27*, 298–305.

[39] Mysels, K.J. The direct measurement of the Debye length, *Phys. Chem.: Enriching Top. Colloid Surf. Sci.* **1975**, 73–86.

[40] Clark, H.; Ross, S. Two-dimensional phase transition of ethane on sodium chloride, *J. Am. Chem. Soc.* **1953**, *75*, 60–81.

[41] Ross, S.; Clark, H. On physical adsorption. VI. Two-dimensional critical phenomena of xenon, methane and ethane adsorbed separately on sodium chloride, *J. Am. Chem. Soc.* **1954**, *76*, 4291–4297.

[42] Ross, S.; Hinchen, J.J. The evaluation and production of homotattic solid substrates of certain alkali halides, in *Clean Surf.: Their Prep. Charact. Interfacial Stud.*; Goldfinger, G., Ed; Marcel Dekker: New York; 1970; pp 115–132.

[43] Ross, S.; Olivier, J.P. *On physical adsorption*; Wiley: New York; 1964.

[44] Morrison, I.D.; Ross, S. The second and third virial coefficients of a two-dimensional gas, *Surf. Sci.* **1973**, *39*, 21–36.

[45] Ross, S.; Morrison, I.D. Computed adsorptive-energy distribution in the monolayer (CAEDMON), *Surf. Sci.* **1975**, *52*, 103–119.

[46] Russel, W.B. *The dynamics of colloidal systems*; University of Wisconsin Press: Madison, WI; 1987; pp 3–9.

[47] Russel, W.B.; Saville, D.A.; Schowalter, W.R. *Colloidal dispersions*; Cambridge University Press: New York; 1989; Chapter 10. (Equilibrium phase behavior).

[48]Fitch loc. cit.; Chapter 9 and esp. initial color plates.

[49]Velev, O.D.; Lenhoff, A.M.; Kaler, E.W. A class of microstructured particles through colloidal crystallization, *Science* **2000**, *287*, 2240–2243.

[50]Velev, O.D.; Kaler, E.W. Structured porous materials via colloidal crystal templating: from inorganic oxides to metals, *Adv. Mater.* **2000**, *12*, 531–534.

[51]Velev, O.D.; Lenhoff, A.M. Colloidal crystals as templates for porous materials, *Cur. Opinion Colloid Interface Sci.* **2000**, *5*, 56–63.

[52]Liu, W.; Zhang, Z.; Chen, S.; Xue, Q. The research and application of colloids as lubricants, *J. Dispersion Sci. Tech.* **2000**, *21*, 469–490.

[53]Hays, D.A.; Morrison, I.D.; Smith, L.S. Role of particles and dispersions in electrophotography, *Part. Sci. Tech.* **1987**, *5*, 39–51.

[54]Scharfe, M. *Electrophotography: principles and optimization*; Wiley: New York; 1984.

[55]Gibbs, W.W. The reinvention of paper, *Scientific American* **1998**, September, 36–37; www.gyricon.com.

[56]Comiskey, B.; Albert, J.D.; Yoshizawa, H.; Jacobson, J. An electrophoretic ink for all-printed reflective electronic displays, *Nature* **1998**, *394*, 253–255. www.eink.com.

[57]Siegel, R.W.; Hu, E.; Roco, M.C. *Nanostructure science and technology, a worldwide study*, published for the National Science Foundation by Intern. Tech. Res. Inst., Loyola College: MD; 1999 available at http://itri.loyola.edu/nano/IWGN.Worldwide.Study/.

[58]Amato, I. *Nanotechnology: shaping the world atom by atom*, published for the National Science Foundation by Intern. Tech. Res. Inst., Loyola College: MD; 1999 available at http://itri.loyola.edu/nano/IWGN.Public.Brochure.

[59]Lyon, L.A.; Peña, D.J.; Natan, M.J. Surface plasmon resonance of Au colloid-modified Au films: Particle size dependence, *J. Phys. Chem. B* **1999**, *103*, 5826–5831.

[60]Shim, M.; Guyot-Sionnest, P. n-Type colloidal semiconductor nanocrystals, *Nature* **2000**, *407*, 981–983.

[61]Hu, E.L.; Shaw, D.T. in Siegel loc. cit. (Ref. 57); Chapter 2: Synthesis and assembly.

[62]Suslick, K.S.; Hyeon, T.; Fang, M. Nanostructured materials generated by high-intensity ultrasound, *Chem. Mater.* **1996**, *8*, 2172–2179.

[63]Brown, K.R.; Natan, M.J. Hydroxylamine seeding of colloidal Au nanoparticles in solution and on surfaces, *Langmuir* **1998**, *14*, 726–728.

[64]Brown, L.O.; Hutchison, J.E. Controlled growth of gold nanoparticles during ligand exchange, *J. Am. Chem. Soc.* **1999**, *121*, 882–883.

[65]Brust, M.; Walker, M.; Bethell, D.; Schiffrin, D.J.; Whyman, R. Synthesis of thiol-derivatised gold nanoparticles in a two-phase liquid–liquid system, *J. Chem. Soc., Chem. Comm.* **1994**, 801–802.

[66]Leff, D.V.; Brandt, L.; Heath, J.R. Synthesis and characterization of hydrophobic, organically soluble gold nanocrystals functionalized with primary amines, *Langmuir* **1996**, *12*(20), 4723–4730.

[67]Yee, C.; Scotti, M.; Ulman, A.; White, H.; Rafailovich, M.; Sokolov, J. One-phase synthesis of thiol-functionalized platinum nanoparticles, *Langmuir* **1999**, *15*, 4314–4316.

[68]Norris, D.J.; Bawendi, M.G. Measurement and assignment of the size-dependent optical spectrum in CdSe quantum dots, *Phys. Rev. B.* **1996**, *53*(24), 16338–16346.

[69]Norris, D.J.; Efros, A.L.; Rosen, M.; Bawendi, M.G. Size dependence of exciton fine structure in CdSe quantum dots, *Phys. Rev. B.* **1996**, *53*(24), 16347–16354.

70. Murray, C.B.; Norris, D.J.; Bawendi, M.G. Synthesis and characterization of nearly monodisperse CdE (E = S, Se, Te) semiconductor nanocrystallites, *J. Am. Chem. Soc.* **1993**, *115*, 8706–8715.

71. Dabbousi, B.O.; Rodriguea-Viejo, J.; Mikulec, F.V.; Heine, J.R.; Mattoussi, H.; Ober, R.; Jensen, K.F.; Bawendi, M.G. (CdSe)ZnS core-shell quantum dots: synthesis and characterization of a size series of highly luminescent nanocrystallites, *J. Phys. Chem. B.* **1997**, *101*, 9463–9475.

72. Haynes, C.L.; van Duyne, R.P. Nanosphere lithography: a versatile nanofabrication tool for studies of size-dependent nanoparticle optics, *J. Phys. Chem. B* **2001**, *105*, 5599–5611.

73. Oldenburg, S.J.; Averitt, R.D.; Westcott, S.L.; Halas, N.J. Nanoengineering of optical resonances, *Chem. Phys. Lett.* **1998**, *288*, 243–247.

74. Schmid, G.; Bäumle. M.; Geerkens, M.; Heim, I.; Osemann, C.; Sawitowski, T. Current and future applications of nanoclusters, *Chem. Soc. Rev.* **1999**, *28*, 179–185.

75. Burst, M.; Bethell, D.; Schiffrin, D.J.; Kiely, C.J. Novel gold-dithiol nano-networks with nonmetallic electronic properties, *Adv. Mater.* **1995**, *7*(9), 795–797.

76. Collier, C.P.; Saykally, R.J.; Shiang, J.J.; Henrichs, S.E.; Heath, J.R. Reversible tuning of silver quantum dot monolayers through the metal-insulator transition, *Science* **1997**, *277*, 1978–1981.

77. Marinakos, S.M.; Brousseau, L.C., III: Jones, A.; Feldheim, D.L. Template synthesis of one-dimensional Au, Au-poly(pyrole), and poly(pyrole) nanoparticle arrays, *Chem. Mater.* **1998**, *10*, 1214–1219.

78. Iijima, S. Helical microtubes of graphite, *Nature* **1991**, *354*, 56–58.

79. Westcott, S.L.; Oldenburg, S.J.; Lee, T.R.; Halas, N.J., Formation and adsorption of clusters of gold nanoparticles onto functionalized silica nanoparticle surfaces, *Langmuir* **1998**, *14*, 5396–5401.

80. Zehner, R.W.; Lopes, W.A.; Morkved, T.L.; Jaeger, H.; Sita, L.R. Selective decoration of a phase-separated diblock copolymer with thiol-passivated gold nanocrystals, *Langmuir* **1998**, *14*(2), 241–244.

81. Leibowitz, F.L.; Zheng, W.; Maye, M.M.; Chuan-Jian, Z. Structures and properties of nanoparticle thin films formed via a one-step exchange-cross-linking-precipitation route, *Anal. Chem.* **1999**, *71*, 5076–5083.

82. Mirkin, C.A.; Letsinger, R.L.; Mucic, R.C.; Stornoff, J.J. A DNA-based method for rationally assembling nanoparticles into macroscopic materials, *Nature* **1996**, *382*, 607–609.

83. Murray, C.B.; Kagan, C.R.; Bawendi, M.G. Self-organization of CdSe nanocrystallites into three-dimensional quantum dot superlattices, *Science* **1995**, *270*, 1335–1338.

84. Kreibig, U.; Genzel, L. Optical absorption of a small metallic particles, *Surf. Sci.* **1985**, *156*, 678–700.

85. Elghanian, R.; Storhoff, J.J.; Mucic, R.C.; Letsinger, R.L.; Mirkin, C.A. Selective colorimetric detection of polynucleotides based on the distance-dependent optical properties of gold nanoparticles, *Science* **1997**, *277*, 1078–1081.

86. Storhoff, J.J.; Elghanian, R.; Mucic, R.C.; Mirkin, C.A.; Letsinger, R.L. One-pot colorimetric differentiation of polynucleotides with single base imperfections using gold nanoparticles, *J. Am. Chem. Soc.* **1998**, *120*, 1959–1964.

[87] Storhoff, J.J.; Mirkin, C.A. Programmed synthesis with DNA, *Chem. Rev.* **1999**, *99*, 1849–1862.

[88] Englebienne, P. Synthetic materials capable of reporting biomolecular recognition events by chromic transition, *J. Mater. Chem.* **1999**, *9*, 1043–1054.

[89] Lyon, L.A.; Musick, M.D.; Natan, M.J. Colloidal Au-enhanced surface plasmon resonance immunosensing, *Anal. Chem.* **1998**, *70*, 5177–5183.

[90] Lyon, L.A.; Holliway, W.D.; Natan, M.J. An improved surface plasmon resonance imaging apparatus, *Rev. Sci. Instr..* **1999**, *70*(4), 2076–2081.

[91] Ridley, B.A.; Nivi, B.; Jacobson, J.M. All-inorganic field effect transistors fabricated by printing, *Science* **1999**, *286*, 746–749.

[92] Goldstein, A.N.; Echer, C.M.; Alivisatos, A.P. Melting in semiconductor nanocrystals, *Science* **1992**, *256*, 1425–1427.

[93] Klimov, V.I.; Mikhailovsky, A.A.; Xu, S.; Malko, A.; Hollingsworth, J.A.; Leatherdale, C.A.; Eisler, H.J.; Bawendi, M.G. Optical gain and stimulated emission in nanocrystal quantum dots, *Science* **2000**, *290*, 314–317.

26 Appendices

26.1 PHYSICAL CONSTANTS

c	2.997925×10^8 m s^{-1}	Speed of light in vacuum
ε_0	8.854188×10^{-12} C V^{-1} m^{-1}	Permittivity of free space
e	1.602176×10^{-19} C	Elementary charge
g	9.807 m s^{-2}	Acceleration due to gravity
h	6.626069×10^{-34} J s	Planck's constant
k	1.380650×10^{-23} J K^{-1}	Boltzmann constant
R	8.314472 J K^{-1} mol^{-1}	Molar gas constant
	1.98717 cal K^{-1}mol^{-1}	
	82.0575 cm^3 atm K^{-1} mol^{-1}	
	0.0820575 L atm K^{-1}mol^{-1}	
	6.23637×10^4 cm^3 torr K^{-1}mol^{-1}	
F	9.648534×10^4 C mol^{-1}	Faraday constant
N_0	6.022142×10^{23} mole^{-1}	Avogadro's number

26.2 UNITS

1 Ångstrom	$= 0.1$ nm
	$= 10^{-10}$ m
1 atmosphere	$= 101.325$ kPa
1 calorie	$= 4.184$ J
	$= 2.613 \times 10^{19}$ eV
0 degrees Celsius	$= 273.16$ K
1 centiPoise	$= 10^{-2}$ dyn s cm^{-2}
	$= 10^{-3}$ N s m^{-2}
	$= 1$ mPa s
1 Debye	$= 3.3349 \times 10^{-30}$ C m
1 dyne	$= 1 \times 10^{-5}$ N
1 dyne per centimeter	$= 1$ erg cm^{-2}
	$= 1$ mJ m^{-2}
	$= 1$ mN m^{-1}
1 dyne per square centimeter	$= 9.869 \times 10^{-7}$ atmosphere
	$= 1 \times 10^{-8}$ bar
	$= 1 \times 10^{-4}$ kPa

1 electron Volt	$= 1.602176 \times 10^{-19}$ J
1 erg	$= 10^{-7}$ J
1 inch	$= 0.0254$ m
1 mile	$= 1.609$ km
1 Poise	$= 0.1$ Pa s
1 Printer's point	$= 3.515 \times 10^{-4}$ m
Radial frequency	$= 2\pi \times$ frequency ($\omega = 2\pi\nu$)
1 statcoulomb	$= 3.336 \times 10^{-10}$ C
1 Stoke	$= 1$ Poise/density of fluid
	$= 1 \times 10^{-4}$ m² s
1 torr	$= 0.1333$ kPa

26.3 MATHEMATICAL FORMULAE USED IN TEXT

$$\exp(x) = 1 + x + \frac{x^2}{2!} + \frac{x^3}{3!} +$$

$$\ln(1+x) = x - \frac{x^2}{2} + \frac{x^3}{3} - \frac{x^4}{4} +$$

$$\sinh(x) = \frac{\exp(x) - \exp(-x)}{2} = x + \frac{x^3}{3!} + \frac{x^5}{5!} +$$

$$\cosh(x) = \frac{\exp(x) + \exp(-x)}{2} = 1 + \frac{x^2}{2!} + \frac{x^4}{4!} +$$

$$\tanh(x) = \frac{\exp(x) - \exp(-x)}{\exp(x) + \exp(-x)} = x - \frac{x^3}{3} + \frac{2x^5}{15} - \frac{17x^7}{315} +$$

$$\int_0^\infty x\,dx \ln[1 - \Delta^2 \exp(-x)] = -\sum_{\nu=1}^\infty \frac{\Delta^2}{\nu^3}$$

The Laplacian operator ∇^2 is

$$\nabla^2 f = \frac{\partial^2 f}{\partial x^2} + \frac{\partial^2 f}{\partial y^2} + \frac{\partial^2 f}{\partial z^2} \quad \text{in Cartesian coordinates}$$

$$= \frac{\partial^2 f}{\partial x^2} \quad \text{in one-dimensional planar coordinates}$$

$$= \frac{\partial^2 f}{\partial r^2} + \frac{2}{r}\frac{\partial f}{\partial r} \quad \text{in one-dimensional spherical coordinates}$$

The principal radii of curvature of a surface in Cartesian coordinates are

$$R_1 = \frac{[1 + (dz/dx)^2]^{3/2}}{d^2z/dx^2} \qquad R_2 = \frac{x\left[1 + \left(\frac{dz}{dx}\right)^2\right]^{1/2}}{dz/dx}$$

26.4 ELECTROSTATIC AND INDUCTION CONTRIBUTIONS TO INTERMOLECULAR POTENTIAL ENERGIES[1]

Charge-charge	Angle and temperature independent	$\dfrac{q_a q_b}{4\pi\varepsilon_0 r}$
Charge-dipole	Averaged over all orientations	$-\dfrac{kT}{3}\dfrac{q_a^2 \mu_b^2}{(4\pi\varepsilon_0)^2 r^4}$
	At the maximum	$-\dfrac{q_a \mu_b}{4\pi\varepsilon_0 r^2}$
Charge-quadrupole	Averaged over all orientations	$-\dfrac{kT}{20}\dfrac{q_a^2 Q_b^2}{(4\pi\varepsilon_0)^2 r^6}$
Dipole-dipole	Averaged over all orientations	$-\dfrac{2kT}{3}\dfrac{\mu_a^2 \mu_b^2}{(4\pi\varepsilon_0)^2 r^6}$
	At the maximum	$-\dfrac{2\mu_a \mu_b}{4\pi\varepsilon_0 r^3}$
Charge-induced dipole	Angle and temperature independent	$-\dfrac{q_a^2 \alpha_b}{8\pi\varepsilon_0 r^4}$
Dipole-induced dipole	Averaged over all orientations	$-\dfrac{\mu_a^2 \alpha_b}{4\pi\varepsilon_0 r^6}$

where q_a = charge on particle a (C)
r = interparticle distance (m)
μ_a = dipole moment of particle a (C-m)
α_a = polarizability of particle a (m^3)
Q_a = quadrupole moment of particle a (C-m^3)
ε_0 = permittivity of free space

Molecular polarizabilities are often estimated from the sum of bond polarizabilities.

[1] Hirschfelder, J.O.; Curtiss, C.F.; Bird, R.B. *Molecular theory of gases and liquids*; John Wiley & Sons: New York; 1954; pp. 25–37.

26.5 ELECTRIC PROPERTIES OF REPRESENTATIVE MOLECULES[2]

	Charge, q (C)	Dipole Moment, μ (C-m)	Polarizability, α (m^3)	Ionization Potential (J)
Na$^+$	1.602×10^{-19}	—	—	—
Cl$^-$	1.602×10^{-19}	—	—	—
N$_2$	0	0	1.74×10^{-30}	2.53×10^{-18}
O$_2$	0	0	1.57×10^{-30}	2.18×10^{-18}
CO	0	3.3×10^{-31}	1.99×10^{-30}	2.29×10^{-18}
NH$_3$	0	5.0×10^{-30}	2.24×10^{-30}	1.87×10^{-18}
H$_2$O	0	6.1×10^{-30}	1.48×10^{-30}	2.01×10^{-18}
C$_6$H$_5$–OH	0	4.83×10^{-30}	—	—
C$_6$H$_6$	0	0	10.3×10^{-30}	1.54×10^{-18}
HCl	0	3.59×10^{-30}	26.3×10^{-31}	2.21×10^{-18}

26.6 LIFETIMES OF CONTRIBUTORS TO COLLOID AND INTERFACE SCIENCE

"Knowledge....

Her ample page, rich with the spoils of Time,"

— Gray's *Elegy*

"Our echoes roll from soul to soul,

And grow for ever and for ever."

— Tennyson

Robert Hook	1635–1702
Sir Isaac Newton	1642–1727
Francis Hauksbee	1666–1713
Janos András Segner	1704–1777
Benjamin Franklin	1706–1790
Pierre Simon de Laplace	1749–1827
Sir John Leslie	1766–1832
Thomas Young	1773–1829
Robert Brown	1773–1858
William Henry	1774–1836
Ferdinand Friedrich Reuss	1778–1852
Claude-Louis-Marie-Henri Navier	1785–1836
Michael Faraday	1791–1867
Pierre Hippolyte Boutigny	1798–1884
Franz Ernst Neumann	1798–1895

[2] Landolt-Bornstein, Volume 1, Part 3; Springer, 1951.

Jean-Louis-Marie Poiseuille	1799–1869
Joseph-Antoine-Ferdinand Plateau	1801–1883
Germaine Henri Hess	1802–1850
Thomas Graham	1805–1869
Francesco Selmi	1817–1881
Jules-Celestin Jamin	1818–1886
John Couch Adams	1819–1892
George Gabriel Stokes	1819–1903
Francis Bashforth	1819–1912
John Tyndall	1820–1893
Hermann von Helmholtz	1821–1894
Mathew Carey Lea	1823–1897
William Thomson, Lord Kelvin	1824–1907
Johann Wilhelm Hittorf	1824–1914
Jakob Maarten van Bemmelen	1830–1911
James Clerk Maxwell	1831–1879
Georg Hermann Quincke	1834–1924
Johannes Diderik van der Waals	1837–1923
Josiah Willard Gibbs	1839–1903
John Aitken	1839–1919
Pierre-Émile Duclaux	1840–1904
Carlo Giuseppe Matteo Marangoni	1840–1925
Ludwig Boltzmann	1844–1906
Osborne Reynolds	1842–1912
John William Strutt, Lord Rayleigh	1842–1919
Sir James Dewar	1842–1923
Joszef Fodor	1843–1901
Ludwig Boltzmann	1844–1906
Walthère Victor Spring	1848–1910
Otto Bütschli	1848–1920
Franz Hofmeister	1850–1922
Friedrich Wilhelm Ostwald	1853–1932
Louis-Georges Gouy	1854–1926
Sir Charles Vernon Boys	1851–1944
Edward Goodrich Acheson	1856–1931
Sir Joseph John Thomson	1856–1940
Percival S. U. Pickering	1858–1920
Pierre-Maurice-Marie Duhem	1861–1916
Wilson Taylor	1861–1923
Agnes Pockels	1862–1935
Henri-Edgard Devaux	1862–1956
Frederick Stanley Kipping	1863–1949
Sir William Bate Hardy	1864–1934
Hermann Walther Nernst	1864–1941
Richard Adolf Zsigmondy	1865–1929

Heinrich Jakob Bechold	1866–1937
Wilder Dwight Bancroft	1867–1953
Georg Bredig	1868–1944
Gustav Mie	1868–1957
Raphael Edward Liesegang	1869–1947
David Leonard Chapman	1869–1958
Jean Perrin	1870–1942
Frederick M. G. Donnan	1870–1956
Marion von Smoluchowski	1872–1917
Bogdan von Szyszkowski	1873–1931
William Draper Harkins	1873–1951
Gilbert Newton Lewis	1875–1946
Leonor Michaelis	1875–1949
Carl Axel Fredrik Benedicks	1875–1958
Jerome Alexander	1876–1959
Frans Maurits Jaeger	1877–1945
Robert Whytlaw-Gray	1877–1958
Maximilian Nierenstein	1878–1946
Eli Franklin Burton	1879–1948
Albert Einstein	1879–1955
Herbert Max Finley Freundlich	1880–1941
Anton Vladimirovich Dumanskii	1880–1967
Irving Langmuir	1881–1957
Hermann Staudinger	1881–1965
James William McBain	1882–1953
Hugo Rudolph Kruyt	1882–1959
Warren Kendall Lewis	1882–1975
Wolfgang Ostwald	1883–1943
Floyd Earl Bartell	1833–1961
Peter Joseph William Debye	1884–1966
Theodore Svedberg	1884–1971
Hendrik Sjoerd van Klooster	1884–1972
Harry Boyer Weiser	1887–1950
Otto Stern	1888–1969
Sir Chandrasekhara Venkata Raman	1888–1971
Sir Hugh Stott Taylor	1890–1974
Sir Eric Keightley Rideal	1890–1974
Sergei Sergeevich Medvedev	1891–1970
Nell Kensington Adam	1891–1973
Michael Polanyi	1891–1976
George Scatchard	1892–1974
Hendrik Gerard Bungenberg de Jong	1893–1977
Sir John Edward Lennard-Jones	1894–1954
Victor Kuhn La Mer	1895–1966
Willis Conway Pierce	1895–1974

Alexander Naumovich Frumkin	1895–1976
Nikolai Albertowich Fuchs	1895–1982
Aladár Buzágh	1895–1962
Per Ekwall	1895–1990
Ernst Alfred Hauser	1896–1956
Erich Armand Arthur Hückel	1896–1980
Petr Aleksandrovich Rehbinder	1898–1972
Katharine Burr Blodgett	1898–1979
Ralph Aionzo Beebe	1898–1979
Foster Dee Snell	1898–1980
John Warren Williams	1898–1988
Georg-Maria Schwab	1899–1984
Harry Herman Sobotka	1899–1965
Fritz London	1900–1954
Curtis R. Singleterry	1900–1971
Frank Thompson Gucker, Jr.	1900–1973
Edward Armand Guggenheim	1901–1970
Mikhaik M. Dubinin	1901–1993
A.S.C. [Stuart] Lawrence	1902–1971
Arne W. T. Tiselius	1902–1971
Boris Vladimirovich Derjaguin	1902–1994
Stephen Brunauer	1903–1986
Wilfried Heller	1903–1982
Jack Henry Schulman	1904–1967
Edward Alison Flood	1904–1979
Pierre van Rysselberghe	1905–1977
Evert Johannes Willem Verwey	1905–1981
William Albert Zisman	1905–1986
Bozo Tezak	1907–1980
Lev Davidovich Landau	1908–1968
Andrei Vladimirovich Kiselev	1908–1984
Winfred Oliver Milligan	1908–1984
Robert Donald Vold	1910–1978
Paul John Flory	1910–1985
Yuli M. Glazman	1911–2001
Marjorie Jean Vold	1913–1999
Albert Ernest Alexander	1914–1970
Myron L. Corrin	1914–1980
Stanley George Mason	1914–1987
Karol Joseph Mysels	1914–1998
William James Dunning	1915–1979
Dmitrii Aleksandrovich Fridrikhsberg	1915–1989
Frederick Mayhew Fowkes	1915–1990
Albert C. Zettlemoyer	1915–1991
Macarius van den Tempel	1919–1990

Alexi Scheludko	1920–1995
Akira Watanabe	1922–1980
Frank Chauncey Goodrich	1924–1980
John W. Vanderhoff	1925–1999
John F. Padday	1927–2000
Geoffrey Derek Parfitt	1928–1985
Russel S. Drago	1928–1997
J. Calvin Giddings	1930–1996
George L. Gaines, Jr.	1930–1996
John Michael Corkill	1931–1974
Raouf Shaker Mikkail	1931–1983
Henry Anton Resing	1933–1988
George Mathew Kanapilly	1934–1982

Living Scientists are not included. The authors are aware that this list has many blatant omissions. Some names deserving entry are missing because vital dates were not found. The authors welcome suggestions for additional names. They request that correspondents provide vital dates.

26.7 AWARDS

ACS Award in Colloid or Surface Chemistry

The ACS Award in Colloid or Surface Chemistry was originally sponsored by the Kendall Company (a subsidiary of the Colgate-Palmolive Co.) to recognize and encourage outstanding scientific contributions to colloid or surface chemistry in the United States and Canada. It is now sponsored by Proctor & Gamble Company.

1954	Harry N. Holmes
1955	John W. Williams
1956	Victor K. La Mer
1957	Peter J. W. Debye
1958	Paul H. Emmett
1959	Floyd E. Bartell
1960	John D. Ferry
1961	Stephen Brunauer
1962	George Scatchard
1963	William A. Zisman
1964	Karol J. Mysels
1965	George D. Halsey, Jr.
1966	Robert S. Hansen
1967	Stanley G. Mason
1968	Albert C. Zettlemoyer
1969	Terrell L. Hill
1970	Jerome Vinograd

1971	Milton Kerker
1972	Egon Matijević
1973	Robert L. Burwell, Jr.
1974	W. Keith Hall
1975	Robert Gomer
1976	Robert J. Good
1977	Michel Boudart
1978	Harold A. Scheraga
1979	Arthur W. Adamson
1980	Howard Reiss
1981	Gabor A. Somorjai
1982	Gert Ehrlich
1983	Janos H. Fendler
1984	Brian E. Conway
1985	Stig E. Friberg
1986	Eli Ruckenstein
1987	John T. Yates, Jr.
1988	Howard Brenner
1989	Arthur T. Hubbard
1990	John M. White
1991	W. Henry Weinberg
1992	David G. Whitten
1993	D. Wayne Goodman
1994	J. Kerry Thomas
1995	Thomas Engel
1996	Theodorus van de Ven
1997	Harden M. McConnell
1998	Eric W. Kaler
1999	Nicholas J. Turro
2000	Darsh T. Wasan
2001	Charles T. Campbell
2002	Charles M. Knobler

Arthur W. Adamson Award for Distinguished Service in the Advancement of Surface Science

1993	David M. Hercules
1994	Gabor A. Somorjai
1995	W. Henry Weinberg
1996	Harden M. McConnell
1997	Robert J. Madix
1998	Kenneth B. Eisenthal
1999	John T. Yates, Jr.
2000	Alvin W. Czanderna
2001	J. Michael White
2002	D. Wayne Goodman

Bibliography

Acton, J.R.; Squire, P.T. *Solving equations with physical understanding*; Adam Hilger: Boston; 1985.

Adam, N.K. *The physics and chemistry of surfaces*, 3rd ed.; Oxford University Press: London; 1941.

Adamson, A.W. *Physical chemistry of surfaces*; Interscience Publishers: New York; 1st ed., 1960; 2nd ed., 1967; 4th ed., Wiley: New York; 1982; 5th ed., 1990; 6th ed. with A.P. Gast; 1997.

Akers, R.J., Ed. *Foams*; Academic Press: New York; 1976.

Alexander, A.E.; Johnson, P. *Colloid science*, in 2 volumes; Oxford University Press: London; 1949.

Alexander, A.L. *Interaction of liquids at solid substrates*; Adv. Chem. Ser. 87; American Chemical Society: Washington, DC; 1968.

Alexander, J. *Colloid chemistry: principles and applications*, 3rd ed.; van Nostrand: New York; 1929.

Alexander, J. *Colloidal chemistry, theory and applications*; Reinhold Publishing: New York; 1944.

Allen, T. *Particle size measurement*; Chapman and Hall: New York; 1st ed., 1968; 2nd ed., 1977; 3rd ed., 1981; 4th ed. 1990; 5th ed. in 2 vols., 1997.

Ariman, T.; Veziroglu, T.N., Eds. *Particulate and multiphase processes*, Vol. 2 Contamination analysis and control*; Hemisphere Publishing: New York; 1987.

Attwood, D.; Florence, A.T. *Surfactant systems, their chemistry, pharmacy, and biology*; Chapman and Hall: New York; 1983.

Aveyard, R.; Haydon, D.A. *An introduction to the principles of surface chemistry*; Cambridge University Press: London; 1973.

Bakker, G. *Kapillarität und Oberflächspannung*; Akademische Verlagsgesellschaft: Leipzig; 1928.

Bancroft, W.D. *Applied colloid chemistry, general theory*, 1st ed.; McGraw-Hill: New York; 1921.

Bancroft, W.D. *Applied colloid chemistry, general theory*, 2nd ed.; McGraw-Hill: New York; 1926.

Bancroft, W.D. *Applied colloid chemistry, general theory*, new ed.; McGraw-Hill: New York; 1932.

Barnes, H.A.; Hutton, J.F.; Walters, K. *An introduction to rheology*; Elsevier: New York; 1989.

Barth, H.G., Ed. *Modern methods of particle size analysis*; Wiley: New York; 1984.

Becher, P. *Emulsions: theory and practice*; Reinhold: New York; 1957.

Becher, P., Ed. *Encyclopedia of emulsion technology,* Vol. 1 *Basic theory*, 1983; Vol. 2 *Applications*, 1985; Vol. 3 *Basic theory, measurement, applications*, 1988; Vol. 4, 1996; Marcel Dekker: New York.

Becher, P. *Dictionary of colloid and surface science*; Marcel Dekker: New York; 1990.

Becher, P.; Yudenfreund, M.N., Eds. *Emulsions, latices, and dispersions*; Marcel Dekker: New York; 1978.

Berg, J.C., Ed., *Wettability*; Marcel Dekker: New York; 1993.

Bhatia, A.B. *Ultrasonic absorption; an introduction to the theory of sound absorption and dispersion in gases, liquids, and solids*; Oxford University Press: London; 1967; Dover: New York; 1985.

Bikerman, J.J. *Surface chemistry, theory and applications*; Academic Press: New York; 1958.

Bikerman, J.J. *Foams: theory and industrial applications*; Reinhold: New York; 1953.

Bikerman, J.J. *The science of adhesive joints*; Academic Press: New York; 1st ed., 1960; 2nd ed., 1968.

Bikerman, J.J. *Physical surfaces*; Academic Press: New York; 1970.

Bikerman, J.J. *Foams*; Springer: New York; 1973.

Bier, M., Ed. *Electrophoresis, theory, methods, and applications*; Academic Press: New York; Vol. 1., 1959; Vol. 2, 1997.

Bingham, E.C. *Fluidity and plasticity*; McGraw-Hill: New York; 1933.

Bird, R.B.; Stewart, W.E.; Lightfoot, E.N. *Transport phenomena*; Wiley: New York; 1960.

Boger, D.V.; Walters, K. *Rheological phenomena in focus*; Elsevier: New York; 1993.

Bogue, R.H., Ed. Vol. 1. *The theory of colloidal behavior,* Vol. 2 *The application of colloidal behavior*; McGraw-Hill: New York; 1924.

Bohren, C.F.; Huffman, D.R. *Absorption and scattering of light by small particles*; Wiley: New York; 1983.

Bohren, C.F. *Clouds in a glass of beer; simple experiments in atmospheric science*; Wiley: New York; 1987.

Boscovitch, R.J. *A theory of natural philosophy, put forward and explained by Roger Joseph Boscovich*; Open Court: Chicago; 1922.

Boothroyd, R.G. *Flowing gas-solid suspensions*; Chapman and Hall: London; 1971.

Boys, C.V. *Soap bubbles, their colors and forces which mold them*; Dover: New York; 1959.

Burk, R.E.; Grummitt, O., Eds. *Frontiers in colloid chemistry*; Interscience Publishers: New York; 1950.

Burke, J.J.; Reed, N.L.; Weiss, V., Eds. *Surface and interfaces I; Chemical and physical characteristics*; Syracuse University Press: Syracuse; 1967.

Burke, J.J.; Weiss, V. *Block and graft copolymers*; Syracuse University Press: Syracuse; 1973.

Buscall, R.; Corner, T.; Stageman, J.T., Eds. *Polymer colloids*; Elsevier: New York; 1985.

Brunauer, S. *The adsorption of gases and vapors,* Vol. 1 *Physical adsorption*; Princeton University Press: Princeton; 1943.

Burdon, R.S. *Surface tension and the spreading of liquids*; Cambridge University Press: London; 1940.

Burton, E.F. *The physical properties of colloidal solutions*; Longmans, Green: London; 1916.

Cadle, R.D. *Particle size determination*; Interscience Publishers: New York; 1955.

Chattoraj, D.K.; Birdi, K.S. *Adsorption and the Gibbs surface excess*; Plenum Press: New York; 1984.

Christian, S.D.; Scamehorn, J.F., Eds. *Solubilization in surfactant aggregates*; Marcel Dekker: New York; 1995.

Chu, B. *Laser light scattering*, 2nd ed.; Academic Press: New York; 1991,

Clark, A. *The theory of adsorption and catalysis*; Academic Press: New York; 1970.

Clayton, W. *The theory of emulsions and their technical treatment*, 3rd ed.; J. & A. Churchill: London; 1935.

Clayton, W., Ed. *Wetting and detergency*; A. Harvey: London; 1937.

Cohen, E.D.; Gutoff, E.B. *Modern coating and drying technology*; VCH Publishers: New York; 1992.

Conley, R.F. *Practical dispersion: a guide to understanding and formulating slurries*; VCH Publishers: New York; 1996.

Copeland, L.E. *Solid surface and the gas-solid interface*; *Adv. Chem. Ser.* 33; ACS: Washington, DC; 1961.

Cunningham, F.F. *James David Forbes: Pioneer Scottish glaciologist*; Scottish Academic Press: Edinburgh; 1990.

Dahneke, B.E., Ed. *Measurement of suspended particles by quasi-elastic light scattering*; Wiley: New York; 1983.

Dallavalle, J.M. *Micromeritics, the technology of fine particles*, 2nd ed.; Pitman Publishing: New York; 1948.

Danielli, J.F.; Pankhurst, K.G.A.; Riddiford, A.C., Eds. *Surface phenomena in chemistry and biology*; Pergamon Press: New York; 1958.

D'Arrigo, J.S. *Stable gas-in-liquid emulsions, Production in natural waters and artificial media*; Elsevier: New York; 1986.

Davies, J.T.; Rideal, E.K. *Interfacial phenomena*; Academic Press: New York; 2nd ed., 1963.

Davis, H.T. *Statistical mechanics of phases, interfaces, and thin films*; VCH Publishers: New York; 1996.

Dean, R.B. *Modern colloids*; D. van Nostrand: New York; 1948.

De Boer, J.H. *The dynamical character of adsorption*; Oxford University Press: London; 1953.

Defay, R.; Prigogine, I.; Bellemans, A. *Surface tension and adsorption*; Everett, D.H., Trans.; Longmans, Green: London; 1966.

De Gennes, P.G. *Soft interfaces — The 1994 Dirac memorial lecture*; Cambridge University Press: New York; 1997.

Derjaguin, B.V.; Churaev, N.V.; Muller, V.M. *Surface forces*; Kisin, V.I., Trans.; Kitchener, J.A., Trans. Ed.; Plenum: New York; 1987.

Derjaguin, B.V. *Theory of stability of colloids and thin films*; Johnston, R.K., Trans.; Plenum Publishing: New York; 1989.

Dickenson, E. *An introduction to food colloids*; Oxford University Press: New York; 1992.

Dickenson, E.; McClements, D.J.; *Advances in food colloids*; Chapman and Hall: New York; 1996.

Dobiáš, B.; Qiu, X.; von Rybinski, W. *Solid-liquid dispersions*; Marcel Dekker: New York; 1999.

Drzaic, P.S. *Liquid crystal dispersions*; World Scientific Publishing: New Jersey; 1995.

Dubin, P.; Tong, P., Eds. *Colloid-polymer interactions*; ACS Symp. Ser., 532; ACS; Washington, DC; 1993.

Du Noüy, P.L. *Surface equilibria of biological and organic colloids*; ACS, The Chemical Catalog Co.: New York; 1926.

Edwards, D.A., Brenner, H.; Wasan, D.T. *Interfacial transport processes and rheology*; Butterworth-Heineman: Stoneham, MA; 1991.

Einstein, A. *Investigations on the theory of the Brownian movement*; notes by Fürth, R.; Cowper, A.D., Trans.; Dover: New York; 1956.

Ekwall, P.; Groth, K.; Runnström-Reio, V., Eds. *Surface chemistry*; Academic Press: New York; 1965.

Elimelech, M.; Gregory, J.; Jia, X.; Williams, R.A. *Particle deposition and aggregation; measurement, modeling, and simulation*; Butterworth-Heinemann: London; 1995.

Elworthy, P.H.; Florence, A.T.; Macfarlane, C.B. *Solubilization by surface-active agents*; Chapman and Hall: London; 1968.

Evans, D.F.; Wennerström, H. *The colloidal domain; where physics, chemistry, biology, and technology meet*; VCH Publishers: New York; 1994.

Everett, D.H.; Ottewill, R.H., Eds. *Surface area determination*; Butterworths: London; 1970.

Everett, D.H. *Basic principles of colloid science*; Royal Society of Chemistry: London; 1988.

Fendler, J.H.; Fendler, E.J. *Catalysis in micellar and macromolecular systems*; Academic Press: New York; 1975.

Fischer, E.K. *Colloidal dispersions*; Wiley: New York; 1950.

Fitch, R.M., Ed. *Polymer colloids*; Plenum Press: New York; 1971.

Fitch, R.M. *Polymer colloids: a comprehensive introduction*; Academic Press: New York; 1987.

Flick, E. W. *Industrial surfactants*; Noyes Publications: Park Ridge, NJ; 2nd ed. 1993.

Flood, E.A., Ed. *The solid-gas interface in 2 Vols.*; Marcel Dekker: New York; 1967.

Flory, P.J. *Principles of polymer chemistry*; Cornell University Press: Ithaca; 1953.

Fort, T.; Mysels, K.J., Eds. *Eighteen years of colloid and surface chemistry*; ACS: Washington, DC; 1991.

Fowkes, F.M., Ed. *Contact angle, wettability, and adhesion*; Adv. Chem. Ser. 43; American Chemical Society: Washington, DC; 1964.

Franks, F. *Polywater*; MIT Press: Cambridge, MA; 1981.

Franks, F. *Water*; Royal Soc. Chem.: London; 1983.

Freundlich, H. *The elements of colloidal chemistry*, Barger, G., Trans.; E.F. Dutton: New York; 1925.

Freundlich, H. *Colloid and capillary chemistry*; Hatfield, H.S., Trans.; Methuen: London; 1926.

Freundlich, H. *New conceptions in colloidal chemistry*; Methuen: London; 1926.

Friberg, S., Ed. *Lyotropic liquid crystals*; Adv. Chem. Ser. 152; American Chemical Society: Washington, DC; 1976.

Fridrikhsberg, D.A. *A course in colloid chemistry*; Leib, G., Trans.; Mir: Moscow; 1986.

Fuchs, N.A. *The mechanics of aerosols*, Daisley, R.E.; Fuchs, M., Trans., Davies, C.N., Trans. Ed.; Pergamon Press: New York; 1964.

Gaines, G.L., Jr. *Insoluble monolayers at liquid-gas interfaces*; Interscience Publishers: New York; 1966.

Garbassi, F.; Morra, M.; Occhiello, E. *Polymer surfaces: from physics to technology*; Wiley: New York; 1994.

Gard, J.A., Ed. *The electron-optical investigation of clays*; Mineralogical society: London; 1971.

Garrett, P.R., Ed. *Defoamimg: theory and industrial applications*; Surfactant Science Series, Vol. 45; Marcel Dekker: New York; 1993.

Gauss, C.F. *Allgemeine Grundlagen einer Theorie der Gestalt von Flüssigkeiten im Zustand des Gleichgewichts* (General basis of a theory of the shape of liquids in the equilibrium state); Wilhelm Engelmann: Leipzig; 1903.

Gibbs, J.W. *Scientific papers of J. Willard Gibbs: Thermodynamics*;Longmans, Green: London; 1906; Dover: New York; 1961.

Goddard, E.D., Ed. *Monolayers*; Adv. Chem. Ser. 144; American Chemical Society: Washington, DC; 1975.

Goddard, E.D.; Vincent, B., Eds. *Polymer adsorption and dispersion stability*; ACS Symp. Ser. 240; American Chemical Society: Washington, DC; 1984.

Good, R.J.; Stromberg, R.R.; Patrick, R.L., Eds. *Techniques of surface and colloid chemistry and physics*, Vol. 1; Marcel Dekker: New York; 1972.

Goodwin, J.W., Ed. *Colloidal dispersions*; Royal Soc. Chem.: London; 1982.

Goodwin, J.W.; Buscall, R., Eds. *Colloidal polymer particles*; Academic Press: New York; 1995.

Goodwin, J.W.; Hughes, R.W. *Rheology for chemists*; Royal Society of Chemistry: Cambridge; 2000.

Gregg, S.J. *The adsorption of gases by solids*; Methuen: London; 1934.

Gregg, S.J. *The surface chemistry of solids*; Whitefriars Press: London; 1st ed., 1951; 2nd ed., 1961.

Gregg, S.J.; Sing, K.S.W. *Adsorption, surface area, and porosity*; Academic Press: New York; 1967.

Gregg, S.J.; Sing, K.S.W. *Adsorption, surface area and porosity*, 2nd ed.; Academic Press: New York; 1982.

Groves, M.J.; Wyatt-Sargent, J.L., Eds. *Particle size analysis*; Soc. Anal. Chem.: London; 1972.

Hair, M.L. *Infrared spectroscopy in surface chemistry*; Marcel Dekker: New York; 1967.

Hair, M.; Croucher, M.D., Eds. *Colloids and surfaces in reprographic technology*; ACS Symp. Ser. 200; American Chemical Society: Washington, DC; 1982.

Hardy, W.B. *Collected scientific papers of Sir William Bate Hardy*; Cambridge University Press: London; 1936.

Harkins, W.D. *The physical chemistry of surface films*; Reinhold: New York; 1952.

Harper, W.R. *Contact and frictional electrification*; Oxford University Press: London; 1967.

Hartland, S.; Hartley, R.W. *Axisymmetric fluid-liquid interfaces*; Elsevier: New York; 1976.

Hatschek, E. *An introduction to the physics and chemistry of colloids*; J. & A. Churchill: London; 1916.

Hatschek, E. *An introduction to the physics and chemistry of colloids*; P. Blakiston's Son: Philadelphia; 1919.

Hatschek, E. *Laboratory manual of elementary colloid chemistry*; P. Blakiston's Son: Philadelphia; 1920.

Hatschek, E., Ed. *The foundations of colloid chemistry, a selection on early papers bearing on the subject*; Ernest Benn: London; 1925.

Hauser, E.A. *Colloidal phenomena, an introduction to the science of colloids*; McGraw-Hill: New York; 1939.

Hauser, E.A.; Lynn, J.E. *Experiments in colloid chemistry*; McGraw-Hill: New York; 1940.

Hedges, E.S. *Colloids*; Edward Arnold: London; 1931.

Heicklen, J. *Colloid formation and growth, a chemical kinetics approach*; Academic Press: New York; 1976.

Helfrich, W.; Heppke, G., Eds. *Liquid crystals of one- and two-dimensional order*; Springer: New York; 1980.

Herdan, G. *Small particle statistics, An account of statistical methods for the investigation of finely divided materials*; Elsevier: New York; 1953.

Hermans, J.J., Ed. *Flow properties of disperse systems*; Interscience: New York; 1953.

Hiemenz, P.C. *Principles of colloid and surface chemistry*; Marcel Dekker: New York; 1977; 2nd ed., 1986; 3rd ed. with R. Rajagopalan, 1997.

Hirschfelder, J.O.; Curtiss, C.F.; Bird, R.B. *Molecular theory of gases and liquids*; Wiley: New York; 1954.

Hirtzel, C.S.; Rajagopalan, R. *Colloidal phenomena, advanced topics*; Noyes Publications: Park Ridge, NJ; 1985.

Holmes, H.N. *Introductory colloid chemistry*; Wiley: New York; 1934.

Holmes, H.N. *Laboratory manual of colloid chemistry*, 2nd ed.; Wiley: New York; 1928.

Hudson, J.B. *Surface science, an introduction*; Butterworth-Heinemann: Boston; 1992.

Hunter, R.J. *Zeta potential in colloid science, principles and applications*; Academic Press: New York; 1981.

Hunter, R.J. *Foundations of colloid science*, in 2 vols.; Oxford University Press: New York; 1987.

Hunter, R.J. *Introduction to modern colloid science*; Oxford University Press: New York; 1993.

Hunter, R.J. *Foundations of colloid science*, 2nd ed.; Oxford University Press: New York; 2001.

Iler, R.K. *The colloid chemistry of silica and silicates*; Cornell University Press: Ithaca; 1955.

Isenberg, C. *The science of soap films and soap bubbles*; 1st ed., Tieto:Avon; 1978; 2nd ed., Dover: New York; 1992.

Israelachvili, J. *Intermolecular and surface forces*; Academic Press: New York; 1st ed., 1985; 2nd ed., 1991.

James, G.V. *Water treatment, a survey of current methods of purifying domestic supplies and of treating industrial effluents and domestic sewage*; CRC Press: Cleveland; 1st ed., 1940; 2nd ed., 1949; 3rd ed., 1965; 4th ed., 1971.

Jaroniec, M.; Madey, R. *Physical adsorption on heterogeneous solids*; Elsevier: New York; 1988.

Jasper, J.J. The surface tension of pure liquid compounds, *J. Phys. Chem. Ref. Data,* **1972**, *1*(4), 841–1010.

Jaycock, M.J.; Parfitt, G.D. *Chemistry of interfaces*; Wiley: New York; 1981.

Jelínek, Z.K. *Particle size analysis*; Wiley: New York; 1970.

Jenson, W.B. *The Lewis acid–base concepts*; Wiley: New York; 1980.

Jirgensons, B. *Organic colloids*; Elsevier: New York: 1958.

Jirgensons, B.; Straumanis, M.E. *A short textbook of colloid chemistry*; MacMillan: New York; 1st ed., 1954; 2nd ed., 1962.

Joos, P. *Dynamic surface phenomena*; VSP: Utrecht, The Netherlands; 1999.

Karsa, D.R., Ed. *Industrial applications of surfactants II*; Roy. Soc. Chem.: Cambridge; 1990.

Kaye, B.H. *Direct characterization of fine particles*; Wiley: New York; 1981.

Kerker, M. *The scattering of light and other electromagnetic radiation*; Academic Press: New York; 1969.

Kipling, J.J. *Adsorption from solutions of non-electrolytes*; Academic Press: New York; 1965.

Kiselev, A.V.; Yashin, Y.I. *Gas-adsorption chromatography*, Bradley, J.E.S., Trans.; Plenum Press: New York; 1969.

Kissa, E. *Dispersions: characterization, testing, and measurement*; Marcel Dekker: New York; 1999.

Kitahara, A.; Watanabe, A., Eds. *Electrical phenomena at interfaces, fundamentals, measurements, and applications*; Marcel Dekker: New York, 1984.

Klinkenberg, A.; van der Minne, J.L. *Electrostatics in the petroleum industry: the prevention of explosion hazards*; Elsevier: New York; 1958.

Krotov, V.V.; Rusanov, A.I. *Physicochemical hydrodynamics of capillary systems*; Imperial College Press: London; 1999.

Kruyt, H.R. *Colloids, a textbook*, van Klooster, H.S., Trans.; Wiley: New York; 1930.

Kruyt, H.R., Ed. *Colloid science*, Vol. 1 *Irreversible systems*; Elsevier: New York; 1952.

Kruyt, H.R., Ed. *Colloid science*, Vol. 2 *Reversible systems*; Elsevier: New York; 1949.

Langmuir, I. *The collected works of Irving Langmuir*, Vol. 9 *Surface phenomenon*; Pergamon Press: New York; 1961.

Larson, R.G. *The structure and rheology of complex fluids*; Oxford University Press: New York; 1999.

Laughlin, R.G. *The aqueous phase behavior of surfactants*; Academic Press: New York; 1994.

Lawrence, A.S.C. *Soap films, a study of molecular individuality*; G. Bell: London; 1929.

Lee, L.-H., Ed. *Adhesive chemistry; developments and trends*; Plenum Press: New York; 1984.

Levich, V.G. *Physicochemical hydrodynamics*; Prentice Hall: Englewood Cliffs, NJ; 1962.

Lewis, W.K.; Squires, L.; Broughton, G. *Industrial chemistry of colloidal and amorphous materials*; MacMillian: New York; 1948.

Lissant, K.J., Ed. *Emulsions and emulsion technology*; Marcel Dekker: New York; Parts 1 and 2, 1974; Part 3, 1984.

Lliboutry, L.A. *Very slow flows of solids*; Kluwer: Dordrecht; 1987.

Loeb, A.L.; Overbeek, J.Th.G.; Wiersema, P.H. *The electrical double layer around a spherical colloid particle*. MIT Press: Cambridge, MA; 1961.

Lowell, S. *Introduction to powder surface area*; Wiley: New York; 1st ed., 1979.

Lowell, S.; Shields, J.E. *Powder surface area and porosity*; Chapman and Hall: New York; 2nd ed., 1984.

Lucassen-Reynders, E.H., Ed. *Anionic surfactants*; Marcel Dekker: New York; 1981.

Lyklema, J. *Fundamentals of interface and colloid science,* Vol. 1: *Fundamentals; 1991,* Vol. 2*: Solid-liquid interfaces; 1995,* Vol 3*: Liquid-fluid interfaces*; 2000; Academic Press: New York.

Macosko, C.W. *Rheology: principles, measurement, and application*; VCH Publishers: New York; 1994.

Mahanty, J.; Ninham, B.W. *Dispersion forces*; Academic Press: New York; 1976.

Malghan, S.G., Ed. *Electroacoustics for characterization of particulates and suspensions*; NIST Spec. Pub. 856; U.S. Gov. Printing Office: Washington, DC; 1993.

Mandelbrot, B.B. *Fractals: form, chance, and dimension*; W.H. Freeman: San Francisco; 1983.

Manegold, E. *Emulsionen*; Straßenbau, Chemie und Technik: Heidelberg; 1952.

Manegold, E. *Schaum*; Straßenbau, Chemie und Technik: Heidelberg; 1953.

Manegold, E. *Kapillarsysteme, Band 1 (Grundlagen)*; Straßenbau, Chemie und Technik: Heidelberg; 1955.

Manegold, E. *Kapillarsysteme, Band 2 (Anwendungen)*; Straßenbau, Chemie und Technik: Heidelberg; 1960.

Maxwell, J.C. *Electricity and magnetism*, in *2* vols.; Clarendon Press: Oxford; 1873.

Maxwell, J.C. *A treatise on electricity and magnetism.*, 3rd ed., Clarendon Press: London; 1891; Vol. 1; Dover: New York; 1954.

McBain, J.W. *The sorption of gases and vapors by solids*; George Routledge: London; 1932.

McBain, J.W. *Colloid science*; D.C. Heath: Boston; 1950.

McCutcheon's: *Emulsifiers & Detergents*, American Edition, MC Publishing: Glen Rock, NJ; (An annual publication.)

McKay, R.B., Ed. *Technical applications of dispersions*; Marcel Dekker: New York; 1994.

McCrone, W.C.; Delly, J.G. *The particle size atlas*; Ann Arbor Science: Ann Arbor, MI; 1993.

Michaelis, L. *Practical physical and colloid chemistry*, Parsons, T.R., Trans.; W. Heffer: Cambridge; 1925.

Michaelis, L. *The effect of ions in colloidal systems*; Williams and Wilkins: Baltimore; 1925.

Mill, C.C., Ed. *Rheology of disperse systems*; Pergamon Press: London; 1959.

Miller, C.A.; Neogi, P. *Interfacial phenomena, equilibrium and dynamic effects*; Marcel Dekker: New York; 1985.

Miner, R.W., Ed. *Properties of surfaces*; New York Academy of Sciences, 58 (6), 721–970: New York; 1954.

Mittal, K.L., Ed. *Adsorption at interfaces*; ACS Symp. Ser. 8; American Chemical Society: Washington, DC; 1975.

Mittal, K.L., Ed. *Colloidal dispersions and micellar behavior*; ACS Symp. Ser. 9; ACS: Washington, DC; 1975.

Mittal, K.L.; Anderson, Jr, H.R., Eds. *Acid-base interactions: relevance to adhesion science and technology*; VSP: Utrecht; 1991.

Mittal, K.L., Ed. *Micellization, solubilization and microemulsions*; Plenum: New York, NY; 1977; Vol. 1 and 2.

Mittal, K.L., Ed. *Acid-base interactions; Relevance to adhesion science and technology*, Vol. 2; VSP: Utrecht; 2000.

Moilliet, J.L.; Collie, B. *Surface activity*; E.& F. N. Spon: London; 1951.

Moilliet, J.L.; Collie, B.; Black, W. *Surface activity, the physical chemistry, technical applications, and chemical constitution of synthetic surface-active agents*, 2nd ed.; van Nostrand: New York; 1961.

Moudgil, B.M.; Somasundaran, P.; Eds. *Dispersion and aggregation, fundamentals and applications*; Engineering Foundation: New York; 1994.

Mukerjee, P.; Mysels, K.J. *Critical micelle concentrations of aqueous surfactant systems*; Nat. Stand. Ref. Data Ser., 36; U.S. Government Printing Office: Washington, DC; 1971.

Myers, A.L.; Belfort, G., Eds. *Fundamentals of adsorption*; Engineering Foundation: New York; 1984.

Myers, D. *Surfactant science and technology*; VCH Publishers: New York; 1988; 2nd ed., 1992.

Myers, D. *Surfaces, interfaces, and colloids: Principles and applications*; VCH Publishers: New York; 1991; 2nd ed.; Wiley-VCH: New York; 1999.

Mysels, K.J.; Shinoda, K.; Frankel, S. *Soap films, studies of their thinning and a bibliography*; Pergamon Press: New York; 1959.

Mysels, K.J.; Samour, C.M.; Hollister, J.H., Eds. *Twenty years of colloid and surface chemistry, The Kendall Award addresses*; American Chemical Society: Washington, DC; 1973.

Mysels, K.J. *Introduction to colloid chemistry*; Robert F. Krieger Publishing: Huntington, NY; 1978.

Napper, D.H. *Polymeric stabilization of colloidal dispersions*; Academic Press: New York; 1983.

Nelson, R.D., Jr. *Dispersing powders in liquids*; Elsevier: New York; 1988.

Niven, W.W., Jr. *Fundamentals of detergency*; Reinhold: New York; 1950.

Niven, W.W., Jr. Ed. *Industrial detergency*; Reinhold: New York; 1955.

Nye, M.J. *Molecular reality: A perspective on the scientific work of Jean Perrin*; Elsevier Publishing: New York; 1972.

Orr, C., Jr; Dallavalle, J.M. *Fine particle measurement, size, surface, and pore volume*; MacMillan: New York; 1959.

Ościk, J. *Adsorption*, Cooper, I.L., Trans.; Ellis Horwood: Chichester; 1982.

Osipow, L.J. *Surface chemistry, theory and industrial applications*; Reinhold: New York; 1962.

Ostwald, W. *An introduction to theoretical and applied colloid chemistry, "The world of neglected dimensions."* Fischer, M.H., Trans.; Wiley: New York; 1917; 8th ed.; 1922.

Ostwald, W.; Wolski, P.; Kuhn, A. *Practical colloid chemistry*, Kugelmass, I.N.; Cleveland, T.K., Trans., 4th ed.; E.P. Dutton: New York; 1924.

Ottewill, R.H., Ed. *Wetting, A discussion covering both fundamental and applied aspects of the subject of wetting and wettability*; Soc. Chem. Ind. Monograph 25; Soc. Chem. Ind.: London; 1967.

Padday, J.F., Ed. *Wetting, spreading and adhesion*; Academic Press: New York; 1978.

Parfitt, G.D. *Principles of the colloidal state*; A monograph for teachers, No. 14; Royal Institute of Chemistry: London; 1967.

Parfitt, G.D. *Dispersion of powders in liquids*; Wiley: New York, 1st ed. 1969; 2nd ed., 1973.

Parfitt, G.D.; Sing, K.S.W., Eds. *Characterization of powder surfaces with special reference to pigments and fillers*; Academic Press: New York; 1976.

Patterson, D., Ed. *Pigments: An introduction to their physical chemistry*; Elsevier: New York; 1967.

Patton, T.C. *Paint flow and pigment dispersion, A rheological approach to coating and ink technology*, 2nd ed.; Wiley: New York; 1979.

Perrin, J. *Les atomes*; 5th ed.; Librairie Félix Alcan; 1914.

Petty, M.C.; *Langmuir–Blodgett films — an introduction*; Cambridge University Press: New York; 1996.

Pillai, V.; Shah, D.O., Eds. *Dynamic properties of interfaces and association structures*; AOCS Press: Champaign, IL; 1996.

Plateau, J. *Statique expérimentale et théorique des liquides soumis aux seules forces moléculaires*; in 2 vols.; Gauthier-Villars: Paris; 1873.

Popiel, W.J. *Introduction to colloid science*; Exposition Press: Hicksville, NY; 1978.

Porter, R.S.; Johnson, J.F., Eds. *Ordered fluids and liquid crystals*; Adv. Chem. Ser. 63; American Chemical Society: Washington, DC; 1967.

Povey, M.J.W. *Ultrasonic techniques for fluids characterization*; Academic Press: New York; 1997.

Provder, T., Ed. *Size exclusion chromatography (GPC)*; ACS Symp. Ser. 138; American Chemical Society: Washington, DC; 1980.

Provder, T., Ed. *Particle size distribution I, assessment and characterization*; ACS Symp. Ser. 332; American Chemical Society: Washington, DC; 1987.

Provder, T., Ed. *Particle size distribution II, Assessment and characterization*; ACS Symp. Ser. 472; American Chemical Society: Washington, DC; 1991.

Provder, T., Ed. *Particle size distribution III, Assessment and characterization*; ACS Symp. Ser. 693; 1998.

Prud'homme, R.K.; Khan, S.A. *Foams: Theory, measurements, and applications*; Marcel Dekker: New York; 1996.

Randolph, A.D.; Larson, M.A. *Theory of particulate processes, analysis and techniques of continuous crystallization*; Academic Press: New York; 1971.

Rayleigh, Foam, in *Scientific Papers by Lord Rayleigh (John William Strutt)*; Cambridge University Press: Cambridge; 1902; also in 6 vols. bound in 3; Dover: New York; 1964.

Rao, S.R. *Surface phenomena*; Hutchinson Educational: London; 1972.

Resing, H.A.; Wade, C.G., Eds. *Magnetic resonance in colloid and interface science*; ACS Symp. Ser. 34; American Chemical Society: Washington, DC; 1976.

Rhodes, M. *Introduction to particle technology*; Wiley: New York; 1998.

Ricca, F., Ed. *Adsorption-desorption phenomena*; Academic Press: New York; 1972.

Riddick, T.M. *Control of colloid stability through zeta potential*; Zeta-Meter, Inc: New York; 1968.

Rideal, E.K. *An introduction to surface chemistry*. Cambridge University Press: Cambridge; 1926; 2nd ed.; 1930.

Rohn, C.L. *Analytic polymer rheology*; Hanser Publishers: New York; 1995.

Rosen, M.J. *Surfactants and interfacial phenomena*; Wiley: New York; 1st ed, 1978; 2nd ed., 1989.

Ross, S.; Olivier, J.P. *On physical adsorption*; Wiley: New York; 1964.

Ross, S., Ed. *Chemistry and physics of interfaces*; American Chemical Society: Washington, DC; 1965.

Ross, S., Ed. *Chemistry and physics of interfaces — II*; American Chemical Society: Washington, DC; 1971.

Ross, S.; Morrison, I.D. *Colloidal systems and interfaces*; Wiley: New York; 1988.

Ross, S. *Nineteenth-century attitudes: men of science*; Kluwer: Dordrecht; 1991.

Rowlinson, J.S.; Widom, B. *Molecular theory of capillarity*; Clarendon Press: Oxford; 1982.

Russel, W.B. *The dynamics of colloidal systems*; University of Wisconsin Press: Madison; 1987.

Russel, W.B.; Saville, D. A.; Showalter, W.R. *Colloidal dispersions*; Cambridge University Press: New York; 1989.

Sanfeld, A. *Introduction to the thermodynamics of charged and polarized layers*; Wiley: New York; 1968.

Sato, T.; Ruch, R. *Stabilization of colloidal dispersions by polymer adsorption*; Marcel Dekker: New York; 1980.

Sauer, E. *Kolloidchemisches praktikum*. Springer: Berlin; 1935.

Sheludko, A. *Colloid chemistry*; Elsevier: New York; 1966.

Schmitz, K.S. *An introduction to dynamic light scattering by macromolecules*; Academic Press: New York; 1990.

Schramm, L.L. *The language of colloid and interface science: A dictionary of terms*; American Chemical Society: Washington DC; 1993.

Schulz, D.N.; Glass, J.E., Eds. *Polymers as rheology modifiers*, ACS Symp. Ser. 462; American Chemical Society: Washington, DC; 1991.

Schwuger, M.J., Ed. *Detergents in the environment*; Marcel Dekker; New York; 1997.

Shaw, D.J. *Electrophoresis;* Academic Press: New York; **1969**.

Shaw, D.J. *Introduction to colloid and surface chemistry*; Butterworths: Boston; 1st ed., 1966; 2nd ed., 1970; 3rd ed., 1980; 4th ed., Butterworth-Heinemann; Oxford; 1992.

Shaw, D.T., Ed. *Fundamentals of aerosol science*; Wiley: New York; 1978.

Sheludko, A. *Colloid chemistry*; Elsevier: New York; 1966.

Sherman, P, Ed. *Rheology of emulsions*; Macmillan Company: New York; 1963.

Sherman, P., Ed. *Emulsion science*; Academic Press: New York; 1968.

Shinoda, K.; Nakagawa, T.; Tamamushi, B-I; Isemura, T. *Colloidal surfactants, some physicochemical properties*; Academic Press: New York; 1963.

Shinoda, K., Ed. *Solvent properties of surfactant solutions*; Marcel Dekker: New York; 1967.

Shinoda, K.; Friberg, S. *Emulsions and solubilization*; Wiley: New York; 1986.

Sjöblom, J., Ed. *Emulsions and emulsion stability*; Marcel Dekker: New York; 1996.

Smith, A.L., Ed. *Particle growth in suspensions*; Academic Press: New York; 1973.

Sobotka, H., Ed. *Monomolecular layers*; American Association for the Advancement of Science: Washington, DC; 1954.

Society of Chemical Industry (Great Britain) *Powders in industry*; Monograph 14; London; 1961.

Somorjai, G.A. *Principles of surface chemistry*; Prentice-Hall: Englewood Cliffs, NJ; 1972.

Sonntag, H.; Strenge, K. *Coagulation and stability of disperse systems*, Kondor, R., Trans.; Wiley: New York; 1972.

Sparnaay, M.J. *The electrical double layer*; Pergamon Press: New York; 1972.

Steele, W.A. *The interaction of gases with solid surfaces*; Pergamon Press: New York; 1974.

Stein, H.N. *The preparation of dispersions in liquids*; Marcel Dekker: New York 1996.

Stockman, J.D.; Fochtman, E.G., Eds. *Particle size analysis*; Ann Arbor Science Publishers: Ann Arbor, MI; 1977.

Stokes, R.J.; Evans, D.F. *Fundamentals of interfacial engineering*; Wiley-VCH: New York; 1997.

Svedberg, T. *Die methoden zur herstellung kolloider lösungen anorganischer stoffe*; Theordor Steinkopff: XXX; 1909.

Svedberg, T. *Colloid chemistry: Wisconsin lectures*; American Chemical Soc.: New York; 1924; 2nd ed.; 1928.

Tadros, Th. F., Ed. *The effect of polymers on dispersion properties*; Academic Press: New York; 1982.

Tadros, Th. F., Ed. *Solid/liquid dispersions*; Academic Press: New York; 1987.

Takeo, M. *Disperse systems*; Wiley-VCH; New York; 1999.

Tanford, C. *The hydrophobic effect: formation of micelles and biological membranes*; Wiley: New York; 1980.

Tanford, C. *Ben franklin stilled the waves*. Duke University Press: Durham and London; 1989.

Taylor, W. *A new view of surface forces*.University of Toronto Press: Toronto; 1925.

Thomas, T.R. *Rough surfaces*; Imperial College Press: London; 1st ed. 1982; 2nd ed., 1999.

Thomas, A. *Colloid chemistry*. McGraw-Hill: New York; 1934.

Thompson, D.W. *On growth and form*, 2nd ed.; Cambridge University Press: London; 1942.

Trapnell, B.M.W. *Chemisorption*; Butterworths Scientific Publications: London; 1955.

Travis, P.M. *Mechanochemistry and the colloid mill including the practical applications of fine dispersions*. The Chemical Catalog Co.: New York; 1928.

Valentin, F.H.H. *Absorption in gas-liquid dispersions*: Some aspects of bubble technology; E. & F. N. Spon: London; 1967.

van de Hulst, H.C. *Light scattering by small particles*; Wiley: New York; 1957; Dover: New York; 1981.

van de Ven, T.G.M. *Colloidal hydrodynamics*; Academic: New York; 1989.

van Klooster, H.S. *Lecture and laboratory experiments in physical chemistry*; The Chemical Publishing Co.: Easton; 1925.

van Olphen, H. *An introduction to clay colloid chemistry*; Wiley: New York; 1963.

van Olphen, H. *Clay colloid chemistry*; Interscience: New York; 1963.

van Olphen, H.; Mysels, K.J., Eds. *Physical chemistry, Enriching topics from colloid and surface science*; Theorex: La Jolla, CA; 1975.

van Oss, C.J. *Interfacial forces in aqueous media*; Marcel Dekker: New York; 1994.

van Waser, J.R.; Lyons, J.W.; Kim, K.Y.; Colwell, R.E. *Viscosity and flow measurement*; Wiley: New York; 1963.

Veale, C.R. *Fine powders, preparation, properties, and uses*; Wiley: New York; 1972.

Verwey, E.J.W.; Overbeek, J. Th. G. *Theory of the stability of lyophobic colloids, the interaction of sol particles having an electric double layer*; Elsevier: Amsterdam; 1948; Dover: Mineola, NY; 1999.

Vold, M.J.; Vold, R.D. *Colloid chemistry, the science of large molecules, small particles, and surfaces*; van Nostrand Reinhold: New York; 1964.

Vold, R.D.; Void, M.J. *Colloid and interface chemistry*; Addison-Wesley: Reading, MA; 1983.

Voyutsky, S. *Colloid chemistry*; Mir: Moscow; 1978.

Walters, K., Ed. *Rheometry: industrial applications*; Wiley-RSP: New York; 1980.

Ware, J.C. *The chemistry of the colloidal state, a textbook for an introductory course*, 2nd ed.; Wiley: New York; 1936.

Weaire, D. *The Kelvin problem; foam structures of minimal surface area*; Taylor & Francis: London; 1996.

Weaire, D.; Hutzler, S. *The physics of foam*; Oxford University Press: New York; 1999.

Webb, P.A.; Orr, C. *Analytical methods in fine particle technology*; Micromeritics Instrument Corp.: Norcross, GA; 1997.

Weber, W.J., Jr.; Matijević, E., Eds. *Adsorption from aqueous solution*; Adv. Chem. Ser. 79; American Chemical Society: Washington, DC; 1968.

Weiser, H.B. *Colloid chemistry (a textbook)*. Wiley: New York; 1939.

Weiser, H.B. *A textbook of colloid chemistry*, 2nd ed.; Wiley: New York; 1949.

Wiersema, P.H. *On the theory of electrophoresis*; Drukkerij Pasmans: Den Haag; 1964.

Williams, J.W., Ed. *Ultracentrifugal analysis in theory and experiment*; Academic Press: New York; 1963.

Williams, RA.; de Jaeger, N.C., Eds. *Advances in measurement and control of colloidal processes*; Butterworth-Heinemann: Boston; 1991.

Williams, RA., Ed. *Colloidal and surface engineering: applications in the process industries*; Butterworths: Oxford; 1992.

Willows, R.S.; Hatschek, E. *Surface tension and surface energy and their influence on chemical phenomena*. P. Blakiston's Son: Philadelphia; 1915.

Willows, R.S.; Hatschek, E. *Surface tension and surface energy and their influence on chemical phenomena*; 3rd ed.; J. and A. Churchill: London; 1923.

Wu, S. *Polymer interfaces and adhesion*; Marcel Dekker: New York; 1982.

Young, D.M.; Crowell, A.D. *Physical adsorption of gases*; Butterworths: London; 1962.

Zsigmondy, R. *Colloids and the ultramicroscope*; Alexander, J., Trans.; Wiley: New York, 1909.

Zsigmondy, R. *The chemistry of colloids. Part I Kolloidchemie, Part II Industrial colloid chemistry*, Spear, E.R., Trans.; Wiley: New York; 1917.

Zsigmondy, R. *Das kolloide gold*; Akademische Verlagsgesellschaft: Leipzig; 1925.

Zubrowski, B. *Bubbles, A children's museum activity book*; Little, Brown: Boston; 1979.

AUTHOR INDEX*†

Abrams, I.M., (80) *497*
Abricossova, I.I., (43) *373*
Absolom, D.R., (12) *372*
Acton, J.R., (19) *47*
Adam, N.K., (11) *244*
Adamson, A.W., (26) *277*
Addison, C.C., (25) *216*
Ademu-John, C.M., (29) *199*
Ahmad, S. I., (42) *495*
Albert, J.D., (56) *551*
Alderfer, J.L., (18) *419*
Alexander, A.E., (15) *276*
Alexandridis, P., (52) *278*
Alivisatos, A.P., (92) *553*
Allen, T., (2) *111*, (18) (30) *112*, (43) (50) *113*, (54) *114*
Allred, R.E., (8) *227*, (6) *419*
Amato, I., (58) *551*
Ambwani, D.S., (12) *216*
Anderson, F.W., (13) 291
Anderson, H.R., Jr., (33) *199*
Anderson, T.F.I., (4) *215*
Ardizzone, S., (7) *348*
Arefi-Khonsari, F., (39) *199*
Arendell, J.P., (16) *453*
Aronson, M.P., (89) *497*
Aronson, S., (10) *307*
Arridge, R.G.C., (64) *351*
Arwin, H., (34) *373*
Asakura, S., (15) *419*
Asianova, M.A., (54) *455*
Audebert, R., (56) *531*
Averitt, R.D., (73) *552*
Axberg, C., (85) *497*
Bäckström, G., (67) *351*
Baglioni, P., (18) *453*
Bailey, F.C., (58) *351*
Bailey, P.A., (53) *455*
Bain, C., (45) *278*
Balard, H., (40) *199*

Balgi, V., (73) *115*
Bancroft, A.R., (64) *496*
Bancroft, W.D., (4) *452*
Bargeman, D., (14) *372*
Barman, B.N., (38) *113*
Barnes, H.A., (1) (2) *46*, (17) *47*
Bartell, F.E., (38) (39) (40) *217*
Barth, H.G., (3) *111*, (35) *113*, (64) *114*
Bascom, W.D., (20) *175*
Basheva, E.S., (47) *495*
Basta, A.H., (37) *530*
Bastos, D., (33) *349*
Bäumle. M., (74) *552*
Bawendi, M.G., (68) (69) *551*, (70) (71) (83) *552*, (93) *553*
Baxter, S., (12) *227*
Becher, P., (9) *453*, (73) *496*
Becker, J.E., (8) *135*
Behringer, R.P., (46) *530*
Belfort, G., (21) *227*
Bellemans, A., (16) *307*
Bennis, R., (12) *175*
Beresford, J., (107) *116*
Berg, J.C., (31) (32) *199*, (16) *227*, (2) *418*, (22) (23) (24) *419*
Berger, E.J., (37) *454*
Berger, P.D., (16) *453*
Bergeron, V., (76) *497*
Bergström, L., (34) (36) *373*
Berman, A.D., (18) *419*
Bernett, M.K., (35) *199*
Bernhardt, C., (14) *112*
Berthelot, D., (3) (4) *197*, (16) (17) *198*
Berti, D., (18) *453*
Bethell, D., (65) *551*, (75) *552*
Bibette, J., (32) (35) (36) *550*
Bigelow, W.C., (33) *217*
Biggs, S., (18) *61*, (16) *419*
Bikerman, J.J., (20) *156*, (91) *497*
Billmeyer, F.W., Jr., (59) *114*

*Authors and Editors in the Bibliography are not listed.
†The number in parentheses is the reference number; the number in italics is the page number on which the reference occurs.

Binks, B.P., (10) *453*
Bird, R.B., (3) *46*, (22) (24) *47*, (1) *60*, (2) *197*, (2) (3) (5) *371*
Biswas, B., (23) *453*
Bjerregaard, S., (44) *454*
Blodgett, K.B., (14) *244*, (16) (17) *245*
Bloom, B.H., (14) *156*, (9) *216*
Bock, E.J., (36) *217*
Boger, D.V., (23) *529*
Bohren, C.F., (3) *14*, (5) *15*, (56) (57) *114*
Boils, D., (10) *381*
Bolles, W.L., (46) *495*
Bolvari, A.E., (5) *419*
Bonini, M., (18) *453*
Boruvka, L., (13) *156*, (8) *215*
Borwankar, R.P., (33) *494*
Boscovitch, R.J., (21) *419*
Bott, S.E., (65) *114*
Boucher, E.A., (21) *156*, (17) (18) (19) (20) *315*
Bowen, P., (31) *113*
Boyd, J., (15) *453*
Boys, C.V., (6) *493*
Brady, A.P., (80) *497*
Brandt, L., (66) *551*
Brendle, E., (40) *199*
Brenner, H., (101) *498*
Briggs, T., (1) *493*
Briggs, T.R., (47) (48) *454*
Brinker, C.J., (28) *176*
Brooks, C.F., (103) *498*
Brousseau, L.C., III, (77) *552*
Brown, F.E, (1) *215*
Brown, K.R., (63) *551*
Brown, L.O., (64) *551*
Brown, M.G., (6) *276*
Brown, R.C., (2) *155*
Broze, G., (17) *453*
Brunauer, S., (93) *116*
Brust, M., (65) *551*
Buboltz, J.T., (16) *216*
Buff, F.P., (4) *175*
Bungenberg de Jong, H.G., (50) *454*
Burcik, E.J., (26) *216*
Burns, J.L., (18) *61*
Burst, M., (75) *552*
Burton, R.A., (98) *498*
Bury, C.R., (14) *276*
Buscall, R., (26) *529*, (30) *550*
Butler, J.N., (14) *156*, (9) *216*
Bütschli, O., (27) *453*
Cadenhead, D.A., (3) *244*
Cahn, J.W., (12) *315*
Calderon, F. L., (36) *550*

Caldwell, K.D., (41) *113*
Canevari, G.P., (55) *455*
Cannon, D., (84) (87) *115*
Cao, Y., (41) *454*
Carpineti, M., (19) *61*
Carroll, B.H., (13) *529*
Caruso, F., (24) (25) *176*
Casamitjana, X., (52) *531*
Casimir, H.B.G., (11) *198*, (27) *372*
Cassie, A.B.D., (12) *227*
Casson, N., (34) *530*
Cates, M.E., (64) (65) *455*
Cayias, J.L., (21) *216*
Cazabat, A.-M., (17) *227*
Centers, P., (59) *496*
Challis, R.E., (78) (81) *115*
Chan, D.Y.C., (38) *373*
Chanamai, R., (20) *529*
Chandar, P., (79) *497*
Chang, K., (37) *454*
Chang, T.M., (5) *276*
Chang, Y.J., (51) *114*
Chaplain, V., (56) *531*
Chapman, D.L., (17) *349*
Chasovnikova, L.V., (83) *497*
Chatterjee, S., (16) *315*
Chehimi, M.M., (39) *199*
Chen. E.S., (6) *307*
Chen, S., (52) *551*
Chen, W., (10) *227*
Chen, X.J., (33) *373*
Christova, C., (70) *496*
Chu, B., (8) *15*
Chuang, K.T., (49) *217*
Chuan-Jian, Z., (81) *552*
Chung, S.I., (21) *453*
Churaev, N.V., (21) *315*, (15) *372*, (46) (47) *373*
Cini, R., (10) *244*, (36) *277*
Cipelletti, L., (19) *61*
Clark, H., (40) (41) *550*
Clausse, M., (34) *454*
Clayfield, E.J., (30) *373*, (12) *382*
Clayton, W., (49) *454*
Clint, J.H., (46) *217*
Cohen, E.D., (26) (27) *176*, (29) (38) (39) *530*
Cohen, G.L., (11) *549*
Cohu, O., (40) *530*
Coles, N.E., (7) *47*
Collier, C.P., (76) *552*
Colomer, J., (52) *531*
Comiskey, B., (56) *551*
Conley, R.F., (4) *528*

Conroy, L.E., (44) *495*
Cook, H.D., (4) *244*
Cooper, P., (68) (70) *496*, (76) *497*
Corkill, J.M., (51) *278*
Corner, T., (30) *550*
Costanzo, P.M., (44) *217*
Cottington, R L., (20) *175*
Cox, J.M., (53) *495*
Crane, M.D., (20) *227*
Cratin, P.D., (21) *198*
Croucher, M.D., (18) *372*, (48) *373*, (11) *381*, (13) (14) *419*, (25) *549*, (26) (27) *550*
Crowley, J.M., (4) *134*
Cummins, H.Z., (41) *350*
Cummins, P.G., (70) *114*
Cunningham, F.F., (5) *47*
Curtiss, C.F., (2) *197*, (2) (3) (5) *371*
Dabbousi, B.O., (71) *552*
Dabros, T., (2) *452*
Daby, E., (11) *549*
Dang, L.X., (5) *276*
Daniel, F.K., (14) (15) *135*
D'Arrigo, J.S., (15) (17) *216*
Dauchot, J., (12) *15*, (13) (14) *549*
Davidson, J., (58) *496*
Davies, D.K., (65) (66) *351*
Davies, J.T., (40) *277*, (13) *453*, (20) *494*,
de Gennes, P.G., (16) (22) *175*, (23) *176*
De Maeyer, L., (67) *114*
De Smet, Y., (18) *156*
Debye, P., (83) *115*
Defay, R., (16) *307*, (10) *315*
Delamar, M., (39) *199*
DelaMaza, A., (48) *495*
Delay, R., (44) *278*
Delly, J.G., (44) *113*
Deminière, B., (36) *550*
Denkov, N.D., (47) *495*, (68) (69) (70) (71) *496*
Depaoli, D.W., (6) *135*
Deriemaeker, L., (18) *156*
Derjaguin, B.V., (7) (8) *155*, (7) *276*, (21) *349*, (59) *351*, (43) (46) (47) *373*, (5) *381*, (8) *493*, (21) *494*, (38) *550*
Dettre, R.H., (30) *216*, (13) (14) *227*
Dewar, J., (51) *495*
D'Haene, P., (26) *529*
Dickinson, E., (74) *115*, (41) *454*
Dinegar, R.H., (60) *531*, (9) *549*
Dippenaar, A., (87) *497*
Dixon, J.K., (7) *236*
Djabbarah, N.F., (41) *495*
Dodge, J.S., (20) (21) *549*

Doremus, R.H., (5) *549*
Doroszkowski, A., (10) *135*
DosRamos, J.G., (34) *113*
Douillard, J.-M., (12) *175*, (41) *199*
Drago, R.S., (25) (26) (27) *198* (28) *199*, (9) *419*
Dreger, E.E., (19) *276*
Drommond, F.G., (36) *454*
Drost-Hansen, W., (1) *155*
Drummond, C.J., (25) *176*, (38) *373*
Drzaic, P.S., (50) *278*
D'Silva, A.P., (101) *116*
Du, Q., (4) *276*
Duits, M.H.G., (21) *529*
Duke, C.B., (72) (73) (74) *351*
Dukhin, S.S., (43) *277*, (21) *349*
Durian, D.J., (28) (29) *494*
Dzyaloshinskii, I.E., (31) *373*
Echer, C.M., (92) *553*
Edser, E., (9) *175*
Edwards, D.A., (2) *493*, (101) *498*
Efros, A.L., (69) *551*
Einarson, M.B., (22) *419*
Einstein, A., (8) *60*, (11) *61*
Eisler, H.J., (93) *553*
El-Aasser, M.S., (27) *277*, (61) *455*, (24) *549*
Elder, M.E., (14) *529*
Eley, R.R., (35) *530*
Elghanian, R., (85) (86) *552*
El-Saied, H., (37) *530*
Elsayad, S.Y., (37) *530*
Emmett, P.H., (93) *116*
Epstein, M.B., (44) (45) *495*
Eres, M.H., (27) *156*
Evans, D.F., (15) *382*
Evans, M.G., (3) *276*
Everett, D.H., (5) *528*
Exerowa, D., (7) *244*, (46) *278*, (49) *495*
Fabish, T.J., (62) (72) (73) (74) *351*
Facemire, B.R., (8) *315*, (61) *496*
Faddis, N., (47) *530*
Fadeev, A.Y., (10) *227*
Fairhurst, D., (47) *350*
Fang, M., (62) *551*
Faraday, M., (2) *548*
Faye, V., (36) *550*
Feigenson, G.W., (16) *216*
Feke, D.L., (16) *529*
Feldheim, D.L., (77) *552*
Fendler, E.J., (3) 291
Fendler, J.H., (47) (48) *217*, (3) *291*
Feng, J.Q., (6) *135*
Ferguson, J., (8) *276*

Fernández, A., (33) *349*
Finsy, R., (71) *115*, (18) *156*
Fiocco, R.J., (55) *455*
Fischer, C., (76) *497*
Fischer, E.K., (8) (10) (18) *47*, (6) *291*, (22) *315*, (25) *529*
Fisher, L., (43) *217*
Fisher, L.R., (15) (16) *156*
Fitch, R.M., (28) (29) (31) (48) *550*
Flautt, T.J., (25) *277*
Fleer, G.J., (52) (53) *114*, (57) *531*
Flory, P.J., (60) *455*
Fochtman, E.G., (11) *112*
Forbes, J.D., (4) *46*
Førdedal, H., (20) *47*, (35) *454*
Forstner, M.B., (10) *61*
Fort, T., Jr., (12) *216*
Fowkes, F.M., (25) *156*, (17) (18) *175*, (13) (14) (18) (19) (22) (23) *198*, (29) (30) (42) *199*, (29) *216*, (35) *217*, (15) *227*, (2) *290*, (7) (8) *291*, (60) (61) *351*, (25) *372*, (29) *373*, (17) *401*, (3) (4) (10) (11) *419*, (26) *494*
Frances, E.I., (47) *278*
Frank, C.W., (103) *498*
Frank, H.S., (3) *276*
Frankel, S., (50) *495*
Franks, F., (19) *156*, (3) *315*
Fraser, A.B, (5) *15*
Freundlich, H., (11) *307*
Freysz, E., (4) *276*
Friberg, S., (53) *278*, (19) (46) *453*, (42) (43) (53) *495*
Frisch, H.L., (21) *175*
Frith, W.J., (26) *529*
Frokjare, S., (44) *454*
Frumkin, A., (8) *307*
Fuchs, N., (3) *400*
Fuerstenau, D.W., (10) *348*
Fuller, G.G., (103) *498*
Furler, G., (23) *291*
Furlong, D.N., (24) *176*
Gaines, G.L., Jr., (5) *215*, (15) *244*
Ganchev, D., (47) *495*
Garner, F.H., (20) (21) *291*
Garrett, P.R., (66) *496*, (84) (86) *497*
Gauss, C.F., (3) *175*
Gavroglu, K., (9) *198*
Geerkens, M., (74) *552*
Genzel, L., (84) *552*
Gerdes, S., (17) *227*
Ghiradella, H., (21) *175*
Ghotani, S., (34) *550*
Gibbons, R., (1) *111*

Gibbs, J.W., (14) *175*, (19) *494*
Gibbs, W.W., (55) *551*
Gibson, H.W., (58) (70) *351*
Giddings, J.C., (42) *113*
Giermanska-Khan, F., (76) *497*
Giese, R.F., (44) *217*
Giglio, M., (19) *61*
Giles, C.H., (100) (101) *116*, (10) *529*
Girifalco, L.A., (6) *155*, (15) *198*
Glatter, O., (63) *114*
Goddard, E.D., (79) *497*
Goia, D.V., (8) *549*
Goldman, P., (14) *135*
Goldstein, A.N., (92) *553*
Gonick, E., (11) *276*
Good, R.J., (6) *155*, (15) *198*, (31) *217*
Goodman, J.F., (51) *278*
Gopal, A.D., (27) *494*
Gopal, E.S.R., (16) *315*
Gorria, Ph., (36) *550*
Götte, E., (32) *277*
Gouy, G., (16) *349*
Grabowski, E.F., (68) (69) *114*, (37) *454*
Grahame, D.C., (18) *349*
Graillat, C., (56) *531*
Gratton-Liimatainen, J., (47) *530*
Gregory, J., (6) *198*, (7) *372*, (55) *531*
Grieser, F., (38) *373*
Griffin,W.C., (11) (12) *453*
Griggs, D.T., (7) *47*
Grist, D.M., (22) *453*, (97) *498*
Grunderbeeck, F., (12) *549*
Gu, J., (1) *236*
Guo, Y., (18) *419*
Gurney, R.W., (1) *348*
Gutoff, E.B., (26) (27) *176*, (29) (38) (39) *530*
Gutop, Yu.V., (38) *550*
Guyot-Sionnest, P., (60) *551*
Guzonas, D., (10) *381*
Haak, R.M., (27) *216*, (37) *495*
Hadamard, J., (19) *291*
Hadjiiski, A., (70) *496*
Hadlington, S., (55) *114*
Hair, M.L., (45) *350*, (62) *351*, (18) *372*, (10) (11) *381*, (14) *419*, (25) *549*, (26) *550*
Halas, N.J., (73) (79) *552*
Hale, A.R., (20) *291*
Halliwell, M.J., (29) *199*
Halverson, F., (50) *530*
Hamada, A., (23) *549*
Hamaker, H.C., (9) *372*
Hammerton, D., (21) *291*

Handon, A., (4) *15*
Hannemann, R.E., (47) *278*
Hardy, W.B., (19) *175*, (10) *401*
Harkins, W.D., (5) *155*, (5) (11) *175*, (1) (4) (7) *215*, (35) *217*, (22) *277*, (65) *496*
Harper, W.R., (71) *351*
Hart, W.H., (65) *114*
Hartenstein, C., (10) *216*, (35) *277*
Hartland, S., (12) *156*, (6) *215*
Hartley, G.S., (17) *276*
Hartley, R.W., (12) *156*
Hasapidis, K., (51) *114*
Hauser, E.A., (1) *400*
Hautecler, M., (12) *549*
Hayashi, K., (26) *291*
Haydon, D.A., (23) *453*
Haynes, C.L., (72) *552*
Hays, D.A., (53) *551*
Haywood, I.R., (62) *496*
Healy, T.W., (15) *349*, (7) *381*, (58) *531*
Heath, J.R., (66) *551*, (76) *552*
Heim, I., (74) *552*
Heine, J.R., (71) *552*
Heitner, H.I., (51) *531*
Helling, J.O., (103) *116*
Hemar, Y., (74) *115*
Hendrix, J., (67) *114*
Heng, Y.M., (90) *497*
Henrichs, S.E., (76) *552*
Henry, D.C., (36) *350*
Hérard, C., (31) *113*
Herb, C.A., (69) *114*, (37) (45) *454*,
Herdan, G., (13) *112*
Hermanie, P.H.J., (10) (9) *291*
Herrmann, N., (80) *115*, (17) *156*, (20) *529*
Herz, A.H., (103) (104) *116*
Hickin, G.K., (22) *112*
Hidalgo-Álvarez, R., (33) *349*
Hielscher, F.H., (61) *351*
Higuchi, W.I., (58) *455*
Hilgenfeldt, S., (25) *494*
Hinchen, J.J., (42) *550*
Hingston, F.J., (6) *529*
Hirelman, E.D., (57) *114*
Hirschfelder, J.O., (2) *197*, (2) (3) (5) *371*
Hofer, M., (63) *114*
Hofer, R., (9) *227*
Hoffman, R.L., (32) *113*
Hollingsworth, J.A., (93) *553*
Holliway, W.D., (90) *553*
Holmes, A.K., (78) (81) *115*
Holmqvist, P., (52) *278*

Homola, A., (58) (59) *531*
Honig, E.P., (5) *400*
Hoogeveen, N.G., (57) *531*
Horne, D.S., (74) *115*
Hórvölgyi, Z., (47) (48) *217*
Hoskins, R., (20) *419*
Hough, D.B., (32) *373*
Hrymak, A., (90) *497*
Hsieh, M.C., (10) *227*
Hsu, C.A., (16) *453*
Hu, E.L., (57) (61) *551*
Huang, J., (16) *216*
Huang, J.S., (7) *549*
Hudson, J.B., (23) *277*
Huethorst, J.A.M., (26) *156*, (45) *217*
Hühnerfuss, H., (10) *244*
Hui, S.W., (18) *419*
Humphreys, C.W., (1) *307*
Humphry-Baker, R., (31) *113*
Hunter, R.J., (88) (89) *115*, (25) (30) (31) (32) *349*, (34) (35) (38) (50) (51) (52) *350*
Hurd, A.J., (28) *176*
Hutchinson, E., (3) (4) (5) *307*
Hutchison, J., (45) *278*
Hutchison, J.E., (64) *551*
Hutton, J.F., (1) *46*
Hutzler, S., (14) *494*
Hyeon, T., (62) *551*
Iijima, S., (78) *552*
Irvine, T.F., Jr., (21) *47*
Isemura, T., (30) *277*
Isenberg, C., (7) *493*
Israelachvili, J., (15) *156*, (10) *276*, (55) (56) *278*, (16) *372*, (45) *373*, (18) *419*
Ivošević, N., (32) (33) *176*
Jacobson, J., (56) *551*, (91) *553*
Jaeger, H., (80) *552*
Jaeger, H.M., (46) *530*
Jakob, C.W., (44) (45) *495*
James, M., (89) *115*
James, R.O., (15) *349*, (7) *381*, (58) (59) *531*
Jameson, J., (18) *61*
Jamieson, A.M., (27) *530*
Jamin, J.C., (18) (19) *227*
Janex, M.L., (56) *531*
Janule, V.P., (18) *216*
Janusonis, G.A., (104) *116*
Jasper, J.J., (4) *155*
Jelínek, Z.K., (4) *111*, (24) *112*
Jensen, K.F., (71) *552*
Jenson, W.B., (24) *198*, (8) *419*
Jeon, S.J., (36) *113*

Jinnai, H., (14) *198*, (13) *291*
Johnsen, E.E., (20) *47*
Johnson, J., (9) *135*
Johnson, R.E., Jr., (30) *216*, (13) (14) *227*
Jones, A., (77) *552*
Jones, E.R., (14) *276*
Jordan, H.F., (7) *215*
Joslin, S.T., (10) *419*
Judson, C.M., (7) *236*
Jura, G., (5) *175*
Kabalnov, A., (63) *455*
Kagan, C.R., (83) *552*
Kagawa, Y., (34) *550*
Kaler, E.W., (49) (50) *551*
Kamath, Y.K., (6) *227*
Karasikov, N., (48) *113*
Karuhn, R., (15) *112*
Käs, J., (10) *61*
Kasuga, K., (47) *495*
Kaye, B.H., (5) *111*, (19) (25) (26) *112*, (45) *113*, (47) (48) *530*,
Keim, G.I., (19) *276*
Keller, W., (53) *531*
Kelvin, *See* Thomson, W.
Kerker, M., (6) (10) *15*, (58) *114*, (11) *549*
Kiely, C.J., (75) *552*
Kim, K.-H., (34) *550*
Kim, Y.-H., (34) *495*, (102) *498*,
Kimball, W.A., (5) (6) *244*
Kipling, J.J., (99) *116*, (8) *529*
Kirkwood, J.G., (8) *198*
Kissa, E., (61) *114*
Klein, B., (22) *529*
Klenerman, D., (45) *278*
Klimov, V.I., (93) *553*
Klinkenberg, A., (4) (14) (15) *291*
Klumpar, I.V., (27) *112*
Koczo, K., (34) *495*, (102) *498*
Koehler, S.A., (25) *494*
Koning, P.A., (5) *419*
Koppel, D. E., (66) *114*
Kornbrekke, R.E., (37) *217*, (17) *291*, (13) (14) (15) *315*, (48) *350*, (96) *498*
Kosmulski, M., (8) *348*
Kourti, T., (62) *114*
Kraft, F., (29) *277*
Krastanka, G., (70) *496*
Kratochvil, P., (7) *15*
Kratochvil, J.P., (11) (18) *549*
Kreibig, U., (84) *552*
Kretzschmar, G., (43) *277*
Krieger, I.M., (20) (21) (22) *549*
Krog, N.J., (1) *452*
Krotov, V.V., (34) *277*, (31) *494*

Kruyt, H.R., (14) *61*, (7) *400*, (17) *549*
Kuhl, T., (18) *419*
Kulkarni, R.D., (79) *497*
Kunieda, H., (14) *453*, (56) *455*
Kurdi, J., (39) *199*
Kurzendoerfer, C.P., (88) *497*
La Mer, V.K., (11) *15*, (60) *531*, (10) (9) *549*
Lacy, L.L., (8) *315*, (61) *496*
Lafuma, F., (56) *531*
Lalchev, Z., (7) *244*, (46) *278*, (49) *495*
Lambert, R.H., (11) *529*
Landau, L., (5) *381*
Lando, J.B., (12) *244*
Langbein, D., (26) *372*
Langevin, D., (76) *497*
Langmuir, I., (13) *244*, (17) *245*, (5) *315*, (3) *381*
Lannigan, L.A., (29) *199*
Lara, J., (33) *530*
Larson, I., (38) *373*
Larson, R.G., (12) (14) *47*, (32) *530*
Larsson, K., (1) *452*
Laskowski, J.S., (22) *529*
Lau, A., (19) *349*
Laughlin, R.G., (48) (54) *278*
Lavaste, V., (39) *199*
Lawson, K.D., (25) *277*
Leal, L.G., (22) *291*
Leatherdale, C.A., (93) *553*
Leckband, D., (18) *419*
Lee, L.-H., (43) *199*
Lee, T.R., (79) *552*
Lee, Y.J., (16) *529*
Leenaars, A.F.M., (26) *156*, (45) *217*
Leff, D.V., (66) *551*
Lehmann, O., (49) *278*
Leibowitz, F.L., (81) *552*
Leng, D.E., (33) *550*
Lenhoff, A.M., (49) (51) *551*
Letsinger, R.L., (82) (85) (86) *552*
Leubner, I.H., (62) (63) *531*
Levi, A.C., (33) *373*
Levich, V.G., (29) *176*, (24) *291*
Levius, H.P., (36) *454*
Lewis, J.A., (19) *419*
Li, J., (41) *113*
Li, M., (9) *135*
Lichtman, V., (16) *135*
Lifshitz, E.M., (10) *372*, (31) *373*
Lightfoot, E.N., (3) *46* (1) *60*
Lin, M.Y., (7) *549*
Lindman, B., (52) *278*
Lindquist, C.G., (80) *497*

Lissant, K.J., (52) *455*
Little, R.C., (20) *276*
Liu, W., (52) *551*
Lliboutry, L.A., (6) *47*
Lok, K.P., (27) *550*
Lombardini, P.P., (10) *244*, (36) *277*
London, F., (1) *197*, (4) *371*
Lopes, W.A., (80) *552*
Lowell, J., (63) *351*
Lowell, S., (91) (92) *115*
Lucassen, J., (37) *277*, (32) *494*
Lucassen-Reynders, E.H., (41) *217*, (37) *277*, (32) *494*
Luckham, P.F., (36) *373*
Lumb, E.C., (30) *373*, (12) *382*
Lumb, P.B., (62) *496*
Lumsdon, S.O., (10) *453*
Lundelius, E.F., (6) *315*
Lunkenheimer, K., (10) *216*
Luzzati, V., (24) *277*
Lyklema, J., (6) *348*, (24) *349*, (44) *350*
Lyon, L.A., (59) *551*, (89) (90) *553*
Mabille, C., (36) *550*
MacEwan, T.H., (10) *529*
MacGregor, J.F., (62) *114*
Mackey, P.H., (30) *373*
Magnin, A., (19) *529*, (40) *530*
Mahale, A.D., (5) (6) *227*
Maham, Y., (49) *217*
Mahanty, J., (19) (39) *372*
Malkina, A.D., (43) *373*
Malko, A., (93) *553*
Ma*ł*ysa, K., (35) *277*
Manas-Zloczower, I., (16) *529*
Mandelbrot, B.B., (16) (17) *61*
Manegold, E., (25) (32) (33) *453*
Mannheimer, R.J., (36) *495*, (98) (99) (100) *498*
Manning-Benson, S., (45) *278*
Manson, J.A., (14) *198*
Marangoni, C., (16) (17) (18) *494*
Maret, G., (72) *115*
Marinakos, S.M., (77) *552*
Marinov, B., (7) *244*, (46) *278*
Marinova, K.G., (70) (71) *496*
Marlnov, B., (49) *495*
Marmo, M.J., (14) *198*
Maron, S.H., (14) *529*, (36) *530*
Martín, A., (33) *349*
Martin, C., (13) *135*
Martin, D., (10) *61*
Martin, J.-Y., (68) *496*
Martínez, F., (33) *349*
Martynov, G.A., (38) *550*

Maru, H.C., (40) *495*
Masliyah, J., (2) (3) *452*
Mason, S.G., (22) *216*
Mason, T.G., (35) *550*
Máté, M., (48) *217*
Matijević, E., (49) *350*, (12) *401*, (54) *531*, (8) (11) (15) (16) (18) *549*
Matsunaga, T., (16) *401*
Matsushita, M., (37) *199*
Mattoussi, H., (71) *552*
Maxwell, J.C., (2) *14*, (24) *156*, (8) *372*, (31) *454*
Maye, M.M., (81) *552*
Maynard, D.F., (1) *276*
McBain, J.W., (13) (9) *276*, (1) (2) *307*, (2) *315*, (38) *495*, (80) *497*
McCarthy, D.C., (17) *175*
McCarthy, T.J., (10) *227*
McClements, D.J., (5) *60*, (80) *115*, (17) *156*, (20) *529*
McCormick, M., (25) *176*
McCoy, M., (5) *236*
McCrone, W.C., (44) *113*
McGuiggan, P., (24) *419*
McKay, R.B., (3) *134*, (1) *528*
McKennell, R., (23) *47*
McKenzie, D.C., (51) *114*
McLaughlin, A., (19) *349*
McLaughlin, S., (19) *349*
McNeil-Watson, F., (47) *350*
Médout-Marère, V., (41) *199*
Meguro, M., (37) *199*
Mellema, J., (21) *529*
Melrose, J.C., (2) *175*
Menon, R., (40) *350*
Menon, V.B., (51) *454*
Mer, V.K., (61) *531*
Meurk, A., (34) (36) *373*
Mewis, J., (26) *529*
Mie, G., (9) *15*, (4) *549*
Mikhailovsky, A.A., (93) *553*
Mikulec, F.V., (71) *552*
Miles, G.D., (19) *276*
Miller, J.F., (46) *350*
Miller, R., (10) *216*, (35) (43) *277*
Milling, A., (16) *419*
Minnaert, M., (15) *15*
Minor, M., (44) *350*
Mirkin, C.A., (82) (85) (86) (87) *552*
Mirnik, M., (18) *549*
Misra, J., (58) *455*
Mitchell, D.J., (16) *276*, (40) (41) (42) *373*, (4) *381*

Mittal, K.L., (33) (38) *199*, (32) *217*, (57) *455*
Mohan, V., (40) *495*
Moore, R.J., (13) *291*
Moran, K., (3) *452*
Morkved, T.L., (80) *552*
Morris, T.R, (17) *112*
Morrison, I.D., (68) (69) *114*, (95) (96) *116*, (11) (12) (17) *291*, (9) *307*, (11) (12) *349*, (48) (54) (55) *350*, (14) (15) *401*, (37) *454*, (11) *494*, (44) (45) *550*, (53) *551*
Morsy, F., (37) *530*
Mostafa, M.A., (17) *175*, (14) *198*, (7) (13) *291*
Motyka, A.L., (18) *529*
Movchan, T.G., (21) *315*
Mucic, R.C., (82) (85) (86) *552*
Mukerjee, P., (15) *307*
Muller, V.M., (47) *373*
Murphy, R.E., (37) *113*
Murray, C.B., (70) (83) *552*
Murray, J.M., Jr., (21) *198*
Musick, M.D., (89) *553*
Mustacchi, H., (24) *277*
Mysels, K.J., (19) *216*, (41) *277*, (22) *494*, (50) *495*, (39) *550*
Nagarajan, R., (21) *453*
Nagel, S.R., (46) *530*
Nakagaki, M., (13) *494*
Nakagawa, T., (30) *277*
Nakamae, K., (37) *199*
Nakhwa, S.N., (100) *116*, (10) *529*
Napper, D.H., (13) *382*, (1) (25) *418*
Natan, M.J., (59) (63) *551*, (89) (90) *553*
Needham, T.E., (25) *198*
Nelson, R.D., Jr., (3) *528*
Németh, S., (47) *217*
Neumann, A.W., (13) *156*, (8) *215*, (12) *372*
Neustadter, E.L., (22) *453*, (97) *498*
Nguyen, N., (34) *199*
Nicoli, D.F., (51) *114*
Ninham, B.W., (16) *276*, (11) (19) (20) (39) *372*, (40) (41) (42) *373*
Nishino, T., (37) *199*
Nishioka, G.M., (7) *227*, (1) *314*, (7) (8) *315*, (12) *494*, (57) (61) *496*, (81) (82) (95) *497*, (96) *498*
Nivi, B., (91) *553*
Nordhage, F., (67) *351*
Norris, D.J., (68) (69) *551*, (70) *552*
Noskov, B.A., (20) (28) *216*, (42) *277* (35) *495*

Novotny, V., (16) *291*
Novotny, V.J., (45) *350*
Nushart, S., (20) *112*
Nye, M.J., (9) *61*
Ober, R., (71) *552*
O'Brien, K.C., (12) *244*
O'Brien, R.W., (85) (86) (87) (88) (89) *115*, (22) *349*, (37) *350*
Ogden, A.L., (19) *419*
Ognyanov, K., (7) *244*, (46) *278*, (49) *495*
O'Hagan, P., (51) *114*
Oja, T., (84) *115*, (17) *291*, (48) *350*
Okahata, Y., (24) *176*
Okazaki, S., (26) *291*
Oldenburg, S.J., (73) (79) *552*
Oliveria, M., (6) *549*
Olivier, J.P., (9) *47*, (22) *112*, (90) *115*, (13) *349*, (24) *529*, (43) *550*
Omenyi, S.N., (12) *372*
Onda, T., (20) *61*, (11) *227*
Oner, D., (10) *227*
Oosawa, F., (15) *419*
Oppliger, P.E., (78) *497*
Orr, C., (10) *112*
Orr, C., Jr., (22) *112*
Ościk, J., (9) *529*
Osemann, C., (74) *552*
Osterhof, H.J., (40) *217*
Österlund, R., (85) *497*
Ostrach, S., (22) *156*
Ostwald, Wa., (24) *453*
Ottewill, R.H., (14) *349*, (55) *495*, (19) *549*
Overbeek, J.Th.G., (10) (12) *198*, (2) *348*, (23) *349*, (28) *372*, (49) *373*, (1) (2) (6) (8) (9) *381*, (4) (8) (11) *400*, (24) *494*
Owen, M.J., (3) *155*
Padday, J.F., (3) *215*, (13) (14) *216*
Palkar, S.A., (37) *113*
Panzer, H.P., (50) *530*
Papir, Y.S., (22) *549*
Papirer, E., (40) *199*
Park, N.A., (21) *47*
Park, S.Y., (47) *278*
Parkinson, C., (15) *453*
Parra, J.L., (48) *495*
Parsegian, V.A., (11) (20) *372*, (37) *373*
Partridge, S.J., (22) *529*
Pashias, N., (23) *529*
Pashley, R.M., (14) *216*, (20) *349*
Patterson, R.E., (15) *175*, (11) *216*
Patton, T.C., (105) (106) *116*, (1) *134*, (28) *530*

Paulussen, N., (67) *114*
Pelssers, E.G.M., (52) (53) *114*
Pelton, R., (90) *497*
Peña, D.J., (59) *551*
Perrin, P., (37) *550*
Peterson, G., (84) *115*
Pétré, G., (44) *278*
Petty, M.C., (18) *245*
Phillies, G.D.J., (59) *114*
Phillips, D.H., (12) *112*
Piau, J.-M., (19) *529*
Pickering, S.U., (8) *452*
Pickett, D.L., (33) *217*
Pieracci, J.P., (21) *227*
Pierce, P.E., (36) *530*, (20) *549*
Pignon, F., (19) *529*
Pike, E.R., (41) *350*
Pine, D.J., (29) *494*
Pinfield, V.J., (6) *60*
Pitaevskii, L.P., (31) *373*
Pitt, A.R., (14) *216*
Pitzer, K.S., (7) *198*
Pochan, J.M., (58) *351*
Pockels, A., (1) *244*
Polder, D., (11) *198*, (27) *372*
Pons, R., (56) *455*
Potter, L., (38) *277*
Pouchelon, A., (76) *497*
Povey, M.J., (2) (3) (4) (5) *60*, (23) *112*, (77) (79) (82) *115*, (38) (40) *454*, (17) *529*
Powers, J.M., (98) *116*
Pratsinis, S.E., (53) *531*
Pravia, K., (1) *276*
Prest, H.F., (4) *493*
Prigogine, I., (16) *307*, (10) *315*
Princen, H.M., (22) (23) *216*, (30) *453*
Prins, M.W.J., (31) *176*
Privat, M., (12) *175*
Provder, T., (6) (7) (8) (9) *112*
Provencher, S.W., (67) *114*
Pugh, R.J., (29) *373*, (16) (17) *401*
Puretz, J., (29) *112*
Quickenden, T., (4) *15*
Quincke, G., (3) (4) (5) *348*
Quinn, M.J., (63) *496*
Quirino, J.P., (28) *277*
Qutubuddin, S., (27) *530*
Rabinovich, Ya.I., (21) *315*, (15) *372*, (46) *373*
Radigan, W., (21) *175*
Rafailovich, M., (67) *551*
Ramsden, J.J., (48) *217*
Ratanathanawongs, S.K., (42) *113*

Rauner, L.A., (77) *497*
Ravey, J.-C., (20) *453*
Rayleigh (17) *112*, (24) *216*, (15) *494*
Reerink, H., (6) *381*
Rehbinder, P.A., (16) (17) (18) *135*
Reiner, M., (8) *175*
Reiter, G., (30) *176*
Ren, L., (6) *215*
Reuss, F.F., (29) *349*
Richardson, E.G., (29) *453*
Richmond, G.L., (6) *276*
Richmond, P., (5) *197*, (41) *373*, (4) *381*
Richter, S.M., (73) *115*
Rideal, E.K., (40) *277*
Rideal, G.R., (16) (17) *112*
Ridley, B.A., (91) *553*
Ries, H.E., Jr., (2) (4) (5) (6) (8) (9) *244*
Riisom, T.H., (1) *452*
Robb, I.D., (20) *419*
Roberts, K., (85) *497*
Robertson, C.R., (103) *498*
Robinson, J.V., (39) *495*, (75) (80) *497*
Roco, M.C., (57) *551*
Rodriguea-Viejo, J., (71) *552*
Roebersen, G.J., (5) *400*
Roessner, D., (75) *115*
Rohn, C.L., (13) (15) *47*, (30) *530*
Rootare, H., (98) *116*
Rose-Innes, A.C., (63) *351*
Rosen, M., (69) *551*
Rosen, M.J., (18) *276*, (33) *277*, (1) *290*, (10) *307*, (54) *495*
Ross, J., (19) *276*, (44) (45) *495*
Ross, S., (90) *115*, (95) *116*, (15) *175*, (34) *199*, (11) (27) *216*, (37) *217*, (23) *277*, (23), (25) *291*, (6) (9) *307*, (1) *314*, (7) (9) (13) (14) (15) *315*, (56) *350*, (5) *452*, (26) *453*, (4) *493*, (9) (10) (11) (12) (24) *494*, (37) (38) *495*, (57) (60) (67) (72) *496*, (80) (81) (82) (93) *497*, (96) *498*, (15) *529*, (40) (41) (42) (43) (44) (45) *550*
Rotenberg, Y., (13) *156*, (8) *215*
Rowcliffe, D.J., (34) *373*
Rowlands, W.N., (87) *115*
Rowlinson, J.S., (13) *175*, (11) *315*
Ruckenstein, E., (6) (7) *452*
Rücker, A.W., (94) *116*
Rusanov, A.I., (34) *277*, (31) *494*
Russel, W.B., (16) *47*, (15) *61*, (14) *382*, (31) *530*, (47) *550*
Rutzler, J.E., Jr., (68) *351*
Rybczynski, W., (18) *291*
Ryde, N.P., (49) *350*

Ryskin, G., (22) *291*
Sacher, R.S., (96) *116*
Sagert, N.H., (63) *496*
Saito, H., (43) *495*
Salley, D.J., (7) *236*
Saltsburg, H.M., (62) *351*
Sanders, J.V., (1) *548*
Sasaki, T., (26) *291*
Satoé, E., (31) *113*
Satoh, N., (20) *61*, (11) *227*
Satoh, S., (47) *495*
Savage, R.L., (68) *351*
Saville, D.A., (16) *47*, (15) *61*, (14) *382*, (31) *530*, (47) *550*
Savitskaya, E.M., (94) *497*
Sawitowski, T., (74) *552*
Sawyer, W.M., (26) *494*
Saykally, R.J., (76) *552*
Scatena, L.F., (6) *276*
Schade, B.E.H., (51) *114*
Schaeffer, H.F., (15) *529*
Scharfe, M., (54) *551*
Schätzel, K., (46) *350*
Schechter, R.S., (21) *216*, (36) *495*, (100) *498*
Schiefer, H.M., (78) *497*
Schiffrin, D.J., (65) *551*, (75) *552*
Schimpf, M.E., (36) *113*
Schmid, G., (74) *552*
Schmidt, H.F., (47) *454*
Schmitt, V., (36) *550*
Schowalter, W.R., (14) *382*, (31) *530*, (47) *550*
Schreiber, H.P., (33) *530*
Schulz, K.F., (18) *549*
Schulze, H., (9) *401*
Schunk, P.R., (28) *176*
Schure, M.R., (37) *113*
Schurz. J., (63) *114*
Schwartz, D.K., (4) *226*
Schwartz, L.W., (28) *156*
Scott, T.C., (6) *135*
Scotti, M., (67) *551*
Scriven, L.E., (23) *156*
Sebba, F., (28) *453*
Segal, D.L., (55) *495*
Sennett, P., (9) *47*, (13) *349*, (24) *529*
Serizawa, T., (24) *176*
Serra, T., (52) *531*
Sevick-Muraca, E.M., (73) *115*
Sharma, A., (30) *176*
Shaw, D.T., (61) *551*
Shawafaty, N., (90) *497*
Shchukin, E.D., (18) *135*

Shedlovsky, L., (19) *276*
Sheehy, D.P., (5) *419*
Shelly, Z.A., (1) *236*
Sheludko, A., (52) *495*
Shen, Y.R., (4) *276*
Sheppard, I.R., (2) *134*
Sheppard, S.E., (11) (12) *529*
Sherman, P., (15) *453*
Shiang, J.J., (76) *552*
Shibuichi, S., (20) *61*, (11) *227*
Shields, J.E., (92) *115*
Shim, M., (60) *551*
Shin, W.-T., (5) (7) *135*
Shinde, R.R., (73) *115*
Shinoda, K., (30) *277*, (14) (19) *453*, (46) *454*, (50) *495*
Shiundu, P.M., (42) *113*
Shortt, D.W., (75) *115*
Showalter, W.R., (16) *47*, (15) *61*
Siebert, P.C., (21) *112*
Siegel, R.W., (57) *551*
Silebi, C.A., (34) *113*
Singleterry, C.R., (20) *175*
Sita, L.R., (80) *552*
Sjöblom, J., (35) *454*
Skinner, S.M., (68) *351*
Skodvin, T., (35) *454*
Skoulios, A., (24) *277*
Slater, J. C., (8) *198*
Smilga, V.P., (8) *155*, (59) *351*
Smith, D., (10) *529*
Smith, D.L., (21) *276*
Smith, E. R., (42) *373*
Smith, F.M., (107) *116*
Smith, L.S., (53) *551*
Smith, N.J., (17) *419*
Smith, W.O., (20) *227*
Söderberg, I., (44) *454*
Sokolov, J., (67) *551*
Solans, C., (56) *455*
Somasundaran, P., (10) *175*
Spagnolo, D.A., (49) *217*
Sparnaay, M.J., (28) *349*, (24) *372*
Spencer, J., (98) *116*
Spencer, N.D., (9) *227*
Spicer, P.T, (53) *531*
Squire, P.T., (19) *47*
Stageman, J.F., (30) *550*
Stainsby, G., (15) *276*
Staples, E.J., (70) *114*
Stenkamp, V.S., (2) *418*, (24) *419*
Stephan, J.T., (74) *496*
Stern, O., (26) *349*
Sternling, C.V., (23) *156*

Stevenson, W.H., (42) *350*
Stewart, W.E., (3) *46*, (1) *60*,
Stiepel, C., (31) *277*
Stockman, J.D., (11) (28) *112*
Stone, H.A., (25) *494*
Storey, J., (17) *112*
Storhoff, J.J., (82) (85) (86) (87) *552*
Ström, G., (17) *227*
Stuart, M.A.C., (52) (53) *114*, (57) *531*
Sugden, S., (2) *215*
Sung, J., (7) *549*
Suslick, K.S., (62) *551*
Suzawa, T., (23) *549*
Suzin, Y., (25) *291*, (93) *497*
Swain, R.C., (2) *307*
Swift, H., (8) *244*
Szalma, J., (48) *217*
Szegvari, A., (9) *135*
Tabor, D., (44) (45) *373*
Tamai, H., (23) *549*
Tamamushi, B-I., (30) *277*
Tanford, C., (2) *276*, (14) *307*
Taniguchi, M., (21) *227*
Tarnawskyj, C.J., (54) (55) *350*
Taylor, P., (62) *455*
Taylor, S.E., (46) *217*
Teller, E., (93) *116*
Ten, P.G., (21) *315*
Terabe, J., (28) *277*
Tester, D.A., (8) *529*
Textor, M., (9) *227*
Tezak, B., (18) *549*
Thiessen, P.A., (3) *548*
Thomas, A.G., (54) *350*
Thomas, T.B., (20) *494*
Thomas, T.R., (3) *226*
Thomas, W.D.E., (38) *277*
Thompson, D.W., (4) *315*, (3) *493*
Thompson, H.D., (42) *350*
Thomson, W., (5) *493*
Thorburn, R.C., (80) *497*
Timm, E.E., (33) *550*
Tischler, D.O., (29) *199*
Titijevskaia, A.S., (43) *373*
Titijevskaya, A.S., (21) *494*
Tomaić, J., (33) *176*
Toporov, Yu.P., (7) *155*
Tordai, L., (39) *277*, (12) *419*
Tosatti, E., (33) *373*
Townsend, D.F., (9) *315*, (60) *496*
Trapeznikov, A.A., (83) *497*
Trasatti, S., (7) *348*
Traube, J., (12) (13) *307*
Trivedi, A.S., (101) *116*

Tronel-Peyroz, E., (12) *175*
Tscharnuter, W.W., (48) *113*, (47) *350*
Tsouris, C., (5) (6) (7) *135*
Tsujii, K., (20) *61*, (11) *227*, (47) *495*
Tyndall, J., (1) *14*
Ueda, Y., (37) *199*
Ugelstad, J., (27) *277*, (59) (61) *455*, (24) *549*
Ulevitch, I.N., (14) *529*
Ulman, A., (67) *551*
Urdahl, O., (20) *47*
Urick, R.J., (82) *115*
van de Ven, T.G.M., (12) *61*
van den Ende, D., (21) *529*
Van den Tempel, M., (32) *494*
van der Linde, A.J., (44) *350*
van der Minne, J.L., (4) (9) (10) *291*
van der Waarden, M., (5) *291*, (42) (43) *454*
van Duyne, R.P., (72) *552*
van Leeuwen, H.P., (24) *349*, (44) *350*
van Oekel, J.J., (26) *156*
van Olphen, H., (41) (43) *530*
van Oss, C.J., (44) *217*, (12) *372*
van Voorst Vader, F., (14) *372*
Vanderhoff, J.W., (27) *277*, (61) *455*, (24) *549*,
Vbranac, M.D., (6) *175*, (31) *199*
Velev, O.D., (49) (50) (51) *551*
Verheijen, H.J.J., (31) *176*
Vermehren, C., (44) *454*
Verwey, E.J.W., (10) *198*, (2) *348*, (23) *349*, (49) *373*, (1) (8) (9) *381*
Vincent, B., (46) *350*
Virden, J.W., (23) *419*
Visser, J., (12) (13) (17) *372*
Vogel, G.C., (25) *198*
Vold, M.J., (33) *113*, (21) (22) (23) *372*, (6) *400*
Vold, R.D., (33) *113*, (6) *400*
von Smoluchowski, M., (2) *400*
von Szyszkowski, B., (7) *307*
Vouk, V.B., (18) *549*
Wade, W.H., (21) *216*
Walker, M., (65) *551*
Walker, R.D., (11) *529*
Walters, K., (1) *46*, (11) *47*
Walton, C.W., Jr., (39) *217*
Wang, G., (90) *497*
Wang, S.Q., (27) *530*
Wang, Y., (12) *135*
Wantke, K.-D., (10) *216*
Ward, A.F.H., (39) *277*, (12) *419*
Ward, T.C., (5) *419*

Warren, P., (20) *419*
Wasan, D.T., (21) *453*, (51) *454*, (2) *493*, (33) *494*, (34) (40) (41) *495*, (101) (102) *498*
Washburn, E.W., (42) *217*
Watanabe, A., (14) *349*
Watillon, A., (12) *15*, (12) (13) (14) *549*
Watkins, R.C., (92) *497*
Watts, J.F., (39) *199*
Weaire, D., (14) *494*
Webers, V.J., (69) *351*
Webster, A.J., (64) (65) *455*
Wedlock, D.J., (78) (81) *115*, (41) *454*
Weerawardena, A., (25) *176*
Weidner, D.E., (27) *156*
Weiner, B.B., (48) *113*
Weiser, H.B., (27) *349*
Weiss, J., (80) *115*, (17) *156*
Weitz, D.A., (29) (30) *494*, (6) (7) *549*
Wennerström, H., (15) *382*
Wenzel, R.N., (1) (2) *226*
Wesson, S.P., (5) (6) (7) (8) *227*, (6) *419*
West, W., (13) *529*
Westcott, S.L., (73) (79) *552*
Whipple, C.L., (78) *497*
Whitcomb, D.L., (13) *529*
White, H., (67) *551*
White, L.R., (22) *349*, (37) *350*, (32) *373*
Whitney, C.E., (38) *217*
Whittingham, K.P., (22) *453*, (97) *498*
Whyman, R., (65) *551*
Widom, B., (11) *315*
Widon, B., (13) *175*
Wiersema, P.H., (39) *350*, (5) *400*
Wiese, G.R., (15) *349*, (7) *381*
Wiglow, H., (29) *277*
Wijmans, C.M., (74) *115*
Willenbacher, N., (45) *530*

Williams, P.A., (17) (20) *419*
Wilson, A., (44) *495*
Winterton, R.H.S., (44) *373*
Wolfe, J.A., (29) *199*
Wolthers, W., (21) *529*
Woo, T.M., (2) *315*
Woodbridge, R.F., (19) *549*
Woods, M.E., (20) (21) (22) *549*
Woods, W.W., (39) *495*
Woodward, J.T., (4) *226*
Wu, J.S., (51) *114*
Wu, S., (20) *198*, (44) *199*
Wyatt, P.J., (75) *115*
Xu, J., (27) *530*
Xu, R., (43) *350*
Xu, S., (93) *553*
Xue, Q., (52) *551*
Yamano, Y., (34) *550*
Yan, Y., (18) *61*
Yates, D.E., (7) *381*
Yee, C., (67) *551*
Yeung, A., (2) (3) *452*
Yiacoumi, S., (5) (7) *135*
Yoshizawa, H., (56) *551*
Young, T., (1) *174*
Youngblood, J., (10) *227*
Zehner, R.W., (80) *552*
Zhang, L., (6) *215*
Zhang, Z., (52) *551*
Zheng, W., (81) *552*
Zia, I.Y.Z., (22) *216*
Zimm, B.H., (12) *276*
Zisman, W.A., (35) (36) *199*, (33) *217*, (57) *351*
Zsigmondy, R., (47) *113*, (3) *548*,
Zuiderweg. F.J., (56) *496*
Žutić, V., (33) *176*

INDEX

Absorption of light, 75, 76, 85
Absorption of sound, 96, 511
Accelerated testing, 443
Acceptor-donor interactions, *see* Acid-base interactions
Acid-base interactions
   charge adjuvants, 542
   componental theory, 179–189
   contributions to interfacial tension, 185–189
   dispersibility map, 510–511
   fiber/fiber interactions, 540
   in nonpolar media, 281–289
   in polymer adsorption, 403–411
   solid/liquid interactions, 166, 190–197
   solid surfaces, 196–197, *table* 196
   spontaneous adsorption, 249, 449, 499
ACN, *see* Alkane carbon number
Acoustic scattering, 53, 95, 106
Adhesion, 87, 179, 343, 482, 540. *See also* Work of adhesion
Adsorption,
   adsorbate/adsorbent definitions, 228
   acid-base nature of solids, 406–408
   Bancroft's rule, 424–426
   block copolymers, 416–418
   in chemical demulsification, 447–448
   chemical processing aids, 130–132
   complex ions, 394–396
   dispersing agents, 499–500
   dye, 105, 504–506
   dynamic, 280
   flocculation, 523–526
   free energy, 257, 304, 407, 424, 448
   gas, 99, 101, 103, 106, 138, 167, 168, 216
   Gibbs, 168, 230, 294, 296, 309, 461
   heats of, 197, 406, 408, 501
   on homotattic substrates, 532, 537–538
   isotherm
      gas, monolayer, 101–102, 239–242, 537
      gas, multilayer (BET), 99–100
      Gibbs adsorption, 294
      Langmuir, 104, *table* 104, 501, 538
   to obtain heats, 408
   spreading pressure, 167–168, 192–195
   values, *table* 195
   standard state free energies, 302–305
   competition with micelles, 250–265
   near phase boundaries, 308–331, 474–480
   negative, 231, 296, 503
   in nonpolar media, 279–287
   polymer, 281, 282, 374, 403–411, 519, 525, 526
   polymer configurations, 405
   positive, 230, 254, 310, 461
   potential, 104, 501, 504
   preferential, 173, 317, 321, 322, 502–504, 547
   rate of, 52, 206, 404, 409, 410, 411, 467
   from solution, 103, 292–307, 402–408, 456, 500–506
   mixed solvent, 412
   surface-tension titrations, 506
   thermodynamics, 292–307
   virial equation of state, 101, 538
   at a water surface, 246–250
Aerosol OT, 131, 259, 269, 485. *See also* AOT
Aerosols
   clouds, 539
   coalescence of, 141
   contamination by, 195
   droplet sizing of
      Fraunhofer diffraction, 10
      photozone detection, 84
      time-of-transition method, 83
   for monodisperse metal oxides, 534
   for nanoparticle synthesis, 546
   Ostwald ripening of, 151
   smokes, 539
   for xerography, 541
AFM, *see* Atomic force microscope
Agglomeration, of suspensions, 69, 421. *See also* Flocculation

Agglutination tests, 532
Aggregate
　definition, 499
　milling equipment used for, 61, 91, 117, 119, 124, 444, 449
Aggregation number of dyes, 104–105, table 104, 274
Alkane carbon number (ACN), 433–434
Alumina, 127, 321, 342, table 360, 391, 403, 534
Aluminum nitrate, 323, 324, 395
Aluminum salts and colloidal stability, 394–396, 524
Aluminum sulfate, 395
Amphipathic, definition, 228
　particles, 426–429
　solutes, 248, 251, 253, 254, 309, 474, 477
Amphipathy, definition, 228
Amphiphile(s), 162, 228, 247, 256, 272, 275, 279, 433
Amphiphilic, definition, 228
Anchoring group(s), 383, 402, 405, 406, 411, 416, 417
Andreason sedimentation pipette, 72
Angle of repose, 522
Anomalous water, 152, 248, 249, 308. *See also* Polywater; Water, anomalous
Antifoams, 308, 477, 478, 482, 483, 484, 486, 488, 489, 491
Antifoaming, 480–489
　agents, 308, 484, 488, 489
　silicones, 486, 488
Antonoff's rule, 164
AOT (sodium di-2-ethylhexylsulfosuccinate or sodium dioctylsulfosuccinate), 259, 260, 319
Applied forces, 48–49
　centrifugation, 50–51
　electric, 51
　gravity, 49–50
　ultrasound, 52–54
　viscous, 51–52
Argon, 101, 168, table 195
Asphalt emulsion, 234, 438
Atomic force microscope (AFM), 220, 224, 368
Attapulgite clay, 509, 510
Attractive potentials
　Hamaker, 355–364
　hydrogen bonding, 182
　Lennard-Jones (6–12), 136, 353
　Lifshitz, 364–369

London constant, 353, 354
London dispersion energy, 182, 353
　metallic bonding, 182
　effect on surface tension, 138, 193
Attritors, 126, 127, 134
Autocorrelation function, 11, 12, 13, 92, 93
Autophobic dewetting, 163, 164

Bactericidal activity, 234
Ball mills, 121, 124, 125, 126, 127, 133, 134
Banbury mixer, 130, 133
Bancroft's rule, 424–426, 427, 429, 432, 437, 446, 555
Barium sulfate, 403
　Barite, table 319
　dispersion energy of, table 195
　preparation of, 444
Bartell cell, 45, 210, 211, 212, 215, 556, 558
Bashforth-Adams tables, 149, 158, 203
Bénard cells, 154
Benzoin, 396
Berthelot's principle, 178, 180, 363
BET (Brunauer, Emmett, and Teller), theory, 99, 100, 105
Betaine, 470, 608
Betel nut, 162
Bilayer, 241, table 275, 488
Bimolecular leaflet, 241, 470–471
Bingham body, 21, 25, 513
　two-dimensional, 468
Biodegradation, 234
Black films, 456, 468, 470, 471
Block copolymers, 233, 272, 406, 411, 416–418, 485, 547
　compositions of, table 417
Blood, 14, 19, 87, 97, 106, 223, 249, 380
Boiling points, table 139
Bond number, 153, 423
Bordeaux mixture, 426
Brandy tears, 154, 474
Breaking of emulsions, 124, 420, 446, 447, 489. *See also* Demulsification
Bridging flocculation, 413, 523–525
Brownian motion
　capillary hydrodynamic chromatography, 77
　diffusion constant, 55
　mean displacement, 55
　particle collisions, 374, 380, 481
　in quasi-elastic light scattering, 10–13, 92–93

random walk, 93
rates of flocculation, 384–386
sedimentation equilibrium, 57–58, 77
thermal FFF, 79–81
Bubbles,
bubble rise, 38, 287–290
Hadamard regime, 288–289
observation, 289, 290
Stokes regime, 289–290
geometry of, 456–459
Jamin effect of, 222–224
lifetime of, 311
maximum bubble pressure, 205–206
pressure increase, see Laplace equation
sessile and pendent, 203
Builders, 235, 236
common builders, table 235
Built-up films (L-B films), 243
BuOH (butanol), table 139
Butter, 420, 421

Cadmium selenide, 534, 547, 548
CAEDMON (Computed adsorptive distribution in the monolayer), 102, 538
Calcite (calcium carbonate), table 319, table 360, table 367
Capacitance, 329, 331
Capillarity, 157, 180, 200, 308, 314, 475, 477, 478, 479, 540. See also Capillary; Jamin effect; Kelvin equation; Laplace equation; Marangoni effect; Marangoni elasticity; Ostwald, ripening; Spreading, Spreading coefficient; Surface tension; Wetting; Work of adhesion; Work of cohesion; Work of desorption; Young-Dupré equation
coalescence, 141–142
cohesion, 144, 146, 161, 202, 240, 246, 353, 468
componental theory, 180, 189
critical surface tension, 190, table 191
curved interfaces, 146–148, table 148, 200
definition, 308
detergency, 170–171
dewetting, 163, 164, 173, 403, 473, 482, 483
encapsulation, 142–144
emersion, 146, 160, 219, 243, 403
entering coefficient, 482
interfacial tensions, table 145

regular solution theory, 311–313
relation to phase diagrams, 308–315
surface free energy, 160, 162, 163, 165, 167, 187, 189, 191, 192, 194, 370, 481
thermodynamic functions, 140, table 140, 302
Zisman plot, 190
Capillary
blood flow, 14
constant, 153, 173
curvatures, 146–148, table 148
drainage, 462
in electric emulsification, 445
in electrokinetics, 335–340
flow, 148, 152–153, 155, 173, 465
in pores, 210–212
in scratches, 221–224
forces, 43, 153, 212, 213, 423
height, 200–201
in lungs, 241
number, 173
Plateau borders, 464–467
pressure, see Laplace equation
rise, 36, 201
stresses in foams, 468
viscosimeters, 34–37
waves, 173, 270
Capillary hydrodynamic chromatography (CHDC), 77–78
Capillary hydrodynamic fractionation (CHDF), see Capillary hydrodynamic chromatography
Capillary viscosimeter, 34–37
Carbon black
adsorption of argon by, 101
dispersed in hydrocarbons, 280–281, 319–321, 342
dispersed in water, 342
dry grinding of, 118
fractal structure of, 59
graphitized, 538
hydrophobic surface of, 403, 409, 426
steric barriers required, 399–400
Carbon disulfide, 249
Casein, 380, 421, 468
Cassie and Baxter equation, 221
Castor oil, 237
Catalysis, 102, 103, 186, 196, 279, 292, 406, 538, 546
CCC, see Critical coagulation concentration
CCD (charge-coupled device), 220
Cellulose, 23, 234, 281, table 346, 499

Centrifugation, *see also* Field flow fractionation
  comparisons, *table* 106
  contact angles on powders, 212
  demulsification by, 446
  disk centrifuge, 75–76
  efficiency of, 74
  line start, 75
  terminal velocity, 50
  ultracentrifuge, 76–77
  uniform start, 76
  for zeta potential, 332
Characteristic times, 32, 36, 49
Charcoal, 249, 503
Charge adjuvant, 542
Charge density, 286, 324, 325, 326, 327, 328, 329, 379, 448, 525
Charge transfer, 345
Charge-to-mass ratio, 287, 343
CHDC, *see* Capillary hydrodynamic chromatograpy
CHDF, *see* Capillary hydrodynamic fractionation
Chemical processing aids, 130
Chemisorption, 195, 394, 418, 547
Chromatography, 77, 78, 80, 196, 264, 406
Clausius-Clapeyron equation, 408
Clay suspensions, 23, 27, 189, 332, 334, 387, 509, 510, 512, 514, 520, 521, 540
Cloud point, 236, 413
Clouds, 1, 177, 344, 353–354, 364, 539, 609
CMC, *see* Critical micelle concentration
Coagulation, 371, 383, 384, 385, 390, 391, 393, 394, 413, 414, 442, 523
  kinetics, 383–387
  rapid, 384, 385, 532
Coalescence, 83, 131, 292, 308, 426, 434, 441, 445, 446, 449, 460, 461, 486, 489
  of droplets, 141–142, 423–424
Coating methods, *table* 521
  dip coating, 172–173
  spin coating, 172
  spray coating, 172
Cohesion, 144, 146, 161, 202, 240, 246, 353, 468
Collectors (flotation), 162
Colligative properties, 58, 251, 252
Colloid mill, 119, 121, 122, 133, 134, 443
Colloid stability, *see* Dispersions, stability of

Colloid vibration potential (CVP), 342
Colloidal crystals, 538–539
Colloidal systems, model, 532–539
  particles of uniform size, 436, 532, 535, 536
  surfaces of uniform composition, 536–538
Comminution of aggregates, 61, 91, 117, 119, 124, 444, 449
Complex ion chemistry, 394–396
Componental theories, Chapter 8, 177–197
  Berg's correction for entropy, 189
Compressibility, 97, 240, 290, 381, 511
Concrete and cement, 103, 513, *table* 540, 545
Conductivity, *see* Electrical conductivity
Constant surface charge density, 379
Constants, physical, Appendix 26.1, 554
Contact angle(s)
  advancing and receding, 166, 167, 218, 221, 223, 225, 226
  apparent, 218, 219, 221
  on autophobic particles, 426–429
  autophobic wetting, 164–167
  capillary flow, 152–153
  capillary height, 200–201
  constancy of, 143, 158
  critical point wetting, 313–314
  definition, 157–158
  degrees of solid/liquid interaction, 160–162
  detergency, 170–173
  variation with electric charge, 173–174
  on fractal surfaces, 60
  hysteresis of, 226
  on irregular surfaces, *see* Wetting of irregular surfaces
  in Laplace equation, *see* Laplace equation
  in mercury porosimetry, 102–103
  methods to measure, 208–215
    on fibers, 209–210
    goniometer, 208–209, 215, 216
    interfacial meniscus, 209
    on irregular fibers, 224–226
    on Langmuir balance, 213
    manufacturers, 214–215
    on powders, 210–214
      Bartell cell, 210
      centrifugation, 212–213
      by heats of immersion, 213–214

with Wilhelmy plate, 45,
201–202–203, 215–216, 219,
224, 242, 270, 423
on solid substrates, 190–197
in Washburn equation, 131
water on solid substrates, *table* 161
in wetting, 131
in Young-Dupré equation, 157,
218–219, 427
Zisman plot, 190
Contact angle goniometer, 208, 215, 216
Contact electrification, 343, 344, 345
Continuous phase, 17, 75, 421, 423, 424,
427, 436, 439, 443, 456
Cosorption lines, 312
Couette flow, 436
Couette rheometer, 16, 42, 513
Coulter counter, 86, 87
Counterions
complex ions, 394–396
in the diffuse double layer, 316, 324,
326, 328, 354, 463
in nonpolar media, 280, 285
in Stern theory, 329–331
in transport phenomena, 253
CPVC, *see* Critical pigment volume
concentration
Creaming of emulsions, 35, 49, 71, 98,
122, 384, 420, 442, 443, 446
Creation, Divine, 270
Creep curve, 44
Creep flow, 21, 32, 40, 44, 132, 436, 521,
609
Critical coagulation concentration (CCC),
391–394, 395
Critical flocculation temperature, 412
Critical micelle concentration (CMC),
250–260
definition, 252
effect of structure, 257–260
equilibrium or phase separation?,
254–257
inverse micelle, 279
Krafft point, 264–265
methods to measure, 253–254
density, 254
examples, *figure* 255
Hittorf transport number, 253–254
refractive index, 254
surface-tension lowering, 254
turbidity, 254
osmotic coefficient, 252

on phase diagram, 272
solubilization, 261–264
surface-tension titration, 506
values, *table* 258
Critical packing parameter, 273–275,
table 275
Critical pigment volume concentration
(CPVC), 517
Critical point, 256, 313–314, 413, 436,
476, 478
Critical point wetting, 313–314
Critical stabilization concentration, 395,
figure 396
Critical surface tension(s), 190
values, *table* 191
Crude oil, 203, *table* 286, 441, 468
Crystal growth, 154, 526, 527
nucleation, 4, 53, 88, 152, 165, 220,
475, 526, 527, 528, 533, 609
CSC, *see* Critical stabilization
concentration
Curvature(s), 146–148
formulae, *table* 148
principal radii of, 556
CVP, *see* Colloid vibration potential
Cyclone mills, 52, 73

Daniel flow point method, 131
Dark field illumination, *see*
Ultramicroscopy
Davies' group numbers, 260, 430
values, *table* 431
Debye forces, 179, *table* 179, 185, 187
Debye length, *figure* 326, 375, 386, 392,
394, 396, 536
definition, 326
variation with ionic strength, 326
Debye-Hückel approximation, 378–379
Deflocculation
of carbon black/cellulose esters, 281
definition, 352
dilatancy, due to, 25–27
kaolin/water with TSPP, 321
phase diagrams, 411–413
preferential adsorption, 321–322
in rheology, 514–516
sample preparation, 510
sedimentation volumes, 19, 508–509
of suspensions, 506–508
titania dispersions, 66
Defoamer, 486, 489
Degrees of liquid/solid interaction,
160–161

Demulsification, 124, 387, 420, 426
  by chemical methods, 447–448
  by mechanical methods, 446
Deoxyribonucleic acid (DNA), 545, 547
Depletion flocculation, 413–414
Depletion stabilization, 413–14
Derjaguin approximation, 377–378, 389
Desorption
  dynamic, 152, 280, 410, 424
  in dynamic surface tension, 269
  free energy, 141, 428–429, 448–449
  of ions, 173, 345
  of polymers, 284, 406
Destabilization of dispersions, 425. *See also* Flocculation
Detergency, 170–171. *See also* Detergents
  adsorption from solution, 103
  use of builders, 236
  example, *figure* 172
  Ferguson effect, 250
  microemulsions, 448
Detergents, *see also* Detergency; HLB scale
  bactericidal activity, 234
  biodegradation, 234
  commercial sources, 230, 431
  common detergents and builders, *table* 235
  demulsification, 447
  dynamic surface tensions, 268–270
  foam stability, 489
  gelatinous surface layers, 465
  Gibbs excess, 300–301
  journals, 230
  Krafft point, 264–265
  Marangoni effects, 467
  oil soluble, 318
  structure, 229
  types, 234, *table* 235
Dewetting, 163, 164, 173, 403, 473, 482, 483
Dielectric spectroscopy, 440–441
Die swell, 33
Diethylene glycol, 310, 311
Differential heat, 408
Diffraction of light, 533, 538
Diffusing wave spectroscopy, 93
Diffusion
  apparent velocity, 55
  Brownian motion, 11, 13, 54–57, 58, 59, 77, 92–93, 374, 481, 523, 532
  coefficient, 12, 55, 92, 261
  in emulsion polymerization, 263, 449, 452

  equation of Ward and Tordai, 269, 410
  in foam lamellae, 466–467
  of gas in foams, 456, 459, 461
  versus Marangoni flow, 154
  mean free path, 55
  mixtures of homologs, 261
  in nucleation and crystal growth, 526–528
  of polymers, 380
  in quasi-elastic light scattering, 92–93
  rate of, 55, 154, 269, 527
    adsorption, 136, 409–411
    flocculation, 384–386
  sedimentation equilibrium, 57
  sedimentation FFF, 78–79
  thermal FFF, 80
Digestion of precipitates, 105, 151, 152
Dilatant flow, 25, 26, 27, 512, 514, 515
Dip coating, 172–173
Dipole-dipole interaction, 177, 179, *table* 179, 354, 403, 510, 552
Disjoining pressure, *See* Disjunctive pressure
Disjunctive pressure, 367, 536
Discontinuous phase, 420, *table* 420
Disk centrifuge, 75, 76, 137, 138
Dispersants, 499–500
  chemical processing aids, 130–132
  common, *table* 235
  Daniel flow-point method, 131
  intermittent milling, 121–122, 444–445
  in milling, 117, 125
  oil soluble, 284–285
  polymer, *table* 417
  published lists, 499–500
    Bernhardt's table, 67
  effect on rheology, 23–25, *figure* 24
  testing, 526
Dispersed phase
  acoustic absorption, 96–98
  definition, 420, *table* 420, 456
  electrical conductivity, 439–441
  electrokinetic phenomena, *table* 333
  emulsion polymerization, 451–452
  emulsions, internal phase, 436–441
  examples, *table* 421–422
  light scattering, 3
  Ostwald ripening, 151–152
  effect on rheology, 17
    dilatancy, 25–27, *figure* 27
    emulsions, 437–439
    plastic flow, 23, *figure* 24
  effect on speed of sound, 53, 97–98
Dispersibility maps, 510–511

Dispersion energy
  componental theory for interfaces, 179–185, *table* 179, *figure* 181
  contributions to DLVO theory, 387–394
  contributions to mercury, *table* 183
  contributions to water, *table* 184
  Hamaker theory, 355–364, 370–371
  Lennard-Jones equation, 353
  Lifshitz theory, 364–369
  London
    equation, 353
    constant, 353–354
  between molecules, 177–179, *table* 179, 353–354
  between particles, 354–371
  of polar liquids, *table* 194
  retardation, 352, 360–361, 366, 400
  of solid substrates, 190–196
    Fowkes' theory, 191–192, *figure* 192
    low energy surfaces, from contact angles, *table* 193
    from π_e, *table* 195
  in steric stabilization, 398–400
  in work of adhesion, 246
  van der Waals
    equation, 353
    forces, 177, 388
Dispersion of light, 178
Dispersions, stability of, 58, 318, 352, 383, 413
  block copolymers, 416–418
  comparison of deflocculated and flocculated, *table* 508
  complex ions, 394–396
  critical coagulation concentration, 391–393
  DLVO theory, 387–394
  electrostatic stability in nonpolar media, 396–398
  electrosteric, 415–416
  polymer stabilization, 402, 411–413
  primary minimum, 394, 414, 415
  secondary minimum, 394
  steric stabilization, 398–400
  steric versus electrostatic, 398–400
Dispersions
  kinetics of coagulation, 383–387
  lyocratic, 383, 398
  lyophilic, 228, 273, 309, 383, 398, 402, 507, 508, *table* 508
  lyophobic, 179, 228, 273, 309, 383, 393, 396, 402, 409, 448, 508, *table* 508, 534
Dissolution, 165, 264, 534

Distributions, 63–65
  cumulative, 64, 71
  Gaussian or normal, 63
  Log normal, 64–65
DLVO (Derjaguin, Landau, Verwey, and Overbeek)
DLVO theory of stability, 173, 284, 316, 333, 376, 386, 387–394, 396
  Derjaguin approximation, 377, 389
  electrocratic, 383, 387, 388, 523, 526
  electrosteric, 388, 415, *figure* 417
DNA, *see* Deoxyribonucleic acid
DN-AN, *see* Donor-acceptor numbers of Gutmann
Donnan equilibrium, 555
Donor-acceptor interactions, 177, 179, *table* 179, 186, 196, *table* 196, 283, 403, 406, 407, 409, 508. *See also* Acid-base interactions
Donor-acceptor numbers of Gutmann, 186
Double layers, *see* Electrical double layer
Drago
  constants, *table* 179, *table* 188
  equation, *table* 186
Dropping mercury electrode, 334
Drops, sessile and pendent, 202–203
  autophobic dewetting, 163–164, *figure* 164
  constancy of contact angle, *figure* 158
  drop weight, 201
  equation for shape of, 149–150
Du Noüy ring, 423
Duke-Fabish model, 346
Duplex film, 164
Dyes, 103, 104, 500, 543
  adsorption, 105, 504–506
  aggregation number, 104–105, 274
  recommended dyes, *table* 104
Dynamic contact angles, 169, 193, 215, 216
Dynamic surface tensions, 154, 208, 266, 268, 269, 270, 288, 467, 481

E & C numbers of Drago, 186–188
  values for organics, *table* 188
  values for solids, *table* 408
EG, *see* Ethylene glycol
Einstein equations, 12, 55, 57, 77, 93, 513, 532, 555
Elasticity
  of fluids, 29, 30, 40,
  of insoluble monolayers, 242,

Elasticity (*Continued*)
  mechanical, 20–21
  stress relaxation, 32, 44, 436
  of surface films, 265–268, 462, 464, 466
  of suspensions, 468, 469, 517, 521, 536
  viscoelasticity, 29, 30, 31, 40, 44, 242, 517
  viscoelastometers, 44
Electrical emulsification, 445
Electrical charges
  in adhesion, 146
  contact electrification, 343, 344, 345
  in dispersions, 316–351
  in dry powders, 343–347
  at liquid interfaces, 173–174
  in nonpolar media, 282–287
  origin of charge
    in aqueous media, 317–318
    in nonpolar media, 318–321
  point of zero charge, 317–318, 319
Electrical conductivity
  CMC determination by, 253, 262
  of emulsions, 439–441
  of nonpolar media, 279–280, 463
    charge-to-mass ratio, 343–344
    electrical charges, 282–287, 319–344
    electrophoresis, 339
    half-value times, *table* 286
    and inverse micelles, 279
    test for emulsion type, 423
Electrical double layer, 324–331, *figure* 330, *figure* 331. *See also* Critical coagulation concentration; Debye length; Ionic strength
  in black films, 470–471
  in colloidal crystals, 538–539
  in contact electrification, 345–346
  counterions, 316, 324, 394–396
  Debye length, 375, 386, 392, 394, 396, 536
  definition, 316
  dielectric spectroscopy, 440–441
  in DLVO theory, 387–391
  electroacoustic effect, 341–342
  in explosions and fires, 285
  in foam stability, 463
  in Hückel theory, 335
  in Lifshitz theory, 369
  O'Brien and White's calculations, 335, *figure* 336, *figure* 337
  Poisson-Boltzmann equation, 316, 324–325, 328, 331, 377

  repulsion, 316, 334, 352, 376, 515, 523–524, 525
    between flat plates, 374–376
    between spheres, 377–379
    in thin films, 463
  Schulze-Hardy rule, 393–394
  Smoluchowski theory, 334–335
  Stern theory, 329–331
  triboelectric series, 345–346, *table* 346
  in Voltaic piles, 345
Electrical properties of dispersions, *see* Electrical conductivity
Electroacoustic sonic amplitude (ESA), 341, *table* 343
Electroacoustics, 98, 341–343, 512
  techniques, *table* 343
Electrocratic, 383, 387–388, 523, 526. *See also* Electrical double layer, repulsion
Electrokinetics, 331–333, *table* 333
  colloid vibration potential, 342
  electroacoustic, 337, 341–343, 512
  electro-osmosis, 332, *figure* 332, *figure* 333, 338, 339–340, *figure* 340
  electrophoresis, *figure* 332, *figure* 334, 334–339, *figure* 339, 543–545
  electrophoretic mobility, 318, 322, *figure* 322, 332, 334, 341
    definition, 334–335
    at low conductivity, 339
    range of applicability of limiting equations, *table* 336
  sedimentation potential, *table* 333
  streaming current, 340–341
  streaming potential, 332, *table* 333, 340–341, 347
Electrolysis, 253, 332, 546
Electrolytes
  effect on surface tension, 230–231
  indifferent, 523
  nonpolar, 285, 295–296, 319
  Schulze-Hardy rule, 393, 523
Electron donor-acceptor interactions, *see* Acid-base interactions
Electron microscope, 81, 83, 93, 532
Electro-osmosis, 332, *figure* 332, *figure* 333, 338, 339–340, *figure* 340
Electrophoresis, *figure* 332, *figure* 334, 334–339, *figure* 339, 543–545
Electrophoretic display (EPID), 285, 543–545
Electrophoretic mobility, 318, 322, *figure* 322, 332, 334, 341
  definition, 334–335

Electrostatic charge, 279, 374, 396. *See also* Electrical double layer
explosions and fires, 285
Electrosteric stabilization, 415–416, 417
Ellipsometry, 168, 169, 270
Elutriation, 73
Emersion, 146, 160, 219, 243, 403
Emulsifier(s), 233–236. *See also* Davies' group numbers; Dispersants, published lists; Ferguson effect; Phase inversion temperature
  amphipathic particles, 426–429
  Bancroft's rule, 424–426
  bibliographies, 431–432
  chemical moieties, *table* 233, 309
  common emulsifiers, *table* 235
  definitions, 420, *table* 420
  effect of structure, 259–261
  HLB scale, *see* Hydrophile-lipophile balance
Emulsion polymerization, 263, 450–452, 479, 506, 528, 534, 535
  surface-tension titration, 506
Emulsion stability, 53, 432, 433, 441, 442
Emulsion(s), Chapter 22, 420–455. *See also* Emulsifier(s); Emulsion polymerization; Ferguson effect; Hydrophile-lipophile balance; Microemulsions; Miniemulsions; Ostwald ripening
  accelerated testing, 443
  asphalt, 234, 438
  Bancroft's rule, 424–426, 432, 437, 446, 555
  breaking, 124, 420, 489
    by chemical methods, 447–448
    by mechanical methods, 446
  coalescence of droplets, 141–142, 423–424
  concentrated, 423, 437, 438, 439, 451, 456, 467
  concentration of internal phase, 426–441
  continuous phase, 17, 75, 421, 423, 424, 427, 436, 439, 443, 456
  creaming, 35, 49, 71, 98, 122, 384, 420, 442–443, 446
  Davies' group numbers, 260, 430
    values, *table* 431
  definitions, 421
  electrical charges, 173–174
  electrical conductivity, 439–441
  emulsification, 121, 123, 132, 250, 285, 287, 387, 421, 424, 426, 444, 445, 451
  external phase, 443, 444, 446, 449
  food, *table* 421–422
  interfacial tension, 423
  internal phase, 423, 424, 436–441, 443, 444–446, 449, 451, 536
  making, 117–124, 443–446
    by condensation, 444
    by electric emulsification, 445
    by intermitten milling, 444–445
    by phase inversion, 443–444
    by special methods, 445–446
  mechanical properties of the interface, 434–436
  monodisperse, 536
  Oil-in-water (O/W), *figure* 96, *figure* 97, *figure* 426, *figure* 430, *table* 431, *figure* 433, *figure* 442, 447
    conductivity, 423, 439–441
    DLVO stability calculation, 390, *figure* 391
    examples, *table* 421–422
  particle sizing
    by electrozone detection, 86–87
    by Fraunhofer diffraction, 90–91
  phase inversion, 432, 437, 441, 443, 476, 511
  phase-inversion temperature (PIT), 432–434, 443
  Pickering, 436
  rheology, 437–439
  silicone foam inhibitors, 486
  speed of sound in, 53, 96–98
  spontaneous emulsification, 259, 424
  stability, 53, 432, 433, 441, 442
    effect of drop size, 390
    freeze-thaw, 392
    Lifshitz constants, 366–367
    measurement of stability, 441–443
    relation to phase diagram, 308–309
  terminology, *table* 420
  Traube's rule, 247, 257, 260, 304, 305–306, 409
  type, 420–423, 432, 439
    determination, 420–423
  ultrasonic techniques to detect separation, 510–512
  W/O (water-in-oil), *figure* 425, *figure* 430, *table* 431, *figure* 433
    conductivity, 423, 439–441
    examples, *table* 421–422

Encapsulation of one liquid by another, 118, 142–144, 308, 545
Entering coefficient, 482
Enthalpy, 140, *table* 140, 160
 acid-base interactions, 186–189, 407–408
Entropy, 141, 160
 Berg's correction for entropy, 189
 role of in hydrophobic effect, 246–248, 304, 448–450
 of polymers on solids, 405
EPID, *see* Electrophoretic display
Equations
 BET, 100
 Casson, 517
 Clausius-Clapeyron, 408
 Coulomb (charge-charge), 556
 Debye
  Dipole-induced dipole, 556
  Double layer thickness, 326
 de Gennes, 168
 Derjaguin, 378
 Drago, 186
 Duhem, *table* 140
 Einstein
  Brownian motion, 55
  diffusion constant, 12
  rheology of dispersions, 513
 Fowkes, 183, 187
 Frumkin, 301
 Fuch, 384
 Gibbs, 251, 254, 294, 296
 Gibbs-Duhem, 293, 294, 297
 Gibbs-Thomson, 151
 Girifalco-Good, 181
 Hamaker, 356, 357–359
 Herschel-Bulkley, 519
 Hooke, 20
 Hückel, 335
 Keesom (Dipole-dipole), 556
 Kelvin, 151
 Kipling and Testor, 503
 Langmuir
  gas adsorption, 100
  solution adsorption, 301, 303, 501
 Laplace
  capillary pressure, 147
  speed of sound, 53
 Lennard-Jones, 353
 Lifshitz, 362
 London, 353
 Marangoni, 153
 Maxwell
  conductivity of emulsions, 439
 speed of light, 2
 viscoelasticity, 31
 Morton, 450
 Neumann, 143
 Newton
  force, 48
  rheology, 17
 Poiseuille, 19
 Poisson-Boltzmann, 325
 Rayleigh, 4
 Reynolds, 52
 Smoluchowski, 334
 Stokes, 48
 Szyszkowski, 301
 Van der Waals, 353
 Ward and Tordai, 410
 Washburn, 131, 210–212
 Wenzel, 218–219
 Young-Dupré (Young-Gauss), 157
Equations of state, 267
 Frumkin, 301–302, 556
 virial, 101, 538
ESA, *see* Electroacoustic sonic amplitude
EtAc (ethylacetate), *table* 139
Ethylene glycol, 476–476, *figure* 475
Ethyl salicylate, 311
Evanescent foam, 270, 467, 468, 479, 480, 489
Evaporation, 153, 154, 169, 212, 465, 528, 533, 537, 539
Excluded volume, 177, 353
Explosions and fires, 285
External phase, definition, 420, *table* 420

Fabric
 of rocks, 21
 woven, 221
Fatty acid(s)
 as antifoams, 484, 485
 Krafft point, 264
 monolayers, 243–244
 rates of adsorption of, 410
Feathering of inks, 222
Ferguson effect, 249–250, 310, 409, 449, 474, 475, 477
Fermi level, 344, 346
Ferric oxide, *table* 195, 408, 508, 526, 528, 534
FFF, *see* Field flow fractionation
Fiber balance, 209, 225
Fick's law of diffusion, 384
Field flow fractionation, 78–81
Film balance, 165

Films, *see also* Detergency;
    Langmuir-Blodgett (L-B) films;
    Laplace equation; Marangoni flow;
    Spreading
  adhesion, 197
  black, 456, 468, 470–471
  Bond number, 153
  elasticity, 265–268
  foam films, 456–459
    stability, 460–480
  insoluble monolayers, Chapter 12,
    237–244
  mechanical properties, 434–436,
    491–493
  monolayer states, 238
  soap, 536–537
  thin-film coatings, 520
  viscoelasticity of monolayer films,
    242–243
Filters and filtration, 285
  crossflow filtration for sample
    preparation, 337
  digestion of precipitates, 105
  filter bed to break emulsions, 446
  filter paper as a Wilhelmy plate, 202
  fractal structure of filter pores, 60
  isolate counterchanges in nonpolar
    media, 287, *figure* 334, 343
  particle-size analysis by filtration,
    69–70
Floating particles, 161–162, 212, 493
Flocculants, 328, 394, 395, 417–418, 523,
  524, 525, 526
Flocculation, *figure* 26, 523–526. *See also*
    Critical coagulation concentration;
    Schulze-Hardy rule
  by block copolymers, 416–418
  compared to deflocculation, *table* 508
  dependence on size, 390, 415
  by depletion attraction, 413, 414
  detection by turbidity, 88
  on dilution, 66
  by electrolytes (double layer
    compression), 388, 523–524
  floc volumes, 508, 509
  freeze-thaw, 337, 392
  heteroflocculation, 523, 526
  by neutral polymers, 524–525
  orthokinetic, 523
  particle thickeners, 520–521
  perikinetic, 523
  by polyelectrolytes, 525
  by polymer bridging, 413, 523–525

  rates of, 67, 89, 361, 371, 384, 385,
    386, 513, 524
  relation to phase diagram, 314,
    411–413
  rheology of flocs, 513–514
  value, 322
  versus stabilization, 525–526
Flotation, 103, 162, 308, 473
  collectors, 162
Fluidity, *table* 34, 272, 273, 465
Fluorapatite (fluorinated calcium
    phosphate), *table* 319
Fluorescence, 489
Fluorite (calcium fluoride), *table* 319
Foam(s), Chapter 23, 456–498
  antifoams, 480–489
    surface and interfacial tensions,
      481–483
  boosters, 470
  defoamers, 486, 489
  drainage, 464
  dynamic surface tension, 268–270
  equation of state, 459–460
  evanescent, 480
  flooding, 479
  froth flotation, 162
  geometry of, 456–459
  Gibbs angle in, 465
  Gibbs elasticity of, 267–268, 466
  Gibbs-Marangoni effect, 462
  ice cream, 421
  inhibition, 479, 483, 488
    aqueous, 483–485
  inhibitors, 478, 481, 484–488
    compound inhibitors, 485
  lamellae elasticity, 466–467
  making (homogenization), 122
  Marangoni flow and stability, 153–155,
    173, 242, 289, 308, 474
  minimum surfaces, 309, 458, 506
  morphology of, 457
  Ostwald ripening of, 151
  and phase diagrams, 273, 311, 474–480
  Plateau borders in, 464–467
  polyhedral, 437, 457, 458
  properties, 456
  stability, 154, 310–311
    electrostatic, 463
    measurement of, 489–491
    theories, 460–480
  surface rheology, 491–492
  unit cell, 458
  yield stress, 438–439
Fog, 539

Food emulsions, *table* 421–422
Forces. *See also* Applied forces
  dispersion, 182, 186, 353, 406
  heteromolecular, 180
  homogeneous, 180, 182
  intermolecular, 138, 146, 177, 178, 179, 180, 182, 308, 352–354, 370, 388
  representative, *table* 179
Forces of attraction
  between particles, 352, 354
Forces of repulsion
  between particles, electrostatic, 374–382
    between flat plates, 374–376
    between spheres, 377–379
    by polymer layers, 379–381
    from rates of flocculation, 386–387
Formazin, 89
Fractal dimension, 60
Fractals, 58–60
Fraunhofer diffraction, 4, 7, 9, 10, 90, 91
Free energy
  of adsorption, 304
  of coalescence, 141, 448
  of desorption, 428
  of double-layer repulsion, 376, 388
  of emersion, 146
  of immersion, 214
  of interaction (flat plates), 362, 390
  of liquid surfaces, 138
  of micellization, 257
  of separation, 145
  of solid substrates, 159, 191
  of spreading, 142
Free meniscus coating, 172
Freeze-thaw cycles
  to flocculate, 392, 447
  to obtain serum, 337
Froth, 162, 309

Gas adsorption, 99, 101, 103, 106, 138, 167, 168, 216
Gelatin, 310, 380, 405
Gelatinous film, 470
Gels, 93, 387, 438, 465, 521
Gibbs adsorption, 168, 230, 292–300, 309, 461
Gibbs angle, 465
Gibbs elasticity, 267, 268, 466
Gibbs excess, 231, 294–302
Gibbs-Marangoni effect, 462–463
Girifalco-Good equation, 181
Glues, 485, *table* 540

Goethite, 501, *figure* 502
Gold
  Lifshitz constants, *table* 358, *table* 360, *table* 361
  mining of, 162
  sols (nanoparticles), 9, 532–548 (*passim*)
  water on, *table* 161
Goniometer, contact angle, 90, 157, 208, 215, 216, 226
Gouy-Chapman theory, 324–329, 330, 331
Grafted polymer, 413
  to hydrated surfaces, 406
Graphite, *table* 161, 162, 168, 189, *table* 195, 354, *table* 358, 403
Gravel, 234
Gravitational sedimentation, 57, 73, 74, 75, 544
Grease(s), 29, 44, 169, 171, 172, 362, 539–540
  cleaning stains, 154–155
Greek and Latin prefixes and suffixes, *table* 229
Greek kalends, 166
Grinding, 117, 118, 124, 125, 126, 127, 128, 129, 132, 134, 411
Gum arabic, 380

Hadamard regime, 288–289, 290
Half-life, 385, 386, 393
Hamaker theory, 356, 364, 369
  criticism of, 363
  different geometries, 357
  Hamaker constants
    definition, 356
    effect of intervening substances, 356–357
    measurement of, 370–371
    values in air, *table* 358–359
    values for heterogeneous interactions, *table* 361
    values in water, *table* 360
    retardation correction, 363
Hard-soft acid-base theory of Pearson (HSAB), 186
Hard-sphere model, 436
HLB scale, 429–432. *See also* Hydrophile-lipophile balance
Heat
  of adsorption, 197, 406, 408, 501
  of immersion, 196, 213, 214, *table* 508
Henry's law, 261, 262, 263, 302, 303, 304

INDEX   601

Herschel-Bulkley equation, 519
Heteroflocculation, 523, 526
Heterogeneity of substrates, 101, 102, 168, 169, 219, 292, 298, 504, 506
Higher order Tyndall spectra (HOTS), 9, 533, 534
High-shear mills, 121
Hittorf transport number, 253, 254
HLB, *see* Hydrophile-lipophile balance
Homogeneous nucleation, 526, 527
Homogenizers, 119, 120, 122, 123, 133, 441, 443
Homomolecular forces, 180, 182
Homotattic surfaces, 537–538
Hooke's law, 20, 30, 31, 456
HOTS, *see* Higher order Tyndall spectra
Hydration, 318, 383
Hydrocarbon moieties, 238, 250, 253, 261, 262, 264, 273, 306, 467, 469
   adsorbed at the air/water interface, 238–239, 302–305
   dispersion forces between, 182–184, 246
   Ferguson effect, 250
   geometric packing factors, 273–275
   micellization of, 250–252
   solubilization in micelles, 261–264
   Traube's rule, 305–306, 409
Hydrodynamic chromatography, 77, 78
Hydrodynamic(s)
   in dip coating, 172–173
   of foam drainage, 464, 473
   instability, 75
Hydrophile-lipophile balance (HLB) 230–232, 257, 429–432, 433, 447, 474, 477, 484, 485. *See also* Ferguson effect; Phase inversion temperature; Traube's rule
   in adsorption, 247–249
   graded series, *table* 232
   homologs, mixtures, 261
   hydrophilic and lipophilic moieties, *table* 233, *table* 417
   hydrophobic bonding, 247–249
   structure, effect of, 259–260
   technical terms, 228, *table* 229
Hydrophilic, definition, 228
Hydrophilic and lipophilic moieties, *table* 233
Hydrophobic, definition, 228
Hydrophobic bonding, 247–249, 251
Hydrophobic particles, 486, 488
Hydroxyapatite (calcium phosphate), *table* 319

Hysteresis
   of contact angle, 218–221, 225–226
   rheological, 29, 40
   of spread monolayer, 241

Ice
   color, 5
   creep, 21
   separation from brine, 392
Ice cream, 421
Ideal equation of state, 267, 300
IGC, *see* Inverse gas chromatography
Immersion
   free energy of, 146
   heats of, 196, 213, 214, *table* 508
Impact mills, 124, 125
Impactors, 52, 73, *figure* 74
Indifferent electrolyte, 523
Indium-tin oxide, 543
Induced dipole, 177, 179, *table* 179, 353, 354, 403, 552
Inhibition of foaming, 479, 488
Initial and final spreading coefficients, 162–167
Inkjet, 161, 222
Inks, 3, *table* 19, *table* 22, 120, 130, 222, 380, 387, 391, 499, 516, 541, 543, 545
   for xerography, 540–542
Insoluble monolayers, Chapter 12, 237–244
Interface, *see also* Adsorption; Capillarity; Contact angles; Electrical charges; Monolayers; Spreading; Surface-active solutes; Thermodynamics; Wetting; Work of adhesion; Work of cohesion
   dynamic equilibrium, 136
   effect of curvature of, 152–153
   theories of interfacial energies, Chapter 8, 177–197
Interfacial viscosity, 432, 448
Interfacial area
   of emulsions, 441, 448–449
   of foams, 467, 490
   of rough solids, 218
   by surface tension titrations, 506
   as a thermodynamic parameter, 140, *table* 140
Interfacial energy
   theories of surface and interfacial energies, Chapter 8, 177–197
   solid/liquid interfaces, Chapter 7, 157–174

Interfacial film, 205, 448, 467
Interfacial free energy, 146, 448
Interfacial phase, 292–300
Interfacial rheology
  mechanical properties, 434–436
Interfacial tension(s)
  against water, table 145
  antifoaming, 481–485
  componental theory, 179–189
  detergency, 170–172
  dispersion forces, 180–185
  experimental methods, 200–208
  function of structure, 259
  microemulsions and miniemulsions, 448–451
  thermodynamics, Chapter 15, 292–306
  ultralow dynamic, 270
  values, table 145
Interfacial thermodynamics, 140–146, table 140
Intermittent milling, 122, 127, 444, 445
Internal phase, definition, 420, table 420
Intermolecular forces, table 179
Interparticle forces, Appendix 26.4, table 556
  attraction between particles, 352, 354
  DLVO theory, 387–394
  electrostatic repulsion between flat plates, 374–376
  electrostatic repulsion between spheres, 377–379
  from rates of flocculation, 386–387
  repulsion between particles, 374–387
  repulsion by polymer layers, 379–381
Inverse micelles, 273, 279, 398, 471
  nonpolar electrostatics, 283–287, 318–321, 542
Inversion of an emulsion, 511
Inverse gas chromatography (IGC), 196, 406–407
Ion adsorption, 523
Ion exchange, 269, 403
Ionic double layer, 335
Ionic solids, PZC, table 319
Ionic strength
  CCC, 392
  chemical demulsification, 447
  colloidal crystals, 538–539
  and Debye length, 326
  definition, 326
  dispersions, stability of, 374, 376, 378, 383
  DLVO theory, 390, figure 391
  double layer compression, 523

inverse micelle, 279
  in Lifshitz theory, 369
  Schulze-Hardy rule, 393–394
  stability ratio, 387
Irregular surfaces, wetting of, 218–227
Isaphroic lines, 476
Isoelectric point, 234, 310, 317, 368, 499, 526. See also Point of zero charge
Isostearic acid, 239, 240
Isotherm, see Adsorption, isotherm; Surface tension, isotherm

Jacob's box, 370, 379, 380
Jamin effect, 222–224, 554
Jar mills, 124, 125, 126
Jaundice, 162

Kady mill, 120, 121
Kaolin, 23, 97, 321, 322, 342, 387, 514, 523
Keesom forces, 179, table 179
Kelvin bubble cluster, 458
Kelvin equation, 150–151, 152, 526
Kinematic viscosity, 38
Kinetic theory, 55, 384
Kinetics, 383, 524, 533
  of flocculation, 533
Kipling and Tester isotherm equation, 503
Krafft point (temperature), 264–265
Krypton, 538

Laminar flow, 70, 79, 288
Langmuir
  adsorption, 104, 501, 538
  approximation, 378, 388
  balance, 213
  equation, 302, 303, 304, 408, 501, 502, 503, 504
  films, 213, 238, 242, 488
  isotherm equation, 301
  trough, 57, 204, 238
Langmuir-Blodgett (L-B) films, 243–244, 547
Lanthanum hydroxide, 540
Laplace equation, 53, 146–149, 153, 200, 211, 213, 450
Laplacian operator, Appendix 26.3, table 555–556
Laponite, 520, 521
Laser Doppler microelectrophoresis, 337–338

Laser light scattering, 5, 609
Laser velocimetry, 13–14, 347, 348
Latex particles, 60, 234, 387, 392, 506, 526, 534, 535
LCFT, *see* Lower critical flocculation temperature
LCST, *see* Lower critical stabilization temperature
LDV (laser Doppler velocimetry), *see* Laser velocimetry
Lecithin, 270, 319, 380
Lennard-Jones potential, 353, 361, 371
Let-down solution, 88
Lewis acid-Lewis base interactions, 186, 352. *See also* Acid-base interactions
Lewis acids, 182, 186, 189, 196, *table* 196, 279, 280, 281, 352, 406, 447, 542
Lewis bases, 182, 186, 189, 196, *table* 196, 280, 281, 406, 447, 542
Lifshitz constant, 364, 365, 366, 368, 369
   across air, *table* 358–359
   across water, *table* 360
   between three materials, *table* 361
Lifshitz theory, 364, 365, 369, 370, 376
   method of Ninham and Parsegian, 365–369
   values of necessary optical constants, 367
Light absorption, 75, 76, 85
Light scattering, 1–14, 62, 87–94, 95–96, *table* 95, *table* 106–108, 385
   Fraunhofer diffraction, 9–10, 90–92, *figure* 91, 87, *figure* 87
   as a detector, 94
   diffusing wave spectroscopy, 93–94
   higher order Tyndall spectra (HOTS), 9, 533, 534
   laser velocimetry, 13–14, 347, 348
   Mie, 7–9, *figure* 87, 90, *figure* 91, 93
   multiple angle light scattering (MALS), 70, 94
   nephelometry, 88, 89, 413
   phase-angle (PALS), 338
   polarization effects, 92
   quasi-elastic (QELS), 10–13, 92–93, 94, 441
   Rayleigh, *figure* 87, 88
   turbidity, 89, 254, 413
   Tyndall effect, 1, 7, 82, 89, 533
Lipid bilayers, 241
Lipids, 55, 57, 241, *table* 275, 369
Lipophilic, definition, 228

Liquid crystals, 271, 272, 273, 436, 469, 471, 472, 538
   middle phase (hexagonal), 271, 272, 442
   neat phase (lamellar), 271, 272
Liquid/liquid interfaces, 157, 209, 279, 292, 427, 428, 500
Liquid/solid interaction, 160, 161
Liquid/solid interface, 279
Liquid-condensed phase, 239
Liquids
   molecular theory of surfaces, 136–140
   surface tension and boiling points, 138
   typical surface tensions, 139
London
   constant, 178, 353, 354, 356, 363
   forces, 182, 186, 353, 406
   equation, 178
Lower critical flocculation temperature (LCFT), 412, *figure* 412
Lower critical stabilization temperature (LCST), 412, *figure* 412
Lubricating greases, 539–540
Lubrication, *table* 19, 103, 169, 287
Lundelius' rule, 249–250, 310, 313, 409, 474, 475
Lung surfactant, 241
Lyocratic, definition, 383
Lyophilic, definition, 228
Lyophilic moiety, 309
Lyophobic, definition, 228

MA, *see* Methyl acetate
Macromolecules, 17, 58, 77, 92, 251, 334, 357, 383, 536. *See also* Polymers
MALS, *see* Multiangle light scattering
Manufacturers
   capillary equipment, 214–215
   electrokinetics, 347–348
   milling equipment, 132–134
   particle sizing equipment, 109–111
   rheometers, 45–46
Marangoni effects, 154, 169, 270, 425, 463, 464, 465, 467, 468, 474, 480, 481
Marangoni elasticity, 267, 466
Marangoni flow, 153–155, 173, 242, 289, 308, 474
   cleaning of surfaces
      removing grease, 154
      thermally induced, 155
   wetting of cloth, 154
Margarine, 422

604 INDEX

Matrix
  in composites, 166
  in greases, 540
  in margarine, 421
Maximum bubble pressure, 205, 215
Mayonnaise, 422
Measurement of
  acid-base character, 406–408, 508–510, 510–511
  adsorption, 99–102, 500–502
  contact angle, 208–215
  critical micelle concentration, 253–254
  dispersion energy, 167–168, 208–215
  dispersions, stability of, 393, *figure* 393, 411–413, *table* 508
  electrophoretic mobility, *table* 333
  emulsion stability, 441–443
  emulsion type, 420–423
  foam stability, 489–491
  Hamaker constants, *table* 358–359, *table* 360, *table* 361, 370–371
  irregular surfaces, 218–227
  Lifshitz constants, 365–366
  molecular weight, 5–6, 36, 57–58, 77
  particle size, *table* 106–108
  required dispersant level, 131
  rheology, 34–45
  spreading pressure, 167–168
  surface area, 99–102, 500–502
  surface excess, 231, 295–296, 297
  surface rheology, 491–493
  surface tension, 200–208
  viscosity, 34–45
  zeta potential, *table* 333, 334–342
Membrane(s)
  cell, 214, 244
  filters, 60
    contact angles on, 224
    demulsification, 446
    positively charged, 285
  nanoparticle, 545, 547
Meniscus
  axisymmetric, 150, 203, 204, 423
  critical point wetting, 314
  dip coating, 172–173
  effect of curvature, 148–151, 200–201
  falling meniscus method, 270
  free meniscus coating, 172
  to measure contact angle, 209–210
Meniscus cleaning, 154
Mercury
  adsorption of vapor from a rose, 163
  in capillary tube, 152

contact angle on glass, 150, 157, *figure* 158, 200
dispersion energy contribution, *table* 183
dropping Hg electrode, 334
interface with aqueous electrolyte, 174, 329
interfacial tension with water, *table* 145
metallic energy contribution, 182
pore-volume analysis, 102–103, 210
surface tension and boiling point, 139
work of adhesion with water
  calculation, 183–184
  measurement, *table* 145
Mercury porosimetry, 102–103, 210
Mesophase, 271, 436
Metal-oxide sols, monodisperse, 534
Metastable, 165, 197, 526
Method of Ninham and Parsegian (Lifshitz theory), 365–369
Methyl acetate, *table* 188, 475–476, *figure* 475
Mica
  flotation, 166
  in Jacob's box, 151, 379–380
  Lifshitz constant, *table* 358, *table* 360, *table* 361
  wetting by water, 163
Micelles, *figure* 252. *See also* Critical micelle concentration
  critical packing parameter, *table* 275
  definition, 251
  dynamic equilibria, 152, 246, 248–249, 251, *figure* 251, 257
  in foam stability, 471–472
  inverse micelles, 279, 398, 471, *figure* 471
  nonpolar electrostatics, 283–287, 318–321, 542
  Krafft point, 264–265, *figure* 265
  molecular weights by sedimentation equilibrium, 77
  phase diagram, *figure* 272
  Rayleigh scatterer, 4
  solubilization, 250, 254, 261–264, 449, 470
Micellization, 250–264, 309, 449, 507
Microelectrophoresis, 335–337, 339, 347, 348
Microemulsions, 250–264, 390, 425, 448–451, 532, 540
Microfluidizer, 122
Micronizer, 118, 134

Microscopy
　Brownian motion, observation of, 54
　comparisons, *table* 106–108
　particle sizing, 81–82
　preferred particle-sizing method, 62
　test for emulsion type, 423
　ultramicroscopy (dark-field
　　illumination), 82–83
Mie theory of light scattering, 7–9, *figure*
　87, 90, *figure* 91, 93
Millbase, 118
Milling
　comminution of aggregates, 64, 91,
　　117, 444, 449
　dry grinding, 118
　equipment
　　attritors, 126, 134
　　ball mills, 121, 124–127, 133, 134
　　colloid mills, 119, 121, 122, 133,
　　　134, 443
　　cyclones, 52, 73, 106, 133, 134
　　dispersers, 120
　　heavy-duty, 130
　　high-shear mills, 121
　　high-speed stirrers, 118–120
　　for high viscosity, 129–130
　　homogenizers, 119, 120, 122, 123,
　　　133, 441, 443
　　impact mills, 124–129
　　impactors, 106
　　jar mills, 124, 125, 126
　　let-down solution, 118
　　micronizers, 118
　　mixers, 119, 120, 129, 130, 133, 134,
　　　443
　　processing aids, 130–132
　　roll mills, 129, 130, 134
　　sand mills, 127–129, 132, 411
　　Sonolator, 123
　　manufacturers, *table* 132–134
　　three-roll mills, 129
　　ultrasonic dispersers, 123, 443
　　vibratory mills, 125–126
　intermittent milling, 121–122, 127, 445
　wet milling, 132
Mines and mining, 334, 344, 387, 410
Mineral oil
　bubble rise in, 289–290
　in greases, 539–540
　Traube's rule revisited, 409
Minerals
　floating fine particles, 161–162
　origin of surface charge, 317
Miniemulsions, 263, 448–451

Minimal tetrakaidecahedron, 458
Minimum surfaces, 309, 456–458, 506
Mixers, *see* Milling, mixers
Model colloidal systems, 532–539
　colloidal crystals, 538–539
　particles of uniform size, 436, 532, 535,
　　536
　surfaces of uniform composition,
　　536–538
Molecular areas of soaps, *table* 507
Molecular interaction, 185, 189, 361, 364,
　448
Molecular weight
　by intrinsic viscosity, 36
　by light scattering, 5–6, 88, 94
　by sedimentation equilibrium, 57–58,
　　77
Monodisperse
　colloidal crystals, 538–539
　by controlled nucleation and growth,
　　528
　emulsions, 536
　latexes, 534–536
　micelles, 254–257
　microemulsions, 448
　nanoparticles, 546–547
　particles, 532–536
Monolayer adsorption, Chapter 13,
　246–275, Chapter 15, 292–306
　adsorption from solution, 103–105,
　　500–502
　equilibrium film thickness, 169
　gas adsorption, 101–102, 537–538
Monolayers
　insoluble, *see* Monomolecular layers
　soluble, *see* Monolayer adsorption
Monomolecular layers, 168, 306, 488,
　Chapter 12, 237–245
　assembly and self-assembly in,
　　547–548
　ideal gaseous state in, 238
　intermediate phase in, 238
　liquid-condensed phase in, 238, 239
　liquid-expanded phase in, 238
　solid-condensed phase in, 238, 239
　viscoelasticity of, 242
Morton equation, 450
Multiangle light scattering (MALS), 70,
　94
Multilayer adsorption, *see* BET theory

Nanoparticles, 4, 9, 379, 533
　applications, 102, 236, 548

Nanoparticles (*Continued*)
  assembly, 547
  self-assembly, 547–548
  synthesis, 546–547
Nanotechnology, 448, 545–548
National Biscuit Dunking Day, 212
Neat soap, 272
Nephelometric turbidity unit (NTU), 89
Nephelometry, 88, 89, 413
Nernst potential, 345
Neumann's triangle of forces, 143–144
Newtonian flow, 23, 27
n-hexatriacontanoic acid, 240
Nickel, 426, 534
NMR, *see* Nuclear magnetic resonance
Non-Newtonian flow, *see* Rheology
Nonpolar media, Chapter 14, 279–290.
  See also HLB; W/O emulsions;
  adsorption in, 280–282
    acid-base interactions in, 281–282,
      figure 282, *figure* 283
  bubble rise in, 287–290
  conductivities of, *table* 286
  electrical charges in, 282–287,
    318–321, 542
  electrocratic dispersions in hydrocarbon,
    396–398
  electrophoretic displays, 543–545
  electrostatic stabilization in, 396–398
  explosions and fires, 285
  half-value times, *table* 286
  hydrocarbons on charged electrodes,
    174
  inverse micelles in, 279, 398, 471,
    figure 471
  liquid crystal phases, 471–473
  method to measure charging capacity,
    286, *figure* 286
  nonpolar polymer colloids, 535–536
  surface potential to stabilize dispersions
    in, *table* 398
  xerographic imaging in, 542
Non-wetting, 102
  autophobic dewetting, 163, 164
NTU, *see* Nephelometric turbidity unit
Nucleation, homogeneous, 526, 527
Nuclear magnetic resonance, 262, 403,
  500
Nucleation and crystal growth, 4, 53, 88,
  152, 165, 220, 475, 526–528, 533,
  609
Nylon, *table* 346

O'Brien and White, calculations, 335,
  figure 336, *figure* 337
Oil-in-water emulsions (O/W),
  definitions, 420, *table* 420
  DLVO theory, example, 390–391,
    figure 391
  electrical conductivity, 439–440
  examples, *table* 421–422, *figure* 426,
    figure 433, *figure* 442
  HLB scale, 429–432
  speed of sound in, 96–98, *figure* 97
OLOA *371*, 319, 399, 400
OLOA *1200*, 319, 320, 321, 399, 400
Opalescence, 533, 538
Opals, 532–533
Optical constants, *table* 367
Optical density, 76, 500
Optical microscopy, 510
Orthokinetic flocculation, 523
Osmotic coefficient, 58, 251–253
Osmotic pressure, 58, 255, 451
Osmotic stabilization, 450–451
Ostwald ripening, 96, 105, 151–152, 263,
  264, 444, 449, 451, 528, 535
O/W, *see* Oil-in-water emulsions
Oxides, *see also* Kaolin; Mica; Silica;
  Titania
  acid-base interactions, 189
  dispersing agents for, 499–500
  E & C numbers for, 408
  as hydrophilic colloids, 403, 506
  iron oxide, 282, 514–515
    ferric oxide, *table* 195, 508
    goethite, 501
  Lifshitz constants, *tables* 358, *table*
    360, *table* 361
  monodisperse, 528, 534
  PZC of, 317–318, *table* 318
  silane treated, 481
  spreading pressure
    of adsorbed gases on oxides, 168
    of organic liquids on oxides, 166
    of water on oxides, 166
  surface charge density of, 379
  zinc oxide, 509

PALS, *see* Phase-angle light scattering
Particle shape, 62, 81–84, 90, 522
Particle size
  distributions, 63–65
    cumulative, 64, 71
    Gaussian or normal, 63
    log normal, 64–65

manufacturers, *table* 109–111
sample preparation, 66–67, 337, 499–500
  magnetic powders, 67
Particle-sizing by
  acoustics, 94–98, *figure* 96, *figure* 97
  Andreason sedimentation, 72, *figure* 72
  centrifugation, *figure* 75
    disk centrifuge, 75–76
    efficiency, 74
    line start, 75
    terminal velocity, 50
    ultracentrifuge, 76–77
    uniform start, 76
  Coulter counter, 86–87, *figure* 86
  cyclones, 52, 73
  electroacoustics, 98, 341–343, 512
    techniques, *table* 343
  electron microscopy, 83
  electrozone detection, 86–87, *figure* 86
  elutriation, 73, 106
  field flow fractionation, 79–81, *figure* 79, *figure* 80
  filtration, 69–70
  Hegman gauge, 105
  hydrodynamic chromatography, 77, 78, *figure* 78
  impactors, 52, 73, *figure* 74
  light scattering, 1–14, 62, 87–94, *figure* 87, 95–96, *table* 95, *table* 106–108, 385
    Fraunhofer diffraction, 9–10, 90–92, *figure* 91, 87, *figure* 87
    Mie, 7–9, *figure* 87, 90, *figure* 91, 93
    multiple angle light scattering (MALS), 70, 94
    polarization effects, 92
    quasi-elastic(QELS), 10–13, 92–93, 94, 441
    Rayleigh, *figure* 87, 88
    turbidity, 89, *figure* 89
    Tyndall effect, 1, 7, 82, 89, 533
  mercury porosimetry, 102–103, 210
  optical microscopy, 81–82
    preferred particle-sizing method, 62
    ultramicroscopy (dark field illumination), 82–83
  photozone detection, 84–86, *figure* 85
  sedimentation, 49–50, 70–72
  sieves, 67–69, *figure* 68
    sieve sizes, *table* 69
  techniques, comparisons, *table* 106–108
  time-of-transition, 83–84, *figure* 83, *figure* 84

ultramicroscopy, 82
ultrasound, comparison to light, *table* 95
Particulates
  in antifoams, 486–489
  Brownian motion, 55
  charge per particle, 287, 334–346
  contact angles on, 210–214
  as emulsion stabilizers, 426–429
  light scattering by, 1, 4
  monodisperse, 526–528, 533–535
  motion of by applied force, 48–54
  nanoparticles, 545–548
  powder flow, 522
    angle of repose, 522
  rate of collisions, 384–386
  Reynolds number, 52
  sedimentation equilibrium, 57
  size distributions, 63–65
  size reduction, Chapter 5, 117–134
  stabilization
    electrostatic, 389–391, 396–398
    electrosteric, 415–416
    steric, 398–400, 416–418
  terminal velocity, 49
PDMS, *see* Silicones
PE, *see* Polyester
PEG, *see* Polyethylene glycol
Pendent drop, 149–150, 202–203, 204
PEO, *see* Polyethylene oxide
Perikinetic flocculation, 523
Permittivity of free space (ε_0), value 554
Petroleum emulsions, 447, 448
Phase-angle light scattering (PALS), 338
Phase diagrams,
  three-component (isothermal)
    figure 272, water/butanol/$E_{100}P_{70}E_{100}$
    figure 274 and *figure* 472 water/hydrocarbon/lipophilic solute
  two-component
    figure 265, SDS/water, concentration/temperature
    figure 310, diethylene glycol/ethyl salicylate, concentration/temperature, cosorption contours
    figure 311, diethylene glycol/ethyl salicylate, concentration/temperature, isaphroic contours
    figure 313, regular solution, temperature/concentration, cosorption contours

Phase diagrams (*Continued*)
  figure 396, aluminum salt
    concentration/pH, solubility
  figure 412, two solvents, temperature
  figure 475, ethylene glycol/methyl
    acetate, bubble
    lifetime/concentration
Phase-inversion temperature (PIT),
  432–434, 443, 444, 447
Phases, *see also* Monomolecular layers
  capillarity and phase diagrams, Chapter
    16, 308–314
  colloid stability and phase diagrams,
    394–396, 411–413
  emulsion types, 420
  foaminess and phase diagrams,
    474–480
  micellar, 265
  phase-inversion temperature, 432–434
  of surface-active solutes, 271–273
  surface phases, 101–102, 164, 231,
    292–300, 308
Photozone counter, 84–86, *figure* 85
Pickering emulsions, 436
PIT, *see* Phase-inversion temperature
Plastic, *see* Plasticity; Pseudoplastic flow,
  Rheology, plastic flow;
  Viscoelasticity
Plasticity, 434, 468, 472, 473, 492
Plateau border, 464–467
Platinum, 201, 545, 546
PMMA, *see* Polymethylmethacrylate
Point of zero charge (PZC), 317, 318, 319.
  See also Isoelectric point
  ionic solids in water, *table* 319
  oxides in water, *table* 318
Poiseuille flow, 19, 211
Poisson-Boltzmann equation, 316, 324,
  325, 328, 331, 377
Poisson equation, 324, 325, 327
Polar forces, 185
Polar organic liquids, *table* 194
Polarizability, 3, 178, 189, 353–357, *table*
  557
Polyacid, 500
Polyacrylamide, 402, 417, 525
Polyacrylate, 500
Polyacrylic acid, 402, 417, 516, 524, 528
Polyacrylonitrile, *table* 346, 417
Polyalkylsuccinimide, *table* 235
Polybutadiene, *table* 358
Polybutene, *table* 358
Polycarbonate, 76, *table* 196, *table* 346,
  547

Polydimethylsiloxane (PDMS), *see*
  Silicones
Polyelectrolyte, 402, 524, 525
Polyester (PE), 477 *figure* 477, 517, 518
Polyethoxyalkenes, 447
Polyethylene, 161, 166, 167, 168, 191,
  table 193, 229, *table* 235, 236, 261,
  table 346, *table* 358, *table* 360, 399,
  414, 417, 424, 447, 500
Polyethylene glycol (PEG), 414, 261
Polyethylene oxide, 229, *table* 233, *table*
  235, 236, *table* 358, 417, 418, 424,
  447, 485, 500
Polyhedron(a), 437, 457, 458
Polyhexafluoropropylene, *table* 191, *table*
  193
Polyhexamethylene, *table* 191
Polyhydric alcohols, 484
Polyion, 499
Polyisobutylene, *table* 358, 399, 400, 417
Polyisoprene, *table* 360, *table* 367, 417
Polymaleate, 500
Polymer(s)
  acidity-basicity, 187, 189, 196–197,
    281–284
  adsorption, 281, 282, *figure* 282, 283,
    figure 283, 403–404, 519, 525, 526
  anchoring group(s), 405, 406, 411, 416,
    table 417
  block copolymers, 405, 416–418
  charges in nonpolar media, 318–321
  coatings, 379, 380, 414
  compounding, 130
  critical surface tensions, *table* 190
  degradation determination of, 80–81
  die swell of, 33
  films, 146, 347, 446, 517
  flocculants, 524–526
  foaming, 477–479
  graft copolymers, 405, 406, 416
  grafted, 413
  homopolymers, 381, 405, 485
  molecular weight
    by light scattering, 5–6
    by sedimentation equilibrium, 77
  monodisperse latexes, 534–537
  phase diagrams, 412
  random copolymers, 405, 406
  silicone foam inhibitors, 486–488
  stabilizing chains, *table* 417
  steric repulsion, 379–381
  steric stabilization, 398–400
    depletion, 414
    by free polymer, 413–414

and its phase diagrams, 411–413
  theory, Chapter 21, 402–418
  thermal FFF, 80–81
  triboelectric series, table 346
  viscoelasticity, 44
  Weissenberg effect, 33
  Zimm plots, 6
Polymer solubility, cloud point, 236, 413
Polymethylmethacrylate (PMMA), 80, table 191, 196, table 196, 197, 281, 282, 342, table 346, table 358, table 360, 417
  adsorption from nonpolar media, 282
Polyoxyethylene, see Polyethylene oxide
Polypropylene oxide (PPO), 417–418
Polysaccharide, 414
Polysiloxane, table 233. See also Silcones
Polystyrene, 7, 11, 12, 13, 60, 79, 90, 173, table 191, table 193, table 346, table 358, table 360, table 361, table 367, 380, 417, 547
Polytetrafluoroethane, 150
Polytetrafluoroethylene, table 191, table 193, table 346, table 360
Polytrifluoromonochloroethylene, table 193
Polytrifluoropropylsiloxane, 486
Polyurethane, 461
Polyvinylchloride (PVC), 282, table 367
  adsorption of from nonpolar media, 283
  post chlorinated, 196–197, table 196
Polywater, 152
Pore size, 224
Pore-volume analysis, 102–103
Pores, 60, 70, 102, 131, 210, 211, 224
Porosity, 102, 210, 212, 517
Porous solids, 102, 537
Post-CPVC, see Polyvinylchloride, post-chlorinated
Potential-determining ions, 322, 326, 327, 379, 395, 523, 524, 534
Powder, see Particulates
PPO, see Polypropylene oxide
Precursor films, 168, 169
Preferential adsorption, 173, 317, 321–324, 502, 503, 547
Prefixes, table 229
Pressure
  across curved surfaces, table 148
  of gas in foams, 459
  osmotic, 58, 239, 251, 263, 375, 376, 379, 413, 444, 450, 451
  surface, 213, 244, 493
Primary minimum, 394, 414, 415

Profoamers, 478
Protective colloids, 380, 393, 405, 447
Protein(s), 379, 380, 414, 421, 442, 464, 468, 469, 489, 532, 546
Pseudoplastic flow, 23, 25, 513
Pseudopod, 162
PVC, see Polyvinylchloride
Pyrophosphate, 23, table 235, 321, 334, 379, 387, 499, 514
PZC, see Point of zero charge

QELS, see Quasi-elastic light scattering
Quantum dots, 546, 548
Quaternary ammonium compounds, table 233, 234, 283, 447, 542
Quartz, 39, 46, 76, table 161, 162, 170, 171, 269, table 358, table 360, table 361, table 367, 403, 409
Quasi-elastic light scattering (QELS), 78, 90, 92, 93, 94, 441

Radius of curvature, 148, 151, 173, 205, 263, 378
Radius of gyration, 6, 88, 94, 381
Rainbow, 136
Random copolymer, 405, 406
Random walk, 93
Rates
  of adsorption, 52, 404, 409, 410, 411, 467
  of flocculation, 67, 89, 361, 371, 384, 385, 386, 513, 524
Rayleigh light scattering, 87, 88
Red blood cells, 87
Regular solution theory, 311–313
Rehbinder effect, 132
Resistazone counter, 86
Retardation
  in bubble rise, 289
  in electrokinetics, 355
  in Hamaker theory, 179, 363, 364, 369, 400
Reverse micelle, see Inverse micelle
Reynolds number, 51–52, 288–289
Rheology, Chapter 2, 16–46. See also Elasticity
  Bingham body, 25, 468, 513
  bubble rise in viscous media, 287–290
  comparison of deflocculated and flocculated dispersions, 508, table 508, 513–516
  Couette flow, 436, 530
  Couette rheometer, 16, 42, 513

Rheology (*Continued*)
  creep flow, 21, 32, 40, 44, 132, 436, 521, 609
  terms defined, *table* 22
  dilatant flow, 514, 515
  of dilute suspensions, 513
  of emulsions, 437–439
  equations, *table* 44
  Hadamard regime, 288–289
  interfacial, 434–436, 467–470
  laminar flow, 70, 79, 288
  instrument manufacturers, *table* 45
  Newtonian flow, 23, 27
  non-Newtonian flow, definition, 18
  Ostwald-Ubbelohde viscosimeter, 34–35, *figure* 35
  of pigments in polymer solutions, 516–520
  plastic flow, 21, 23, 27
  plasticity, 434, 468, 472, 473, 492
  plug flow, 25
  Poiseuille flow, 211
  power-law flow, 25
  pseudoplastic flow, 23
  Reynolds number, 51, 52, 288, 289
  rheopexy, 29, 517
  shear rate, 16–19
    typical, *table* 19
  shear stress, 16–19
  shear thinning, 25, 27, 456, 512, 513, 517, 519
  shear viscosity, *see* Viscosity
  slump height, 514
  Stokes regime, 289
  stress relaxation, 32, 44, 436
  surface rheology, 456, 491–493
  surface viscosity, 464, 468, 469
  of suspensions, 512–523
  thickeners, 520–521
  thixotropy, 29, 30, 40, 515, 517
  units, *table* 34
  viscoelasticity, 29, 30, 31, 40, 44, 242, 517
  viscoelastometers, 44
  viscosity
    definition of, 16, 17
    typical, *table* 18
  yield point, 21, 23, 25, 43, 394, 434, 456, 466, 492, 513, 514, 517, 521, 539, 540
Rheometers, 23, 25, 33–46, 215, 493, 513
  cone-and-plate, 42, 43, 44
Rheopexy, 29, 517
Ring tensiometer, 216

Roll mills, 129, 130, 134
Ross-Nishioka effect, 434, 444, 475, 477
Roughness, 60, 90, 218, 219, 221, 222, 224, 362, 522
Rubber, 20, *table* 22, 130, 346, *table* 346, *table* 358, 387, 396, 451, 468, 499, 534

Salad dressing, 422
Sand, 26, 27, 332, 411, 447, 522
Sand mills, 127–132, 411
SBR, *see* Styrene-butadiene rubber
Schulze-Hardy rule, 393, 523
Scratches, 221, 222
SDS, *see* Sodium dodecyl sulfate
Secondary minimum, 391, 394, 414
Sedimentation, 539
  Andreason sedimentation pipette, 72
  for a dispersibility map, 510–511
  equilibrium, 57, 76, 77
  Hadamard, 288, 290
  potential, *table* 333
  rates, 50
  Stokes law, 539
  Stokes regime, 290
  terminal velocity, 38, 48, 49, 50, 51, 62, 288, 289
  volumes, *table* 508, 509
    example (ZnO), 509
Sedimentation field flow fractionation (SF^3), 78–79
Sedimentation FFF, *see* Sedimentation field flow fractionation
Selenium, 7, 285, *table* 358, *table* 360, *table* 361, 403, 508, 528, 533, 609
Self-assembly, 250, 257, 536, 547, 548
Sessile drops, 147, 149–150, 160, 164, 190, 194, 200, 202–203, 204, 210, 226
  mercury on glass, 158
SF^3, *see* Sedimentation field flow fractionation
Shear rate, 16–19
  typical values, *table* 19
Shear stress, 16–19
Shear thinning, 21–25
  of flocculated suspensions, 512–515, 517–520
  of foams, 456
Shear viscosity, *see* Viscosity
Sieves, 63, 67, 68, 69, 73, 106, 137, 138, 285, 545
  sizes, *table* 69

Silica
  acidity-basicity, 196, table 196,
     281–282, 408
  in antifoams, 486–489, 508
  coating on titania, 391
  cobalt adsorption on, 342
  dispersion free energy of, table 195
  as flow agent, 522
  hydrophilic interface, 403, 506
  hydrophobic silica, 483, 487, 488, 489
  in kaolin, 321
  Lifshitz constants of, table 360–361
  nanoparticle silica, 547
  in opals, 532–533
  optical constants, table 367
  polymer grafted on, 406
  stearylated, 413
  surface dissociation of, 317
  as thickener, 540
  as turbidity standard, 89
  in water-butanol mixtures, 314
  water on, 249
     no contact angle with, table 161
     spreading pressure, table 161, 168
     work of adhesion, table 161
Silica gel(s), 403, 409
Silicate(s), 197, 234, table 235
Silicone(s)
  in electrophoretic displays, 544
  as foam inhibitors, 470, 483–489
  polydimethylsiloxane (PDMS)
     antifoam in oils, 477–479
     in block copolymers, 417
     in depletion repulsion, 413
     interfacial tension with water, 194
     Lifshitz constant, table 358
     low surface tension and low
        volatility, 169
     in micellization, 249
     surface activity in oils, 189, 289–290
     surface tension of, 139
     used to treat silica, 508
  in Silly Putty, 27
  spreading in grooves, 222
  as viscosity standards, 37
Silly Putty, 27
Silver
  dispersion energy of, table 195
  Lifshitz constant of, table 358, table
     360, table 361
  nanoparticles of, 546, 547
  tray, 463
Silver bromide, 504, 505, 534
Silver chloride, 319

Silver halide, 317, 318, 380, 504, 534
Silver halide sols, monodisperse, 534
Silver iodide, 319, 322, 329, 330, 379,
   394, 395, 396, 444, 524, 534
Size distributions, 63–65
  cumulative, 64, 71
  Gaussian or normal, 63
  log normal, 64–65
Sky, blue of, 5, 53
Slow coagulation, 384
SLS, see Sodium lauryl sulfate
Slump height, 514
Smectic phase, 273. See also Liquid
   crystals, lamellar
Smokes, 539
Smoluchowski equation, 334, 335, table
   336, 384, 385, 386, 555
Soaps, alkali salts of fatty acids, 233–234.
   See also Detergents; Detergency
  in black films, 456, 468, 470, 471
  bubbles of, 40
  in chemical demulsification, 447
  constant surface-charge density of, 379
  effect of structure on CMC, 257–260
  film(s), 456, 461, 532
     as model surfaces, 536–538
  grease making, metal soaps in, 539–540
  Krafft point (temperature) of, 264
  manufacturers of, 230
  micelles in emulsion polymerization,
     451–452
  molecular areas of, table 507
  phases of, 271
  preferential adsorption of, 321
  sodium laurate, 264
  sodium oleate, 424, 429, 430, 444, 445,
     470
  solubilization in micelles, 261–262
  surfaces of uniform composition,
     536–537
  surface-tension titration, 506
Sodium dioctylsulfosuccinate, 483. See
   also AOT and Aerosol OT
Sodium dodecyl sulfate (SDS), 96, figure
   252, 253, 255, table 275, 290, 323,
   473. See also Sodium lauryl sulfate
  CMC, 255
Sodium dodecyl sulfonate, 265
Sodium laurate, 264
Sodium lauryl sulfate (SLS), 295, 469,
   472. See also Sodium dodecyl sulfate
Sodium oleate, 424, 429, 430, 444, 445,
   470
Soil, 102, 170, 234, 259

Solid/liquid interface, 214, 317, 345, 409, 499, 500, 506. *See also* Chapter 7, 157–174
Solid substrates, Chapter 10, 218–226. *See also* Solid/liquid interface
  acid or base character, *table* 196, 196–197, 406–408
  acid-base values, *table* 408
  dispersion energies, 190–195, *table* 195
  gas adsorption on, 100
  homotattic, 537–538
  low energy, *table* 193
Solubilization in micelles, 250–264, 390, 425, 448, 449, 470, 532, 540
Solutes, *see also* Surface-active solutes
  amphipathic, 248, 249, 251, 253, 254, 309, 474, 477
  anionic, 234–235, *table* 235
  cationic, 234–235, *table* 235
  nonionic, 234–235, *table* 235, 296, 429, 433, 447, 448, 465, 523
Solution adsorption isotherm, 504
Solvation, 141, 309, 318, 383, 404
Sonolator, 123
Sorbitan ring, *table* 431
Sorbitol(s), 234, 309
Specific heat, 54, 197
Specific surface area, 62, 98, 99, 100, 101, 102, 103, 160, 501, 504, 508, 509
Speed of light, 2, 179, 363
Speed of sound, 53, 71, 96–98, 441–442, 511–512
Spherical double layers, 377
Spin coating, 172
Spinning drop method, 206, 207, 215
Spontaneous adsorption, 248, 249, 280, 294, 306, 309, 449
Spontaneous association, 250
Spontaneous emulsification, 259, 424
Spray coating, 172
Spraying, *table* 19, *table* 22, 23, 124, 268
Spreading, *see also* Spreading pressure
  coefficient, 142, 143, 146, 159, 160, 170, 194, 371, 402, 482
  dynamics, 168–169
  final, 169
  initial and final, 162–167
  monolayer, *see* Chapter 12, 237–244
  negative, 143, 173, 222
  one liquid on another, 142–144
  positive, 165, 166, 194, 357, 371, 402, 479, 507
  precursor film, 168, 169
Spreading pressure
  of adsorbed vapor, 101, 159, 165–168, 195, *table* 195
  of insoluble monolayers, 238–242
  by surface tension lowering, 265, 300
Squalane, 168
Stability ratio, 284, 334, 384, 386, 393, 397
Stabilization by polymers, Chapter 21, 402–418
  anchoring groups, *table* 417
  block copolymer, 416–418
  electrosteric stabilization, 414–416, *figure* 417
  free polymer, 413–414
  lower critical flocculation temperature (LCFT), 412
  lower critical stabilization temperature (LCST), 412
  repulsion by polymer layers, 379–381
  stabilization and the phase diagram, 411–413, *figure* 412
  stabilizing chains, *table* 411, 416, 417, 418
  steric versus electrostatic stabilization, 414–415, *figure* 415, *figure* 416
  theory of polymer stabilization, 398–400
  theta conditions, 412–414
  upper critical flocculation temperature (UCFT), 412
  upper critical stabilization temperature (UCST), 412
Stabilizing chain, 411, 416, 417, 418
Standard states, 302–305
Starch, 27, 332, 380, 421
Starlings, plague of, 162
Stearic acid, 230, 237, 239, 240, 241
Steric repulsion, 334, 379, 402, 427, 432, 463, 523
Steric stabilization, *see* Polymer stabilization
Stern layer, 284, 324, 325, 329–331, 333, 440, 556
Stirred media mill, 126, 127
Stokes law, 48, 70
Stokes regime, 290
Stokes sedimentation, 37, 49, 50, 63, 70–72, 539
Streaming current, 340–341, 347, 348
Streaming potential, 332, *table* 333, 340–341, 347
Stress relaxation, 32, 44, 436
Styrene-butadiene rubber, *figure* 396

INDEX   613

Sucrose, 230, *figure* 231, *table* 232, *table* 421, 482, 506
Sugar, 53, 98, 421, 511
Sulfate(s), 230, *table* 233, 234, *table* 235, 253, 260, 264, 323, 395, 426, 534. *See also* Sodium lauryl sulfate
  barium, *table* 195, 319, 403, 444
  determination by turbidity, 88
  flotation of, 162
  sodium, *table* 232
Sulfonate, 230, *table* 233, 234, *table* 235, 258, 259, 279, 443, 468, 473, 514, 515
Sulfur, 161, 162, 403, 408, 491, 528
  sols, 7, 533
Supersaturation, 151, 152, 165, 526, 527, 533
Surface-active solutes, Chapter 11, 228–236, Chapter 13, 246–275, Chapter 15, 292–306. *See also* Critical micelle concentrations; Emulsifiers; Ferguson effect; Lundelius' rule; Phase inversion temperature; Traube's rule
  adsorption, 246
  autophobic wetting, 164
  biodegradation, 234
  colligative properties, 251–252, 255
  definitions, 228
  dynamic equilibria, 246, 250–253
  effect of packing, 273–275
  entropy changes, 247–249
  Greek prefixes and suffixes, *table* 229
  hydrophobic effect, 247–249
  phase behavior, 271–273
  preferential adsorption, 321–324
  Rehbinder effect, 102
  solubilization, 261–264
  types
    amphoteric, 234
    anionic, 234
    cationic, 234, 261, 409
    examples, *table* 235
    hydrophilic and lipophilic moieties, *table* 233
    nonionic, 296, 429, 433, 447, 448, 465, 523
Surface activity, Chapter 11, 228–236, Chapter 13, 246–275, Chapter 14, 279–290, Chapter 15, 292–306, Chapter 16, 308–314. *See also* Lundelius' rule; Ross-Nishioka effect
Surface area
  of foams, 457, 459–460, 489–491
  by gas adsorption
    BET method, 99–100
    monolayer, 101–102
  by solution adsorption, 103–105, 500–506
  by surface tension titration, 506
Surface charge
  calculated from zeta potentials, 326–328
  change during collisions, 379
  effect on adhesion, 146
  origin of charge
    aqueous media, 317–318
    nonpolar media, 318–321
  Stern theory, 329–331
Surface concentration, *see also* Gibbs excess
  effect on bubble rise, 288–290
  regular solution theory, 311–313
Surface conductivity, 341
Surface elasticity, 265, 268, 468, 469
Surface free energy
  componental theory, Chapter 8, 177–197
  degrees of solid/liquid interaction, 160
  electrical charges, 173–174
  Helmholtz free energy for solids, 158–160
  irregular surfaces, Chapter 10, 218–226
  low energy solids, *table* 193
  polar organic liquids, 194
  Rehbinder effect, 132
  two-body interactive energies
    attraction, 335–371
    repulsion, Chapter 19, 374–381
  Young-Dupré equation, 161
  Young-Dupré equation for solids, 159
Surface excess, *see* Gibbs adsorption; Gibbs excess
Surface films, 197, 265, 266, 267, 268, 465, 468, 469
  elasticity of, 65–268
Surface heterogeneity, 218, 219, 504
Surface ionization, 317
Surface molecular theory, 136–140
Surface phase, 101, 164, 231, 292–300
Surface pressure, *see* Spreading pressure
Surface region, 136, 137, 230, 231
Surface rheology, 456, 491
  monolayer viscoelasticity, 242
  viscosimeter, 492, 493
Surface roughness, 90, 218, 219, 221, 222, 224, 362

Surface tension, *see also* Componental theories; Gibbs-Marangoni effect
  dispersion forces, 362–363
  dynamic, 154, 208, 266, 268–270, 288, 467, 481
    ultralow, 270
  elasticity, 265–268
  gradient, 154, 155, 173, 242, 463, 467
  isotherm, 300–301
  lowering by adsorption, 230–231
  methods to measure
    capillary height, 200–201
    drop weight, 201
    Du Noüy ring, 216, 202, 423
    maximum bubble pressure, 205–206, 215, 216
    maximum pull on a rod, 203–205
    pendent drop, 202–203
    sessile drop, 202–203
    spinning drop, 206–207, 215
    vibrating jet, 207–208
    Wilhelmy plate, 201–202
  origin, 136–138
  of solid substrates, 157–160
  thermodynamic variable, 140–155, table *140*
  values, *table* 139
Surface-tension isotherm, 300–301
Surface-tension titration, 506
Surface thermodynamics, 140–146
Surface viscosity, 464, 468, 469
Surface waves, 242
Surface-enhanced Raman scattering (SERS), 547
Surfactant, definition, 228. *See also* Surface-active solutes
Suspensions
  adsorption from solution, 292, 500–502
  adsorption of dyes, 504–506
  charging by preferential adsorption, 321–324
  deflocculated, 506–508
  deflocculated, concentrated, 515–516
  dispersions, stability of, Chapter 20, 383–400
  electroacoustics, 341–343
  electrokinetic phenomena, *table* 333
  flocculated, 513–514
  flocculated versus deflocculated, 506–508
  flocculation, 523–526
    by electrolyte, 523–524
    by neutral polymers, 524–525
    by polyelectrolytes, 525
    versus stabilization, 525–526
  heteroflocculation, 526
  pigments in polymer solution, 516–520
  rheology, 512–523
  rheology of dilute suspensions, 513
  sedimentation, 70–72
  surface-tension titrations, 506
  techniques, 500–512
  technology, Chapter 24, 499–528
  thin-film coatings, 520
  ultrasonic techniques, 510–512
Swelling, 450, 451, 535, 544
Szyszkowski equation, 301, 304

Teflon®, 222
  foam inhibitor, 481
  Lifshitz constant, *table* 360
  rod for measuring surface tension, 205
  triboelectric series, *table* 346
  water contact angle on, *table* 161
TCP, *see* Tri-*p*-cresyl phosphate
Terminal velocity, 38, 48, 49, 50, 51, 62, 288, 289
Ternary systems, 272, 314
Tetrasodium pyrophosphate, 23, *table* 235, 321, 334, 379, 387, 499, 514
Textiles, *table* 22, 221, *table* 235
Thermal field-flow fractionation (Thermal FFF), 79–81
Thermally induced Marangoni flow (TIM), 155
Thermodynamic functions, 140–141, *table* 140, 302
Thermodynamics
  adsorption from solution, 292–307
  critical point wetting, 313–314
  curved interfaces, 146–148
  degrees of liquid/solid interaction, 160–161
  Kelvin equation, 150–151
  liquid/solid interfaces, 157–160
  Ostwald ripening, 151–152
  standard states, 302–305
  surfaces and interfaces, 140–146
  Young-Dupré equation, 157, 218, 219
Thermophoresis, 80
Theta conditions, 412, 535
Theta temperature, 412
THF (tetrahydrofuran)
Thin-film coatings, 171–173, 520
Thixotropy, 29, 30, 40, 515, 517
TIM flow, *see* Thermally induced Marangoni flow

INDEX   615

Tiselius' method, 556
Titania, 1, 3, 323, 324, 342, *table* 360, 391, 392, 399, 403, 499, 543, 544
 anatase, *table* 195, *table* 318
 optical constants, *table* 368
 rutile, *table* 318, *table* 358, *table* 360, 368, *table* 368, 408, 510, 511, 518
Transport number, 253–254
Traube's rule, 247, 257, 260, 304–306, 409
Triboelectric series, 345, 346
 values, *table* 346
Tri-*p*-cresyl phosphate, 194, 239, 240
Triton X-100, 229
TSPP, 23, *table* 235, 321, 322, 334, 514.
 See also Tetrasodium pyrophosphate
TU, *see* Turbidity units
Turbidity, 88, 89, 106, 137, 254, 413
Turbidity units, 89
Turbulence, 52, 119, 122, 123
Twin-screw extruders, 130
Tyndall effect, 1, 7, 9, 82, 89, 533, 534, 554, 609

UCFT, *see* Upper critical flocculation temperature
UCST, *see* Upper critical stabilization temperature
Ultracentrifuge, 75, 76, 77, 532
Ultrafiltration, 224
Ultramicroscopy, 82–83
Ultrasonic dispersers, 123, 443
Ultrasonic vibrations, 123
Ultrasound, 52, 53, 95, *table* 95, 123, 133, 134, 152, 341, 443, 510, 511
 comparison to light, *table* 95
Units, Appendix 26.1, 554–555
Upper critical flocculation temperature (UCFT), 412, *figure* 412
Upper critical stabilization temperature (UCST), 412, *figure* 412

Van der Waals forces, 177, 388
Venn diagram, 536
Vibrating jet, 206–207, 269, 467
Virial equation of state, 101, 538
Viscoelasticity, 29, 30, 31, 40, 44, 242, 517
Viscoelastometers, 44
Viscosimeters, 34–46
 interfacial, 435
 silicone standards, 37
 surface, 492, 493

Viscosity, *see also* Rheology, viscosity
 coating methods, *table* 521
 kinematic, 38
 values, *table* 18
Void volumes, 103, *table* 508
Volta potential, 344

Ward and Tordai equation for diffusion, 410
Water
 anomalous, 152, 248, 249, 308
 contact angles, *table* 161
 dispersion forces, *table* 184
Water-in-oil emulsions, (W/O)
 definitions, 420, *table* 420
 electrical conductivity, 439–440
 examples, *table* 421–422
 HLB scale, 429–432
 lipophilic amphipathic particles, 426
 petroleum emulsions, 447, 448
 speed of sound, 97
Washburn equation, 131, 210–212
Washing, 162, 171
Water/hydrocarbon, 230, 247, 248
Waterproofing, 221
Water-repellent, 60, 221
Waves, damping, 154, 237, 242
Weissenberg effect, 33
Wenzel equation, 218–219
Wet milling, 132
Wetting, *see also* Wetting of irregular surfaces
 agents, *table* 235, 259. *See also* AOT
 chemical processing aids, 130–132
 of cotton yarn, 154
 critical point, 313–314
 electrical effects in, 173–174
 HLB, *table* 431
 mixing dry powders, 117
 pore-volume analysis, 102
 preferential wetting, 162
 sediment volumes, 508–509
 Washburn equation, 210–212
Wetting of irregular surfaces, 218–227
 Jamin effect, 222–224
 pores, 220–221
 roughness, 218–219
 scratches, 221–222
 surface heterogeneity, 219–220
 Wenzel equation, 218–219
White blood cells, 87
Wicking, 196, 211, 212
Wilhelmy plate, 45, 201–202, 203, 215, 216, 219, 224, 242, 270, 423

W/O, *see* Water-in-oil emulsions
Work function, 344, 345, 346
Work of adhesion
　acid-base character, 196–197, *table* 196
　amphipathic particles, 427–429
　componental theories, 180–190
　definition, 144–145
　deflocculated versus flocculated, *table* 508
　degrees of liquid/solid interaction, 160–161
　direct measurements, 368, 370. *See also* Atomic force microscope; Jacob's box
　dispersion energy, 363
　electrical charges, 146
　hydrocarbon/water, 246–249
　liquid/solid, 160
　values, *table* 145
Work of cohesion, 144–146, 166, 180, 246, 362
Work of desorption, 448, 449

Xanthates, 162
Xerography, 343, 540–543
　inks, 540–542

Yield point, 21, 23, 25, 43, 394, 434, 456, 466, 492, 513, 514, 517, 521, 539, 540
Young-Dupré equation, 157, *figure* 158, 218, 219, 427

Zeolite, 403
Zeta potential
　carbon/benzene, 283
　in DLVO theory, 388–390
　electrical forces, 51
　electroacoustics, 512
　electrokinetic phenomena, 332–333, *table* 333
　electrosteric stabilization, 415–416
　foam films, 463
　Hückel equation, 335
　measurements, 334–344
　in nonpolar media, 284, 319–321
　O'Brien and White calculations, 335–337
　preferential adsorption, 321–324
　Smoluchowski equation, 334
Zeta-meter, 348
Zinc oxide, 509
Zisman plot, 190